DATE DUE

MAY 0 2 2003	

CREATING QUALITY

Process Design for Results

William J. Kolarik
Texas Tech University

WCB McGraw-Hill

Boston • Burr Ridge, IL • Dubuque, IA • Madison, WI • New York • San Francisco • St. Louis
Bangkok • Bogotá • Caracas • Lisbon • London • Madrid
Mexico City • Milan • New Delhi • Seoul • Singapore • Sydney • Taipei • Toronto

WCB/McGraw-Hill

*A Division of The **McGraw-Hill** Companies*

CREATING QUALITY: PROCESS DESIGN FOR RESULTS

This book is printed on acid-free paper.

1 2 3 4 5 6 7 8 9 0 DOC/DOC 9 4 3 2 1 0 9

ISBN 0-07-036309-9

Vice president/Editor in Chief: *Kevin T. Kane*
Publisher: *Thomas Casson*
Executive editor: *Eric M. Munson*
Marketing manager: *John T. Wannemacher*
Project manager: *Carrie Sestak*
Production supervisor: *Michael R. McCormick*
Freelance design coordinator: *JoAnne Schopler*
Supplement coordinator: *Jennifer L. Frazier*
Compositor: *GAC/Indianapolis*
Typeface: *10/12 Times Roman*
Printer: *R. R. Donnelley & Sons Company*

Library of Congress Cataloging-in-Publication Data

Kolarik, William J.
   Creating quality : process design for results / William J. Kolarik.
      p.   cm.
   Includes index.
   ISBN 0-07-036309-9
   1. Production engineering.   2. Quality control.   I. Title.
TS176.K635   1999
658.5—dc21   98-46222

http://www.mhhe.com

McGraw-Hill Series in Industrial Engineering and Management Science

Barnes
Statistical Analysis for Engineers and Scientists:
A Computer-Based Approach

Bedworth, Henderson, and Wolfe
Computer-Integrated Design and Manufacturing

Black
The Design of the Factory with a Future

Blank
Engineering Economy

Blank
Statistical Procedures for Engineering,
Management, and Science

Bridger
Introduction to Ergonomics

Denton
Safety Management: Improving Performance

Ebeling
Reliability and Maintainability Engineering

Grant and Leavenworth
Statistical Quality Control

Hillier and Lieberman
Introduction to Mathematical Programming

Hillier and Lieberman
Introduction to Operations Research

Juran and Gryna
Quality Planning and Analysis: From Product
Development through Use

Kelton, Sadowski, and Sadowski
Simulation with ARENA

Khoshnevis
Discrete Systems Simulation

Kolarik
Creating Quality: Concepts, Systems, Strategies,
and Tools

Law and Kelton
Simulation Modeling and Analysis

Marshall and Oliver
Decision Making and Forecasting

Moen, Nolan, and Provost
Improving Quality through Planned Experimentation

Niebel and Freivalds
Methods, Standards, and Work Design

Nash and Sofer
Linear and Nonlinear Programming

Nelson
Stochastic Modeling: Analysis and Simulation

Pegden
Introduction to Simulation Using SIMAN

Riggs, Bedworth, and Randhawa
Engineering Economics

Sipper and Bulfin
Production: Planning, Control, and Integration

Steiner
Engineering Economic Principles

Taguchi, Elsayed, and Hsiang
Quality Engineering in Production Systems

Wu and Coppins
Linear Programming and Extensions

PREFACE

Processes are essential. We can do nothing, and accomplish nothing, without a process—ad hoc or carefully defined, designed, controlled, and implemented. Therefore, it seems reasonable to study processes systematically, from perspectives of definition/redefinition, control, and improvement.

BACKGROUND

Evidence is available to suggest that we sometimes focus on the glitter and promise of the latest technical and business initiatives to the exclusion of fundamental core technical/business issues and practices. Examination of textbooks and professional reference books suggests that this is so, as do direct observations of the manner in which we conduct our affairs. It is disheartening to observe an organization where more emphasis is placed on command of the latest buzzwords, initiatives, and tools than on fundamental processes and their products, relative to customer outcomes and business results. *Creating Quality: Process Design for Results* serves as a compass in redirecting our energy and creative abilities toward understanding and mastering fundamental processes.

Creating Quality: Process Design for Results aligns itself with both academics and professional practice. It addresses the fundamental processes used to create quality. These processes were originally introduced in Chapter 1 of the author's earlier book, *Creating Quality: Concepts, Systems, Strategies, and Tools.* This present book, *Creating Quality: Process Design for Results,* presents a detailed view of processes in terms of (1) process definition/redefinition, the conceptual essence of a process, (2) process control, in terms of both monitoring and adjustment, (3) process improvement in terms of continuous improvement, and (4) transformation to process-based organizations.

Creating Quality: Process Design for Results and *Creating Quality: Concepts, Systems, Strategies, and Tools* are complementary works. They can be used together or separately. Each is capable of standing on its own merits; both together provide wider perspectives.

PURPOSE

The purpose of *Creating Quality: Process Design for Results* is to encourage/address natural means of enhancing competitive advantage in a production system. We stress scientifically based, process-related principles and creative thinking, as opposed to checklist and anecdotal approaches. *Creating Quality: Process Design for Results* focuses on processes—the fundamental means available to us to define, design, develop, produce, deliver, sell, use, and dispose or recycle products, and in general create quality and productivity for our customers and prosperity for our stakeholders and ourselves.

Creating Quality: Process Design for Results places the process—definition/redefinition, control, and improvement—in the foreground, and places initiatives and tools in the background. Initiatives and tools are not without value; to the contrary, they are invaluable when we need them and can use them to help us enhance our core processes in physical, economic, timeliness, and customer service performance. However, when initiatives and tools begin to drive organizations, a true focus/bearing is lost and ineffectiveness and inefficiency follow. A focus on creating value for customers and stakeholders—through value-created processes and their resulting products—is maintained throughout the book.

CONTENTS

The materials in *Creating Quality: Process Design for Results* have been defined, designed, and developed with both academic and professional practice requirements in mind. They encourage a holistic view/understanding of a production system and its customers, yet provide for analytical detail in design, control, and implementation. The text points out that optimizing processes in a production system, one at a time, does not typically provide an optimal production system as a whole. All sections together offer a balanced, process-based organizational structure squarely positioned to address this critical issue.

All materials are sectioned to allow and encourage instructors to build a significant hands-on design project element into their instruction. Through design projects, students experience a living, growing, case-based environment. This environment encourages instructors and students to develop the meaningful dialogs necessary to hone cases to the point of mastery. Students stand to gain professional practice-based experience through planning and executing open-ended projects. In addition, the text supports professional practice settings (e.g., workshops), where we use hands-on projects to directly address organizational goals and objectives in real-time, yielding immediate organizational learning and improvement.

Creating Quality: Process Design for Results materials allow instructors to deliver a comprehensive course, centered in process definition/redefinition, control, and improvement. The text material allows several options to instructors for building either a one- or two-course sequence relating to production systems and processes. Course materials can be adapted for a wide range of college students, ranging from sophomores to seniors. The course materials, with some supplements, support graduate studies.

Materials are available to present a conceptual course, without significant mathematical prerequisites by focusing on Sections 1, 3, 5, 6, and the conceptual process control elements in Section 4. Highly quantitative approaches can be taken by focusing on Sections 1, 2 and 4, with side trips through Sections 3, 5, and 6. Regardless of course orientation, a case-based approach in any, or all, of the three areas—definition/redefinition, control, and improvement—is capable of producing impressive project portfolios. Trial usage of the materials has produced results well beyond the author's original (ambitious) expectations.

These materials work best in the context of extended team projects that last essentially the entire term, with oral and written/story-board reporting two or three times per term. Heavy emphasis is placed on graphical depictions, with supporting words, rather than the reverse. When we use this approach, we always cover Section 1 first. Then, we survey Chapter 18 with its teaming, leadership, and creativity topics. With the teaming, leadership, and creativity knowledge, we move into the technical sections and our projects. We use the initiatives and tools chapters, Chapters 19 and 20, as needed to support our projects.

A solid, two-course sequence is developed by incorporating all sections, in depth. The first course of this sequence serves as an introduction to process-based thinking and organizations, mostly on a conceptual and project-intensive basis. We use the materials to support a combination of lectures and team-based projects. The concept-based course can appear as early as the sophomore year. It serves to help students grasp the nature of processes and their criticality in our modern production systems, as they obtain the quantitative basis for in-depth process control work.

The second course in this sequence takes on a quantitative nature, focusing primarily on process characterization and control, e.g., Sections 2 and 4. This course requires a basic probability and statistics background, and can appear as early as the junior year. Here, we place the technical aspects of process characterization/exploration and control within the context of the production system, as a complement to process definition/redefinition and improvement.

Throughout the two-course sequence, we emphasize the importance of implementation, in addition to design and planning. The overall effect sought is one of balance, compatible with what students will ultimately experience in professional practice. Provided project portfolios are developed, students can impress potential employers with tangible evidence of their knowledge of and expertise in process issues.

ACKNOWLEDGMENTS

The *Creating Quality: Process Design for Results* project has drawn deeply from several areas of expertise. These areas are supported by theory and practice. Searching out this information in both literary sources as well as professional practice required considerable time and effort. Many contributors helped and guided the project along.

The final result has taken several twists and turns along the ideation and development paths. Three primary resource groups deserve formal recognition. First, many students, ranging from freshmen to doctoral students, have participated in development work—reading, evaluating, and contributing to these materials. Traditional undergraduate and graduate students, as well as off-campus students, and practicing engineers and managers, shaped the product—many of the cases developed in the book are a direct result of contributions from students. Explicit case contributions are denoted with initials, [XX], at the end of each contribution.

A second set of contributors have expedited development through their hard work and diligence in helping to write several sections and review the writing of other sections. Specifically, Babu Chinnam, Iulian Gherasoiu, Mehmud Karim, Huitian Lu, Shuxia Lu, Sanjuka Patro, Nawshaba Rahman, Michael Sanders, and Beverly Wiley have contributed many hours to the project. Valuable strategic guidance was provided through several project reviews. Specifically, Pirooz Vakili, *Boston University;* Karl D. Majeseke, *KM Consulting;* Gary S. Wasserman, *Wayne State University;* John R. English, *University of Arkansas;* D. L. Kimbler, *Kimbler Associates Inc.;* Jeremy D. Semrau, *University of Michigan;* Robert R. Safford, *University of Central Florida;* Diane Schaub, *University of Florida;* Ali A. Houshmand, *University of Cincinnati;* Suraj M. Alexander, *University of Louisville;* Karl J. Arunski, *Texas Instruments;* Hoang Pham, *Rutgers University;* K. N. Balasubramanian, *California Polytechnic State University–San Luis Obispo;* and M. Jeya Chandra, *Pennsylvania State University* deserve credit for supplying guidance and support for the project. My deepest appreciation goes to Eric Munson and Ken Case for their executive direction, regarding essence and organization, throughout the *Creating Quality: Process Design for Results* project.

The third set of contributors includes family and friends, who have supplied encouragement during project development. Especially noteworthy is the support of my wife, Yvonne, and sons, William II, Charles, and Franklin.

William J. Kolarik

CONTENTS

SECTION 1

PRODUCTION SYSTEMS AND PROCESS PERFORMANCE 1

Chapter 1

PRODUCTION SYSTEMS—THE BASICS 2

1.0 Inquiry 2
1.1 Introduction and Overview 2
1.2 Basics 4
Quality 5
Productivity 8
Quality-Productivity Connection 11
1.3 Production System Linkages 13
1.4 Cooperative Efforts 14
1.5 Organizational Optimization and Synergy 15
Review and Discovery Exercises 21
References 23

Chapter 2

SYSTEMS THINKING—CONCEPTS AND DEVELOPMENT 24

2.0 Inquiry 24
2.1 Introduction 24
2.2 Analytical and Systems Philosophies 24
2.3 Models and Modeling 26
2.4 Systems Theory and Thinking 26
System Classification 27
Systems Thinking 28
Adaptations of General Systems Thinking 29
2.5 Production Systems 30
Craft Production System Paradigm 33
Factory Production System Paradigm 33
Mass Production System Paradigm 34
Lean Production System Paradigm 35
Agile Production System Paradigm 36
Common Ground 36
2.6 Systems Thinking—Science and Engineering 37
Review and Discovery Exercises 40
References 41

Chapter 3

PROCESS FUNDAMENTALS 43

3.0 Inquiry 43
3.1 Introduction 43
3.2 Process Features and Synergy 43
Direction—Production Systems 44
Process Definition/Redefinition Concept 45
Process Control Concept 45
Process Improvement Concept 45
Scope—Define/Redefine, Control, Improve 45
Synergy 46
3.3 Purpose—Goals, Objectives, Targets, Tolerances—and Action 47
Location and Dispersion 48
3.4 Process Structures 51
3.5 Production System Views 53
3.6 Robust, Mistakeproof, and Benchmark Performance 55
Robustness 55
Mistakeproofing 57
Benchmarks 58
Review and Discovery Exercises 60

SECTION 2

PROCESS CHARACTERIZATION, EXPLORATION, AND RESPONSE MODELING 62

Chapter 4

PROCESS CHARACTERIZATION 63

4.0 Inquiry 63
4.1 Introduction 63
4.2 Process Understanding 63
4.3 Process Models 65
4.4 Process Measurement Scales 67
4.5 Process Levers and Leverage 69
4.6 Physical Characterization 69
4.7 Statistical Characterization 73
Data Collection 74
Graphical Assessment 75
Numerical Assessment 76

Review and Discovery Exercises   87
References   89

Chapter 5

PROCESS EXPLORATION   90

5.0   Inquiry   90
5.1   Introduction   90
5.2   Experimental Protocol   91
5.3   Single-Factor Experiments   92
      Single-Factor CRD Models   92
      Random Effects Model   96
      Fixed Effects Model   100
      Prediction and Residual Calculations   102
      Treatment Level Interval Analysis (Fixed Effects
          Factors)   106
      Treatment Mean Comparisons (Fixed Effects
          Factors)   107
5.4   Model Adequacy   109
5.5   Multiple-Factor Experiments   114
      Multiple-Factor Analysis of Variance
          (ANOVA)   114
      Pairwise Treatment Comparisons (Fixed Effects
          Factors)   122
5.6   Summary   124
Review and Discovery Exercises   124
References   125

Chapter 6

PROCESS RESPONSE MODELING   126

6.0   Inquiry   126
6.1   Introduction   126
6.2   Least Squares Estimation   126
      Least Squares Estimators   127
      Normal Equations   127
6.3   Regression Analysis   128
      Regression ANOVAs   129
      Response Surface Structure   131
      Model Simplification   133
      Model Fit and Adequacy   135
6.4   Response Surface Designs   146
      2^f with a Center Point   146
      Central Composite Design   149
Review and Discovery Exercises   157
References   158

SECTION 3

PROCESS DEFINITION AND REDEFINITION   159

Chapter 7

PROCESS DEFINITION/REDEFINITION— OUTPUT PERSPECTIVES   160

7.0   Inquiry   160
7.1   Introduction   160
7.2   Timing, Personnel, and Exposure   163
7.3   Critical Elements   164
7.4   Production System Level Results
      Definition   164
7.5   Process Level Results Definition   175
      Customers   175
      Outcomes   177
      Concepts   179
Review and Discovery Exercises   185
References   187

Chapter 8

PROCESS DEFINITION/REDEFINITION— TRANSFORMATION AND INPUT PERSPECTIVES   188

8.0   Inquiry   188
8.1   Introduction   188
8.2   Process Means   188
      Options   189
      Evaluation   194
      Plan   198
8.3   Process Creation   210
      Resources   210
      Schedule   210
      Action   213
Review and Discovery Exercises   217

SECTION 4

PROCESS CONTROL **219**

Chapter 9

PROCESS CONTROL—CONCEPTS AND OPTIONS **221**

9.0 Inquiry 221
9.1 Introduction 221
9.2 Critical Elements of Process Control 222
9.3 Process Control Options and Growth 224
9.4 Process Control Models Overview 227
9.5 Introduction to SPC Models 229
Review and Discovery Exercises 232
References 233

Chapter 10

PROCESS MONITORING—VARIABLES CONTROL CHARTS FOR GROUPED MEASUREMENTS **234**

10.0 Inquiry 234
10.1 Introduction 234
10.2 SPC Model Rationale for Variables Data 234
 SPC Concept 235
 Subgrouping Rationale 240
10.3 Notation for Subgrouped SPC Models 240
 General Control Chart Symbols 240
 R, S, X-bar Chart Symbols 241
 Exponential Weighted Moving Average and Deviation
  (EWMA and EWMD) Chart Symbols 241
 Cumulative Sum (CuSum) Chart Symbols 241
10.4 Shewhart X-bar, R, and S Control Chart Concepts and Mechanics 242
 Normal Model 242
 X-bar Control Chart Mechanics 242
 R and S Control Chart Mechanics 244
10.5 Interpretation of Shewhart Control Charts 248
 Basic Interpretation 249
 Pragmatic Interpretation 252
10.6 Shewhart Control Chart OC Curves and Average Run Lengths 255
10.7 Probability Limits for Shewhart Control Charts 256

10.8 EWMA and EWMD Control Charts 259
10.9 CuSum Control Charts 263
10.10 Limited Duration Process Runs 269
 Deviation from Target Charts 269
 Standardized Charts 270
 Limited Duration Process Run Summary 270
Review and Discovery Exercises 270
Appendix: OC Curve Construction and ARL
  Calculations 273
References 279

Chapter 11

PROCESS MONITORING—VARIABLES CONTROL CHARTS FOR INDIVIDUAL MEASUREMENTS AND RELATED TOPICS **280**

11.0 Inquiry 280
11.1 Introduction 280
11.2 SPC Model Rationale for Individuals Data 281
11.3 Notation for Individuals SPC Models 282
 General Control Chart Symbols 282
 $X, \overline{X}_M$, and R_M Chart Symbols 282
 Exponential Weighted Moving Average and Deviation
  (EWMA and EWMD) Chart Symbols 283
 Cumulative Sum (CuSum) Chart Symbols 283
11.4 $X, \overline{X}_M$, and R_M Control Chart Concepts and Mechanics 283
11.5 EWMA and EWMD Control Charts 291
11.6 CuSum Control Charts 294
 Two-Sided CuSum Charts 295
 One-Sided CuSum Charts 298
 CuSum Charting Variations 298
11.7 Sampling Schemes 299
11.8 Production Source Level Control 300
11.9 Target-Based Control Charts 302
11.10 Process Capability 303
 Process Capability Indices 303
 Interpreting Capability Indices 306
11.11 Gauge Studies 309
Review and Discovery Exercises 313
Appendix: Additional Capability Indices 317
References 318

Chapter 12

PROCESS MONITORING—ATTRIBUTES CONTROL CHARTS FOR CLASSIFICATION MEASUREMENTS   320

12.0 Inquiry   320
12.1 Introduction   320
12.2 Defects and Defectives   321
12.3 SPC Model Rationale for Attributes Data   322
12.4 Notation for Attributes SPC Models   325
    General Symbols   325
    P-Chart Symbols   325
    C-, U-Chart Symbols   326
    Binomial Model   326
    Poisson Model   327
12.5 P Control Chart Concepts and Mechanics   327
    P-Chart Mechanics   328
12.6 C and U Control Chart Concepts and Mechanics   334
    C-Chart Mechanics   335
    U-Chart Mechanics   337
12.7 Process Logs and Pareto Charts   341
Review and Discovery Exercises   344
Reference   345

Chapter 13

PROCESS MONITORING—NONTRADITIONAL SPC CONCEPTS AND MODELS   346

13.0 Inquiry   346
13.1 Introduction   346
13.2 SPC Model Performance Evaluation   347
13.3 Performance Assessment with *Nid/Iid* Data Streams   348
13.4 Performance Assessment with *Non-Nid* Data Streams   354
13.5 Introduction to Multivariate SPC Models   365
    Multivariate SPC Characterization   365
    χ^2 Multivariate Location Chart—Subgrouped Data   368
    Hotelling T^2 Multivariate Location Chart—Subgrouped Data   368
    Sample Generalized Variance Multivariate Dispersion $|S|$ Chart—Subgrouped Data   375
    Hotelling T^2 Multivariate Location Chart—Individuals Data   379

Interpretation of Multivariate SPC Charts   380
    Other Multivariate Models   381
Review and Discovery Exercises   381
References   384

Chapter 14

PROCESS ADJUSTMENT—INTRODUCTION TO AUTOMATIC PROCESS CONTROL, CONVENTIONAL MODELS   385

14.0 Inquiry   385
14.1 Introduction   385
14.2 Classical Control Concepts   386
14.3 Discontinuous/Discrete Control Action   392
14.4 Continuous Control Action   397
    Proportional Control   397
    Integral Control   401
    Derivative Control   405
    PID Control   407
14.5 Controller Tuning   407
14.6 Transfer Functions and Block Diagram Representation   409
Review and Discovery Exercises   415
References   417

Chapter 15

PROCESS ADJUSTMENT—INTRODUCTION TO AUTOMATIC PROCESS CONTROL, UNCONVENTIONAL MODELS   418

15.0 Inquiry   418
15.1 Introduction   418
15.2 Advanced Concepts in Conventional APC   419
    Cascade Control   419
    Ratio Control   419
    Feedforward Control   419
15.3 Process Identification and Nonparametric Models   420
    Mathematical Models   420
    Dynamic Process Modeling   421
    Nonparametric Models   422
    Artificial Neural Networks   422
    Neural Network Architectures   424
    Feedforward Networks   424
    Recurrent Networks   424

Learning in Neural Networks   425
Applications of Neural Networks in Control   425
Expert Systems   426
Evolutionary Computation   428
15.4 Self-Tuning Control   429
15.5 APC/SPC Model Combinations   431
Review and Discovery Exercises   433
References   434

SECTION 5
**PROCESS ANALYSIS AND
IMPROVEMENT   436**

Chapter 16
**PROCESS IMPROVEMENT—QUESTIONING
PERSPECTIVES   437**

16.0 Inquiry   437
16.1 Introduction   437
16.2 Critical Elements   439
16.3 Process Improvement Opportunity   443
Observation   443
Concepts   445
Options   451
Review and Discovery Exercises   454

Chapter 17
**PROCESS IMPROVEMENT—ANALYSIS AND
IMPLEMENTATION PERSPECTIVES   456**

17.0 Inquiry   456
17.1 Introduction   456
17.2 Process Change Description   456
Alternatives   457
Evaluation   462
Plan   466
17.3 Process Change   471
Resources   471
Schedule   473
Action   477
Review and Discovery Exercises   479

SECTION 6
**PROCESS-BASED TRANSFORMATIONS,
INITIATIVES, AND TOOLS   481**

Chapter 18
PROCESS-BASED TRANSFORMATIONS   482

18.0 Inquiry   482
18.1 Introduction   482
18.2 Organization   483
Structure and Channels   484
Information Exchange and Archives   486
Reward Structure—Awards and Recognition   488
18.3 Creativity   492
Creative Thinking   493
Knowledge Base—Domain   498
Environment—Field   500
18.4 Leadership   500
Direction   502
Teamwork   503
Empowerment   508
Review and Discovery Exercises   509
References   510

Chapter 19
PROCESS-COMPATIBLE INITIATIVES   511

19.0 Inquiry   511
19.1 Introduction   511
19.2 Benchmarking   512
19.3 Concurrent Engineering   515
19.4 Continuous Improvement   516
19.5 Cycle Time/Waste Reduction   518
19.6 Fifth Discipline   521
19.7 Function-Value Analysis   522
19.8 ISO 9000   523
19.9 Mistakeproofing (Poka-Yoke)   525
19.10 Quality Awards   527
19.11 Quality Function Deployment   530
19.12 Reengineering   532
19.13 Robust Design   534
19.14 Six Sigma   535
19.15 Theory of Constraints   538
19.16 Total Quality Management   539
Discovery Exercises   542
References   542

Chapter 20

PROCESS-COMPATIBLE TOOLS   544

20.0   Inquiry   544
20.1   Introduction   544
20.2   Activity/Sequence List   545
20.3   Break-Even Analysis   545
20.4   Capability Analysis   546
20.5   Cash-Flow Analysis   547
20.6   Cause-Effect Diagram   548
20.7   Check Sheet   550
20.8   Control Chart   550
20.9   Correlation/Autocorrelation Analysis   551
20.10 Critical Path Method (CPM)   551
20.11 Experimental Design   553
20.12 Failure Mode and Effects Analysis
       (FMEA)   554
20.13 Fault Tree Analysis   556
20.14 Flowchart   557
20.15 Force Field Analysis   559
20.16 Gantt Chart   560
20.17 Histogram   561
20.18 Matrix Diagram   561
20.19 Pareto Analysis   562
20.20 Process Value Chain Analysis   563
20.21 Relations Diagram   564
20.22 Root Cause Analysis   565
20.23 Runs Chart   566
20.24 Scatter Diagram   567
20.25 Stratification Analysis   568
Discovery Exercises   569
References   569

SECTION 7

PROCESS CASES—DESCRIPTIONS AND DATA   571

VII.1 Introduction   571
VII.2 Data Extensions   572
VII.3 Cases   573
      Case VII.1   AA Fiberglass   573
      Case VII.2   Apple Core—Dehydration   574
      Case VII.3   Apple Dehydration Exploration   575
      Case VII.4   Back-of-the-Moon—Mining   576
      Case VII.5   Big City Waterworks   578

Case VII.6   Big Dog—Dog Food Packaging   578
Case VII.7   Bushings International—Machining   579
Case VII.8   Door-to-Door—Pizza Delivery   580
Case VII.9   Downtown Bakery—Bread Dough   582
Case VII.10 Downtown Bakery—pH
             Measurement   583
Case VII.11 Fix-Up—Automobile Repair   584
Case VII.12 Hard-Shell Aquaculture   585
Case VII.13 Health Assist—Service   586
Case VII.14 High-Precision—Collar Machining   587
Case VII.15 High-Precision—Collar
             Measurement   588
Case VII.16 Link-Lock Chain   589
Case VII.17 LNG—Natural Gas Liquefaction   590
Case VII.18 M-Stick Manufacturing   591
Case VII.19 Night Hauler Trucking   593
Case VII.20 PCB—Printed Circuit Boards   593
Case VII.21 Punch-Out—Sheet Metal Fabrication   594
Case VII.22 Reuse—Recycling   596
Case VII.23 Reuse—Sensor Precision   597
Case VII.24 Silver Bird—Baggage   599
Case VII.25 Snappy—Plastic Injection Molding   600
Case VII.26 Squeaky Clean Laundry   601
Case VII.27 Rainbow—Paint Coating   601
Case VII.28 Sure-Stick Adhesive   603
Case VII.29 TexRosa—Salsa   604
Case VII.30 Tough-Skin—Sheet Metal Welding   606

SECTION 8

STATISTICAL TABLES   608

Table VIII.1 Cumulative Standard Normal Distribution
              Table   609
Table VIII.2 t Distribution Table—Critical Values   611
Table VIII.3 Chi-Squared Distribution Table—Critical
              Values   612
Table VIII.4 F Distribution Tables—Critical
              Values   613
Table VIII.5 X-bar, R, and S Control Chart—3-Sigma
              Limit Constants   617
Table VIII.6 X-bar, R, and S Control Chart—Probability
              Limit Constants   618
Table VIII.7 EWMA and EWMD Control Chart Limit
              Constants   619
Table VIII.8 Tabled Pseudo-Standard Normal Random
              Numbers   619
Table VIII.9 Normal Probability Plotting Paper   622

1

PRODUCTION SYSTEMS AND PROCESS PERFORMANCE

The purpose of Section 1 is to introduce both the essence and nature of process-based organizational concepts.

PART OUTLINE

Chapter 1: Production Systems—The Basics

The purpose of Chapter 1 is to introduce the process-based concepts and their relationships to organizational synergy in terms of effectiveness (quality) and efficiency (productivity).

Chapter 2: Systems Thinking—Concepts and Development

The purpose of Chapter 2 is to give an overview of both analytical and systems thinking in the context of production system evolution.

Chapter 3: Process Fundamentals

The purpose of Chapter 3 is to introduce whole-process thinking through process purpose, definition/redefinition, control, and improvement elements.

1

PRODUCTION SYSTEMS—THE BASICS

1.0 INQUIRY

1. What is a production system?
2. How do production systems work?
3. What is quality? What is productivity? How are they related?
4. What is a cooperative effort? How do they start? How are they sustained?
5. How are organizations focused? Optimized?

1.1 INTRODUCTION AND OVERVIEW

Every product in existence—hard-good, perishable, or service—**can be traced back to a production system.** Some production systems depend primarily on nature's bounty, e.g., agriculture, mining, and petroleum. Some depend on intricate machinery, e.g., electronics, automobiles, and textiles. Some depend on personal attention to consumer needs, e.g., education, banking, retail, and food service. In short, our lives are impacted and sustained by production systems. **A production system is an integrated collection of people and processes that together transforms resources into products.**

Our purpose in this textbook is to explore the fundamental nature of production systems, specifically their constituent working parts: people, products, and processes. Figure 1.1 provides a graphical overview depicting and relating the three constituent elements of any production system. Customer needs, demands, and expectations drive the intricate network of production systems that surround us today. People participate in many of these for-profit and not-for-profit production systems simultaneously as consumers or external customers, producers or internal customers, suppliers of affiliated production systems, or stakeholders—owners, creditors, and so on.

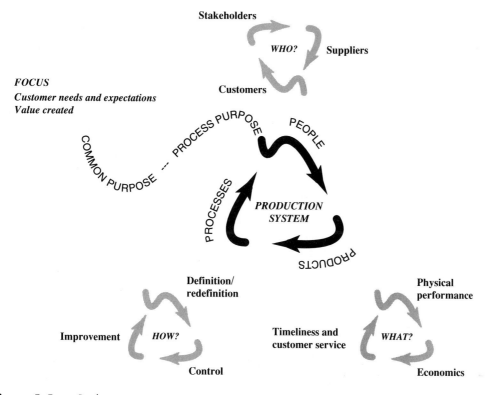

Figure 1.1   Production system overview.

Products are ultimately judged by customers in terms of the benefits they generate against the burdens incurred. More specifically, we describe these benefits and burdens in terms of physical, economic, timeliness, and customer service performance. Here, physical performance encompasses function (how it works for the customers), form (how it looks to the customers), and fit (how it addresses specific customer applications). Economic performance involves both the price that a customer pays up front, in money, and the cost to sustain the product, as well as revenues, if any, ultimately generated by the product. Timeliness performance includes the time it takes to produce a product, the time we must wait to obtain a good or service, or delays we encounter during the course of product usage later in the product life cycle. Customer service performance includes how we treat our customers, in terms of customer perception of our attention to their needs and responses to their demands.

Every endeavor associated with a product involves a process—planned and practiced, or improvised and executed in an ad hoc manner. There are no exceptions. **All processes require some level of definition, control, and improvement.** This process triad is our primary focus and is expanded in the course of our discussions in all seven sections.

1.2   BASICS

A production system in its broadest perspective transforms a set of resources into a set of products and by-products; see Figure 1.2*a*. Figure 1.2*b* depicts a process where a variety of resources serve as inputs and are transformed into products and by-products, the outputs. Typically, we see a wide variety of inputs/resources transformed into a limited number of products and by-products. The transformation literally acts as a funnel. It is critical that this funneling effect add value. Added value requires that customer benefits increase faster than customer burdens.

Here, we define **value** broadly as

$$\text{Value} = \frac{\text{customer benefits}}{\text{customer burdens}} \qquad\qquad \textbf{[1.1]}$$

where customer benefits constitute fulfillment of customer needs and expectations, and customer burdens constitute what the customer gives up to obtain the product: money, time, and so on. The definition in Equation (1.1) is sometimes expressed as a ratio of worth to cost.

The process concept is fundamental to all production systems. Every production system is literally an integrated series of processes working in harmony to serve both internal and external customers with products that meet their needs and expectations.

There are any number of ways that resources can be transformed into products and by-products. These ways are referred to as **process configurations.** The best of these process

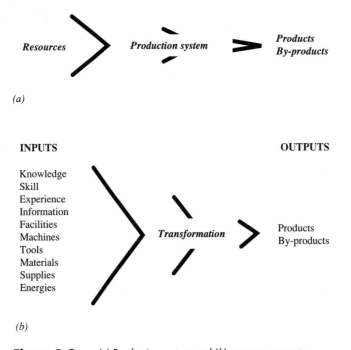

(a)

(b)

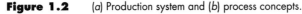

Figure 1.2   (*a*) Production system and (*b*) process concepts.

configurations usually lead to a competitive edge or advantage for their owners in extracting benefits and/or suppressing burdens. In order to develop and maintain a competitive edge, we must address the effectiveness and efficiency of our processes. In general, **the concept of quality addresses effectiveness, while the concept of productivity addresses efficiency.**

QUALITY

The quality concept is complicated. A number of authors have put forth definitions based on both customer benefits as well as customer burdens (primarily regarding products). Some definitions are expressed in a rigid manner:

Quality is meeting and exceeding customer needs and expectations; common expression.

Quality is fitness for use; Juran [1].

Quality is conformance to requirements (clearly stated); Crosby [2].

Quality should be aimed at the needs of the consumer, present and future; Deming [3].

Quality is the total composite product and service characteristics of marketing, engineering, manufacture, and maintenance through which the product and service in use will meet the expectations of the customer; Feigenbaum [4].

Quality is the loss (from function variation and harmful effects) a product causes to society after being shipped, other than any losses caused by its intrinsic functions; Taguchi [5].

Quality is the totality of features and characteristics of a product or service that bear on its ability to satisfy stated or implied needs; ISO 9000 [6].

Other quality definitions are stated in a more flexible manner:

Quality, as applied to the products turned out by industry, means the characteristic or group or combination of characteristics which distinguishes one article from another, or the goods of one manufacturer from those of competitors, or one grade of product from a certain factory from another grade turned out by the same factory; Radford [7].

There are two common aspects of quality. One of these has to do with the consideration of the quality of a thing as an objective reality independent of the existence of humans. The other has to do with what we think, feel, or sense as a result of the objective reality; this subjective side of quality is closely linked to value; Shewhart [8].

The extent of quality is determined by how well the true quality characteristics (customer needs, expressed in customer language) match substitute quality characteristics (product specifications, expressed by a producer in technical language); Ishikawa [9].

The Shewhart and Ishikawa definitions lead us to view quality through the customer's eyes. **True quality characteristics echo customer needs and set up subjective customer expectations. We translate these expectations into substitute quality characteristics** that are defined in technical terms sufficient to design and produce products. Ultimately, **customer satisfaction results from the degree of correspondence between the customer's true quality characteristics and our substitute characteristics.**

Example 1.1	List a true quality characteristic for (*a*) a microcomputer system and (*b*) a knife. Then, list three substitute quality characteristics pertaining to each true quality characteristic.

Solution

a. One example for the microcomputer system is

True quality characteristic: A monitor that is easy on my eyes.

Substitute quality characteristics: Screen contrast level, dot pitch, refresh rate.

b. One example for the knife is

True quality characteristic: Cuts well.

Substitute quality characteristics: Blade material, blade angle, surface roughness of cutting edge.

In order to design and build quality into products in a systematic manner, we pursue a course of both knowledge and action in a scientific manner. The scientific nature of quality, based on the dynamic definitions of quality above, is addressed.

Two fundamental elements address the science of quality: the customer's experience of quality and the producer's creation of quality. The experience of quality is a function of the fulfillment of human needs and expectations and results in some degree of customer satisfaction or dissatisfaction. Quality is created through processes: definition, design, development, production, delivery, sales and customer service, use, and disposal or recycle. Kolarik [10].

Value as perceived by the customer is related to the experience of quality. Customers judge benefits and burdens through experience. We know that **the customer's experience of quality is based on human needs.** The individual quality experience (IQE) model, as described by Kolarik [10] and depicted in Figure 1.3, provides a human needs–based experience model to describe quality in terms of the customer's experience in assessing value.

Our human needs structure is represented by a simplification of Maslow's [11] description of human needs. Here, **we see five basic human needs:**

1. **Physiological:** Basics such as food, water, and sleep

2. **Safety-security:** Freedom from threats to our health and well-being

3. **Social:** Our needs to interact with other humans

4. **Esteem:** Power, prestige

5. **Self-actualization:** Our desire to be all we can be personally and professionally

   In the IQE cycle, customer experience involves:

1. **Observing:** Sensing and perceiving within our physical and social environments

2. **Assessment and interpretation:** Thinking in both the cognitive (logical) and affective (emotional) dimensions

3. **Attitudes and behavior:** Shaped by decisions and initiatives

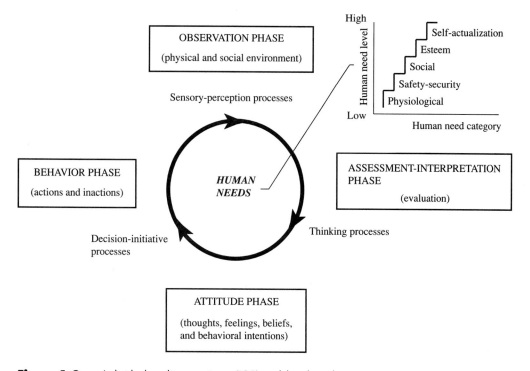

Figure 1.3 Individual quality experience (IQE) model and needs structure.

The IQE model helps us associate the customer with our product and production system. For example, customers experience/perceive physical, economic, timeliness, and customer service performance, and then shape their attitudes and behaviors accordingly.

Customers always buy on the basis of perceived benefit/burden ratios, rather than on the basis of the products themselves. For example, we buy a computer for aid in developing accounting records, engineering drawings, entertainment, and so on, as opposed to buying a sculptured lump of plastic, silicon, and metal.

Customers in turn use processes in order to gain product related benefits. These customer usage processes ultimately determine the extent of benefits and burdens that our customers reap and bear, respectively, and ultimately establish the degree of customer satisfaction or dissatisfaction experienced. With regard to our shoes, for example, we decide where to wear our shoes, how to put them on and take them off, how to take care of them, and so on, all process related.

Hence, we conclude that every endeavor involves a process—planned and practiced, or improvised and executed in an ad hoc manner. What remains to be developed are systematic and repeatable means to both better understand and create positive quality experiences for our customers.

Example 1.2

Describe the experience of quality as it relates to a pair of shoes. Describe the creation of quality as it relates to a pair of shoes.

Solution

For a pair of shoes, as external customers, we desire an experience of quality that includes foot comfort, good looks, long wear, good traction, no black marks on our floors, and so on. We may also add to our list friendly and courteous service in the retail outlet store and prompt attention to any complaints we may have after we buy our shoes.

For a pair of shoes, the creation of quality involves:

- Listening to customers' needs and demands relative to their experience of quality
- Translating these true quality characteristics into meaningful substitute quality characteristics so that we can define our product and production processes in the best interest of our customers
- Designing the product and its associated processes
- Developing the product and production processes
- Manufacturing the product
- Delivering the product to our customers
- Helping our customers acquire our product
- Helping our customers use our product to their best advantage
- Finally, helping our customers to dispose/recycle our product in an acceptable manner

The experience/creation approach to quality captures both the product nature (in the experience of quality) as well as the process nature (in the creation of quality). Hence, we can see a clear, sequential relationship between the experience and creation of quality.

If we have a reasonably good understanding (relative to our competition) of the experience of quality our customers need and expect, then we can structure our processes (aimed at the creation of quality) accordingly and reduce our risk of business failure. Kolarik [10].

PRODUCTIVITY

The productivity concept—production system efficiency—is straightforward:

Productivity is concerned with the efficient utilization of resources (inputs provided) in producing goods and/or services (outputs generated). Sumanth [12].

With reference to Figure 1.2, productivity associates quantities/costs of resources at the input to product at the output. Here, the transformation process determines our level of productivity.

We can express productivity as a ratio metric or measurement of process inputs to outputs:

$$\text{Productivity ratio} = \frac{\text{output produced per a given time period}}{\text{resource(s) consumed in production per a given time period}} \qquad \textbf{[1.2]}$$

Since we typically see a number of different resource classes as inputs to our production systems, we define two basic productivity ratios.

Partial productivity is the ratio of output to one class of input. Sumanth [12].

For example, labor productivity is the ratio of output to labor input; capital productivity is the ratio of output to the capital input; energy productivity is the ratio of output to energy input; material productivity is the ratio of output to material input; and so on.

Total productivity is the ratio of total output to the sum of all input factors. Sumanth [12].

The numerators and denominators of **partial productivity ratios may be expressed in natural units** (e.g., units of product, time, energy) or expressed in economic units (e.g., dollars). **Total productivity ratios must be expressed in economic units** in order to combine the resource classes in the denominator.

A small plastic injection molding plant last month produced 50,000 acceptable piece parts worth $300,000 in a 15,000 square foot facility (rented for $5000 per month). They have 10 presses (including tooling, maintenance, and so on) that were charged out at $120,000 for the month. The company used 200,000 kilowatt-hours of electricity during the month at a cost of $10,000. A total of 7000 labor hours was recorded for the month. The payroll including benefits for the month was $70,000. The 25,000 pounds of materials and supplies consumed for the month amounted to $50,000. Define and calculate both partial and total productivity metrics for the molding plant.

Example 1.3

Solution

A set of representative productivity calculations for partial productivity is shown below.

For labor,

$$PP_{labor} = \frac{50{,}000 \text{ parts}}{7000 \text{ hours}} = 7.14 \text{ parts/hour}$$

$$PP_{labor\$} = \frac{50{,}000 \text{ parts}}{\$70{,}000} = 0.71 \text{ parts/dollar of labor}$$

or, in terms of dollars,

$$PP\$_{labor\$} = \frac{\$300{,}000}{\$70{,}000} = \$4.29/\text{dollar of labor}$$

$$PP\$_{labor} = \frac{\$300{,}000}{7000 \text{ hours}} = \$42.86/\text{labor-hour}$$

For facility,

$$PP_{facility} = \frac{50{,}000 \text{ parts}}{15{,}000 \text{ sq ft}} = 3.33 \text{ parts/sq ft}$$

$$PP_{facility\$} = \frac{50{,}000 \text{ parts}}{\$5000} = 10 \text{ parts/facility dollar}$$

or, in terms of dollars,

$$PP\$_{facility\$} = \frac{\$300,000}{\$5000} = \$60/\text{dollar of rent}$$

$$PP\$_{facility\$} = \frac{\$300,000}{\$15,000 \text{ sq ft}} = \$20/\text{sq ft}$$

For equipment,

$$PP_{press} = \frac{50,000 \text{ parts}}{10 \text{ presses}} = 5000 \text{ parts/press}$$

$$PP_{press\$} = \frac{5000 \text{ parts}}{\$120,000} = 0.042 \text{ parts/dollar of machinery and tooling}$$

or, in terms of dollars,

$$PP\$_{press\$} = \frac{\$300,000}{\$120,000} = \$2.50/\text{dollar of machinery and tooling}$$

$$PP\$_{press} = \frac{\$300,000}{10 \text{ presses}} = \$30,000/\text{press}$$

For energy,

$$PP_{energy} = \frac{50,000 \text{ parts}}{200,000 \text{ kWh}} = 0.25 \text{ parts/kWh}$$

$$PP_{energy\$} = \frac{50,000 \text{ parts}}{\$10,000 \text{ energy}} = 5 \text{ parts/dollar of energy}$$

or, in terms of dollars,

$$PP\$_{energy\$} = \frac{\$300,000}{\$10,000} = \$30/\text{dollar of energy}$$

$$PP\$_{energy} = \frac{\$300,000}{200,000 \text{ kWh}} = 1.5 \text{ \$ parts/kWh}$$

For material,

$$PP_{material} = \frac{50,000 \text{ parts}}{25,000 \text{ pounds}} = 2 \text{ parts/pound of material}$$

$$PP_{material\$} = \frac{50,000 \text{ parts}}{\$50,000} = 1 \text{ part/dollar of material}$$

or, in terms of dollars,

$$PP\$_{material\$} = \frac{\$300,000}{\$50,000} = \$6/\text{dollar of material}$$

$$PP\$_{material} = \frac{\$300,000}{25,000 \text{ pounds}} = \$12/\text{pound}$$

We calculate total productivity in terms of dollars in both the numerator and denominator:

In total,

$$TP = \frac{\$300,000}{\$70,000 + \$5000 + \$120,000 + \$10,000 + \$50,000} = 1.18$$

Here, obviously, we must see ratios greater than 1.0 in the long term to sustain a for-profit organization.

QUALITY-PRODUCTIVITY CONNECTION

The literature points out that quality and productivity are closely connected. For example, the concept of total quality (a broad view of quality) tends to encompass productivity in general; see Juran [1]; Deming [3]; Ishikawa [9]; Kolarik [10]. Contemporary views of productivity also tend to encompass quality in general; see Christopher and Thor [13].

The external customer judges our products before, during, and after acquisition in four basic performance dimensions: (1) physical, (2) economic, (3) timeliness, and (4) customer service. Although all four dimensions pertain to both quality and productivity results, physical and customer service performance are more closely associated with classical definitions of quality, while economic and timeliness performance are more closely associated with classical definitions of productivity.

When we extend our quality-productivity connection, we see that **the creation of both quality (effectiveness) and productivity (efficiency) is a result of fundamental processes: market/definition, design/development, production, distribution/marketing/sales/service, use/support, and disposal/recycle.** An additional fundamental process is necessary to deal with the integration and support of the first six fundamental processes. **The seventh process contains leadership and management support services** that include planning, budgeting, coordination, communications, data processing, billing and payments, and so on. These support subprocesses are essential, and add value to the organization in its quest to serve external and internal customers and stakeholders. **Through our seven fundamental processes we create an experience of quality for internal and external customers as well as stakeholders.**

In summary, we make several critical observations about the quality-productivity connection:

1. External and internal customers, as well as stakeholders, are impacted by both product and process effectiveness and efficiency.

2. Competitive edge in a production system can be gained through any combination of a product or process edge developed through effectiveness or efficiency gains, with the most competitive systems typically developing competitive edges in both products and processes.

3. The working definitions of either quality or productivity typically contain both effectiveness and efficiency provisions. For example, the concept of total quality includes economics and timeliness as a part of product performance. In addition, productivity usually considers only product output that conforms to customer requirements.

Example 1.4

Studies are sometimes performed to estimate the cost of finding and correcting a specific quality-productivity problem in different stages of the product life cycle. The following breakdown of cost is typically encountered.

Life cycle phase	Cost of correction
Definition	$ 1
Design	$ 10
Development	$ 100
Production	$ 1,000
Delivery, sales, and service	$ 10,000
Use (in the field by customers)	$100,000

Explain why the costs tend to increase in order of magnitude as we go through the product life cycle.

Solution

Several major points are relevant, dealing with both quality and productivity issues: First, we are committing more and more resources to our product as we move from product definition to field use. Hence, changes require much more effort as we move through the life cycle. For example, a change can be made in product definition with the stroke of a pencil or "mouse," while at the manufacturing level, we have materials, machinery, tooling, and so on to deal with relative to the change.

Second, the life cycle is sequential in nature. For example, if we encounter a serious quality problem in the field use phase, we must flow the change back up to the top, encountering cost in each prior level, as well as in field repair.

Third, our product leaves our control after the production phase. Once our customers take delivery of our product, it is difficult and expensive to locate our products. And, once we do locate our products in the field, we must get our customers to allow us access to the product, thus interrupting our customer's activities by recalling and retrofitting the product. Or, worse yet, we might expose our customers to a potential safety hazard and possible loss of limb or life, and our organization to resulting litigation.

1.3   P<small>RODUCTION</small> S<small>YSTEM</small> L<small>INKAGES</small>

Production systems do not work in isolation. We see a universe of production systems—people, products, and processes—linked together to form integrated chains of organizations. Figure 1.4 depicts a product chain flanked on each side by a people chain and a process chain, respectively. Here, a product is created in each step through a value-added transformation and passed on to the next "organization" in the chain. For example, scrap metal enters the foundry (as $input_n$, from $supplier_n$ to $customer_{n+1}$) where a casting is produced (as $output_n$ from $supplier_{n+1}$). Here, we see $customer_{n+1}$ receive scrap metal that is transformed into a casting; thus $customer_{n+1}$ now becomes $supplier_{n+1}$ to the equipment manufacturer. This casting then enters the equipment manufacturer's process as $input_{n+1}$. Value is added by transforming the casting to a functional bulldozer part and placing it on the bulldozer ($output_{n+1}$). Here, we see $customer_{n+2}$ receive a casting and attach it to a bulldozer product, thus becoming $supplier_{n+2}$ to the dirt moving contractor. And so it goes.

Examination of this linkage points to the fact that **the product chain is a result of the process chain.** In other words, without the process chain, we would not have a product. The effectiveness and efficiency of the product and processes, together, determine the level of competitive edge each supplier-customer organization in the chain will realize. Hence, **the process as well as the products—through quality and productivity—ultimately dictate business success.**

Our purpose in subsequent chapters is to focus on the process chain and its basic structure through process definition, control, and improvement. Likewise, fundamental product issues are captured in detailed discussions of the individual quality experience and quality chain model described by Kolarik [10].

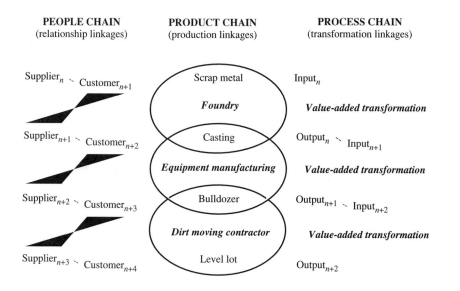

Figure 1.4   People, product, and process linkages.

1.4   COOPERATIVE EFFORTS

The production of goods and services, through processes, requires cooperative efforts. **A cooperative effort is a joint endeavor involving two or more people.** Cooperative efforts are built with personal contributions. According to Barnard, **there are three fundamental requirements to establish cooperative efforts** [14]: **(1) common purpose, (2) individual ability and willingness, and (3) interpersonal communication.** In addition, **cooperative efforts must be effective and efficient over time in order for an organization to survive.**

A common purpose, accepted by all participants, supplies direction and focus. A purpose hierarchy starts with a broad vision and mission, broken up into goals, subgoals, objectives, and finally, specific targets. Varying levels of abstractness and detail of purpose exist; for example, not-for-profit religious, political, and educational purposes tend to be much more abstract than for-profit business purposes.

Purpose may be dynamic or it may be relatively static. Nevertheless, a meaningful common purpose must exist at all times and be accepted with enthusiasm in order to generate high performance in a cooperative effort. An individual's acceptance of purpose is related to personal attitudes and manifested in behavior.

Individual ability and willingness are the second necessity. Purpose is accepted by an individual; ability and willingness spring forth from within an individual. Willingness is the commitment to cooperate, and it is attitude- and behavior-related. Ability is the potential to successfully carry out commitment.

Interpersonal communication is the third prerequisite for the formation of a cooperative effort. Communication links all cooperators so that they can function as a unit or team. There are many forms of communication: verbal (words and expressions), written (words, expressions, numbers), graphical (pictures, drawings), physical expressions (voice tone, body language), and actions (role modeling).

By definition, organizational effectiveness is the ability of an organization to accomplish, to some degree, its common purpose. This organizational accomplishment must be recognized by the cooperators and considered meaningful in terms of their individual need structures. When purposes are totally accomplished, new purposes must take their place to perpetuate the cooperative effort over time; otherwise, an organization will falter and disband. For example, sports teams/ players focus their purpose on winning the next game/match.

We desire to keep individuals who possess the ability to contribute toward accomplishing common purpose engaged in our cooperative efforts. Organizational efficiency is defined here as a ratio of benefits to burdens—a value concept.

$$\text{Organizational efficiency} = \frac{\text{benefits obtained by an individual}}{\text{burdens encountered by an individual}} \qquad \textbf{[1.3]}$$

If this ratio is far greater than 1, then a high degree of willingness is generated. Otherwise, at a ratio of about 1, an individual will fail to contribute in the cooperative effort. With respect to the organization as a whole, individuals may come and go, but the organization must retain a net positive efficiency. For example, a sports team must attract and retain enough players to make a full team; otherwise it cannot compete.

Benefits and burdens are multidimensional—usually, we do not receive benefits in like kind to the burdens we endure. For example, we give our time and ideas in return for salary, interpersonal recognition, and/or intrapersonal feelings of warmth, or perhaps power. In other words, organizational effectiveness and efficiency are determined relative to the need structures of individual contributors, e.g., physiological, safety-security, social, esteem, and self-actualization needs. See Kolarik [10] for a discussion of these needs related to product quality, and see Maslow [11] for a detailed discussion of human needs.

1.5 ORGANIZATIONAL OPTIMIZATION AND SYNERGY

Purpose or direction in general is expressed in three basic dimensions: (1) vision, (2) mission, and (3) core values statements. A vision is a construct of the mind that allows us to see beyond today. A vision has value in what it allows/inspires us to do, rather than what it is as an abstract entity. A vision statement sets forth a position that the organization is expected to attain, e.g., "to be recognized by our customers as the best. . . ." Sometimes a vision will contain a time frame for accomplishment, e.g., ". . . by the year. . . ."

A mission statement typically declares what business or customer needs the organization serves, e.g., "to bring the highest value . . . solutions to our . . . customers." Or, for example, "we produce . . . using innovative processes such that we create satisfied customers for our products, satisfied and empowered employees . . . , and satisfied stakeholders. . . ."

A core value statement typically identifies and addresses customers, employees, stakeholders, products, service, and satisfaction to one degree or another. Key words may include customer focus or expectations, personal responsibility, teamwork, improvement, costs, product value, and so on. For example, "We deliver high performance . . . products and services at a fair price through innovative technical and business processes, carried out in an ethical manner. . . ." Or "We will establish and maintain employee leadership and creativity by providing education, training, and opportunity for all employees to define, design, and implement best practices that yield customer, employee, and stakeholder satisfaction."

The exact titles, words, and details vary widely for the vision, mission, and core value statements. However, several generalizations can be made: (1) the statements are brief so they will be read and remembered; (2) the statements are well publicized both within the organization to employees and outside the organization in advertisements addressed to customers, stakeholders, and the general public; and (3) the statements usually appear together to complement and reinforce the message of each.

The leadership and managerial challenges within an organization are substantial. Declaring a common purpose and direction is not in itself a difficult task. Gaining acceptance of common purpose and implementing effective, efficient, and sustained action toward accomplishing the common purpose are a significant leadership and managerial challenge. **Leadership and management are different;** see Kolarik [10]:

Leadership deals with vision and the subsequent defining and delineating of purpose as well as cultivating belief in purpose and motivating people to act accordingly.

Management entails providing, metering, and monitoring the resources necessary to sustain the action sparked by leadership.

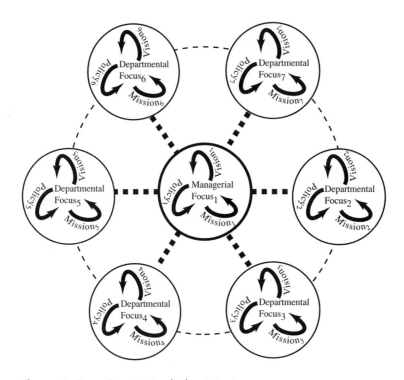

Figure 1.5   Organizational suboptimization.

In a departmental structure there is a tendency for each department to "optimize" internally. In Figure 1.5 we see internal focus and internal "optimization." For example, the people within an identifiable functional unit, such as a purchasing department, a design engineering department, or an accounting department, may attempt to optimize their specific unit without, or with limited, regard for the remainder of the organization. This internal optimization is natural, considering the multiplicity of functional/departmental visions and missions.

Several "optimal" departments typically do not create an optimal organization. It is not unusual to see competition between departments or functional units for resources to further their own functional purposes, rather than a focus on a broad customer-centered organizational purpose—one clear, explicit vision, mission, and core values expression relevant to the entire organization.

| Example 1.5 | **AIRCRAFT FLEET SUPPORT CASE—STRONG DEPARTMENTS/WEAK PROCESSES**   The customer service department of a large U.S. transport aircraft manufacturer receives all incoming correspondence from the company's airline customers. The customer service organization provides a "firewall" buffer between the company's many technical units and its customers. All customer communications are intercepted, dissected, and disseminated to the appropriate internal departments regarding any issue under question. The customer service department collects the internal responses from the involved departments, prepares the company response, and provides it to the original point of inquiry. Customer service |

representatives determine which departments are consulted concerning a given issue; they may choose to avoid consultation with a particular department if they feel that it impedes the timeliness of their response, or if they anticipate an answer that they believe will be displeasing to the customer.

A particular Pacific Rim airline was experiencing problems with cracking in the leading edge flap panels on several of their aircraft. The flap panels in question were made of fiberglass and were sustaining cracks in the area adjacent to an engine pylon. Replacement of these panels became an annoying and costly problem that resulted in a continuing dialog with the aircraft manufacturer. The manufacturer's customer service department worked the problem with the help of the company's structural engineering department. They decided that the cracks occurred in the end of the flap panels because of physical contact with the engine pylon, which, they reasoned, must have been occurring as the engine swayed on the pylon during flight operations.

As a solution, the manufacturer recommended that the fiberglass flap panels be trimmed—in hopes of avoiding the assumed contact with the engine pylon. The airline complied and modified its fleet accordingly. Continued flight operations proved that this fix did not solve the problem, and cracks continued to develop, resulting in continued costly repairs.

Again, it was reasoned that the pylons must be contacting the flap panels to cause the cracking, and again, the manufacturer recommended that the airline trim the panels back a little farther. For good measure, it was also recommended that a few plies of fiberglass be added in certain areas to reinforce the panels. The airline again complied, trimming more from the flap panels and adding fiberglass to the back of the panels to strengthen them.

At this time, the structural engineering department decided to collect flight data. This test was the first attempt to acquire engineering data. A flight test was planned with the intent of getting load data on the flap panels. In addition, a video camera was installed in the flap cavity to record any contact between the flap panels and the engine pylon. As a courtesy, an aerodynamics staff representative was invited along to observe this test, since the problem flap panels were, in fact, aerodynamic devices.

The results of the structural flight test were twofold. First, there was no physical contact occurring between the flap panels and the engine pylons. In fact, the measured pylon movement was very close to the original design predictions and was far less than what would be required to cause contact with the flap panels. Second, it was apparent that the integrity of the aerodynamic seal between the flap panel and the pylon was lost due to the increased gap between the flap panels and the engine pylons, a problem overlooked by the structural engineers but noticed by the aerodynamicist who was "brought along for the ride."

The flight test uncovered the fact that flap panel cracking was not associated with the panels contacting the engine pylons in flight. Unfortunately, the prior fix (widening the space between the flap panel and the pylon) of the nonexistent contact problem resulted in significant aerodynamic changes to the aircraft that could not be ignored. Compounding the problem, the airline had been instructed to modify its entire fleet, which by this time it had completed. Now, the aerodynamic issue existed across the airline's entire fleet of aircraft. Fortunately, it was noted that aircraft that had their flap panels reinforced at the time of the flap panel trimming appeared to show no further signs of cracking. The haste in correcting the cracking problem had introduced an aerodynamic problem.

At this point, the aircraft manufacturer was forced to conduct further testing to determine if the aerodynamic changes caused by the initial flap panel trimming could be tolerated. Laboratory tests and then a very expensive series of fully instrumented performance flight tests were ultimately required. The results of these additional tests were that the aerodynamic changes brought about by the structural fix caused unacceptable effects on the airplane's performance and handling qualities. As a result, the manufacturer took corrective action to restore the customer airline's fleet of aircraft to the original aerodynamic configuration.

Epilog

While these events are unique, they are symptomatic of the type of events that seem to happen frequently in many organizations because of overspecialization, organizational barriers, lack of communication, decision making without data, and overzealousness. The modifications made to the airline customer's fleet, while well intentioned, were made in haste, by structural experts, without engineering data as to the cause of the original cracking and without understanding the aerodynamic consequences of what appeared to be an inconsequential external change to the aircraft. This hasty action, taken without proper investigation into the root causes of the problem, resulted in an expensive and embarrassing incident for the manufacturer. Fortunately, the added panel stiffening solved the original cracking problem. [KC]

A process-based organization, as depicted in Figure 1.6, **aligns processes through a series of process purposes.** Each process purpose is aligned with and supports the common purpose expressed in our vision, mission, and core values. **We see a natural customer/product focus in the process basis, as opposed to the natural functional focus in the departmental basis. This customer/product focus nature encourages value-added activity and discourages non-value-added activity.**

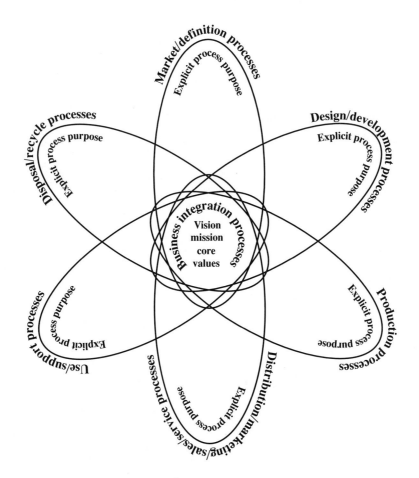

Figure 1.6   Organizational optimization.

In Figure 1.6 we see a natural alignment, converging in the center area of the figure, so that coordination and communication can be established in the form of cross-sectional business integration processes that optimize the organization as a whole. Here, strategically defined processes, operating in a state of control and improved on an ongoing basis, provide synergistic effects.

Each of the fundamental processes, as depicted in Figure 1.6, is composed of subprocesses, to form an integrated process hierarchy/system. Table 1.1 identifies several subprocesses associated with the basic processes. These labels are generic, and aligned with manufacturing production systems. Service production system subprocesses are similar but contain subtle differences in each fundamental process. In all cases, subprocesses and their purposes are tailored to support our specific vision, mission, and core values, which are in turn tailored to meet customer needs and expectations and to create value for our customers and stakeholders.

Table 1.1   Process hierarchy example (manufacturing).

Fundamental Process	Selected Subprocesses
Market/definition	Market characterization, e.g., customer identification; customer needs and expectations Product and production system characterization, e.g., product/production concepts
Design/development	Product and production system design, e.g., targets, requirements, and specifications determination Product and production system proof of concept and capability
Production	Materials acquisition Fabrication Assembly Materials handling
Distribution/marketing/sales/service	Promotion/customer information Packaging Transportation/loading/unloading Sales/selling Product installation Product warranty and returns
Use/support	Product setup User training Usage Service and maintenance
Disposal/recycle	Collection Transportation Disposition/recycling
Business integration (support services)	Business planning/budgeting Process coordination Training Information/communication Record keeping and reporting

Example 1.6	**SOFTWARE DEVELOPMENT CASE—STRONG PROCESSES**　　Project E was a large software development program. Project E's charter was to upgrade the current customer software baseline with faster processing time and leading-edge client/server technology under a tight schedule. Project E based all of the program processes on existing company process documents. The integrated product team concept was used to implement the project. Each engineering team had its own set of process documents that included quality checks of the products and artifacts. Each phase of the project had a quality control gate for delivery to the next phase.

The integrated product team consisted of members from each of the engineering disciplines. Each member had a well-defined role in the project development, design, and implementation as a result of company process documentation. Members of the integrated product team were quality engineering, systems engineering, integration and test engineering, software engineering, and configuration management. Members were matched to a project segment based on their knowledge and skills in that area. With the integrated product team in place, Project E was started.

Project E started with customer requirements. The system team, with participation of the test team, addressed the requirements. The requirements were allocated and broken down to software design requirements and placed in a relational database for ease in traceability from requirement down to the verification matrix. The quality control gate was a requirements and verification review by the customer and the integrated product team. Approval by the customer and the integrated product team was the entry key to the next phase, the software development phase. Quality engineering reviewed and signed off the requirements for accuracy in traceability, allocation, and proper documentation.

The next phase was the software development. The software development was a difficult task. The client/server technology was not easy to implement. A change in computer hardware and operating systems was warranted to accomplish the task. Software design inspections and software code inspections by the integrated product team identified design and implementation defects. These defects were noted in quality metrics and fixed. The client/server software had performance problems when it was first installed in the test system. Per the software and systems process documents, a "tiger" team was assembled to work the performance problems. It was determined that the hardware and operating systems had to be tuned to accommodate the performance requirements of the client/server interface. The software baseline also was tuned. After a short rework time, the performance problems were solved.

The test team was responsible for the early detection of the performance problems as well as other project impact problems. The integration and test phase began with the start of the software development phase. Early detection of the problems saved the project many costly hours of rework later in the project schedule. The integration and test phase tested the software and hardware baseline under configuration management control. All test plans and procedures were inspected and reviewed by the integrated product team for requirement traceability, verification traceability, and test design accuracy. Quality engineering provided a detailed checklist for the test reviews and inspections. Changes to the test requirements were made in order to fix several disconnects in requirements and verification traceability and software design. The integrated product team concept worked effectively and efficiently. Because of the integrated product team concept, Project E installation at the customer's site was very successful.

Epilog

Project E benefited from well-disciplined, integrated teams. Each team had defined processes in place to handle risk management, quality, schedule, rework, and other project impact issues. Each team member was trained and understood his/her role and responsibilities. The integrated product team concept, along with a systematic quality engineering process, provided successful definition, design, implementation, and installation of a challenging, leading-edge product, under a tight schedule, for a discerning customer. The customer was so satisfied with the results of Project E that each member of Project E received a monetary bonus for a job well done. [RC]

Each subsequent chapter of this text builds on a process-based theme, ultimately offering the reader alternative concepts and guidance in organization transformation. **The process-based structure offers several natural advantages over the departmental-based structure:**

1. **The process basis focuses primarily on outputs,** e.g., outcomes.

2. **The departmental basis focuses primarily on inputs,** e.g., resources.

The distinction of focus has a profound impact on both quality and productivity. From the quality perspective, the process basis naturally orients itself toward the customer in terms of process outputs/outcomes. From the productivity perspective, the process basis pulls resources as needed to create outputs/outcomes. Hence, waste in the form of excessive resources is exposed/prevented in process-based structures.

In contrast, the departmental basis tends to focus on inputs/resources so that customers and customer outcomes become secondary. The resource-focused departmental perspective allows hiding places for waste/non-value-added activities in the form of excessive resources and product features that may be of little importance to customers.

Departmental structures can and do develop competitive edges in quality and productivity. However, the natural tendency in departmental structures works against these efforts, whereas natural tendencies in process-based structures work with these efforts. **In head-to-head competition, process bases have distinct advantages due to their natural customer outcome orientations and ability to expose waste/non-value-added activities.**

Recognition of the criticality of process structures and performance is not new. Engineering disciplines such as industrial and chemical are developed around processes. Recent major business-related initiatives involving heavy process emphases include the Malcolm Baldrige National Quality Award [15] and ISO 9000 [6]. These and many other process-based initiatives are described in Chapter 19. Other efforts stressing process-related attributes in the form of process maturity/evaluation efforts exist [16, 17].

REVIEW AND DISCOVERY EXERCISES

REVIEW

1.1. Identify the fundamental components of a process. What purpose does each serve? Explain.

1.2. For a production system of your choice: (*a*) Identify a product and one true quality characteristic. Then, identify two related substitute quality characteristics. (*b*) For the product in part *a,* identify a related process and one true process quality characteristic. Then, identify two related substitute quality characteristics.

1.3. Identify and describe the need levels in Maslow's hierarchy of human needs.

1.4. How does the productivity concept differ from the quality concept? Explain.

1.5. Given the following basic production statistics, determine a set of partial and total productivity statistics.

Production: 250,000 units, worth $1,000,000

Facility: 40,000 sq ft, monthly rental of $15,000

Utilities: 500,000 kWh, at $0.06 per kilowatt-hour, 20,000 mcf (thousand cubic feet) of natural gas at $1.75 per mcf, 200,000 gallons of water at $2.50 per 100 gallons

Labor: 12,000 direct labor hours at $12.50 per hour (includes all benefits), 5000 hours of indirect labor at $15.00 per hour (includes all benefits)

Management: 3 person-months of management at a total of $24,000 (includes all benefits)

1.6. Develop a sequence of production system linkages for a product of your choice. Refer to Figure 1.4 before formulating your response.

1.7. Explain why organizations prosper. Explain why organizations die. What is the functional difference? Explain.

1.8. Explain the difference between the concepts of management and leadership.

1.9. Explain the difference between the concepts of a department and a process.

1.10. What is the difference between the two organizational structures depicted in Figures 1.5 and 1.6, respectively? Can both structures coexist in the same organization? Explain.

DISCOVERY

1.11. From the variety of definitions of quality presented in Chapter 1, develop a set of key words from each, and identify common elements.

1.12. Select a product of your choice. Describe the physical performance, economic performance, timeliness performance, and customer service performance that you need/expect.

1.13. The quality concept is described in many ways, e.g., definitions of quality. We see definitions that imply the maximization of benefits, and we see at least one definition that implies the minimization of losses. Will both approaches lead to the same basic results/outcomes? Is one approach better than the other? Should some processes/products be approached from maximization and others from minimization? Explain.

1.14. Why does such a variety of human needs and human need priorities exist?

1.15. For decades people believed that productivity and quality were traded off. In other words, the conventional wisdom postulated that in order to enhance one, we had to give up the other (to some degree). Today, we believe that we can simultaneously enhance both. What has changed? Explain.

1.16. Identify an organization in which you have enthusiastically participated. Describe the common purpose, willingness, and communication elements of that organization. Identify the organizational effectiveness and efficiency that resulted during the time period of your association.

1.17. Identify an organization that you have quit in disgust. Describe the common purpose, willingness, and communication elements of that organization. Identify the organizational effectiveness and efficiency that resulted during the time period of your association.

1.18. Critique the aircraft fleet support case. Structure your critique along lines of departmental versus process bases. Could a process basis have prevented the outcomes experienced in this case? Explain why.

1.19. Critique the software development case. What elements helped to provide positive results/outcomes? Explain.

REFERENCES

1. Juran, J. M., *Juran on Leadership for Quality*, New York: Free Press, 1989.

2. Crosby, P. B., *Quality Is Free*, New York: McGraw-Hill, 1979.

3. Deming, W. E., *Out of the Crisis*, Cambridge, MA: MIT Center for Advanced Engineering Studies, 1986.

4. Feigenbaum, A. V., *Total Quality Control*, 3rd ed., New York: McGraw-Hill, 1983.

5. Taguchi, G., *Introduction to Quality Engineering: Designing Quality into Products and Processes*, White Plains, NY: Kraus International, UNIPUB (Asian Productivity Organization), 1986.

6. *ISO International Standards for Quality Management*, Geneva, Switzerland: International Organization for Standards, 1992.

7. Radford, G. S., *The Control of Quality in Manufacturing*, New York: Ronald Press, 1922.

8. Shewhart, W. A., *Economic Control of Quality Manufactured Product*, New York: Van Nostrand, 1931.

9. Ishikawa, K., *What Is Total Quality Control? The Japanese Way*, Englewood Cliffs, NJ: Prentice-Hall, 1985.

10. Kolarik, W. J., *Creating Quality: Concepts, Systems, Strategies, and Tools*, New York: McGraw-Hill, 1995.

11. Maslow, A. H., *Motivation and Personality*, 2nd ed., New York: Harper and Row, 1970.

12. Sumanth, D. J., *Productivity Engineering and Management*, New York: McGraw-Hill, 1984.

13. Christopher, W. F., and C. G. Thor, eds., *Handbook for Productivity Measurement and Improvement*, Portland, OR: Productivity Press, 1993.

14. Barnard, C. I., *The Functions of the Executive*, Cambridge, MA: Harvard Press, 1938.

15. "Malcolm Baldrige National Quality Award—1998 Award Criteria," Washington, DC: U.S. Department of Commerce, 1997.

16. Bate, R., et al., "A Systems Engineering Capability Maturity Model," Report SECMM-95-01, Carnegie Mellon University/U.S. Department of Defense, 1995.

17. Paulk, M. C., "How ISO 9000 Compares with the CMM," *IEEE Software,* 0740-7459, pp. 74–83, January 1995.

2

SYSTEMS THINKING—CONCEPTS AND DEVELOPMENT

2.0 INQUIRY

1. How do analytical and systems philosophies differ?
2. What do analytical and systems procedures offer process engineering?
3. What are models? Why are they useful? How are they used?
4. Why do we see different production system paradigms?
5. What forces drive production system evolution/change?

2.1 INTRODUCTION

The pragmatic process basis described in Chapter 1 utilizes scientific/analytical concepts as well as general system theory concepts. The purpose of this chapter is to provide a background regarding these two fundamental procedural approaches, and to justify the pragmatism utilized throughout the remaining chapters. Our pragmatism is due to the inability of either the analytical or systems concepts to provide a sufficient basis for understanding the nature of a production system and its processes. **By combining both analytic and systems thinking, we can enhance both quality and productivity results: physical, economic, timeliness, and customer service.**

2.2 ANALYTICAL AND SYSTEMS PHILOSOPHIES

The process basis draws on two recognizable cognitive thinking philosophies: (1) the analytical procedures of science and (2) the holistic procedures of general systems theory. We

resort to a pragmatic combination of these two basic procedures because neither one is complete enough to assess or model complex production systems.

The analytical procedures of science stress the resolution of a whole into isolatable causal units, cascading down to fundamental units—atomic/subatomic units. This philosophy breaks the whole into parts, which are then studied and explained in detail. These parts are then "reassembled" to explain the whole. **As long as these parts are basically independent in operation and system impact—possess no or insignificant interacting characteristics—the analytical philosophy produces good results.**

The scientific approach has evolved over many centuries. Beginning with Greek science in about 600 B.C., science developed from logical argument to the emergence of observation and empiricism. Major contributions are many—the Arabic numbering system, the concept of inductive science, the concept of experimental science, and the methods of controlled experimentation are but a few highlights.

The acceptance of observation (in addition to reasoning) as a legitimate form of gaining knowledge required about 2000 years, spanning from Hippocrates's time, about 400 B.C., to Bacon's time, about 1600 A.D., eventually yielding modern experimental inquiry. Three essential characteristics of experimental inquiry are cited by Checkland [1]: (1) reductionism, (2) repeatability, and (3) refutation. The complexity of our world is reduced in order to perform experiments; we simplify the world by selecting a few variables to manipulate and control (hold constant) the remainder as best we can in our experiments. Experimental results are validated through repeatability, by independently running the experiment and comparing results for consistency. Finally, hypotheses are established, and refutation of hypotheses is used to support scientific conjecture. We structure null and alternative hypotheses; if we see sufficient evidence, we reject the null hypotheses in favor of the alternative hypothesis.

In contrast to the specificism of analytical science, **classical systems theory takes the approach that interrelationships in the whole preclude or limit the analytical approaches.** In other words, general systems theory counters scientific specificism and reductionism, suggesting the study and examination of a system as a whole, rather than by parts. General systems theory is a manifestation of the twentieth century.

> The system problem is essentially the problem of the limitations of analytical procedures in science. . . . Analytical procedure means that an entity investigated be resolved into, and hence can be constituted or reconstituted from, the parts put together, these procedures being understood both in their material and conceptual sense. Ludwig von Bertalanffy [2].

The point is that complex systems are more than the sum of the properties of the components taken separately. Complex systems are explained not only by their components, but also by the relationships between their components—hence, the holistic view of the system in its entirety, rather than by basic parts.

Our perspectives in this book stress engineering more than science. **Science is based on the "urge to know" whereas engineering is based on the "urge to do."** Many great inventions preceded scientific inquiry and the scientific method. For example, pottery, paper, and gunpowder were not products derived from scientific inquiry. However, **when modern scientific methods are coupled with the need to accomplish practical ends, a synergy transpires,** yielding new and improved products and processes.

Our engineering-oriented view recognizes the likelihood that production systems cannot be broken into "atomiclike" units and totally explained in this most basic context. But, because of complexity, on the other hand, neither can they be totally defined, controlled, and improved on the basis of the whole alone. According to Boulding [3], if we rely only on specificism, we risk a loss of meaning, and if we rely only on generalism, we risk a loss of content. Hence, we encourage both specificism and generalism to understand and master production systems.

2.3 MODELS AND MODELING

The basis for our definition, control, and improvement of production systems lies in modeling. We model specific relationships. We propose models to deal with general interrelationships. We use models that rely on simple verbal logic. We use models that involve equations that predict values of random variables. These models play critical roles in helping us to understand and predict physical and social behavior without resorting to full-scale experiments. Today, engineering and business models are at the core of our understanding.

For our purposes, **we classify models into two basic categories: (1) physical and (2) abstract.** Physical models can be divided into iconic and analogic models. Iconic models are scaled models. Here, we usually see miniature, partial-scale models, but we could just as well see an enlarged scale model. Iconic models differ from the real system only by a scale factor. Analogic models are physical models, but they reproduce representative system behavior, not the actual system elements. For example, voltages and liquid levels are used to represent temperature levels in temperature sensors, e.g., thermocouples and thermometers, respectively.

Abstract models can be subclassified as quantitative or qualitative. A quantitative model uses mathematical and/or logical relationships relative to basic laws or empirical model structures. Physical laws and response surface models are examples of quantitative models. Qualitative models utilize sequences of symbols and words to describe system components and the interrelationships between components. **Typically, we begin our modeling efforts with qualitative models and then proceed to quantify them as far as possible.** Both quantitative and qualitative models are used frequently in subsequent chapters.

2.4 SYSTEMS THEORY AND THINKING

Systemwide thinking is certainly not new. People since the beginning thought in terms of system outcomes. With the advent of our recognition of, and attempt to explain, complex systems—physical, biological, and social—we evolved the concepts of modern analytical techniques. This evolution required several thousand years and tended to focus on breaking the whole down to fundamental pieces, observing causality, and explaining the whole as the sum of the pieces.

Inadequacies with analytical methods and classical closed-system models were pointed out by several people over the years. However, it was not until the 1920s that Ludwig von Bertalanffy described modern systems theory with the formulation of basic concepts regarding a biological organism as an open system. He later consolidated his biological theory into a general systems theory with interdisciplinary applications (Laszlo [4]).

A living, open system is defined as a system in exchange of matter with its environment, presenting import and export, building-up and breaking-down of its material components. Von Bertalanffy [2].

The profound nature of an open system (as opposed to a closed system) is important.

The system remains constant in its composition, in spite of continuous irreversible processes, import, and export, building-up and breaking-down, taking place. The steady state shows remarkable regulation characteristics which become evident particularly in its equifinality. If a steady state is reached in an open system, it is independent of the initial conditions, and determined only by the system parameters, i.e., rates of reaction and transport. This is called equifinality as found in many organismic processes, e.g., in growth. In contrast to closed physico-chemical systems, the same final state can therefore be rendered equifinally from different initial conditions, and after disturbances of the process.

According to the second principle of thermodynamics, the general trend of physical processes is towards increasing entropy, i.e., states of increasing probability and decreasing order. Living systems maintain themselves in a state of high order, or may even evolve toward increasing differentiation and organization as is the case in organismic development and evolution. Von Bertalanffy [2].

Since its inception, systems thinking has been applied to virtually every endeavor known to humanity. It has seen application to biology/life sciences, engineering, psychology, sociology, and economics/commerce, for example. Both proponents and critics abound. See Ellis et al. [5] for more details.

Perspectives in general systems theory stress holism and synergy. For example, a production system is more than the sum of several basic processes. Market/definition, design/development, production, distribution/marketing/sales/service, use/support, and disposal/recycle each serve a process purpose; however, the impact on society as a whole is more than the simple sum of the processes. Here, the simple sum of the processes results in a product for sale. But, if we consider the benefits/burdens that our product brings to society, the impact is typically far greater (for better or worse) than the sum of the parts.

SYSTEM CLASSIFICATION

Systems abound in our physical, economic, and social environments. They are highly varied in their general nature and structure. **Checkland [1] identifies four basic system classes:**

1. **Natural systems:** Systems whose origin is in the origin of the universe. These systems are a result of the forces and processes of the universe, e.g., the planets and the atmosphere.

2. **Designed physical systems:** Systems whose origin is human. These systems are designed as a result of some human purpose, e.g., bridges and automobiles.

3. **Designed abstract systems:** Systems whose structure is nonphysical. These systems represent the ordered conscious product of the human mind, e.g., mathematics and poetry.

4. **Human activity systems:** Systems whose structure is based on human activities more or less consciously ordered as a result of some underlying purpose. These systems may include other physical and/or abstract systems; for example, sports activities represent a human activity system played on a physical system—the field of play.

For our purpose, **a complex production system is composed of combined designed physical, designed abstract, and human activity systems.** This combination is apparent throughout our subsequent discussions of process definition/redefinition, control, and improvement.

SYSTEMS THINKING

Checkland [1] and Wilson [6] describe **hard and soft systems** and systems thinking. **The term "hard" is applied to systems with clear missions/goals.** These systems tend to be highly structured with clear purposes. **Soft systems tend to lack clear (agreed-upon) missions/goals.** Here subjectivity tends to produce a lack of structure. Examples of hard systems include buildings and transportation networks. Examples of soft systems include managed health care and education.

Checkland [1] **concludes that hard system thinking methodologies are a special case of soft system thinking methodologies.** Checkland offers a general systems model, as depicted in Figure 2.1, which includes seven stages. Stages 1 and 2 address the expression of the situation wherein the perception of a problem, or opportunity, lies. Regarding a hard system, e.g., a physical system, this expression may be explicit, whereas regarding a soft system, e.g., most human activity systems, such expression may be in terms of a multitude of personal perspectives. These perspectives present different pictures of the situation. For example, we might present many diverse internal or external customer perspectives as well as stakeholder or competitor perspectives within a soft production system situation regarding managed health care.

Stage 3 involves developing root definitions. This expression defines the resolution level of what is taken to be the system, relevant to the situation expressed in Stage 2. The root definition

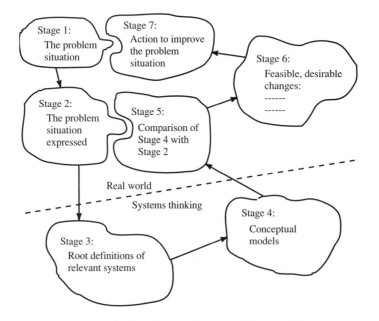

Figure 2.1   Checkland's general systems thinking model.

addresses relevant system properties. A clear root definition typically focuses on "what" rather than "how." Relative to a production system, considerations such as customers, stakeholders, transformation processes, interaction with other systems including the environment, and the general meaningfulness of the definition itself are all included. In general, any root definition relates the transformation of system inputs to outputs.

Conceptual models are developed in Stage 4. A conceptual model represents the minimum, necessary sets of activities that the system must encompass in order to meet the root definition, e.g., transform the inputs to outputs. Conceptual models refer to activity systems and are therefore described with verb phrases, e.g., describe customers. The result is a set of verb phrases with regard to any specific human activity system. The root definition and the conceptual model lie in the systems thinking world, e.g., the abstract world of models.

Entry into Stage 5 relies on the model builder's judgment. This comparison stage brings intuitive perceptions of the situation from Stage 2 together with system constructs from Stage 4. This overlay/comparison may require returning to the root definition and/or modifying the conceptual model. Such comparisons may lead to extended debate regarding either.

Stages 6 and 7 focus on feasible changes/alternatives and action, respectively. Hard systems thinking is straightforward as to system design and implementation, once alternatives are developed and evaluated and a decision for implementation is rendered. Soft systems are less straightforward in both changes and implementation.

According to Checkland [1], the hard system approach addresses the question, "What system, exactly, is to be engineered to address the situation or meet a need?" In contrast, the soft system must allow for unexpected answers to emerge at later stages. Hence, Stage 5 becomes critical in the soft approach in order to allow/encourage these unexpected answers as to both need and proposed systems.

ADAPTATIONS OF GENERAL SYSTEMS THINKING

The concept of systems thinking is straightforward; however, the application is complex. In analytical thinking, we have an advantage in application in that we break the whole into pieces, subdividing until we find a manageable chunk to study. In systems thinking, we consider the whole, and hence, typically, focus on large multifaceted entities. **A broad focus on a complex entity challenges us in developing mental images of each of the parts and their interrelationships simultaneously.** Nevertheless, systems thinkers persist in experimenting with such imagery. Two contemporary applications of systems thinking, relative to production systems, include the Fifth Discipline, Senge [7], and the Theory of Constraints, Goldratt [8] and Dettmer [9].

Senge proposes a combination of systems thinking, personal mastery, mental models, shared vision, and team learning in his initiative aimed at building the "learning" organization. A learning organization is viewed as an entity of people, focused on a common vision, working collectively in both mental and physical dimensions to grow and evolve over time. Here, systems thinking includes building cause-effect circles to model interrelationships as well as observing/seeing processes of change over time embedded in these circles, as opposed to seeing the system as a snapshot, at one instant of time. In general, the learning organization's system context is viewed as an "intuitive" closed-loop control system, with feedback and feedforward characteristics. Chapter 3 provides background and Chapters 14 and 15 provide details regarding closed-loop systems and process control.

The purpose of Goldratt's Theory of Constraints (TOC) is to expose and address production system limitations—the things that prevent a production system from attaining its goals or accomplishing its objectives. The primary premise in TOC is that we can readily identify our overall goal, e.g., profits. This goal then drives the decisions we make and everything we do as a result of the decision. The TOC initiative focuses on locating constraints and breaking them down so that they are no longer constraining factors. Basic principles underlying TOC include:

Systems thinking is preferable to analytical thinking.

An optimal system solution deteriorates over time as the system's environment changes.

The system optimum is not the sum of the local optima (of the parts).

Systems are analogous to chains. Each has a weakest link. Strengthening of any link in a chain, other than the weakest link, does nothing to improve the strength of the chain as a whole.

Knowing what to change requires an understanding of the system's current reality and its goal, and the difference between the two.

Most of the undesirable effects within a system are caused by a few core problems.

System constraints can be of a physical or a policy nature.

The TOC principles are practiced through a sequence of tree/relationship diagrams. Both Fifth Discipline and Theory of Constraints initiatives are described further in Chapter 19.

2.5 PRODUCTION SYSTEMS

The wide array of production systems that surround us today has been shaped by many people and events over the last 4 centuries. This shaping and transformation process is still at work and will continue to change the ways in which we define, design, develop, operate, and improve our production systems. In this section we briefly describe this shaping and transformation, characterizing major breakthroughs in production system technology wrought by the pioneers of production systems. This shaping and transformation involves people, products, and processes. **People are the key components; they shape products and processes through leadership and creativity-based endeavors, in response to customer needs, demands, and expectations.**

Production systems and the processes that configure them are constantly changing. Over extended periods of time, major changes have occurred. **Five major production system paradigms are observed: (1) the craft system, (2) the factory system, (3) the mass production system, (4) the lean production system, and (5) the agile production system.** Each is summarized in Table 2.1.

None of these paradigms was devised and created through perfect planning. Rather, they evolved over long periods of time, with the aid of shifts in economic and market conditions as well as technological advances. We can use general systems theory to help explain this evolution and clearly see that the pioneers of these paradigms were thinking in systemwide terms. For example, the Ford Motor Company's vision in the early 1900s of automobiles for the masses clearly looked beyond the automobile as an assembly of parts. Likewise the views of proponents for telephones and microcomputers also looked beyond their physical boundaries, in terms of systems. We can

Table 2.1 Production system paradigms and characteristics.

Paradigm and General Nature	General Requirements	Competitive Advantage
Craft production system: Product 　Custom configuration 　(tailored to customer needs) Process 　Custom configuration 　(manual, whole-process 　skill basis)	Limited product demand Skilled craftsperson Basic hand tools	Performance **Physical** (function, form, fit) |+++++++++++++++++++ | low　　　　　　　　high **Economic** (product cost) |++　　　　　　　　| expensive　　　inexpensive **Timeliness** (time to market) |++++　　　　　　| slow　　　　　　　　fast **Customer service** (customer assistance) |+++++++++++++++++++ | low　　　　　　　　high **Flexibility** (product/process changeovers) |+++++++++++++++++++ | difficult　　　　　　easy
Factory production system: Product 　Common core configuration 　(selected options, customized 　details) Process 　Standard configuration 　(mechanized production, 　job specialization)	Expanded product demand Semiskilled labor pool Basic hand tools Factory power/machines Large amounts of capital Specialization of labor and management	Performance **Physical** (function, form, fit) |+++++++++++ | low　　　　　　　　high **Economic** (product cost) |++++++++++++++ | expensive　　　inexpensive **Timeliness** (time to market) |++++++++ | slow　　　　　　　　fast **Customer service** (customer assistance) |++++++ | low　　　　　　　　high **Flexibility** (product/process changeovers) |++++++++ | difficult　　　　　　easy
Mass production system: Product 　Common configuration 　(selected detail options) Process 　Standard configuration 　(mechanized production and 　mechanized assembly lines, 　rigid labor and management 　specialization)	Mass product demand Unskilled labor pool Factory power/machines Very large amounts of capital Specialization of labor and management Interchangeable parts Mechanized/integrated production lines Hard selling	Performance **Physical** (function, form, fit) |+++++++ | low　　　　　　　　high **Economic** (product cost) |++++++++++++++++++++ | expensive　　　inexpensive **Timeliness** (time to market) |++++ | slow　　　　　　　　fast **Customer service** (customer assistance) |++++ | low　　　　　　　　high **Flexibility** (product/process changeovers) |++++ | difficult　　　　　　easy *(continued)*

Table 2.1   Production system paradigms and characteristics *(concluded).*

Paradigm and General Nature	General Requirements	Competitive Advantage

Lean production system:
  Product
    Several configurations
    (product mix with basic
    options)
  Process
    Standard configuration
    (mechanized/automated line
    flow, empowered work teams,
    whole process focus)
	Expanded product choice demand
	Educated employees
	Factory power/machines
	Large amounts of capital
	Interchangeable parts
	Statistical methods
	Empowered employees
	Customer relationships
	Supplier partnerships
	Computer-aided design
	Computer-aided manufacturing
	Market segmentation
		Performance
		
		Physical (function, form, fit)
		I+++++++++++++++++ I
		low high
		Economic (product cost)
		I++++++++++++++++++++ I
		expensive inexpensive
		Timeliness (time to market)
		I+++++++++++++++ I
		slow fast
		Customer service (customer assistance)
		I+++++++++++++++ I
		low high
		Flexibility (product/process changeovers)
		I+++++++++++++ I
		difficult easy

Agile production system:
  Product
    Custom configuration
    (tailored to customer needs)
  Process
    Custom configuration
    (high-tech communications,
    design, production, flexible
    teams, whole-process
    partnerships)
	Custom product demand
	Very well-educated employees
	Factory power/machines
	Large amounts of capital
	Interchangeable parts
	Statistical methods
	Empowered employees
	Customer partnerships
	Supplier partnerships
	Interorganizational production
	partnerships
	Computer-aided design
	Computer-aided manufacturing
	Computer-integrated
	communications, definition,
	design, production
		Performance
		
		Physical (function, form, fit)
		I++++++++++++++++++++ I
		low high
		Economic (product cost)
		I++++++++++++++++ I
		expensive inexpensive
		Timeliness (time to market)
		I++++++++++++++++++++ I
		slow fast
		Customer service (customer assistance)
		I++++++++++++++++++++ I
		low high
		Flexibility (product/process changeovers)
		I++++++++++++++++++++ I
		difficult easy

also clearly establish that these pioneers and their peers used analytical methods to devise strategies and plans to build and operate these magnificent production systems. Hence, pragmatic combinations of general systems and analytical thinking produce synergistic impacts.

    The five systems have evolved in a sequential manner, as a function of technological, economic, and sociological factors. The later paradigms have not entirely replaced the earlier paradigms, primarily because of both the variety and nature of the products customers need, demand, and expect. For example, some products naturally lend themselves to specific paradigms and others do not. However, innovations incorporated in the later paradigms find their way back into the earlier paradigms in a pragmatic manner, in order to increase the effectiveness and efficiency of the former. For example, a modern factory production system paradigm might incorporate employee empowerment practices created in the lean production paradigm development. Or, a contemporary craft system paradigm might incorporate computer-aided design technology.

In practice, this transfer of technology from the later paradigms to the earlier paradigms tends to create healthier but somewhat hybrid production systems. In our discussions, we must always keep in mind that the long-term purpose of any production system is to serve the customers and stakeholders effectively and efficiently, not to adhere rigidly to any particular paradigm.

CRAFT PRODUCTION SYSTEM PARADIGM

The craft production system paradigm dates back many centuries. Its products are custom-configured and tailored to individual customer needs, demands, and expectations. Here, we see individual craftspeople developing one-of-a-kind products in close "consultation" or direct contact with their customers. Customers obtain the function, form, and fit they need and desire. However, the economic cost and production volume available limit the customers that can be served. For example, if we used the craft paradigm to produce automobiles today, then we would be able to produce only a limited number of rather crude vehicles, even if we put a sizable proportion of our population to work in these production systems. In addition, the initial cost of our product would be very high. Hence, our market would be severely limited on both the supply and demand sides.

However, many modern products do lend themselves to this paradigm. Modern hard-good products include handmade custom clothing, furniture, buildings, and so on. Modern service products include hair styling, legal advice, medical diagnosis and prescription, engineering work, and so on. Here, unique product creations in goods or services are the telling mark of the craft paradigm.

FACTORY PRODUCTION SYSTEM PARADIGM

The inherent disadvantage of craft production systems is low productivity—low production efficiency. This disadvantage leads to a relatively high price and limited product volume. In response to growing customer bases and to this limitation of the craft production system, an opportunity presented itself in the late eighteenth century. This opportunity, in short, challenged would-be producers to find the means to produce essentially the same basic products, such as textiles, shoes, and so on, that were being produced in craft systems, but at higher production efficiencies. In short, craftspeople were producing as much product as fast as they could with their current technology, but the product was not adequate in price and volume to satisfy the growing population of customers.

The first innovation that initiated the factory system came from the specialization of labor. Here, people were organized and encouraged to specialize in only a part of a whole production process. In addition, simple tools and fixtures for forming and holding parts were introduced to aid craftspeople to more fully utilize their talents. This innovation led to higher production efficiency due to workers becoming more proficient in fabrication and assembly techniques. It made sense to organize these people into groups and locate them in adjacent areas to create a simple product flow. Hence, small factories were born.

A further outcome of this challenge was to develop technology that would allow for the transfer of craft skills to machines. This transfer essentially required the development of factory power through water, steam, and finally electric engines. In addition, it required mechanical means of building as much craft skill as possible into machines. People were then used to bridge the gaps in the production systems that resulted. For example, machine operations and materials handling, as well as machine repair, were accomplished by people. In addition to the introduction of machinery, specialization of labor, management, and ownership emerged.

In order to facilitate the factory production system, product variety was restricted to a core configuration, with selected options and customized details. For example, core shoe designs were selected, with a limited number of standard size and color combinations available. Hence, factory-produced products and custom craft-produced products allowed customers a choice. Typically, by choosing the factory-produced products, customers could retain primary product function, while trading off product form and fit to some degree for a significant reduction in economic cost.

Many products are produced today by using essentially a factory paradigm. For example, high-performance sports cars and firearms use basic core configurations, but involve customized details. Service products, such as dental and medical services and procedures, food services, banking services, and so on, customize basic core products to customer needs.

MASS PRODUCTION SYSTEM PARADIGM

The factory production system provided a breakthrough in production efficiency, compared to the craft system. However, several innovations were developed in the late nineteenth and early twentieth centuries that in turn yielded new breakthroughs in production efficiency. The concept of interchangeable parts, along with the machine tooling and gauging technology essential to actually produce interchangeable parts on a mass scale, paved the way for the mass production system. Here, labor and management specialization were further developed and became more highly specialized. The net result was the capability to produce products in a mechanized line, where materials and parts flowed in at one end and finished products flowed out at the other end. This moving production line essentially allowed the materials and product components to flow to the workers, replacing the necessity for the workers to collect and move the materials and product components to a common area for fabrication and assembly—a characteristic of the factory system.

Products are restricted to common configurations, with some selection as to details, e.g., paint color. Mass production processes are capital-intensive due to their mechanized flow lines of special-purpose machines and transfer mechanisms. Rigid labor and management specialization was used to attend to operations that could not be mechanized, because of economic or technical constraints, and to plan and coordinate the work.

In this paradigm, unskilled labor is used for many operations; specialization and mechanization allow for short training periods—on the order of hours or days, rather than years for craft apprenticeships. These abbreviated training periods allowed for the rapid introduction of large numbers of workers in large production systems. As a counterpart to labor specialization, management specialization by functions, and within each function, developed. For example, accounting, engineering, marketing, production control, and so on emerged and were segmented into highly specialized departments. Here again, training time was minimized, but layer after layer of management was required for coordination purposes, leading to bureaucratic structures. These structures were justified for coordination but resulted in substantial overhead costs and limited communications effectiveness within the organization as a whole. Hence, production system flexibility to respond to challenges from changes in customer needs, demands, and expectations became a disadvantage from both cost and timeliness perspectives.

In the mass production systems, product function is typically addressed, but form and fit are traded off for substantial cost reductions. The results are commoditylike products that are generally functional, and cost-effective in terms of initial costs, but do not have the customer appeal in form and fit that products from the craft or even factory production systems possess. Classical examples of mass production products include many automobiles and automotive components, basic food products and beverages, chemicals and fuels, bulk electronic components, bulk fasteners such as bolts and nails, and so on. Example service products include fast food, automatic car washes, and customer billing systems.

LEAN PRODUCTION SYSTEM PARADIGM

About midway into the twentieth century, the large production volumes necessary, and expensive special-purpose machines and tooling they demanded, created barriers in serving all but large markets. At about the same time, the narrow commoditylike product selection available to customers from mass production systems stimulated questions regarding how well customer needs, demands, and expectations were being met. Also at about the same time, other issues with human resource utilization, resulting from strict labor-management specialization, were surfacing.

In this same period of time, a higher level of education was emerging in industrialized countries. New technologies were emerging in materials, machines, tooling, and process control. Opportunities for product effectiveness gains, in addition to process efficiency, were being discovered. At the same time, midsized markets in their own right as well as strata within larger markets were emerging. These midsized markets and market strata, while reasonably large, were not large enough to support the massive capital investments in plants and highly specialized equipment that were characteristic of mass production systems.

During this period of time the seeds of the quality revolution—and a renewed emphasis on a customer focus, common to the craft paradigm—sprouted and began to grow. In short, enhanced human resources developed, coupled with (1) new production technologies (machines, tools, and process control) and (2) unconventional technologies (communications through customer surveys and focus groups, supplier networks and involvement). The term "lean production" was applied to this integrated system of people and machines by Womack, Jones, and Roos [10]. Other labels such as total quality control (TQC) and total quality management (TQM) have been applied, although in a slightly narrower sense; see Feigenbaum [11] and Ishikawa [12].

Here, we see products tailored to customer needs, demands, and expectations. We do not see custom configurations as in the craft system, but we do see distinctly different product configurations developed to appeal to market segments or niche markets. Although, on the surface, we see many process characteristics similar to those in mass production, we also see distinct differences in the manner in which processes are defined, controlled, and improved. The lean production paradigm reversed the specialization trend in machinery and tooling by utilizing human creative potential to develop quicker changeovers, giving added flexibility in production systems.

Lean production systems have a profound process emphasis in the sense of addressing both effectiveness (quality) and efficiency (productivity) together, rather than separately in a trade-off sense. Workplace teams and empowered employees seek out continuous improvements in both processes and products. Technology in terms of computer-aided design and computer-aided manufacturing allows lean production systems to reduce design-to-production cycles in both time and cost. Communications with, and involvement of, suppliers allow more effective and efficient processes.

Lean production systems attempt to optimize the entire production system as a whole by effectively and efficiently using all resources, including human resources. At the same time, they are customer-focused, aiming to define products that customers truly want and need, and then go on to produce these products in an efficient manner.

AGILE PRODUCTION SYSTEM PARADIGM

A transformation beyond the lean production system is described in terms of agility by Goldman, Nagel, and Preiss [13]. The agile production system is difficult to describe for two reasons. First, it is an emerging production system, and therefore limited experience and information exist. Second, it is a production system that takes the form of a contingency or virtual organization, crafted through partnerships with customers, suppliers, and other producing organizations, each with necessary core competencies and experiences. This crafting is done to define, design, develop, produce, deliver, and service customer-tailored products, many of which contain substantial service elements designed to help customers get the most benefit from their products.

In agile production processes we see custom process configurations. We see high-tech communication systems used to craft together the components of the contingency or virtual organization. Teams composed of personnel from a number of organizations—customers, suppliers, and other partnering producers—are empowered to define, design, develop, and produce products and to aid in implementation and operation of these products to serve specific customer needs. The advent of almost instantaneous global communications and computer-aided and integrated systems for design, development, and production have laid the foundation for virtual production organizations. Leadership and management present new challenges in coordination and multicultural communications.

In agile production systems both effectiveness and efficiency are pushed to the limit. Here, function, form, and fit are all captured while holding cost advantages and customer value. Timeliness, customer service, and flexibility are gained through the integrated partnerships in the contingency organizations. Products may range from reasonably simple, such as clothing, to reasonably complicated, such as communications or information processing services.

COMMON GROUND

From the broad observations of the production system transformations we have seen, several points are obvious. First, **historical accounts generally point toward the formation of a customer need–based vision (purpose) as the initiating event in production system formation** and transformation. Next, **we tend to see a leadership element emerge to champion and communicate this vision to others,** building a willingness to pursue the common purpose in the face of technical, economic, and social adversity. **This pursuit is typically guided by a pragmatic strategy** that evolves to resolve technical, economic, and social bottlenecks that develop during the action phase, **rather than a profound plan** that guides the activities and assures a smooth journey toward a successful outcome.

It is of great interest to note that some products have moved through the paradigms and actually stimulated their own transformation. The textile, automobile, and computer industries are characteristic of this transformation. Here, we have seen craft-based production systems originating in cottages and garages, transformed to small factories, then to mass production, on to lean production, and now entering the agile production paradigm. **In all cases of production system innovation, we see processes being defined/redefined, controlled, and improved.**

2.6 SYSTEMS THINKING—SCIENCE AND ENGINEERING

Both analytical (reductionism) and systems (holism) thinking help us describe, understand, design, and operate production systems. Understanding is essential to effectively and efficiently define/redefine, control, and improve our production systems. Effectiveness and efficiency are essential to allow competitive edge in our systems.

It is possible to engineer production systems without the aid of analytical or systems thinking, i.e., without explicit scientific methods. History holds rich accounts of great inventions and production systems before science, as we know it, was established, e.g., Roman highways and Egyptian pyramids. However, with the advent and use of science, even greater production systems have been engineered, e.g., modern health care systems, high-speed computers, and high-speed transportation systems.

RAINMAKER CASE—SYSTEMS THINKING   Rainmaker is a small manufacturer of agricultural irrigation systems. Over the years Rainmaker employees have produced product innovations that have allowed customers to conserve precious groundwater. One innovation resulted in low-pressure flexible nozzles that yield approximately a 40% savings in costs of water volume and pumping and pressurizing energy. Recently, a new reusable and relayable drip irrigation system was introduced to save about 75% in water and associated pumping costs.   **Example 2.1**

These innovations have attracted the attention of the international community, thus creating what appears to be a demand for Rainmaker products worldwide. Rainmaker currently perceives itself as a small regional manufacturer (about 300 people and about $100 million in sales), with one facility located in the semiarid region of the United States, near the present customers.

Everything, other than its products, at Rainmaker is traditional. It is a privately held corporation, utilizing relatively low-paid employees, operating mostly manual machines, using a traditional command and control (boss-subordinated) management/leadership structure. The one exception is that Rainmaker has computer-aided design (CAD) capability, based on high-end PC CAD packages. This capability is used for product design work only.

We desire to use systems thinking, specifically the Checkland model for systems thinking, Figure 2.1, to assess Rainmaker's situation regarding possible expansion to national and international markets.

In Stages 1 and 2 we address discovery and expression regarding Rainmaker's situation, respectively. A basic description table, Table 2.2, depicts the most critical systems and environments impacting the Rainmaker production system. Here, we have identified and described systems that deal with world food supply, markets, irrigation applications, financial systems, Rainmaker's community, and Rainmaker's production systems.

This first-level situational expression lacks details but is provided to illustrate the wide variety of systems relevant to Rainmaker's possible expansion beyond its current regional markets. It is important to note that in these stages everything is described or expressed "as is," as opposed to how we would like things to be.

Table 2.2   Rainmaker's situation expression overview.

System	Characteristic
World food supply	Poor crops
	Poor agricultural practices
	Lack of rain in semiarid regions
	Lack of irrigation technology
Worldwide irrigation markets	Southwest region of United States
	Australia
	Western Canada
	Middle East
	Africa
	South America
Irrigation applications (methods)	Flood
	Ditch
	Sprinkler
	Drip
Irrigation applications (crops)	Alfalfa
	Cotton
	Fruit and nut trees
	Grapes
	Small grains
	Vegetables
Financial system	Cash
	Conventional credit
	Regional farm credit
	Foreign cash/credit
	International guaranteed loans/credit
Rainmaker community/plant	Only industry in town
	Limited utilities
	Limited transportation facilities
	No postsecondary trade education in area
	Limited number of workers
	Loyal work force
	Limited technical skills in work force
Rainmaker production system	Field engineering applications
	Test laboratory
	Customer field tests
	CAD facilities
	Conventional machine shop
	Regional suppliers
	Command and control management

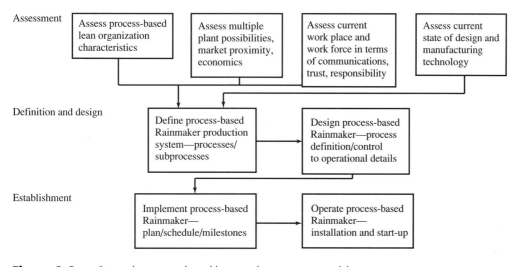

Figure 2.2   Rainmaker process-based lean production system model.

Stage 3 involves developing root definitions for relevant systems. Here, we coalesce the situation as it is, and extend our outlook to linkages that allow us to expand to 3 times our current level of production. On the basis of benchmarking knowledge of other manufacturers and their expansion experiences, several major systematic issues are expressed in our root definitions:

1. Define, design, and establish a process-based production system, within the lean production system paradigm, to triple Rainmaker capacity.

2. Define, design, and establish an international marketing and product support network system of dealers and service providers to assure customer satisfaction and national and international receivables.

3. Define, design, and establish an international product research and development system to maintain and increase product performance leadership in irrigation technology.

Each of the three root definitions above addresses critical "whats." We have not addressed the "hows" at this point, nor have we focused on system constraints. To do so would, at this point, limit our creativity and flexibility in defining, designing, and establishing the "best" systems possible.

Stage 4 allows us to develop conceptual models for our root definitions. Here, in the name of brevity, we present a partial but informative strategic model to illustrate this critical step. A model overview relating to the first root definition is depicted in Figure 2.2. We have included basic elements ranging from detailed assessment to definition, design, and implementation/action. These elements provide us a starting point for a full systems-level development.

Although a good deal of model refinement is essential, our model establishes definite business strategies that appear to be technically and economically reasonable. For example, the lean production process–based strategy, as opposed to a mass production strategy, is feasible for our expansion to at most a midsized manufacturer, with possibly two or three plants. A strategy of forging a dealer network with existing dealers, exporters, and importers is reasonable, considering our expected size of about $300 million in annual sales. A product research and development model might take advantage of the extensive government-sponsored agricultural research facilities both in the United States and internationally.

Each strategy requires extensive study and discussion. These studies and discussions follow from the root definition stage into the modeling stage. They carry over to, or resume again, at Stage 5 when we assess/compare our models with Rainmaker's situation. Eventually, we either commit to the expansion or carry on at present capacity.

Stage 6 consists of developing detailed plans for our production system. Here, we specifically address the "hows." We develop and assess feasible alternatives, choose, and develop formal/detailed plans for Rainmaker's expansion. Finally, in Stage 7, once we are satisfied that our production system plan possesses the characteristics and potential to meet our expectations, we schedule and take action.

The Rainmaker case has introduced us to general systems theory through a partial example. Several particularly noteworthy issues appear in this case: (1) thinking beyond the plant toward other high-impact systems that affect Rainmaker, (2) focusing on the "whats" before the "hows," and (3) a graded application of resources for thinking and doing, from root definitions, to strategies, to plans, to schedule, to action. Here we keep adding resources and detail so long as the endeavor makes technical and business sense.

General systems theory is applicable to any problem/opportunity situation. Overall, general systems theory, expressed through the Checkland model, is adequate to guide production system definition, design, and implementation. However, in such an explicit endeavor, we develop a more tailored approach to help us focus specifically on production systems. Such an approach, with a process basis, was introduced in Chapter 1, and is expanded and detailed fully in subsequent sections.

REVIEW AND DISCOVERY EXERCISES

REVIEW

2.1. Compare and contrast the analytical and systems philosophies in their approaches to assessing and exploring our physical, economic, and social environments.

2.2. What constitutes a model? What forms of modeling exist? Explain.

2.3. Why is modeling stressed in every branch of engineering? How do models help us? Explain.

2.4. Systems can be classified as natural, designed physical, designed abstract, or human activity systems. How is a production system, e.g., a manufacturing or service organization, classified? Explain.

2.5. Explain the difference between a hard and a soft system.

2.6. Explain and critique Checkland's general systems model. What steps are involved? What is the point of using this model?

2.7. The advocates of general systems theory claim that it is applicable to any system, anywhere, any time. Is this an accurate statement? Explain your position.

DISCOVERY

2.8. It appears that engineering predates science in many respects. For example, we see rather amazing projects built over the centuries in societies that were not solidly grounded in science. Why do we stress both science and engineering today? Explain.

The following questions are structured in the context of the five production system paradigms in Table 2.1.

2.9. Why is there not one dominant production system paradigm? Explain.

2.10. What allowed each production system to evolve, e.g., from craft to factory, from factory to mass, from mass to lean, from lean to agile?

2.11. Provide a list of five typical products, one for each of the five production system paradigms.

2.12. Identify a product that has been produced under (or is compatible with) all five production system paradigms.

2.13. Draw five "radar" charts, one for physical performance, one for economic performance, one for timeliness performance, one for customer service performance, and one for flexibility. Each chart should have five axes, one for each of the five production system paradigms. Using the radar charts, summarize the trends in performance realized from the evolution of production systems. *Note:* A radar chart is a multidimensional chart where we project each dimension out from a center point. Using a radar chart, we can place any number of dimensions on the chart. The axis scale is the same in each direction. Hence, a circle on the perimeter indicates that all dimensions are of the same magnitude.

The following questions are focused primarily on the Rainmaker Case—Systems Thinking.

2.14. How do the three root definitions for Rainmaker relate to the situation expression overview in Table 2.2? Explain.

2.15. How does Figure 2.2 correspond to a process-based organization, as discussed in Chapter 1? Explain.

REFERENCES

1. Checkland, P., *Systems Thinking, Systems Practice*, New York: Wiley, 1981.

2. Von Bertalanffy, L., *General System Theory*, New York: Braziller, 1968.

3. Boulding, K. E., "General Systems Theory—The Skeleton of Science," *Management Science*, vol. 2, no. 3, 1956.

4. Laszlo, E., "The Origins of General Systems Theory in the Work of von Bertalanffy," *The Relevance of General Systems Theory*, E. Laszlo, ed., New York: Braziller, 1972.

5. K. Ellis et al., eds., *Critical Issues in Systems Theory and Practice*, New York: Plenum Press, 1995.

6. Wilson, B., *Systems: Concepts, Methodologies and Applications*, New York: Wiley, 1984.

7. Senge, P. M., *The Fifth Discipline*, New York: Currency Doubleday, 1994.

8. Goldratt, E. M., *It's Not Luck*, Great Barrington, MA: North River Press, 1994.

9. Dettmer, H. W., *Goldratt's Theory of Constraints: A System Approach to Continuous Improvement*, Milwaukee: ASQC Press, 1997.

10. Womack, J. P., D. T. Jones, and D. Roos, *The Machine That Changed the World*, New York: Harper Perennial, 1990.

11. Feigenbaum, A. V., *Total Quality Control*, Third Edition, New York: McGraw-Hill, 1983.

12. Ishikawa, K., *What Is Total Quality Control? The Japanese Way*, Englewood Cliffs, NJ: Prentice-Hall, 1985.

13. Goldman, S. L., R. N. Nagel, and K. Preiss, *Agile Competitors and Virtual Organizations*, New York: Van Nostrand Reinhold, 1995.

PROCESS FUNDAMENTALS

3.0 INQUIRY

1. How do process definition, control, and improvement impact competitive edge?
2. How is process synergy generated?
3. What two parameters are used to characterize process performance?
4. How do open-loop and closed-loop processes differ?
5. How is process performance judged?

3.1 INTRODUCTION

Competitive edge in organizations, either for-profit or not-for-profit, is impacted both by unique physical characteristics of processes and by common features and constraints. It is reasonable for us to identify this commonality and address it in generic terms, while at the same time, we intend to identify and exploit unique process characteristics. This chapter focuses on an understanding of common process fundamentals, such that we can build a foundation to systematically develop competitive edges in all processes, throughout our entire organization.

3.2 PROCESS FEATURES AND SYNERGY

Effective and efficient production systems and their processes share several fundamental features; see Figure 3.1. They are directed through common purpose—vision, mission, and core values—and specific process purposes. They are customer-focused and value-created in nature. They are systematically structured in a hierarchical manner. Furthermore, **processes are created and operated through sequences of definition/redefinition, control, and improvement efforts.**

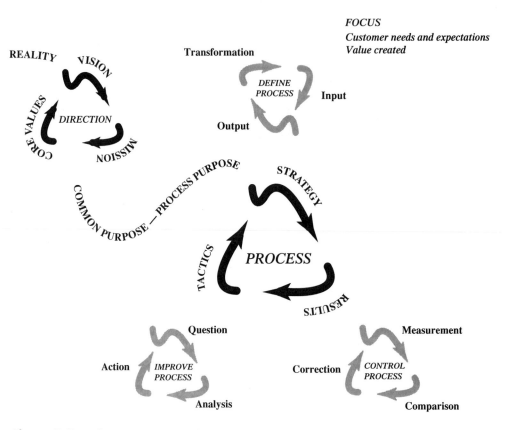

Figure 3.1   Process structure overview.

DIRECTION—PRODUCTION SYSTEMS

Direction is a result of our awareness of impacting systems and intentions of dealing within them—an awareness of our reality. The direction triad, Figure 3.1, is critical. First, we assess our situation as it is now. Any systems that impact our production system are relevant in this assessment. For example, economic/financial systems, competing production systems, environmental systems, and so forth are relevant.

    Vision is a complex, abstract picture of our current situation, with extensions to the future. From our situation reality we mentally project our production organization into the future, adding characteristics we desire regarding physical, economic, timeliness, and customer service performance, and ultimately business success. For the most part, we focus on outcomes.

    From our vision, we verbalize and set out our mission statement. Here, we utilize the general systems theory and root definition concept. **A mission statement crisply verbalizes our intentions for our production system as a whole.**

    Our core values reflect our mission-related intentions and our ethics, thus providing a verbalized description of both. **Together our vision, mission, and core values develop a directional essence for our production system.** This directional essence is applicable to each and every person in the production system, therefore serving as the core common purpose.

PROCESS DEFINITION/REDEFINITION CONCEPT

Process definition/redefinition includes three elements: output, transformation, and input. It represents a strategic means by which the organizational vision and mission are addressed and supported within the core values set out by the organization. If the definition is off the mark, no amount of control and improvement is likely to create competitive edge; only redefinition will help.

Output refers to and describes the products and by-products of the process. Input refers to the resources, such as materials, supplies, energy, information, and so on, needed to execute the process. Transformation refers to the systematic way the resources are changed into products and by-products. In process definition/redefinition we select or create the best technology for our desired transformation, besides defining best practices. We set the explicit goals and targets for the process in terms of outputs. We define and design for the "whole process" picture. Point-to-point process definition/redefinition subelements are developed in Section 3.

PROCESS CONTROL CONCEPT

Process control focuses on results. **Process control includes three elements: measurement, comparison, and correction.** The control elements of the process fundamentals triad involve measurement of both strategic and tactical parameters, e.g., calibrating real-time or recent operational positions. These measurements are compared to their respective target values. Some measures are upstream or leading metrics/indicators and some are downstream or lagging metrics/indicators. Appropriate action is taken, when necessary, to bring the parameter being measured back to its target. Section 4 is devoted to detailed discussions of process control issues.

PROCESS IMPROVEMENT CONCEPT

Process improvement involves three elements: question, analysis, and action. Process improvement is a tactical means of enhancing process effectiveness and/or efficiency that questions our current process regarding transformation, resources input, and/or products output. Here, we analyze our process operations by using facts, figures, and enlightened opinions. Then, we generate and analyze possible alternatives/countermeasures. Finally, action is taken to improve the process in terms of final and intermediate results, thus providing incremental improvements in accomplishing purpose. Process improvement is continuous; we continually question, analyze, and act to enhance process effectiveness and efficiency. In contrast, process definition/redefinition is intended to impact process strategy and occurs infrequently, but on a radical scale. Process improvement subelements are discussed in Section 5.

SCOPE—DEFINE/REDEFINE, CONTROL, IMPROVE

The process definition, control, and improvement structure shown in Figure 3.1 is applicable at any process level, e.g., to all seven fundamental processes and their subprocesses. For example, we seek to define or redefine, control, or improve the entire production system in a global sense. We seek to define or redefine, control, or improve a fundamental process, or we seek to define or redefine, control, or improve a subprocess. In any case, **we define or redefine, control, and improve down to the lowest level necessary in order to** provide a systematic and thorough effort that will **yield higher organizational effectiveness (quality) and efficiency (productivity)** as judged by our internal and external customers and stakeholders in terms of the benefits they receive and the burdens they bear.

SYNERGY

The business integration process, the seventh fundamental process, Figure 1.6, **serves to align the other fundamental processes.** In addition, it facilitates coordination and communication, and hence, nurtures an environment wherein we can continuously enhance our process effectiveness and efficiency. The business integration process (1) counters fragmentation with process alignment toward common purpose, (2) counters conflicting directives with coordination, and (3) counters confusion with communication.

We gain process synergy, in the presence of organizational complexity, by systematically moving through the process definition/redefinition, control, and improvement elements. This synergistic effect is depicted in Figure 3.2. Here, we begin with the initial process definition. At first, we realize only part of the possible gain (the solid line), with unrealized potential remaining (the dashed line). Using process control, over time, we stabilize our process and obtain additional limited gain (the diagonal line). Using process improvement, we can realize additional gain (the

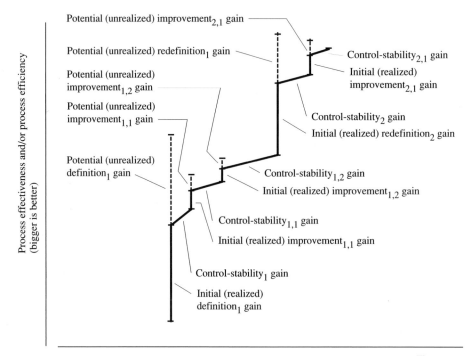

Figure 3.2   Conceptual definition, control, improvement process performance synergy effects.

short vertical line), with some potential/gain remaining (the short vertical dashed line). Some of this potential gain can be realized through process control by targeting and stabilizing our process. When new, feasible options develop, we can redefine our process and continue with our control and improvement efforts. Hence, **each process-related issue—definition/redefinition, control, improvement—has a distinct role to play.**

Confusion between/within roles or the omission of any of the three roles creates disharmony and frustration in the production system, which ultimately limits production system effectiveness and efficiency. Sometimes, e.g., in the presence of confusion, it is possible that effectiveness and efficiency (Figure 3.2) may decrease. In this situation, we hope to learn from our failures and subsequently produce improvement trends in the long term.

3.3 PURPOSE—GOALS, OBJECTIVES, TARGETS, TOLERANCES—AND ACTION

Organizational direction and focus are derived from the common purpose, which includes vision, mission, and core values, as described in Chapter 1. The common purpose is delineated through a purpose hierarchy including a single mission, broad process purpose statements, and specific goals, objectives, targets, and tolerances. Hence, **we use a hierarchy of purpose to tie our processes into our organizational vision, mission, and core values;** see Figure 1.6. Furthermore, a cohesive purpose hierarchy prevents "hiding places" and exposes existing waste: nonquality and nonproductivity.

A process/subprocess purpose statement is a broad statement that describes key process/subprocess performance results and support for the production system. For example, it may contain implications relative to physical, economic, timeliness, and/or customer service results and support that are critical to the production system. Purpose statements tie each process/ subprocess to the production system, i.e., justify the process's or subprocess's existence. **A process or subprocess without a legitimate purpose or role is subject to elimination.**

A goal is an action-oriented expression of purpose that is typically qualitative. While visions and missions are rather fuzzy, goals are more crisp and definite. For example, a vision might "see" the organization as a world-class competitor in a specific industry or product line. In contrast, a goal might be expressed as "to be recognized by our customers worldwide as the industry leader in superior product performance."

An objective is a specific end result we seek to realize. It is narrower and more tightly defined than a goal. Objectives allow us to break down goals into pieces that we can address in terms of measurable results. For example, if we state a goal regarding a productive physical environment for our workplace, we may state one (of perhaps many) related objectives as establishing and maintaining the workplace temperature.

A target addresses a specific benchmark, typically quantitative in nature. A target defines specific, measurable result-based expectations. **A tolerance or engineering specification is the most detailed level in the purpose hierarchy.** Here, we specify process or product limits, which clearly define acceptable performance and unacceptable performance. **Most process/product targets and specifications are explicit delineations of substitute quality characteristics.**

In short, we delineate common purpose when we see a vision, define a mission, state our core values, state process purposes, set goals, state objectives, and establish targets and specifications. Likewise, we accomplish common purpose when we work within our defined mission and process purpose, pursuing our goals, as we meet specifications, hit targets, and accomplish our objectives.

Example 3.1 | **RAINMAKER CASE—PURPOSE HIERARCHY**   Common purpose, as expressed through a purpose hierarchy, supplies direction for an organization. An effective purpose hierarchy is brief and crisp, flowing from the production system level, through the process level, down to the subprocess levels. A sample of the purpose hierarchy that directs activities in the Rainmaker organization appears below.

> *Vision (one of one)*: To lead the international markets in irrigation solutions for semiarid regions of the world.
>
> *Mission (one of one)*: To provide effective and efficient irrigation solutions to regional, national, and international customers through a lean, process-based organization.
>
> *Core values expression (one of one):* To use innovative technical and business practices, and establish and maintain employee leadership and creativity by providing education, training, and opportunity for all employees to define, control, and improve best practices that yield customer, employee, and stakeholder satisfaction.
>
> *Production process purpose statement (one of several)*: To fabricate flat and tubular parts that meet or exceed design specifications on time, every time.
>
> *Goal (one of many)*: To provide a productive and safe physical environment for our employees.
>
> *Objective (one of many)*: To maintain a productive workplace temperature.
>
> *Target (one of many)*: To maintain 75ºF in the workplace.
>
> *Specification (one of many)*: To keep the workplace temperature between 72ºF and 78ºF.

Once we develop and delineate our common purpose, we take action to accomplish our purpose. We devise strategies and action plans to initiate the action phase. A strategy is a general course of action that we intend to pursue to address a goal or objective, one step or phase at a time. A strategy is flexible in that we restrategize after we see the results from each phase. A change in strategy might result in a minor change in direction or perhaps a different direction entirely.

Within our strategy, we formulate action plans that provide who, what, when, where, and how details. Plans are more detailed and rigid than strategies. However, we change our plans when necessary. Such a change alters the who, what, when, where, and how details.

LOCATION AND DISPERSION

Targets are generally of three types: (1) smaller is better, (2) nominal is best, or (3) bigger is better. Smaller is better refers to production time, product variation, operating costs, warranty costs, defect counts and percentages, impurities, accidents, and so forth. Nominal is best refers to a specific point (the target) such as a dimension, chemical content level, operating temperature, and so forth. Bigger is better refers to process yield, service life, time to failure, and so forth.

Performance measurement involves two critical parameters: (1) location, relative to a specific point, such as the center of the bull's eye on a target, and **(2) dispersion,** relative to the

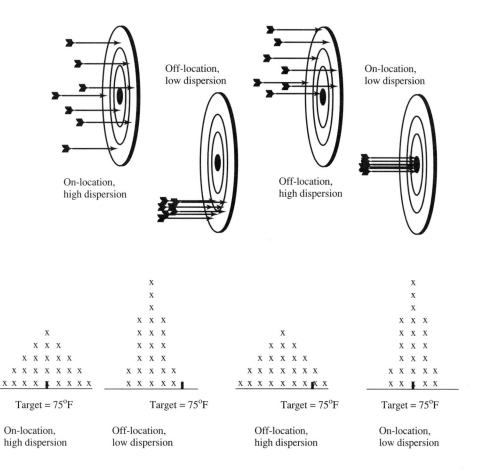

(a)

(b)

Figure 3.3   Location and dispersion depiction. (a) Archery target example of location and dispersion (two-dimensional physical analogy). (b) Check sheet (histogram) example for workplace temperature.

center of several measurements, such as an arrow pattern center. Figure 3.3a illustrates possible outcomes relative to four location-dispersion possibilities in an archery context, i.e., a nominal is best case. The on-location, low-dispersion case is the most desirable case—the highest performance. It is on target (located in the bull's eye) and has relatively little dispersion or scatter relative to the center of the arrow pattern. Hence, we would declare it to have higher performance than the other three alternatives shown.

Regarding the workplace described in Example 3.1, we can illustrate location and dispersion in the context of the temperature-maintenance target by using check sheets. Figure 3.3b depicts our data. Here, we conclude that the on-location, low-dispersion case provides the highest, i.e., most desirable, process performance, as judged against the nominal is best target criterion. The check sheet tool is described in Chapter 20.

Example 3.2 | **AMPLIFIER ACQUISITION CASE—PRODUCT VARIATION**   We were designing the next generation for a data acquisition system. We were quite concerned about reducing cost as well as improving performance and expanding functionality. One of our first activities was to reduce system cost. We performed a Pareto analysis (see Chapter 20) on the overall system cost of the previous system and discovered an unusually high cost associated with a small operational amplifier. This amplifier was by far the highest-cost item in the system, a complete surprise to the design engineers.

The system design could not use a "production run" amplifier, but instead required screening of the parts received from the one supplier capable of providing this part. A fundamental concept of this system was "low noise"; low-noise operational amplifiers were quite critical and unavoidable for this design concept to function correctly. The screening essentially sorted the production run amplifiers into two groups, the low-noise amplifiers and the high-noise amplifiers.

We discovered that our manufacturing operation was performing the noise screening and scrapping the unusable, but still functional (at a higher noise level), parts that would not perform adequately for our system. Scrapping of the functional amplifiers resulted since the supplier was unwilling to accept the return of the screened-out amplifiers. In other words, the high costs associated with this amplifier included the cost of scrapping the screened, but still functional, units in addition to the units that were actually used in the system.

We contacted the manufacturer to explore possible alternatives to our time-consuming and costly situation. What we discovered was that the screening we were performing was technically beyond the capability of the supplier to perform. We had developed a special piece of test equipment that performed extensive computerized analysis associated with the device, while the device was operating in a special test circuit.

We asked the supplier if they would be willing to perform the screening if we gave them the screening equipment, its software, and training in its use, and also if they would be willing to absorb their future screening costs in a higher price per part. Their response was very enthusiastic because this no-risk, no-cost (to supplier) approach also gave them this low-noise capability on a nonexclusive basis—they could offer these specially screened parts on the open market as well.

A simple comparison of our existing costs, including the scrap versus the cost to build and "give away" another test unit, immediately convinced our management that building the test unit was a very good investment. We built the additional test unit and negotiated an agreement with the supplier on its use, ownership, and data rights, as well as pricing of their screened operational amplifiers.

Once our supplier commenced screening at their plant, we were able to immediately install their parts into our systems. Hence, now we could avoid the scrap cost and labor and time associated with screening in our plant, lowering our existing production costs immediately.

Epilog

This case represents an example of how product/process variability can affect the overall cost, performance, and even schedule of the end item. A common technique in the semiconductor industry is to place devices into performance categories. For memory devices, a key is speed. For operational amplifiers, it is the noise floor. Lower-performing product is recognized as having good quality, but just not "fast" or "low power" or "high speed" or whatever the key attributes of the devices are.

The unacceptably wide noise variation (for this particular customer) of the parts provided by the supplier was narrowed through the screening process, thus achieving a very tight dispersion on the noise characteristic. However, the process used to achieve this physical performance enhancement was expensive and suboptimal. Fortunately for this supplier, the competition was poor and the customer base was willing to pay high prices for the high-performance data acquisition products ultimately provided.

The cost reduction and process improvement method employed here was a unique experience for the customer. This experience reduced cost while maintaining the customer's reputation for providing high product performance. More importantly, it demonstrated the importance of process improvements, in addition to product improvements, something they were not, at the time, focusing on. [BF]

3.4 PROCESS STRUCTURES

In general, **we see two types of process structures: (1) open loop and (2) closed loop.** The **open-loop process** is characterized by operations "blind" to present conditions regarding inputs and - outputs, as well as surroundings. The **closed-loop process** can operate in either a feedback or feedforward manner, whereby the process's present output condition (the feedback case) is related to current operations, or input/surrounding conditions (the feedforward case) are related to current operations. Figure 3.4 depicts a workplace space environment example where we are utilizing an open-loop process (Figure 3.4*a*) and two closed-loop feedback control processes (Figure 3.4*b* and *c*).

In the open-loop process in Figure 3.4*a* we see a person feeding fuel logs to the stove without any influence from the process results/output affecting the fueling activities. For example, the fueler deposits one log in the fuel chute each hour, and the temperature inside varies accordingly. It may or may not be comfortable inside. The inside temperature (or any other input or surrounding condition) does not influence the fueler's activities; that is, activities are preprogrammed and the program is followed regardless of circumstances.

In Figure 3.4*b,* we see a person "closing the loop" by sensing, i.e., subjectively measuring, the temperature and then judging or comparing the temperature in the workplace to his or her idea of too hot, too cold, or acceptable. Here, a person is sensing, measuring, comparing, and correcting the process. This situation is an intuitive and subjective form of a closed-loop process control. When the heat level is considered too low, the operator puts another log in the stove and allows the heat level to increase. When the heat level is too high, the operator stops fueling the fire, allowing the workplace area to cool. This cycle is repeated over time.

If we define a continuous metric regarding temperature level within the workplace, e.g., room temperature, then we can plot the measurements as shown in Figure 3.4*b*. We can summarize the measurements, i.e., calculate an average, to obtain an indication of the temperature's location over the time period. We can also assess the dispersion or variation, i.e., variance, in the temperature measurements over the same time period.

If we assess the level of heat by assessing how hot, or cold, the room is at the present time, then we have a feedback control system. If we look forward rather than backward in time to inputs coming in and/or anticipated external conditions, then we have a feedforward control system. The information fed forward is typically extracted from the process environment and/or from the nature of the resources input. Here, if perhaps the stove operator obtains a weather forecast for a blizzard or cold front that will arrive in the next few hours, then the operator could use this advance information to trigger a heat buildup in order to counteract this anticipated environmental effect before or as it arrives, rather than during or after its arrival. If, on the other hand, the operator senses that the incoming wood is wet (the input will change, and the wet wood will not burn the same as the dry wood) then through this feedforward information, countermeasures to properly heat with wet wood, or prevent the wood from getting wet, can be taken. Hence, the difference in feedback versus feedforward control has to do with reacting versus proacting, respectively, in the execution of the process transformation.

Depending on our organization's common purpose (see Example 3.1 and Figure 3.4) we may choose to define our process using different technology. Here, depicted in Figure 3.4*c*, we use a gas-fired stove, a temperature sensor, a gas flow control valve, a quantitatively defined comfort level target of 75°F, and a temperature specification interval of 72°F through 78°F. Our hardware

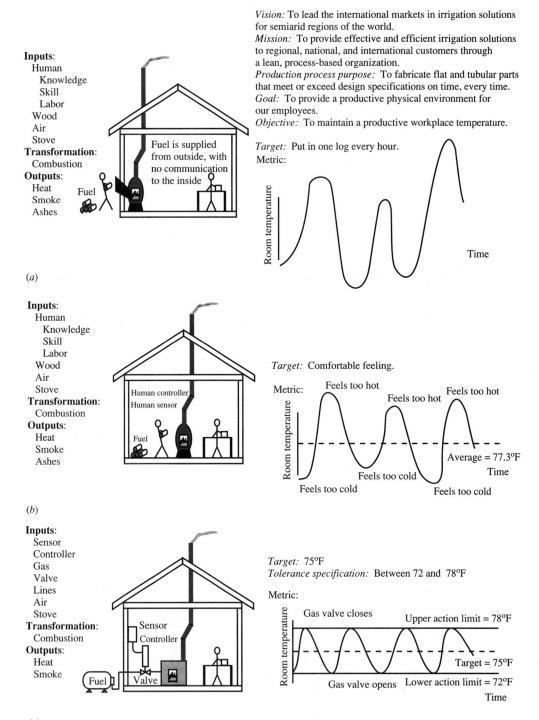

Inputs:
  Human
    Knowledge
    Skill
    Labor
  Wood
  Air
  Stove
Transformation:
  Combustion
Outputs:
  Heat
  Smoke
  Ashes

Fuel

Fuel is supplied
from outside, with
no communication
to the inside

Vision: To lead the international markets in irrigation solutions
for semiarid regions of the world.
Mission: To provide effective and efficient irrigation solutions
to regional, national, and international customers through
a lean, process-based organization.
Production process purpose: To fabricate flat and tubular parts
that meet or exceed design specifications on time, every time.
Goal: To provide a productive physical environment for
our employees.
Objective: To maintain a productive workplace temperature.

Target: Put in one log every hour.
Metric:

Room temperature

Time

(a)

Inputs:
  Human
    Knowledge
    Skill
    Labor
  Wood
  Air
  Stove
Transformation:
  Combustion
Outputs:
  Heat
  Smoke
  Ashes

Human controller
Human sensor

Fuel

Target: Comfortable feeling.

Metric:

Room temperature

Feels too hot

Feels too hot

Feels too hot

Average = 77.3°F
Time

Feels too cold

Feels too cold

Feels too cold

(b)

Inputs:
  Sensor
  Controller
  Gas
  Valve
  Lines
  Air
  Stove
Transformation:
  Combustion
Outputs:
  Heat
  Smoke

Sensor
Controller

Fuel

Valve

Target: 75°F
Tolerance specification: Between 72 and 78°F

Metric:

Room temperature

Gas valve closes

Upper action limit = 78°F

Target = 75°F

Gas valve opens   Lower action limit = 72°F

Time

(c)

Figure 3.4   Open and closed control loop process depictions. (*a*) Open (manual) loop space heating
process. (*b*) Closed (manual) feedback control loop space heating process. (*c*) Closed
(automatic) feedback control loop space heating process.

and knowledge are integrated together to automatically (without human assistance) measure the shop temperature, compare it to the upper and lower action limits, and take action to turn the gas flow and stove on and off accordingly. This depiction represents a closed-loop, automatic feedback control system.

Most processes are initiated as open-loop or closed-loop manual processes; we introduce loop-closing and/or feedback or feedforward information as we begin to understand the technical nature of the process and its surroundings, i.e., the cause-effect relationships. In other words, we tend to use subjective rules of thumb in closed-loop processes before, and as, we learn about the process. The simpler the process, the simpler it is to close the loop and/or incorporate feedback or feedforward information. Section 4 revisits this topic and provides details.

3.5 PRODUCTION SYSTEM VIEWS

The world around us that encapsulates our production systems is one of cause and effect— but the causes and effects do not manifest themselves immediately; we see delayed reactions to our action or inaction. Typically, we do not entirely understand cause-effect linkages in production systems. However, we can usually identify the basic effects we seek and piece together several fundamental causes. As an example, in the context of Example 3.1 and Figure 3.4, controlled combustion in a stove, located in a cold workplace, affects the heat level in a building, which affects workplace temperature, which affects workers, which affect product quality and productivity, which affect cost of production and finally affect business profits.

A process value chain (PVC) diagram helps us map relationships between primary variables and all process/subprocess results so as to understand action-counteraction characteristics. A generic process value chain for an entire organization is a very complex thing. We may take an analytical view, studying the parts, as depicted in Figure 3.5*a*. Or we may take a general systems view, as depicted in Figure 3.5*b*. Here, we see business results downstream on one end, and primary controllable and uncontrollable variables upstream on the other end. In the middle, we see process and subprocess causes and effects—all interrelated, acting simultaneously across time. We can trace from inputs to outputs, or from outputs to inputs—through our subprocesses to processes, or processes to subprocesses, respectively.

The uncontrollable and controllable variables in Figure 3.5 constitute our physical, economic, and social environments, surrounding and permeating our production system. These variables all interact to form lines of influence in our cause-effect structures over time. These structures are usually highly dynamic and nonlinear. We see sequences of dashed lines in Figure 3.5, representing feedback and feedforward characteristics of our process counteractions. Hence, we have depicted a complex set of relationships, which we seek to understand and manipulate in order to produce results—accomplish our mission, goals, objectives, and targets.

Each production system possesses its own unique process value chain, complete with time lags between causes and effects. We use our process definition/redefinition, control, and improvement sequences to tailor the PVC in terms of quality and productivity within each distinct process/subprocess. **We calibrate our degree of success with metrics, in terms of performance—physical, economic, timeliness, and customer service.** Figure 8.8 depicts a partially developed PVC for the Downtown Bakery case developed in Section 3.

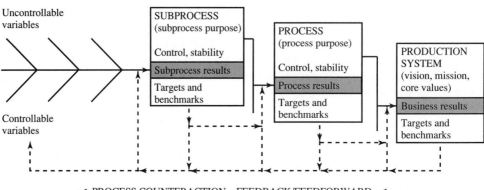

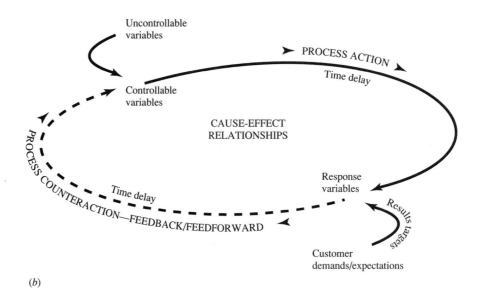

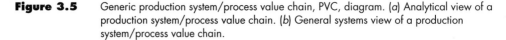

(a)

(b)

Figure 3.5   Generic production system/process value chain, PVC, diagram. (a) Analytical view of a production system/process value chain. (b) General systems view of a production system/process value chain.

The PVC can be approached in either direction, from cause to effect or from effect to cause. The point is to clearly illustrate to people in operational roles and people in leadership roles the necessity of creativity and process discipline to bottom-line results. In other words, **the PVC helps us to link cause and effect relationships between the physical, economic, and social variables to our business results in terms of customer and stakeholder satisfaction.** The challenge is that these variables and their results are complex, probabilistic, and dynamic. No exact science for modeling complex PVCs exists at this time. However, the use of systematic/scientific methods is appropriate for dealing with these components because we know that the concepts of location and dispersion are ever-present in our variables and chosen metrics.

3.6 ROBUST, MISTAKEPROOF, AND BENCHMARK PERFORMANCE

Ultimately customers buy products in anticipation of benefits. Customers expect consistently high performance with little variation, regardless of their environment, application, configuration, and method of operation. Lack of effectiveness and efficiency in product and production performance results in lost work time, environmental damage, wasted energy and materials, late delivery, excessive maintenance, and other costs, inconveniences, and personal aggravations. The manner in which we define, design, develop, produce, distribute, and market/sell/service our products and the manner in which customers use and dispose/recycle our products matter. Together, they determine customer benefits/burdens experienced.

ROBUSTNESS

Robustness in performance refers to our ability to deliver consistent high performance to all users, over reasonable ranges of application, environment, configuration, and utilization/ operation methods. The robust performance concept is illustrated through a performance variance "stack-up" model depicted in Figure 3.6. Here, we see four sources of input variation converging to produce or accumulate variation in the output. By controlling/reducing the input variations, we seek to shrink/control the output variation. We use engineering and training to impact the inputs so as to tailor the outputs to yield the highest level of customer satisfaction possible, i.e., high performance in the eyes of external and internal customers. Process characterization and tools such as experimental design, described in Section 2, support our engineering efforts.

From the external customer's perspective, product performance variation stack-up results from four primary factors; see Figure 3.6a. In the case of product performance, variation or noise in the field environment, product application, product configuration, and product utilization/ operation method all stack up to produce variation in the performance our external customers experience through their product usage-based processes. The performance variance stack-up counterpart that our internal customers experience, shown in Figure 3.6b, includes the production environment, the production configuration and equipment, the production materials and supplies, and the production/operation method itself.

In many cases, the product utilization/operation method or production/operation method is expected to compensate for variations in environments and applications as well as shortcomings

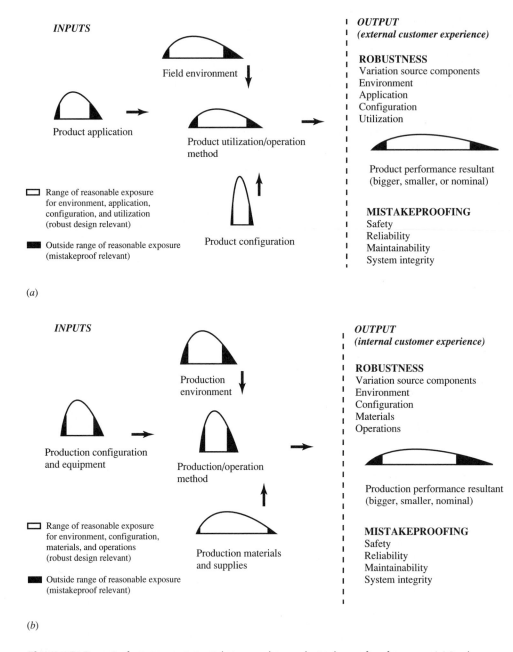

Figure 3.6   Performance variation relative to robust and mistakeproof performance. (*a*) Product performance variation (from the external customer's perspective). (*b*) Production performance variation (from the internal customer's perspective).

in configurations to help shrink the output variation. Hence, in both cases, Figures 3.6*a* and 3.6*b*, operators are often called on and expected to sense and adapt to various conditions arising out of the three other inputs.

The most desirable outcome possible is to simultaneously reduce process or production performance variation and "raise" the performance level, without using tighter tolerances or more expensive materials. Hence, we seek both better location effects to targets, i.e., bigger is better, smaller is better, or nominal is best, as well as smaller dispersion effects, i.e., less spread, in performance outputs, given variation in our process inputs and transformations. A related but more specific initiative, robust design, has emerged to explore and model product/process variation. It is described in Chapter 19.

MISTAKEPROOFING

The concept of mistakeproofing is tightly coupled with that of robust performance. **Mistakeproofing emphasizes safety, reliability, maintainability, and system integrity issues associated with both external and internal customer experiences,** rather than the performance variation emphasis of robustness. The point is that reasonable ranges or bounds exist for environmental, application, configuration, and utilization/operation parameters. Within these bounds, robust performance is relevant; outside these bounds, mistakeproofing is relevant.

Mistakeproofing contains the same general factors as robust performance. The mistakeproofing focus is depicted in Figures 3.6*a* and 3.6*b* for product and production performance, respectively. Mistakeproofing focuses on inputs at unintended levels, beyond the range considered in robust performance. The shaded tail areas in Figure 3.6 on the input bars represent inputs that are considered to be outside the environmental, application, configuration, and utilization parameter ranges. The mistakeproof performance focus tends to be safety, reliability, maintainability, and system integrity to avoid system damage or degradation, production losses, production inefficiency, and possible environmental impact.

Utilization/operation method and production/operation method–related activities are typically expected to compensate for anticipated or unanticipated inputs in the three other input categories. In Figure 3.6*a,* we may expect operators to sense and adapt product utilization to unusual or unexpected combinations of field environment, product application, and product configurations. In Figure 3.6*b,* we may expect operators to sense and adapt production operations to unusual or unexpected combinations of production environments, configurations and equipment, and/or materials and supplies.

It can be argued that inappropriate application, unforeseen environments, unexpected operating conditions, and unauthorized equipment configurations are induced by customers and acts of nature, and therefore, producers are not responsible for the consequences. However, most customers challenge the preceding argument. Simply put, we do all we can through definition and design to protect the integrity of our customers and their property. Mistakeproofing has emerged as an effective initiative (see Chapter 19) to eliminate or prevent mistakes through engineering means.

Mistakeproofing strategies encourage creative efforts to identify potential mistakes resulting from human as well as physical sources. Mistakeproofing combined with robust performance efforts should always be considered in any and every process. **Robustness and mistakeproofing**

are effective in preventing product dissatisfaction as well as in yielding product satisfaction. Products and processes that are tolerant of application and environmental differences and mistakes, either accidental or deliberate, are always favored over those that are not robust and mistake-tolerant, other benefits and burdens being about equal.

Example 3.3 | **RAINMAKER CASE—VARIABLES IDENTIFICATION**   At this point, we revisit the Rainmaker irrigation production system, with a focus on identification of the robustness and mistakeproofing of its primary product, the low-pressure, circular irrigation system. This system is typically about 400 yards long. It moves, on its own power, around its circular path, and must withstand outdoor elements, over all four seasons, throughout the year, both when in service and when out of service. The product itself is expected to last over 10 years, with many customers expecting 20 years of service with only minor service and maintenance. Customers may desire to use the product on several different crops, requiring different water flow rates during the course of a year. In addition, various types of customers buy and use these irrigation systems. Some customers are meticulous in their service and maintenance, and some are not. Several languages are native to users.

Using the format of Figure 3.6, we have listed the basic variables that fall in the input and output categories. A summary is presented in Figure 3.7. Here, we provide ranges for most variables, so that the reader can visualize the distribution and shaded tails that were abstractly introduced in Figure 3.6.

Rainmaker personnel involved in all processes, e.g., market/definition, design/development, . . . , distribution/marketing/sales/service, must be aware of these variables and ranges in order to serve their customers. The ranges depicted represent aggregates over the entire customer base. Typically, any one customer will not experience the entire range, especially in the environmental category, but Rainmaker as a production system must plan for its products in the field to see these varied environments and conditions. Sometimes, Rainmaker can furnish specific customers with options to deal with extreme input conditions, but first, the company attempts to define and design to the spectrum of input conditions and use production options as a last resort.

BENCHMARKS

Competition is an ever-present factor that has a profound influence on production system outcomes. **Benchmarks allow us to calibrate our performance,** e.g., physical, economic, timeliness, and customer service. Our benchmarks bring a degree of realism to our vision and mission declarations and help us to set targets for both products and processes.

Benchmarking initiatives (see Chapter 19 for details) offer a formal means to establish product/process benchmarks. **Benchmarking involves the careful observation, measurement, and description of what others are doing and the results they obtain.** We hold these results up to similar products/processes within our organization.

Benchmarks may be based on average performance and/or best practices. Each basis serves the purpose of locating our performance level relative to the performance of others. This information is critical to establish our direction from two standpoints. First, the benchmark metrics, e.g., average and/or best, introduce a sense of realism. Second, knowledge of best practices in terms of product/process functionality offers ideas, although not original, for helping us to define/redefine, control, and improve our processes. Both standpoints support awareness; they help integrate the reality of competition into common/process purpose, as depicted in Figure 3.1.

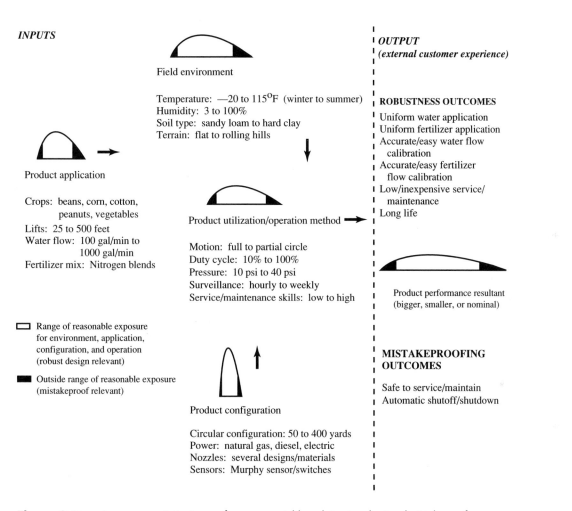

INPUTS

Field environment

Temperature: —20 to 115°F (winter to summer)
Humidity: 3 to 100%
Soil type: sandy loam to hard clay
Terrain: flat to rolling hills

Product application

Crops: beans, corn, cotton,
      peanuts, vegetables
Lifts: 25 to 500 feet
Water flow: 100 gal/min to
      1000 gal/min
Fertilizer mix: Nitrogen blends

☐ Range of reasonable exposure
   for environment, application,
   configuration, and operation
   (robust design relevant)

■ Outside range of reasonable exposure
   (mistakeproof relevant)

Product utilization/operation method ➤

Motion: full to partial circle
Duty cycle: 10% to 100%
Pressure: 10 psi to 40 psi
Surveillance: hourly to weekly
Service/maintenance skills: low to high

Product configuration

Circular configuration: 50 to 400 yards
Power: natural gas, diesel, electric
Nozzles: several designs/materials
Sensors: Murphy sensor/switches

OUTPUT
(external customer experience)

ROBUSTNESS OUTCOMES

Uniform water application
Uniform fertilizer application
Accurate/easy water flow
   calibration
Accurate/easy fertilizer
   flow calibration
Low/inexpensive service/
   maintenance
Long life

Product performance resultant
(bigger, smaller, or nominal)

**MISTAKEPROOFING
OUTCOMES**

Safe to service/maintain
Automatic shutoff/shutdown

Figure 3.7 Low-pressure irrigation performance variables relative to robust and mistakeproof performance.

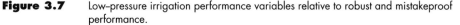

RAINMAKER CASE—BENCHMARKS   Initially, benchmark data is quite fragmented and somewhat incomplete. Eventually, we sort, arrange, and orient our benchmarking information to support our process purposes, e.g., to aid in setting explicit process goals, objective, and targets.

    Rainmaker, in its quest for expanding the organization, is seeking information relative to competitive forces in the environment, specifically the manufacturing environment. Sources of benchmarking information include industry publications, organizational records, and special studies such as plant visits, trade show tours, and so on. Table 3.1 presents a summary of several metrics/indices that impact Rainmaker's direction—vision, mission and process/subprocess purposes.

Example 3.4

Table 3.1   Summarized benchmarks for the Rainmaker case.

	Rainmaker	Industry Average	Industry Best	Best of Best*
Physical Performance Index				
Energy consumption (lift and distribution)	100	75	100	N/A
Uniformity of application (water/fertilizer distribution)	95	80	100	N/A
Economic—Cost per Unit				
Machining (Al alloys— labor, tooling, equipment)	$ 5k	$ 6k	$4k	$2k
Fabrication (labor, tooling, equipment)	$7.5k	$7.5k	$5k	$3k
Timeliness				
Design to production	18 months	20 months	12 months	1 month
Production cycle	60 days	40 days	30 days	5 days
Order delivery	75 days	45 days	35 days	7 days
Customer Service Index				
Sales and setup satisfaction	95	75	98	N/A
Maintenance satisfaction	85	70	95	N/A

*Similar processes—nonirrigation manufacturers.

REVIEW AND DISCOVERY EXERCISES

REVIEW

3.1. Process definition/redefinition, control, and improvement are broken out for separate treatment in the text. See Figure 3.1. What distinguishes one from the other? Explain.

3.2. Define synergy.

3.3. How can process synergy be gained through the integration of process definition/redefinition, control, and improvement? Refer to Figure 3.2. Explain.

3.4. Develop a partial common purpose hierarchy for one of the following organizations:

   An educational organization, e.g., a university
   A service organization, e.g., a bank, laundry, educational institution, medical clinic
   A manufacturing organization, e.g., electronics, automotive
   A basic production organization, e.g., oil and gas, agricultural, mining

3.5. Provide one example each for a bigger is better, a smaller is better, and a nominal is best target from your everyday life.

3.6. What is a process value chain? What is its purpose? Explain.

3.7. Why is it particularly challenging to build a complete process value chain diagram? Explain.

3.8. Explain the difference between an open-loop and a closed-loop process. What advantages/disadvantages does each enjoy? Explain.

3.9. Explain the difference between feedforward and feedback closed-loop process control.

3.10. Provide two examples each of open-loop and closed-loop processes in our everyday lives.

3.11. How are robust performance and mistakeproofing performance related? Explain.

DISCOVERY

3.12. How do the concepts of robust performance and mistakeproofing relate to the concepts of quality and productivity? Explain.

3.13. Although the process synergy figure, Figure 3.2, was developed with production organizations in mind, relate/map the concepts of process definition/redefinition, control, and improvement to a sports activity of your choice, e.g., baseball or volleyball.

3.14. Referring to the Amplifier Acquisition Case, how were location and dispersion associated with the amplifiers' performance characteristics? How conducive is this screening process to mass production? What implications exist for maintenance of products built using a component screening process as described in this case? Explain.

3.15. Develop an example analogous to the space heating example in Figure 3.4. Include open-loop, manual closed-loop, and automatic closed-loop depictions.

3.16. Using the Rainmaker Case—Variables Identification (Example 3.3) as a basis, develop an analogous diagram for a product or process of your choice.

3.17. Using the Rainmaker Case—Benchmarks (Example 3.4) as a basis, develop an analogous table for a process of your choice.

Process Characterization, Exploration, and Response Modeling

The purpose of Section 2 is to introduce physical and statistical process characterization, as well as to provide tools for process exploration and response models.

PART OUTLINE

Chapter 4: Process Characterization

The purpose of Chapter 4 is to provide systematic means for physical and statistical process characterizations, focused on critical process leverage points and related performance metrics.

Chapter 5: Process Exploration

The purpose of Chapter 5 is to describe fundamental experimental design tools that are useful for exploring possible cause-effect relationships.

Chapter 6: Process Response Modeling

The purpose of Chapter 6 is to introduce the regression concept for building process response surface models from empirical data.

4

PROCESS CHARACTERIZATION

4.0 INQUIRY

1. How are process input-output linkages described?
2. What is a process lever? What is process leverage?
3. How is process performance measured?
4. What is process inertia?

4.1 INTRODUCTION

The focus of this section is process understanding. Our approach is to systematically build on existing process knowledge. With new processes, our knowledge base may be rather meager. With ongoing processes, our knowledge base may be extensive. In either case, the more and faster we learn and upgrade our process knowledge, the more competitive we become. **We have two primary options for process learning. We can gain our knowledge in a piecemeal fashion,** observing things as they happen. **Or we can aggressively pursue process knowledge through systematic means.** Systematic means include physical reasoning, designed experiments, and response surface modeling.

4.2 PROCESS UNDERSTANDING

Process understanding is critical in all phases of process work: definition/redefinition, control, and improvement. Understanding helps us to shape our processes toward process purpose and then control our processes relative to our process purpose, goals, objectives, targets, and specifications. In definition and redefinition we usually start with a rather speculative knowledge base—one we infer from past experiences, including the experiences of other production systems.

In process control, we build extensive experience bases. Provided we document and communicate this information, we gain understanding that leads to process improvement. The specific purpose of this chapter is to provide guidance in both physical and statistical characterization of our processes—so that we gain cause-effect knowledge as rapidly as possible and so that, later, we do not attempt to control strategies that simply are not compatible with our process.

In understanding processes, two fundamental issues are explored simultaneously: (1) general principles and (2) specific variables. These issues are depicted in sequence in Figure 4.1. We initiate our principles and variables analyses by identifying the effects we desire and the technologies we employ to obtain them. We apply process flow diagrams, cause-effect (C-E) diagrams, and basic statistical analysis tools to help us identify and study the relevant principles and variables, respectively. Primary tools are overviewed in Chapter 20.

Any given process usually contains several technologies. Each technology usually contains several basic principles. **In the principles analysis we explore how our process technology works.** Here, we begin with a description of basic principles. Each basic principle may involve several process mechanisms, encompassing several variables and metrics and their associated

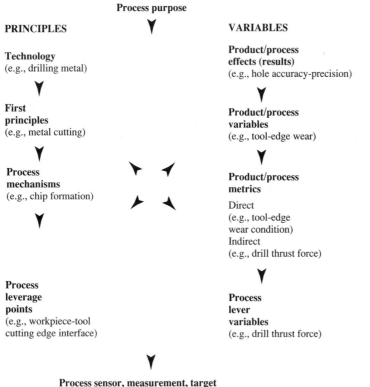

Figure 4.1   Process lever and leverage identification structure.

targets, i.e., bigger is better, smaller is better, or nominal is best. A simplified example is provided for a metal drilling technology in Figure 4.1.

At each step of our variables analysis, we relate to the counterpart step in our principles analysis; see Figure 4.1. These relations track back and forth so that **each principle is described by variables, and each variable applies to one or more principles.** As a result of this crisscrossing, we develop a working knowledge of, and technical insights into, our process. We also identify a comprehensive hierarchy of variables, both controllable and uncontrollable, that impact our process mechanisms.

We work through these process details so as to identify our most critical variables and their associated process leverage points. **Process leverage points are cause-related points where small changes in cause variables provide large impacts on the effect or results variables.** By moving systematically from process technology and effects to process leverage and levers, we gain additional insights into process mechanics and cause-effect. In these analyses, we seek out indirect metrics, as opposed to direct metrics (Figure 4.1), that yield more timely process information at reduced cost.

An indirect metric corresponds to a variable that closely follows/measures the actual variable of interest, but is faster and/or cheaper to track. In the drilling example of Figure 4.1, we are using thrust force to track tool wear. Sometimes several indirect metrics can be identified for a given variable. For example, tool time in service is a second possible indirect metric that can be used to track tool wear. The point is to identify a cost-effective and timely indirect variable that is highly correlated with the variable of interest, and then define meaningful metrics that correspond to these variables.

4.3 PROCESS MODELS

As an extension of our discussion of robustness, illustrated in Figure 3.6, Figure 4.2 depicts process performance in a mathematical format. In this idealistic depiction, we see input vectors, X_i, i.e., sets of controllable variables, along with vectors of uncontrollable variables, Z_i, on the left-hand side. Here, **controllable variables are variables that we are able and willing to manipulate. Uncontrollable variables are variables that we are unable or unwilling to manipulate** because of physical and/or economic constraints. We see corresponding input vectors addressing environment E, configuration C, materials and supplies M, and transformation methods P.

The output results on the right-hand side, **Y, are complex variables—functions of the input side variables. Ideally, the output characteristics are "optimized" when we set our input variables at their respective targets.** Optimization presents several challenges. The first challenge is to identify appropriate output targets, and then to determine corresponding input targets. The second challenge is to adjust the process inputs to their targets in operations. The third challenge is to monitor the process so as to determine when it changes, and take appropriate action.

Output variation is affected by both common cause variation in X_i and Z_i, as well as special cause variation in our X_i and Z_i variables. Common cause variation is that variation we expect to see as a result of natural physical, economic, and/or social fluctuations. Special cause variation is variation resulting from unexpected changes or disturbances.

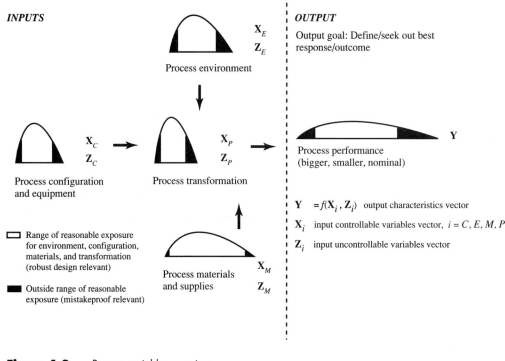

Figure 4.2   Process variables overview.

Our goal in targeting is to define best or at least acceptable output performance. Our initial effort in defining desirable outputs usually takes place at the conceptual level, in process definition; details are described in Section 3. Essentially, we express the results we seek from a process and then detail corresponding results from related subprocesses. We use whatever information we have initially, plus newly acquired process understanding, and eventually define detailed targets/requirements for every critical product and process output characteristic. Each one of these output characteristics makes up one element of the **Y** vector. Each element will be of a smaller is better, larger is better, or nominal is best nature.

The identification of the $\mathbf{X}_i$ and $\mathbf{Z}_i$ vector components follows from process understanding. In Figure 4.2, these vectors pertain to controllable and uncontrollable variables, respectively, associated with the process environment, process configuration and equipment, process materials and supplies, and process transformation. Here, cause-effect logic as well as experimental design technology and response surface modeling play major roles.

We seek out our most critical input variables. Typically, many input variables act to influence output characteristics, **Y** elements. However, some variables possess more influence than other variables. Chapter 5 introduces the discipline of experimental design, which provides a systematic means to explore **X, Z, Y** relationships and determine the extent of influence present.

Once we establish our critical input variables, both controllable and uncontrollable, we seek to relate them in a quantitative sense to our output, *Y*, variables. Given that each $\mathbf{X}_i$ and

$\mathbf{Z}_i$ can take on several, possibly an infinite, number of values, we seek out target or best values for the $\mathbf{X}_i$ elements. The use of response surfaces helps us model our $\mathbf{X, Z, Y}$ relationships so that we can develop targets in a systematic manner. Chapter 6 focuses on response surface modeling.

In summary, any given process presents significant challenges in targeting. Here, we want to determine the "best" values for our controllable variables $\mathbf{X}_i$ so that the response, $\mathbf{Y}$, is "optimized." **In practice, targeting requires a high degree of process understanding.** To address these challenges in a systematic manner and gain process understanding, we use experimental design and response surface modeling in conjunction with basic tools such as flowcharts and C-E diagrams. Examples of these combinations are provided in Chapters 5 and 6. The concepts of adjustment and monitoring are taken up in Section 4.

4.4 PROCESS MEASUREMENT SCALES

Most process results are multidimensional, with physical, economic, timeliness, and customer service performance components. Each critical output result is usually impacted by several input and transformation or conversion variables. Furthermore, variables typically do not act independently of each other. We usually see interaction to one degree or another. As long as we obtain some degree of understanding of our critical variables and their impact on our process, we can set targets at reasonable levels and then move toward "optimality" as we learn from our process experience.

Once we identify our critical process results and leverage points, we identify our process levers—associated controllable variables. **For each result and lever, we consider an appropriate metric.** Here, we have several choices. **Four primary technical scales of measurement exist: (1) nominal, (2) ordinal, (3) interval, and (4) ratio.** Table 4.1 describes each scale and presents several examples.

In process-related work, two working scales are usually used, the attributes scale and the variables scale. When a process characteristic is assessed as being "go" or "no-go" or acceptable or unacceptable (one of two, or perhaps more, classes), it is termed an **attributes characteristic,** subject to an attributes measure. A set of such assessments is termed **attributes data.** A process characteristic assessed by means of a measurement (limited only by the resolution of the measuring device) is termed a **variables characteristic.** We term such a measure a variables measure. A set of such measurements is termed **variables data.**

Table 4.1   Measurement scales.

Scale	Characteristic	Examples
Nominal	Number or label distinguishes between categories (the weakest scale)	Acceptable, unacceptable
Ordinal	Number or label provides rank or order	Small, medium, large
Interval	Order with defined intervals between measurements—an arbitrary zero point	Temperature, hardness
Ratio	Order with defined intervals between measurements—a natural zero point (the strongest scale)	Length, time, yield, profits

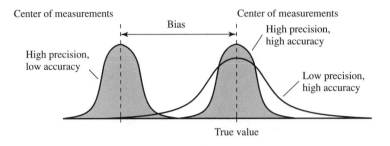

Figure 4.3   Measurement precision and accuracy depiction.

Attributes and variables data can be mapped against the four basic technical measurement scales. The nominal and ordinal scales describe attributes measures. For example, an acceptable or unacceptable product metric generally follows a nominal scale. **The interval and ratio scales describe variables measures.** Examples include temperatures and process yields.

After we define appropriate metrics, according to a scale of measure, we identify or develop an appropriate sensor. Our sensor requires calibration so as to convert what it "sees" in the physical world to a meaningful scale value. Sensors range from complex physical devices, rendering very accurate and precise readings on interval or ratio scales, to humans rendering subjective judgments on nominal or ordinal scales.

Several characteristics of sensors are critical in providing measurements for process control.

1. **Accuracy:**   The absence of bias in the measurements

2. **Precision:**   The dispersion of the measurements

3. **Timeliness:**   The time it takes to obtain a measurement

4. **Economics:**   The cost in personnel, equipment, materials, and supplies necessary to obtain a measurement

Accuracy is location-related. Precision relates to dispersion about a central value, usually the mean of the measurements. Figure 4.3 illustrates these concepts. The measurement process can be one of relatively high precision with low accuracy, high accuracy with low precision, and so on; see Figure 3.3 for the basic concepts of location and dispersion.

Regarding accuracy, in statistical terms, the estimated value of a parameter should approach the true value of the parameter as the sample size increases. In practical terms, the average of our measured values should approach the "true" dimension or true value of our product or process as we make repeated measurements.

If we are dealing with biased measurements, and if we can determine the bias, then we can adjust the bias out of our measurements. For example, if we are using a sensor that reports 10% high, we can simply decrease our recorded measurements by the appropriate amount. Hence, we can readily adjust for bias and derive unbiased measurements. The point is, we must know or estimate the bias in order to make the appropriate adjustment. Instrument calibration is an effective means to minimize, but not totally eliminate, bias. As long as our bias is small relative to the magnitude of our measurement, we usually tolerate its presence.

Precision, on the other hand, is a more complex issue. Dispersion in measurements results from the failure of our sensor to exactly repeat itself. Depending on the nature of the sensor, poor precision can result from electrical, chemical, mechanical, and human causes, anywhere from the point of sensing, through data conversion and transmission, all the way to data recording. Specific methods to quantify and measure measurement precision will be discussed in Chapters 5 and 11.

In general, we know that our recorded measurements differ from the actual characteristic we are measuring. Differences are attributable to both inaccuracy and imprecision. As long as inaccuracy and imprecision are relatively small, relative to the actual product/process characteristic measurement, they can be tolerated.

In many cases, we have a choice of sensors. For example, we might use meticulous laboratory tests or crude on-line samplers. Here, timeliness and economics play large roles. The laboratory analysis may be more accurate and precise, but may require an order of magnitude more time. Cost per measurement varies with the sampling rate; frequent sampling requirements usually favor automatic high-speed sensors, whereas very infrequent samples may favor off-line general laboratory tests. The exact physical nature of the process characteristic and the sampling rate, accuracy, and precision necessary for effective and efficient process control all play a role in sensor selection. In general, **each sensor is justified on the basis of benefits** (how it helps us to produce our product and please our customers) **and burdens** (how it draws on our physical resources, time, and money).

4.5 PROCESS LEVERS AND LEVERAGE

Our critical variables, within the X_i vectors, **are termed process levers.** They allow us to manipulate the process results, the **Y** effects. **The points at which these critical X_i variables act in the process mechanisms are termed process leverage points.**

Our highest leverage variables are upstream variables that, when manipulated, heavily influence response variables downstream, finally impacting our business results. This hierarchy is expressed through the process value chain, PVC (Figure 3.5). In the PVC, and related process flowcharts and C-E diagrams, we mark high-leverage variables, by process location, to expedite our subsequent process control and improvement efforts. The higher the leverage any given variable provides, the more critical that variable becomes in targeting and process control. Our ultimate goal is to identify and then exploit our process levers so as to meet our process output targets.

Given a potential leverage point—typically located in a subprocess—we explore the downstream nature of the leverage, toward the business results side. Then we focus as best we can on the upstream linkages, with respect to related controllable and uncontrollable variables. In other words, **we explore both downstream and upstream from a leverage point in order to determine both business and physical impacts, so that we can more effectively and efficiently control our process.**

4.6 PHYSICAL CHARACTERIZATION

Our physical characterization/examination has three basic facets: (1) upstream physical performance relationships and characteristics, **(2) downstream** leverage in terms of economics and timeliness, and **(3) proactive and reactive counteraction**—diagnosis, prescription, and corrective actions. We explore each facet systematically and thoroughly.

Table 4.2   Physical characterization elements.

Element	Comments
Upstream performance	Many basic variables affect process results. We seek to identify all critical variables by systematically assessing our physical processes—technologies, principles, mechanisms. Tools such as designed experiments are described in Chapter 5.
Downstream leverage	We project the influence of upstream variables to downstream business results. From this projection, we establish the degree of leverage upstream variables exert on downstream business results. Time delays may occur between changes in critical upstream variables and changes in process results. Tools such as response surface modeling are available for more detailed studies; see Chapter 6.

Counteractions
    Proactive
    Reactive
	Counteractions refer to control actions that may be useful at our process leverage points. Counteraction involves diagnosis, prescription, and action:
	Process diagnoses range from fully deterministic (certain) to judgmental (uncertain). Counteraction prescriptions range from algorithmic (fully prescribed) to judgmental (several possible courses of address).
	Corrective action itself involves implementing the prescription.

Our three-point physical characterization is detailed in Table 4.2. Here, we focus on cause-effect regarding specific variables with the intent of distinguishing the few critical variables from the many variables that exert some influence on our process leverage metric. Furthermore, we want to describe the cause-effect response dynamics in quantitative terms. We want to assess the time delay between changes in upstream and downstream variables. We want to estimate physical bounds on process responses and upstream causal variables, in order to obtain a feel for how fast/far our process might respond for better or for worse. This information will help us to eventually work out a reasonable process control sampling/timing scheme.

At this point, we piece together a general process value chain (PVC) model, so that we can clarify and communicate the need for timely monitoring and corrective action when necessary. High-speed/volume production encourages rapid detection/corrective action schemes. From an economics standpoint, we consider the impact of timeliness under "correct" detection and corrective action as well as "false alarms" and "failures to detect." Our work here and our resulting estimates are usually only rough approximations. However, this information proves invaluable in assessing our process control options, described in Section 4, Chapters 9 to 15.

Preventive/proactive and remedial/reactive means include three basic subelements: diagnosis, prescription, and action. Each of these subelements helps determine or refine our process control counteraction options and activities. Diagnosis refers to detection of a process upset in general, and isolating its source and nature of manifestation. Here, we classify the special cause disturbance and describe its impact on the process. Our diagnosis may be fully determined or certain at one extreme, or it may be rather judgmental or uncertain on the other extreme. The more certain the diagnosis, the more straightforward the prescription and action elements.

Next, we prescribe. Prescription differs from diagnosis in that we are defining and designing counteraction. We encounter a spectrum ranging from fully prescribed counteractions to judgmental counteractions. In other words, in some cases our prescription is obvious, and in other cases some level of interactive inquiry in terms of further observation-diagnosis-represcription is necessary.

Action to offset process upsets is the final counteractions issue. Here, we implement the prescription. In simple processes, action in terms of what to do and how much to do is straightforward. In other cases, action is fully planned, but the amount or extent of "adjustment" is determined "on the fly." Feedback/feedforward mechanisms of various sorts may be used in this case. In yet other cases, some sort of ad hoc action strategy is used; e.g., actions are changed/adjusted as responses are measured and assessed. In this case, a high level of judgment along with accompanying rediagnosis and replanning may be necessary. Physical characterization initiatives and tools are described in Chapters 19 and 20.

METER STICK PRODUCTION CASE—PHYSICAL CHARACTERIZATION   The Meter Stick Pro- | **Example 4.1**
duction Case describes an upstream-downstream production process relationship for a meter stick product. Here, we expect to encounter a sequence of processes, each with physical principles and variables relevant to our selected technology. We expect to see a clear upstream-downstream relationship, where we add cost and value to our product as it flows from raw materials to a finished product.

A meter stick is a useful but somewhat primitive measuring device. The product is simple in both concept and production technology. Essentially, we are dealing with a calibrated piece of material that is useful for several tasks, two of which include measuring linear dimensions and providing a straight edge for marking. The usual meter stick is made of wood or metal, one meter long, give or take a little, with calibration to the nearest millimeter. Typically, when the stick is used for measuring, we expect to be able to measure to about the nearest millimeter. We also expect to be able to produce a reasonably straight line when it is used in the marking/guideline mode.

Several subprocesses involved in a meter stick production process are depicted in Figure 4.4a. Here, we start with our raw material—metal in a coiled condition. Our first subprocess consists of feeding material from the coil, straightening it, and shearing the meter stick blank. Other subsequent subprocesses include deburring, cleaning and drying, calibrating and marking, and wrapping and packaging. The production process output is a packaged meter stick, ready for our customer.

Observation and physical description are critical for physical characterization. Figure 4.4b provides a simple physical depiction of the feed-shear and deburr subprocesses. Here, we begin to develop/describe physical details regarding subprocess technologies, principles, and mechanisms, with the aim of finding process leverage points.

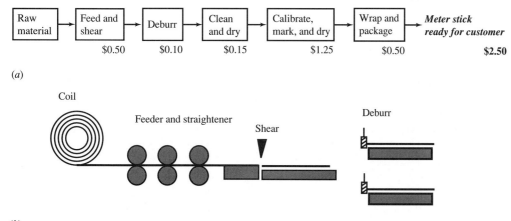

(a)

(b)

Figure 4.4   Meter stick production process overview with product dimension leverage featured. (a) Meter stick production subprocesses, with approximate production costs per unit. (b) Physical depiction of feed-shear and deburr subprocesses. *(continued)*

Activities	Technology	Principle	Mechanism	Leverage	Levers/Variables
Material handling	Coil	Rotate, unwind	Energy release, cold work	Surface condition, residual stresses	Coil wrapping, storage conditions
Feeder and straightener	Rollers, guides	Tension to unroll, compression to grip, timing to feed	Friction, cold work	Thickness, flatness, straightness	Roll force, roll speed, length index control
Shear	Blade	Fracture force	Deformation to fracture	Raw dimensions, sheared surface	Blade condition, force, speed, lubrication
Deburr	Grinding	Abrasion	Chip formation	Finished dimensions (1000 mm)	Wheel characteristics, size, speed, depth of cut

(c)

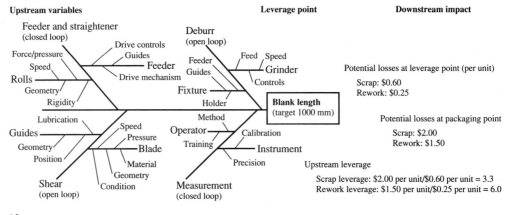

(d)

Counteractions (maintenance related)

Proactive: Off-line service and preventive maintenance—scheduled, typical cost $15 per hour plus supplies and parts

Reactive: On-line feeder-straightener roll pressure adjustments, shear adjustments, $3000 per hour plus parts
            On-line deburr adjustments, $1500 per hour per station plus parts

Counteractions Leverage
    Shear leverage: $3000 per hour/$15 per hour = 200
    Deburr leverage: $1500 per hour/$15 per hour = 100

(e)

Figure 4.4   *(Concluded)* (c) Meter stick feed and shear subprocess leverage and lever analysis (partial).
                   (d) Upstream-downstream leverage analysis (partial). (e) Counteractions leverage analysis
                   (partial).

Figure 4.4*c* provides a partial subprocess leverage analysis. Here, a number of process leverage points and process levers are identified. For example, we have isolated four basic leverage points: (1) the material-related surface condition and residual stresses; (2) feeder and straightener-related thickness, flatness, and straightness; (3) shear-related dimensions and shear surface condition; and (4) final overall dimensions, immediately after the deburring subprocess.

Once our basic physical analysis has isolated potential process levers and leverage points, we continue our physical characterization. We develop essential details with respect to our potential process levers.

For illustrative purposes, we focus on blank length immediately after the deburring subprocess. This process point is the earliest point where we can assess the product's final dimension with a high degree of accuracy and precision. The shearing operation creates a very rough surface, which we must clean up and deburr before we can determine a final product length dimension. Figure 4.4*d* provides a simplified C-E diagram related to our blank length process lever. Here, we have taken the finished overall length and worked further back upstream to identify relevant variables that may have some influence on blank length outcome. We have identified a target of 1000 mm for our characteristic. Our scrap and rework leverages between final product and the deburr-stage product are estimated to be 3.3 to 1 and 6.0 to 1, respectively. These estimated leverage statistics indicate that final length control makes sense early in the process.

Figure 4.4*e* provides a brief leverage overview in terms of proactive versus reactive counteractions. Here, we have estimated the costs associated with providing preventive service and maintenance as opposed to corrective repair for our feed and shear and deburr subprocesses. For example, we estimate a 200 to 1 advantage by substituting proactive off-line (when the line is not operating) preventive maintenance as opposed to stopping production for unscheduled repairs on the shear line. For each deburr station the comparable leverage is 100 to 1.

In summary, a good, high-leverage control point for overall length is located immediately after the deburr subprocess, rather than later in the production process—say after we have gone to the time and expense to clean and calibrate the product. We focus on each process leverage point and its associated process lever, so that we can find upstream metrics to monitor, measure, and compare—all with the objective of eventually producing a stream of zero-defect products. Later, in Chapter 5, we will provide a more detailed process exploration case that focuses further upstream at the shear.

The exact form and extent of physical characterization vary from one process to another. Each characterization usually involves process flow diagramming and C-E diagramming. Tools to portray physical, timeliness, economic, and customer service leverage are usually improvised. Typically, we develop our structure to include graphical communication as far as possible, so as to involve or engage operational personnel in this exploratory exercise.

4.7 STATISTICAL CHARACTERIZATION

Once we have explored the physical characteristics of our processes, we assess statistical characteristics. **Our statistical characterization aids us in developing a suitable modeling strategy;** we want to avoid ill-suited models, which may mislead us in our process understanding and/or control efforts. **Statistical characterization,** introduced in Table 4.3, **contains three basic elements: (1) data collection, (2) graphical assessment**—descriptive data analysis, and **(3) numerical assessment**—quantitative data analysis.

Table 4.3   Statistical characterization.

Element	Comments
Data collection	
Process result variable (Y)	Given the process result metric selected and the related upstream variables (if any), carefully put sensors and data collection equipment in place.
Related variables (X's and Z's)	Take care to assure unbiased data are collected. Document process conditions in existence during data collection so as to explain/support the integrity of the data.
Graphical assessment	
Process result variable/metric plot	Plot a runs chart of the data collected. Develop a histogram or stem and leaf plot.
Related variables/metrics plots	Devise and plot X-Y scatter plots as warranted by the physical characterization (to detect relationships between variables).

Time-modified plots
      X-Y scatter plots
 	
      Y_{t-k}-Y_t scatter plots
	Consider time lags between X and Y variable metrics, and between the response itself, Y_{t-k} versus Y_t.

Numerical assessment	
Goodness of fit	Develop a goodness of fit test/analysis.
X-Y correlation	Develop a correlation study between the upstream variables and the process result metric.
Y-Y autocorrelation	Develop an autocorrelation study and associated correlogram. Plot confidence intervals on the correlogram to assess randomness/independence in the data.

DATA COLLECTION

First, we identify a meaningful process variable and identify a scale of measure. We may want to simultaneously collect data on associated variables. Associated variables are typically upstream variables, X's and Z's, that we believe may hold some relationship/clue to help explain the process response metric, Y. For each variable we provide a sensor and data collection/recording equipment. Sensors range from human observation to electromechanical devices. Likewise, data collection/recording equipment ranges from paper and pencil devices to high-speed digital signal processing and recording devices.

    It is critical to preserve the time order/sequence of our data and to document physical conditions/circumstances surrounding their collection period. The time order sequencing in data collection poses challenges for data sets that include Xs, Zs, and Y. Some sort of master clock or production unit counter is necessary. Data integrity is critical.

    Process logs are used to record and document production conditions and circumstances across our time sequence. **A fully documented data set allows us to identify periods of process upsets or transitions and explain them;** it is critical that we find, explain, and neutralize special causes that disrupt our processes. Otherwise, we would proceed to characterize a data set that is not representative of our process, operating under common cause variation, and thereby obtain a distorted picture of our process metrics. If this distorted picture is allowed to wash through our entire sequence of process understanding activities, our understanding will be compromised.

The measurement process itself introduces variation into the data collected. It is important to ensure both accuracy and precision in our measurements. In general, accuracy/bias is addressed through sensor calibration. The random effects models in Chapter 5 and the gauge study techniques in Chapter 11 provide means to estimate measurement precision and isolate components of variation responsible for measurement imprecision.

GRAPHICAL ASSESSMENT

Graphical assessment through data displays allows us to see a data set as a whole, rather than just a sequence of numbers. In a properly designed data display, we see location, dispersion, trends, and targets. When we consider our process log notes, along with the data display, we look for correspondence between the events we have documented and the responses we recorded in the data.

One of the most effective data displays is the simple runs plot. Here, we see a time-sequenced picture of our data. We assess our plots visually, looking for basic location, dispersion, trends, patterns, and so on. This information, along with our physical reasoning, provides help in understanding our processes.

Another basic graphical data analysis tool is the histogram. Histograms do not preserve the time order of our data stream, but they do allow us to visualize the "shape" of our data set as a whole, e.g., location and dispersion together. For example, a normally distributed data stream appears "bell" shaped, with lower density in the tails and higher density in the center portion. When assessing normality, we also look for symmetry. If we do not see these characteristics, then models that assume normality such as process control models, described in Chapters 10, 11, and 13, may prove to be misleading.

A histogram essentially depicts a data set by assigning each observation in the data set to one of several cells or predefined categories, and then depicting associated cell counts and relative frequencies. Typically, a minimum of 20 to 30 observations is necessary to produce a meaningful histogram. Histograms are constructed in several steps:

1. Determine the range of the data. Here we calculate the difference between the largest and smallest values.
2. Determine the number of cells or categories. Usually we develop from 5 to 15 cells.
3. Determine cell midpoints and boundaries. Cells of equal width are usually defined; however, cells can vary in width.
4. Place each observation in a cell. Each observation must fall in only one cell. A check sheet format is useful to classify each observation.
5. Display the cells. The frequency and/or relative frequency of each cell determine the cell's relative heights in vertical histograms.

Computer aids are available to develop histograms. Modern spreadsheets as well as dedicated statistical analysis packages can be used to produce and display histograms.

Based on clues from our physical analysis, we pursue relationships between our X's, Z's, and Y. Here, a series of X-Y scatter plots may be useful. Correlations tend to show up as patterns on these plots. We may also investigate time-order lags/leads in our data. For example, an X might be related to Y, but Y_t movement may lag X_t movement by several time units. Also, we may investigate time-order relationships in our process result response, Y. Here, we may plot Y_{t-k} versus Y_t to graphically assess autocorrelation in our response variable, provided our data were collected at equally spaced intervals of time.

Basic tools for descriptive data analysis are overviewed in Chapter 20. Other descriptive statistical analysis tools used to assess location and dispersion include stem and leaf plots and box plots; see introductory texts such as Walpole, Myers, and Myers [1] for details.

NUMERICAL ASSESSMENT

We use numerical assessment/analysis to complement and supplement our graphical analyses. Here, three primary technologies are described: (1) goodness of fit, to a hypothesized model; (2) correlation analysis between two variables; and (3) autocorrelation analysis in a time sequence within a single variable.

Goodness of Fit   **Goodness of fit (GOF) essentially measures the degree of fit between an empirical data set and a hypothesized probability density model/function (pdf).** Here, we are quantifying what we see visually in comparing our histogram to a probability density function trace.

A variety of GOF tests exists. Some methods are applicable to any form of probability density model. These methods are labeled nonparametric. Parametric methods are applicable only to selected probability density models. Many computer-aided statistical analysis packages contain GOF capabilities. We describe two GOF tests: (1) the nonparametric chi square GOF test and (2) the parametric Geary's GOF test, applicable to a normal probability density model. Details, in addition to the overview presented here, can be found in Walpole, Myers, and Myers [1].

In the chi square GOF method, we divide our hypothesized pdf into cells. In order to utilize this method, we need 30 or more observations. Several steps are involved:

1. Develop a set of hypotheses based on a hypothetical pdf. For example,

    H_0: A normal pdf adequately describes the data.

    H_1: A normal pdf does not adequately describe the data.

2. Collect a sample of $n \geq 30$ data points from the process.

3. Divide the hypothetical pdf into m cells/portions of about equal probability. For example, we might divide a normal pdf into $m = 6$ cells/portions, centered about the mean, as illustrated in Table 4.4. In this case, we can use the standard normal distribution (see Table VIII.1 in Section 8) to determine the expected probabilities.

4. Determine the empirical cell boundaries appropriate for the cells selected above. See Table 4.4.

5. Sort each observation into its appropriate cell. Be sure each observation falls into only one cell.

6. Calculate the χ^2 test statistic:

$$\chi^2 = \sum_{i=1}^{m} \frac{(o_i - e_i)^2}{e_i}$$

[4.1]

Table 4.4   Chi square GOF test format for a test of normality.

Description	Cell Number					
	1	2	3	4	5	6
Area under the normal pdf	Beyond $\mu - 1\sigma$	From $\mu - 0.5\sigma$ to $\mu - 1\sigma$	From μ to $\mu - 0.5\sigma$	From μ to $\mu + 0.5\sigma$	From $\mu + 0.5\sigma$ to $\mu + 1\sigma$	Beyond $\mu + 1\sigma$
Expected probability	0.1587	0.1498	0.1915	0.1915	0.1498	0.1587
Expected frequency	$e_1 = 0.1587n$	$e_2 = 0.1498n$	$e_3 = 0.1915n$	$e_4 = 0.1915n$	$e_5 = 0.1498n$	$e_6 = 0.1587n$
Observed cell boundaries	Below $\hat{\mu} - 1\hat{\sigma}$	Between $\hat{\mu} - 0.5\hat{\sigma}$ and $\hat{\mu} - 1\hat{\sigma}$	Between $\hat{\mu}$ and $\hat{\mu} - 0.5\sigma$	Between $\hat{\mu}$ and $\hat{\mu} + 0.5\hat{\sigma}$	Between $\hat{\mu} + 0.5\hat{\sigma}$ and $\hat{\mu} + 1\hat{\sigma}$	Above $\mu + 1\hat{\sigma}$
Observed frequency	o_1	o_2	o_3	o_4	o_5	o_6

where m = the number of cells

o_i = the observed count frequency for the ith cell

e_i = the expected count frequency

7. Compare the calculated test statistic with a tabulated value for a selected significance level α. Tabulated χ^2 values appear in Table VIII.3, Section 8. Here, the degrees of freedom are calculated as

    df = (number of cells used) − (number of parameters estimated from the data set) − 1

8. If $\chi^2_{calc} \leq \chi^2_{\alpha,df}$, then we do not reject H_0 on the basis of the evidence presented.

    Note: This test requires expected cell counts of at least 5 before it is credible. If a large number of cells is used, and some contain expected counts of less than 5, adjacent cells can be combined. Hence, for our cells in Table 4.4, we should collect about 35 or more data points.

Geary's test is a parametric test for testing normality in data. Here, we should have a large data sample, say n ≥ 30, for meaningful results. Several steps are involved:

1. Develop a set of hypotheses based on a hypothetical normal pdf. For example,

    H_0: A normal pdf adequately describes the data.

    H_1: A normal pdf does not adequately describe the data.

2. Collect a sample of $n \geq 30$ data points from the process.

3. Calculate the Geary test statistic:

$$u = \frac{\sqrt{\pi / 2} \; \sum_{i=1}^{n} |y_i - \bar{y}|/n}{\sqrt{\sum_{i=1}^{n} (y_i - \bar{y})^2/n}} \qquad \textbf{[4.2]}$$

4. Determine a tabular comparison value for the Geary statistic. Here, if H_0 is true, then the calculated Geary statistic u should be approximately equal to one; i.e., for normally distributed data, the numerator and denominator of Equation (4.2) should be approximately equal.

5. Now, we standardize the Geary statistic u as

$$z_{\text{calc}} = \frac{u - 1}{0.2661 / \sqrt{n}}$$ [4.3]

so as to compare with standard normal tabulated values.

6. We select a significance level α and obtain associated $\pm z_{a/2}$ tabulated values from our standard normal table, Table VIII.1, Section 8.

7. If $- z_{a/2} \le z_{\text{calc}} \le z_{a/2}$, then we do not reject H_0, based on the evidence presented.

Manual GOF tests are effective but not highly efficient in terms of our time. If we have access to a computer-aided statistical analysis package, then we can save a considerable amount of time and calculations. However, we must be sure we understand what test our package uses and how to interpret the results.

Correlation   Samples subject to correlation analysis must be collected in sets. For example, for an X, Y correlation analysis, we must collect X's and Y's in bivariate pairs, observed together. **Correlation for our purposes of analysis is a scalefree (unitless) measure of the strength of linear association between two variables.**

$$\rho = \frac{C_{xy}}{\sqrt{\sigma_x^2 \sigma_y^2}}$$ [4.4]

where $C_{xy} = E[(X - \mu_x)(Y - \mu_y)] =$ covariance

$\sigma_x^2 = E[(X - \mu_x)^2] =$ variance of random variable X

$\sigma_y^2 = E[(Y - \mu_y)^2] =$ variance of random variable Y

$\rho =$ the linear correlation coefficient $(-1 \le \rho \le 1)$

$E[.] =$ the expected value of $[.]$

The covariance C_{xy} is a measure of linear association between random variables X and Y, where the sign of C_{xy} indicates the general nature of the relationship. Positive indicates high X is associated with high Y or low X is associated with low Y; negative indicates high X with low Y or low X with high Y; and zero indicates no linear association between X and Y. The unit of measure is the product of the units of X and Y.

In terms of sample data, we estimate ρ as

$$r = \hat{\rho} = \frac{c_{xy}}{\sqrt{s_x^2 s_y^2}} = \frac{\sum_{j=1}^{n} (x_j - \bar{x})(y_j - \bar{y})}{\sqrt{\sum_{j=1}^{n} (x_j - \bar{x})^2 \sum_{j=1}^{n} (y_j - \bar{y})^2}}$$ [4.5]

where n is the number of samples in the data set.

We can set up a test of hypotheses regarding the linear correlation coefficient

$$H_0: \rho = 0$$

$$H_1: \rho \neq 0$$
$$\rho > 0$$
$$\rho < 0$$

where

$$t_{calc} = \frac{r}{\sqrt{(1 - r^2) / (n - 2)}}$$ [4.6]

We reject H_0 at the α level of significance as follows:

For $H_1: \rho \neq 0$, if $|t_{calc}| > t_{n-2,\,\alpha/2}$, then reject H_0

For $H_1: \rho > 0$, if $t_{calc} > t_{n-2,\,\alpha}$, then reject H_0

For $H_1: \rho < 0$, if $t_{calc} < t_{n-2,\,\alpha}$, then reject H_0

Introductory probability and statistics texts, such as Walpole, Myers, and Myers [1], provide further details regarding the derivation and general nature of linear correlation and related inferences. Typically, we use computer-aided statistical analyses for assessing correlation since calculations for all but the smallest data sets become laborious.

In cases where a time delay is suspected between two variables, we can use a standard correlation analysis format, as described above. However, we shift one variable by our chosen time difference and proceed as usual. For example, a two-time-unit shift (lag) between X input and Y output can be assessed by systematically shifting our X variable back by two time units and then calculating the correlation coefficient between the two variables, X_{t-2} and Y_t. This time-shifting procedure is used to obtain a corresponding X-Y plot, e.g., X_{t-2} versus Y_t. A word of caution is in order: If we choose to study time lags in our data, then we must be careful to collect our data at equally spaced time intervals.

By developing both the X-Y plot and the correlation analysis, we can obtain a more comprehensive picture of any possible data relationships that exist. Also, **linear correlation measures do not generally pick up nonlinear patterns.** Hence, the X-Y plot becomes extremely valuable if we encounter second-order effects.

We must remember that statistical relationships may support a cause-effect conjecture, but they do not establish one—that is where the physical analysis comes into play. Both physical and statistical analyses, together, provide invaluable insights to process understanding and subsequent model formulation.

Autocorrelation   **Autocorrelation, as a concept, refers to serial or order correlation in a univariate data stream.** In a physical sense, a process with little or no physical inertia or momentum should produce a response across time or order of production whereby each observation is independent—does not depend on any other physical result/measurement. In the case of no process inertia, in a statistical sense, we would expect that no correlation would exist between measurements taken across any time delay sequence or lag we consider.

Autocorrelation provides a scalefree measure of the strength of linear association in a time series of data taken k time units apart. Its general form is similar to that of correlation, Equation (4.4):

$$\rho_k = \frac{C_{Y_t Y_{t-k}}}{\sqrt{\sigma_{Y_t}^2 \sigma_{Y_{t-k}}^2}} = \frac{\text{Cov}\,(Y_t, Y_{t-k})}{\sqrt{\text{Var}\,(Y_t)\,\text{Var}\,(Y_{t-k})}} \qquad \text{[4.7]}$$

where $\text{Cov}\,(Y_t, Y_{t-k}) = E[(Y_t - \mu_t)(Y_{t-k} - \mu_{t-k})] = $ autocovariance for time lag k
$\text{Var}\,(Y_t) = E[(Y_t - \mu_t)^2] = $ variance of Y_t
$\text{Var}\,(Y_{t-k}) = E[(Y_{t-k} - \mu_{t-k})^2] = $ variance of Y_{t-k}
$\rho_k = $ autocorrelation coefficient for time lag k, $-1 \le \rho_k \le 1$

In its most general form, Equation (4.7) is difficult to work. For example, each t might index a unique mean, variance, and autocovariance, which would be difficult to estimate. However, if we assume our time series Y_t is weakly stationary,

1. $E[Y_t] = \mu$ for all t
2. $\text{Var}\,(Y_t) = E[(Y_t - \mu)^2] = \sigma_Y^2$ for all t
3. $\text{Cov}\,(Y_t, Y_{t-k}) = E[(Y_t - \mu)(Y_{t-k} - \mu)]$ for all t

Then

$$\rho_k = \frac{\text{Cov}\,(Y_t, Y_{t-k})}{\sigma_Y \sigma_Y} \qquad \text{[4.8]}$$

is workable (e.g., $\rho_0 = 1$, $\rho_{-k} = \rho_k$). In this context, we define white noise as a purely random process, where

$$\text{Autocorrelation}\,(Y_i, Y_j) = 0 \qquad \text{for all } i \ne j$$

Thus, a white noise signature yields

$$\rho_0 = 1$$
$$\rho_k = 0 \quad \text{for } k = 1, 2, \ldots$$

In order to develop autocorreletion estimates from samples, we use

$$r_k = \frac{c_k}{s_Y^2} = \frac{\displaystyle\sum_{t=k+1}^{n}(y_t - \bar{y})(y_{t-k} - \bar{y})}{\displaystyle\sum_{t=1}^{n}(y_t - \bar{y})^2} \qquad k = 1, 2, \ldots \qquad \text{[4.9]}$$

where n represents the sample size.

In order to calculate a full set of r_k terms, we use computer aids. For example, each r_k requires a considerable effort relative to hand calculations. We display our sample autocorrelations, r_k's, on a correlogram, in a graphic format. We use correlogram "signatures" to assess the nature of our data relative to white noise. From the correlogram, we obtain clues as to appropriate time series models we can use to model the "signal" portion of the data. For example, given a sample stream of a process metric, we produce a correlogram. Then, we look to see how close we are to a white noise pattern. Provided we see approximately white noise, we see evidence that suggests the data are likely independent.

Many process control models, such as the statistical process control (SPC) models, Chapters 10 and 11, **require independent data to perform well.** If we detect significant inertia, nonindependence, we may choose to develop or fit a time series model or response surface, extract the residuals from the model, produce a correlogram of the residuals, and assess the randomness of our residuals. If they appear more or less as white noise (random) we can then apply our classical SPC tools. Here, a residual refers to the "unexplained" portion of the response after applying our model. This model-fitting protocol is described in Chapters 6 and 13.

An approximate test for white noise, according to Newbold and Bos [2], considers

$$\text{Var } (r_k) \cong \frac{n - k}{n(n + 2)} \qquad \textbf{[4.10]}$$

Var (r_k) is approximately normally distributed for large sample sizes. Here, an approximate 95% confidence interval is set at

$$\text{CI}_{0.95} \cong \pm\, 2 \sqrt{\frac{n - k}{n(n + 2)}} \qquad \textbf{[4.11]}$$

This approximate confidence interval can be superimposed on our correlogram over $k = 1, 2,$ ..., to provide a graphical assessment of the hypotheses over the lags, k's.

H_0: The data stream is random (white noise).

H_1: The data stream is not random (other than white noise).

Provided our correlogram points r_k, for $k = 1, 2, \ldots$, fall inside the graphical bounds, we have support for randomness, e.g., signs of independence and low process inertia. Here, we should note that $r_0 = 1$. In other words, each observation is perfectly correlated with itself.

METER STICK BLANK CASE—LENGTH DATA CHARACTERIZATION   The physical nature of the **| Example 4.2** meter stick production line was described in Figure 4.4. Here, we identify the blank length after the deburr subprocess as a critical leverage variable. Once we identify our leverage metric, i.e., length, we assess its statistical nature. This assessment helps us to understand this particular results variable, adding to our ability to develop a valid control strategy.

For example, we might be interested in monitoring the length metric using an SPC model. SPC models are explained in detail in Section 4. When we consider a metric for SPC, it is critical to assess the validity of the data independence assumption. In this case, we rely on a graphical analysis. First, we start with a runs chart. Figure 4.5*a* provides 100 length data points continuously sampled from our process. We assume that we sampled every stick produced in sequence. From our runs chart, we notice that our values appear across time or order of production without any noticeable patterns. Our plot tends to support our independence assumption.

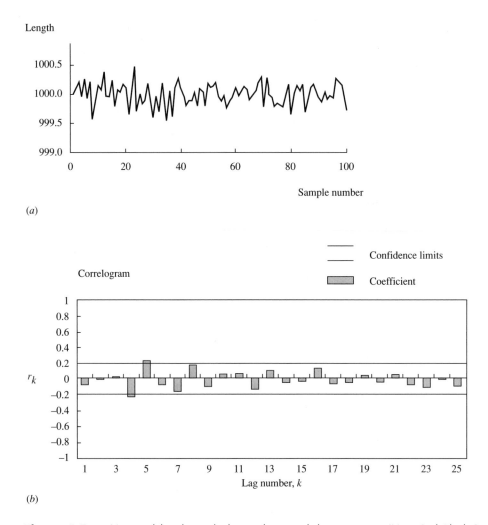

Figure 4.5   Meter stick length sample data and statistical characterization (Meter Stick Blank Case, Example 4.2). (*a*) Length runs plot (100 samples). (*b*) Length autocorrelation correlogram.

Second, we perform an autocorrelation analysis on the sample of 100 data points shown in the runs chart. For this test to be meaningful, we are associating our order of production with a time series. This association is approximate at best. In a case such as this, we use our autocorrelation characterization in conjunction with physical characterization to assess process inertia.

We resort to a computer aid and generate a correlogram, as previously described. Our correlogram is depicted in Figure 4.5*b*. We see the r_k falling pretty much within the confidence limits, which provides support for the use of SPC. In general, our correlogram appears to indicate a white-noise-like signature—an active correlogram with plus and minus values across the lags—staying generally within the confidence limits. Hence, we conclude that we see little if any statistical evidence of process inertia.

Now, we finish our data characterization by testing the data stream for normality. Here, we demonstrate with the first 50 data points, as displayed in Table 4.5.

Table 4.5 First 50 meter stick measurements.

Sample	Length, mm	$\|y_i - \bar{y}\|$	$(y_i - \bar{y})^2$	Sample	Length (mm)	$\|y_i - \bar{y}\|$	$(y_i - \bar{y})^2$
1	1000.007	0.01476	0.00022	26	999.834	0.15858	0.02515
2	1000.120	0.12712	0.01616	27	999.891	0.10127	0.01026
3	1000.214	0.22106	0.04887	28	1000.186	0.19359	0.03748
4	999.943	0.04933	0.00243	29	999.858	0.13472	0.01815
5	1000.279	0.28679	0.08225	30	999.593	0.39992	0.15993
6	999.913	0.08015	0.00642	31	999.988	0.00459	0.00002
7	1000.235	0.24266	0.05889	32	999.695	0.29774	0.08865
8	999.560	0.43281	0.18733	33	1000.195	0.20238	0.04096
9	999.837	0.15523	0.02410	34	999.981	0.01179	0.00014
10	1000.166	0.17308	0.02996	35	999.541	0.45155	0.20390
11	1000.063	0.07005	0.00491	36	1000.076	0.08379	0.00702
12	1000.391	0.39824	0.15860	37	999.606	0.38667	0.14951
13	999.957	0.03523	0.00124	38	1000.135	0.14189	0.02013
14	999.950	0.04310	0.00186	39	1000.272	0.27934	0.07803
15	1000.236	0.24333	0.05921	40	1000.092	0.09923	0.00985
16	999.788	0.20430	0.04174	41	1000.010	0.01720	0.00030
17	1000.078	0.08483	0.00720	42	999.816	0.17628	0.03108
18	1000.031	0.03789	0.00144	43	999.898	0.09425	0.00888
19	1000.167	0.17467	0.03051	44	999.898	0.09450	0.00893
20	1000.106	0.11308	0.01279	45	1000.054	0.06145	0.00378
21	999.659	0.33339	0.11115	46	999.808	0.18501	0.03423
22	1000.097	0.10405	0.01083	47	1000.110	0.11711	0.01372
23	1000.490	0.49736	0.24737	48	1000.066	0.07286	0.00531
24	999.702	0.29042	0.08434	49	999.805	0.18776	0.03525
25	1000.024	0.03105	0.00096	50	1000.212	0.21971	0.04827
				Averages	999.9927	0.17234	0.04539

Our hypotheses are as follows:

H_0: A normal pdf adequately describes the data.

H_1: A normal pdf does not adequately describe the data.

We test these hypotheses with the Geary GOF test. From Equations (4.2) and (4.3),

$$u = \frac{\sqrt{\pi/2} \sum_{i=1}^{n} |y_i - \bar{y}| /n}{\sqrt{\sum_{i=1}^{n} (y_i - \bar{y})^2/n}} = \frac{0.2160}{0.2131} = 1.014$$

Now, we standardize the Geary statistic u as

$$z_{calc} = \frac{u - 1}{0.2661/\sqrt{n}} = \frac{1.014 - 1}{0.2661/\sqrt{50}} = 0.367$$

Here, we use $\alpha = 0.05$; therefore, $z_{0.025} = \pm 1.96$. In this case,

$$-z_{0.025} < z_{\text{calc}} < z_{0.025}$$

We can therefore safely assume that the data are sampled from a normal population, and hence use our SPC models, described in Chapter 10 and 11, with confidence.

The meter stick case involved only one response variable, the blank length. The analysis was straightforward in the sense that the data stream possesses very little, if any, process inertia, is normally distributed, and is therefore suitable for classical SPC models directly. In other cases, we see reasonably strong process inertia. The natural gas case that follows is one such case.

Example 4.3 | **NATURAL GAS PRODUCTION CASE—PHYSICAL AND STATISTICAL CHARACTERIZATION**

Process vigilance is a critical part of process control. Knowing where to look, what to look for, and how to measure or calibrate our observations is critical. If we are not vigilant, or if we are focused only on down-stream process/product characteristics, we forgo substantial gains in process control. Merely *thinking* we know what is happening, and actually *knowing* what is happening—truly understanding the process—are two different perspectives. The major difference is that in the former case we are guessing, and in the latter case we are measuring and projecting our downstream results from upstream activity. Attention to upstream process levers along with careful measurement pays large dividends.

In producing natural gas liquids for sale to large-volume customers, the naturally occurring gas, usu-ally a coproduct with oil, must be separated and processed. Typical product/process-related characteristics monitored are chemical and energy content of the input feed material stream and volume throughput.

The entire production system from ground reservoir to pipeline consists of several processes, each of which is semiautomated. The physical processes have several subprocesses, depending on the primary source of the raw materials. A process sequence of machinery is sized and set up to handle a target volume through-put. The machinery is controlled primarily by the interaction of sensors and control panels—switches and actuators—monitored by both computers and humans. Different but similar technology is used in power plants, feed mills, steel mills, and so on.

Engineers build elaborate models to improve and optimize these processes. Plant operators spend their time observing the process through control panel displays as well as through direct observation, turning switches on and off at appropriate times. Typical product output goals are "faster," "better," and "cheaper." Product output, liquid natural gas, is always a major focus. It is monitored accurately and precisely as a result of supplier/customer protocols. Managers naturally focus on these output characteristics and set production goals, profit goals, and so on. Many of these goals are established historically; they are based on past per-formance under reasonably favorable conditions.

At a field gas plant, results were obtained in exactly this manner. Historically, plant performance was considered adequate. In this case, the gas plant served as a bottleneck, as the field could deliver raw materi-als faster than the gas plant could process them. But building additional capacity is extremely expensive, and all the while on-site natural resources are being depleted. An aggressive plant operations team began to focus on their processes and work upstream looking for process leverage that could lead to improved plant throughput without significant new capital investments.

Their protocol was to use a process flow diagram as a starting point. They built C-E diagrams, starting with the natural gas liquid volume as the effect, working back upstream. In less than an hour they isolated chiller temperature as one critical process lever. The process technology in place used a chillant that demanded a nominal-is-best target; too high created low throughput, and too low created a freeze-up and resulting production problems.

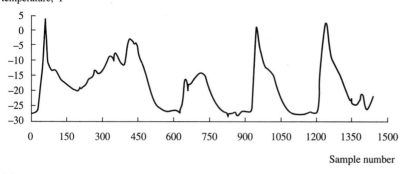

Chiller temperature, °F

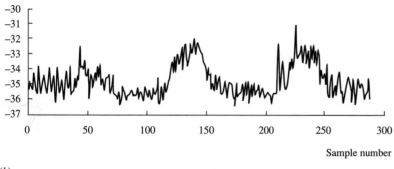

Chiller temperature, °F

(a)

(b)

Figure 4.6 Chiller temperature performance characteristics. (a) Chiller temperature before vigilance, sampled every minute. (b) Chiller temperature after vigilance, sampled every 15 minutes.

They programmed their control console to monitor chiller temperature and plotted the results. Figure 4.6a provides the "before" picture of the upstream process variable. The chiller temperature metric displayed very erratic up-down cycles. Reasons for the cycles varied; the big cycles were due to production problems, while the small cycles were due to input flow and ambient conditions.

The empowered operators immediately decided to pursue a two-pronged strategy: (1) to carefully monitor the temperature and take counteraction when the temperature begin to drift, using experience as their guide, i.e., a judgment-intuition form of process controls; (2) to develop a classical SPC charting approach to address process stability around their nominal-is-best target temperature.

The first strategic action of vigilance produced positive results immediately. Figure 4.6b provides a plot of the process lever metric after the vigilance action was taken. Here, we see cycles in the data, but the data are all falling between −30 and −37°F, as opposed to the +5 to −30°F cycle data we observe in Figure 4.6a. Success here was due to experienced operators setting action limits on chiller temperature, taking counteractions, and communicating these results to less experienced operators. Experience-based engineering judgment, extracted from past performance observations, was critical in setting action limits. For example,

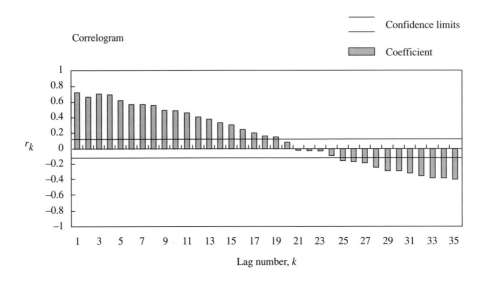

Figure 4.7   Chiller temperature performance inertia characterization, after vigilance (Natural Gas Production Case, Example 4.3).

seasonal temperatures, day-night cycles, inlet mass flow, and inlet gas temperatures were affecting, to one degree or another, chiller temperature dynamics.

In this case, we see a combination of physical reasoning, measurement, documentation, and training, through process vigilance, produce significant results, within the confines of a very dynamic process. In the end, the second, SPC, strategy was at best partially successful—the process dynamics and high process inertia limited the effectiveness of traditional SPC tools.

At this point, we assess our chiller-related process data for process inertia and influence from other upstream variables. This assessment may help us to develop a statistical analysis protocol that we can use to develop strategies to exploit our SPC tools, rather than settle for less than adequate performance. Visual examination of the runs chart in Figure 4.6b indicates a high level of process inertia. We see what appear to be cycles, composed of runs up and down, across time. We can support this argument physically, i.e., from a thermodynamic standpoint.

We can also support the process inertia argument from a statistical standpoint. Figure 4.7 depicts a correlogram obtained from the chiller data after vigilance. The correlogram indicates a high degree of autocorrelation. The correlogram plot shows the bars at each lag far beyond the confidence limits. Hence, it is no real surprise that the standard SPC efforts of the process team were frustrating. SPC did not fail the team— the independence assumption was violated, and the result was a misapplication of a perfectly good process control aid. We will learn more about this topic in Section 4.

Provided we want to use SPC on the chiller, we must figure out a way to transform our data, to produce a more white-noise-like signature. One way we can approach this challenge is to find a covariable that may "explain out" the patterns that produced the high autocorrelations. In this case, we reason from a physical perspective that the chiller temperature is impacted by the ambient temperature surrounding it.

Since the chiller is located outdoors, we measure the ambient temperature along with the chiller temperature, i.e., both are measured at the same point in time. We call these bivariate data pairs. Then, we develop a scatter diagram; see Figure 4.8. Here, we see what appears to be a positive correlation. For example, the scatter plot appears on an upslope as the ambient temperature rises. Our data, using our correlation calculations, produce a correlation coefficient of 0.79, which supports our visual observations and makes

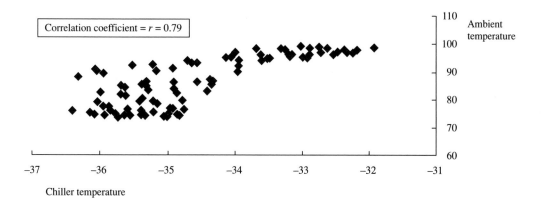

Correlation coefficient = $r = 0.79$

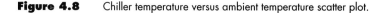

Figure 4.8 Chiller temperature versus ambient temperature scatter plot.

sense physically. In a similar manner, although not presented here, we might assess mass flow and temperature of the mass inflow as other covariables for our chiller. In this case, we would measure all of our covariables at the same time we measured our chiller temperature. We might use a response surface model, as described in Chapter 6, to model and explain our data. Finally, we might use the model residuals to help control the process. Chapter 13 describes such an approach.

Our objective here is to characterize our process—later, in subsequent sections, we pursue more detailed treatments of process definition/redefinition, control, and improvement. **Process characterization serves as a gatekeeper for keeping ineffective process definition/redefinition, control, and/or improvement from robbing us of time and resources.**

REVIEW AND DISCOVERY EXERCISES

REVIEW

4.1. Explain how physical principles and variables are addressed together in order to gain process understanding sufficient for process control. Refer to Figure 4.1 to support your discussion.

4.2. What is the difference between a "direct" metric and an "indirect" metric? Why do we resort to indirect metrics? Explain.

4.3. Identify each of the four fundamental measurement scales and provide an example of each. Which scale offers the most information? Explain.

4.4. Describe the concepts of accuracy and precision. How are they related to the concepts of location and dispersion? Explain.

4.5. What major considerations are associated with the choice of a sensor? Explain.

4.6. Explain the difference between open-loop and closed-loop processes. Provide an example of a system that relies on each.

4.7.   Compare and contrast the concepts of feedback and feedforward control. Provide examples of each in the process of driving/controlling an automobile.

4.8.   What is a process control lever? What is process control leverage? Explain.

4.9.   How does upstream/downstream position impact process control leverage? Explain.

4.10.   What is process inertia and where does it come from? Explain.

4.11.   Conceptually, describe the purpose, methods, and limitations of correlation analysis.

4.12.   Conceptually, describe the purpose, methods, and limitations of autocorrelation analysis.

4.13.   Interpret the runs and correlogram plots shown in Figures 4.5, 4.6, and 4.7. What can we conclude?

Data Characterization: The following questions pertain to the case descriptions located in Section 7.

4.14.   Select one or more of the following cases:

   Case VII.6:   Big Dog—Dog Food Packaging
   Case VII.7:   Bushings International—Machining
   Case VII.9:   Downtown Bakery—Bread Dough
   Case VII.11:   Fix-Up—Automobile Repair
   Case VII.13:   Health Assist—Service
   Case VII.21:   Punch-Out—Sheet Metal Fabrication
   Case VII.24:   Silver Bird—Baggage
   Case VII.27:   Rainbow—Paint Coating
   Case VII.29:   TexRosa—Salsa
   Provide:
   a. A physical assessment
        i. A basic process characterization
        ii. A basic leverage characterization
   b. A statistical assessment
        i. An appropriate graphical characterization
        ii. An appropriate numerical characterization

DISCOVERY

4.15.   Select a process of your choice and develop a set of vector elements and targets analogous to the generic development depicted in Figure 4.2.

4.16.   In industrial processes, typically, countermeasures that rely on shifting location to target are more easily carried out than those associated with reducing dispersion. Why? Explain.

4.17.   The ideal situation in process control is for a zero-defects outcome. Which concept, feedback control or feedforward control, is more conducive to zero defects? Why? Here, zero defects is related to in-process correction that totally prevents rework and scrap.

4.18.   Select a product/process of your choice and develop a process overview analogous to that depicted in Figure 4.4. Be sure to locate and describe the process lever and leverage.

4.19.   How does physical characterization differ from statistical characterization? Consider purpose and methods in your response. Explain.

The following questions pertain to the Natural Gas Production Case (Example 4.3).

4.20. What was primarily responsible for the process performance improvement regarding the chiller? Explain.

4.21. Interpret the statistical characterization of the chilling process data shown in Figures 4.6 and 4.7. What signs of process inertia are present? How was process inertia manifested physically? How was it detected statistically? Explain.

More challenging data characterization: The following questions pertain to the case descriptions located in Section 7.

4.22 Select one or more of the following cases:
Case VII.2:   Apple Core—Dehydration
Case VII.4:   Back-of-the-Moon—Mining
Case VII.8:   Door-to-Door—Pizza Delivery
Case VII.14: High-Precision—Collar Machining
Case VII.17: LNG—Natural Gas Liquefaction
Case VII.20: PCB—Printed Circuit Boards
Case VII.22: Re-Use—Recycling
Case VII.25: Snappy—Plastic Injection Molding
Case VII.30: Tough-Skin—Sheet Metal Welding

Provide:
a. A physical assessment
  i. A basic process characterization
  ii. A basic leverage characterization
b. A statistical assessment
  i. An appropriate graphical characterization
  ii. An appropriate numerical characterization

4.23. Select a process of your choice. Locate a meaningful process lever and collect 50 to 100 data points.

Provide:
a. A physical assessment
  i. A basic process characterization
  ii. A basic leverage characterization
b. A statistical assessment
  i. An appropriate graphical characterization
  ii. An appropriate numerical characterization

REFERENCES

1. Walpole, R. E., R. H. Myers, and S. L. Myers, *Probability and Statistics for Engineers and Scientists*, 6th ed., Upper Saddle River, NJ: Prentice Hall, 1998.

2. Newbold, P., and T. Bos, *Introductory Business and Economic Forecasting*, 2nd ed., Cincinnati: South-Western Publishing, 1994.

PROCESS EXPLORATION

5.0 INQUIRY

1. What constitutes process exploration?
2. What is a designed experiment?
3. What advantages do designed experiments offer?
4. How are experiments designed to assess location? To assess dispersion?
5. How are designed experiments interpreted?

5.1 INTRODUCTION

Process characterization is a critical first step in gaining process understanding. Initial physical and statistical characterization provides both qualitative and quantitative insight as to what is currently happening, e.g., identification of controllable variables, uncontrollable variables, and responses—the X's, Z's, and Y's, as depicted in Figure 4.2. **We use systematic exploration tools to follow up basic process characterization so that we can learn more about our processes, faster.**

In the long term, systematic methods in the exploration and modeling of cause-effect pay huge dividends, but they require a disciplined protocol. Figure 5.1 depicts a logical and systematic sequence of inquiry and activity. Questioning is the logical initiation point for this cycle. We question relationships in our physical, social, and economic environments. We formulate relevant research questions and hypotheses. We apply experimental methods to explore the nature of these relationships. We confirm our preliminary findings and produce appropriate models and plans. Our models and plans lead to the action phase, where we advance quality and productivity performance.

The purpose of this chapter is to introduce and demonstrate useful tools for process exploration. We introduce designed experiments as an efficient means of exploration. Chapter 6

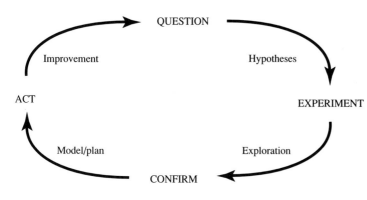

Figure 5.1   Process exploration and modeling discipline.

follows, featuring regression modeling as a means to develop process response surfaces. Both sets of tools allow us to systematically explore and model critical process relationships in order to provide quantitative understanding to support process decisions.

5.2   EXPERIMENTAL PROTOCOL

A well-designed experiment involves six basic steps; see Table 5.1. Each experiment is guided by a well thought-out and stated purpose. Relevant variables are clearly identified and defined. An appropriate randomization and replication structure is developed within the context of the selected model, before actual experimentation. Care is taken in the execution of the experiment to avoid mistakes and prevent biases, so that the validity of results can be defended. Three basic forms of analysis are used: (1) descriptive, e.g., summary statistics and graphs; (2) inferential, e.g., formal

Table 5.1   Fundamental steps in planned experimentation.

1. **Establish the purpose:**   Clearly define and state the problem at hand. Set goals, objectives, and targets relative to what is needed and expected from each experiment.

2. **Identify the variables:**   Clearly delineate response, controllable, and uncontrollable variables. Here we select or determine our responses, select our factors, and establish factor levels.

3. **Design the experiment:**   Our experimental design is based on the way we randomize our experiment. The randomization and the replication structure we use determine the appropriate experimental model and analysis.

4. **Execute the experiment:**   Here, we physically do what we have planned above, and record and document our observations carefully. We are careful to avoid mistakes and biases in our physical setups and execution.

5. **Analyze the results:**   Our analysis may consist of descriptive, inferential, and predictive components, based on the theoretical structure of our model and the physical nature as reflected in our observed measurements.

6. **Interpret and communicate the analysis:**   Here, we put our findings in perspective and defend our methodology (from above) in terms of its conformance to good statistical practices as well as its physical relevance.

hypothesis tests; and (3) predictive, e.g., models used to predict future responses. Finally, the experimental process and results are interpreted and communicated.

Before we present experimental designs, it is important to understand several basic terms. These terms, summarized in Table 5.2, refer to explicit parts and features of designed experiments. Each of these terms is used in the development of subsequent cases.

Each experiment is driven by purpose. These purposes are of two basic orientations—location-oriented or dispersion-oriented. Location-oriented purposes seek to identify a best average level of a response relative to the levels of its causal variables. Here, we define what are termed **fixed effects models.** We hand-select our factors and the levels we wish to explore on the basis of our process experience and purpose. Then, we make inferences only to these specifically selected factor levels, in terms of mean values.

On the other hand, provided we seek to explore dispersion characteristics, we use what are termed **random effects models.** Here, we seek to estimate components of variation associated with our process factors/variables. Random effects factor levels are selected randomly from a population of possible levels. Then, estimates of variation are produced.

In general, we seek to discover and understand factor effects relative to our chosen response. We know that not all possible factors produce equal influence on our response. In both fixed and random effects factors, large influences are more important to us than small influences. Both fixed and random effects factors are illustrated in subsequent sections.

5.3   SINGLE-FACTOR EXPERIMENTS

Single-factor experiments limit our process exploration to only one factor during the course of a single experiment. They are of limited value, since most processes are complex cause-effect structures involving many variables. Nevertheless, the understanding of single-factor experiments is critical, before expansion to multiple factors. We provide understanding through example, exposing limited amounts of theory as we develop our cases.

First, we describe the **completely randomized design (CRD)** using a single factor. Then, we demonstrate the random effects model. Next, we demonstrate the CRD with a single-factor fixed effects model. In these two developments we address both purpose, strategy, and experimental models and their analysis and inferences.

SINGLE-FACTOR CRD MODELS

The CRD gets its name from the nature of its randomization requirements. **In the CRD, initially, each eu stands an equal chance of being selected for any treatment level.** The single-factor CRD, or one-way analysis of variance (ANOVA), can be represented in model form as

$$Y_{ij} = \mu + \tau_i + \epsilon_{ij} \qquad \text{[5.1]}$$

where Y_{ij} = response observation, treatment i, observation j, $i = 1, 2, \ldots,$
$\quad\quad\quad a, j = 1, 2, \ldots, n$

$\quad\mu$ = overall mean

$\quad\tau_i$ = treatment effect

$\quad\epsilon_{ij}$ = random error associated with observation ij

TABLE 5.2 Experimental design terms and features.

Term	Definition	Comment
Designed experiment (controlled experimentation)	A systematic plan of investigation based on established statistical principles and practices, whereby we allow one or more factors to vary, holding the remaining variables as constant as possible.	The purpose is to offer more efficient and effective means to explore our environments—as compared to trial and error methods.
Factors	Independent variables that are deliberately manipulated in the course of the experiment.	Represents cause in a cause-effect context—controllable and/or uncontrollable variables, e.g., X's and Z's, respectively.
Factor level (treatment or treatment level)	The level of a factor that is to be examined.	Sometimes called treatments or treatment-level combinations.
Response (observation)	A measured or observed result obtained from a given factor setting or treatment level.	Represents effects in a cause-effect context, i.e., the Y's.
Hypotheses	Formal conjectures expressed prior to physical experimentation—stated in terms of a null hypothesis, H_0, of no difference and a corresponding alternative hypothesis, H_1, of a difference.	The H_0 is assumed to hold until significant evidence is presented to reject H_0.
Experimental unit (eu)	The largest collection of "experimental material" to which a single independent application of a treatment is made at random.	EU's are defined in the context of the experiment and constitute the fundamental building blocks of an experiment.
Balance	A balanced experiment contains the same number of observations for each treatment/treatment combination.	Refers to the amount of information available regarding responses from eu's.
Experimental error	The measure of variation that exists among responses from eu's that have been treated alike.	Experimental error cannot be measured directly unless replication is present in the experiment.
Replication	Assignment of multiple eu's to a given treatment combination, treating the eu's independently, and observing/measuring the results.	One response/observation/replicate is obtained from each eu.
P value	The probability of observing a more extreme test statistic (e.g., t, F, χ^2, etc.) given that H_0 is true.	P values represent observed significance levels.
Statistical errors (probabilities)	Type I (α) Type II (β)	$\alpha = P$ (rejecting H_0 | H_0 is true) $\beta = P$ (not rejecting H_0 | H_0 is false)
Statistical significance	Refers to the rejection of H_0 at some given α level of significance. It is related to the probability of a type I error.	Statistical significance can be forced out of a physical experiment by increasing the sample size, provided H_0 is not absolutely true.
Practical significance	Refers to how much difference or influence is necessary to impact a process decision.	Practical significance is used to make process action/inaction decisions. It is distinctively different from statistical significance. It is related directly to the physical world and our process/experimental purposes.

The CRD model includes an overall mean as well as a treatment effect and an experimental error. The error term serves as a "residual" term to capture natural variation between the experimental units as well as variation associated with any uncontrolled variables that may be affecting the responses. An expansion, relative to the data, is shown in Equation (5.2). Notice that the equality is preserved and that the terms on the right-hand side correspond to the model terms in Equation (5.1). The geometric counterparts of these terms, the overall mean, the treatment means, and the individual observations, are identified in each subsequent case presented.

$$Y_{ij} = \bar{Y}_{..} + (\bar{Y}_{i.} - \bar{Y}_{..}) + (Y_{ij} - \bar{Y}_{i.}) \qquad \text{[5.2]}$$

Equation (5.3) can be developed by rearranging, squaring, and summing the terms in Equation (5.2). Hence, the term sum of squares (SS) and, thereafter, the ANOVA partitioning concepts are developed. The SS development is the basis for the ANOVA technique. The simplified result is shown in Equation (5.4). The left-hand side of Equation (5.4) represents what we call the "total corrected sum of squares," SS_{tc}, where the sum of squares is corrected for the mean. **In the CRD, we partition the total corrected sum of squares into two parts: (1) the sum of squares associated with the treatments,** i.e., between treatments, and **(2) the sum of squares associated with experimental error,** i.e., variation within the treatments.

$$\sum_{i=1}^{a} \sum_{j=1}^{n} (Y_{ij} - \bar{Y}_{..})^2 = \sum_{i=1}^{a} \sum_{j=1}^{n} [(\bar{Y}_{i.} - \bar{Y}_{..}) + (Y_{ij} - \bar{Y}_{i.})]^2 \qquad \text{[5.3]}$$

$$\sum_{i=1}^{a} \sum_{j=1}^{n} (Y_{ij} - \bar{Y}_{..})^2 = n \sum_{i=1}^{a} (\bar{Y}_{i.} - \bar{Y}_{..})^2 + \sum_{i=1}^{a} \sum_{j=1}^{n} (Y_{ij} - \bar{Y}_{i.}) \qquad \text{[5.4]}$$

Summarized generic ANOVA tables, along with basic assumptions and notations for random and fixed effects factors, are presented in Table 5.3a and b, respectively. These ANOVAs pertain to balanced experiments with a treatment levels and n observations per treatment. The sums of squares shown in Table 5.3 are the partitioned terms of Equation (5.4), which have been algebraically simplified to expedite hand calculations. The expected mean square terms (EMS) are used to develop appropriate hypothesis tests, discussed later in this chapter. Typically, computer-aided analysis packages are used in constructing ANOVA tables.

Essentially, we see three rows, total corrected, treatment, and experimental error, in the single-factor CRD ANOVA. **Each ANOVA row is associated with a measurement.** Clues to each measurement appear in the sum of squares equation components, Equation (5.4). **The total corrected row measures the failure of each eu to respond the same.** The $SS_{tc} = 0$ is possible only if each observation is equal to the grand mean. **The treatment row measures the failure of eu's treated differently to respond the same, on the average.** A $SS_{trt} = 0$ is obtained only if each treatment mean is equal to the grand mean. **The experimental error row measures the failure of each eu within a treatment to respond the same.** A $SS_{err} = 0$ is possible only if, for all treatments, each observation within a treatment is equal to its respective treatment mean.

Table 5.3   Single-factor CRD models and assumptions.

a. Random Effects Model*

Model:
$Y_{ij} = \mu + \tau_i + \epsilon_{ij}$ $i = 1, 2, \ldots, a$
                                            $j = 1, 2, \ldots, n$

Source	df	SS	MS	F	P	EMS
Total corrected	$an - 1$	$\sum_{ij} Y_{ij}^2 - \dfrac{Y_{..}^2}{an}$				
Treatment (between trt's)	$a - 1$	$\dfrac{\sum_{i=1}^{a} Y_{i.}^2}{n} - \dfrac{Y_{..}^2}{an}$	$MS_{trt} = \dfrac{SS_{trt}}{df_{trt}}$	$\dfrac{MS_{trt}}{MS_{err}}$	P_{trt}	$\sigma_\epsilon^2 + n\sigma_\tau^2$
Exp. error (within trt's)	$a(n-1)$	$SS_{tc} - SS_{trt}$	$MS_{err} = \dfrac{SS_{err}}{df_{err}}$			σ_ϵ^2

*Assumptions and hypotheses, random effects:

1. We randomly select each of the a treatments. Selection is based on a random sample from the treatment factor's distribution of values.
2. The τ_i's are independent, distributed normally, $\mu = 0$, $\sigma^2 = \sigma_\tau^2$.
3. The Σ_{ij}'s are independent, distributed normally $\mu = 0$, $\sigma^2 = \sigma_\epsilon^2$.
4. H_0: $\sigma_\tau^2 = 0$, no treatment related variation is present.
   H_1: $\sigma_\tau^2 > 0$, treatment related variation is present.

b. Fixed Effects Model†

Model:
$Y_{ij} = \mu + \tau_i + \epsilon_{ij}$ $i = 1, 2, \ldots, a$
                                            $j = 1, 2, \ldots, n$

Source	df	SS	MS	F	P	EMS
Total corrected	$an - 1$	$\sum_{ij} Y_{ij}^2 - \dfrac{Y_{..}^2}{an}$				
Treatment (between trt's)	$a - 1$	$\dfrac{\sum_{i=1}^{a} Y_{i.}^2}{n} - \dfrac{Y_{..}^2}{an}$	$MS_{trt} = \dfrac{SS_{trt}}{df_{trt}}$	$\dfrac{MS_{trt}}{MS_{err}}$	P_{trt}	$\sigma_\epsilon^2 + n\Phi_\tau$
Exp. error (within trt's)	$a(n-1)$	$SS_{tc} - SS_{trt}$	$MS_{err} = \dfrac{SS_{err}}{df_{err}}$			σ_ϵ^2

†Assumptions and hypotheses, fixed effects:

1. We nonrandomly select each of the a treatments. Selection is based on engineering judgment.
2. $\sum_{i=1}^{a} \tau_i = 0$
3. The Σ_{ij}'s are independent, distributed normally, $\mu = 0$, $\sigma^2 = \sigma_\epsilon^2$.
4. H_0: All $\tau_i = 0$, no treatment related difference is present.
   H_1: At least one $\tau_i \neq 0$, treatment related difference is present.

NOTE: $\Phi_\tau = \sum_{i=1}^{a} \tau_i^2 / (a - 1)$

ANOVA Notation:

df: degrees of freedom

SS: sum of squares

MS: mean square

EMS: expected mean square

F: F test ratio

P value: Probability of observing a more extreme test statistic given H_0 is true

P_{trt}: P value associated with the ANOVA F test

SS_{tc}: total corrected sum of squares

SS_{trt}: treatment sum of squares

SS_{err}: experimental error sum of squares

MS_{trt}: treatment mean square

MS_{err}: experimental error mean square

RANDOM EFFECTS MODEL

The random effects model is appropriate when we want to study the variation associated with a factor. Table 5.3a lists the model, ANOVA, EMS, and assumptions for the CRD random effects model. The overall F test is based on an EMS ratio. For the treatment row, the EMS is

$$\text{EMS}_{\text{trt}} = \sigma_\epsilon^2 + n\sigma_\tau^2 \text{ and } \text{MS}_{\text{trt}} = \hat{\text{E}}\text{MS}_{\text{trt}} \qquad \textbf{[5.5]}$$

and the EMS for the experimental error row is

$$\text{EMS}_{\text{err}} = \sigma_\epsilon^2 \text{ and } \text{MS}_{\text{err}} = \hat{\text{E}}\text{MS}_{\text{err}} \qquad \textbf{[5.6]}$$

The EMS terms for the numerator and denominator are developed separately by summing the mean square term, i.e., the sum of squares divided by the degrees of freedom, over all possible values.

The ratio of MS_{trt} over MS_{err} forms the basis for the overall F test. Here, we are in essence comparing the magnitude of the variation associated with the treatment factor relative to the magnitude of the experimental error present. Our hypotheses are

H_0: $\sigma_\tau^2 = 0$ (no treatment factor variation is present)

H_0: $\sigma_\tau^2 > 0$ (treatment factor variation is present)

Relatively large F values (obviously greater than 1) support rejection of H_0. If we reject H_0, we proceed to estimate σ_τ^2:

$$\hat{\sigma}_\tau^2 = \frac{\text{MS}_{\text{trt}} - \text{MS}_{\text{err}}}{n} \qquad \textbf{[5.7]}$$

If we do not reject H_0, we need not attempt to estimate σ_τ^2 because, in the extreme case when $F < 1$, we obtain $\hat{\sigma}_\tau^2 < 0$, which makes no sense at all. Hence, if the F test is not significant, we typically state that we cannot distinguish σ_τ^2 from 0 in the presence of the experimental error σ_ϵ^2 at the α significance level, based on the evidence at hand (b.o.e.). In other words, if H_0 is not absolutely true, and we fail to reject H_0, our experimental error is large enough to obscure our ability to detect the presence of σ_τ^2. If a true H_0 does not seem physically feasible, and we desire to estimate σ_τ^2, we attempt to obtain more precise estimates by increasing our sample size, tightening our control, and/or being more careful in our physical laboratory procedures. **Our conclusions drawn from designed experiments are always relative to the evidence we collected and assessed. Hence, we always state our conclusions based on the evidence at hand (b.o.e.).**

Example 5.1 | **ROLL PRESSURE CASE—RANDOM EFFECTS MODEL**   In the previous chapter we characterized a meter stick production process, focusing on meter stick length as a critical process characteristic. We physically characterized the process, breaking it into subprocesses, with the intent of finding leverage points. We identified the length leverage point upstream at the shear and deburr subprocesses; see Figure 4.4.

At this point in our case, we notice that deburr can do only so much for our length. For example, long lengths can be trimmed, but short lengths must be scrapped. Our examination now focuses on the feed and shear subprocess. A qualitative cause-effect characterization of the physical nature, Figure 4.4d, leads us to explore the role of the rollers used to advance/index and straighten the coiled raw material before the shear cuts the blank.

We have observed that the rolls are adjusted into the material, creating friction to supply indexing movement and to work-harden and straighten the material. Since the friction is a result of roll pressure, it stands to reason that roll force/pressure is a critical upstream variable. For example, inadequate pressure may lead to slippage and erratic blank lengths. On the other hand, excessive pressure may provide undesirable cold work to the blanks. Roll force/pressure is set manually and reset periodically; however, even with periodic resets roll pressure tends to vary.

Experimental purpose. Explore the relationship of roll force/pressure and blank lengths.

Results sought: Estimate the variation in the blank lengths, before deburr, created by roll pressure and compare it to total variation at the same point in the process.

Possible impact. If the influence on blank length is found to be excessive, we will rethink our roll force/pressure setting and control means. We may then seek to reduce product length variation at the roller source.

Model selection. We will use a single-factor, CRD, random effects experiment.

The CRD requires that each eu stand an equal chance of being assigned to each treatment. Here, we define our eu as the amount of coiled material that is indexed and sheared. We measure the sheared length using a ratio scale. The units of measure are millimeters. For each eu, we independently set/reset our roll pressure. Roll pressure levels are randomly selected from the possible universe of roll pressures. Three levels, of 100, 110, and 130 N/cm^2, are randomly selected to represent pressures that are encountered. Five replicates for each pressure level are planned, yielding a total of 15 observations in a balanced CRD with random effects. Basic descriptive statistics are shown in Figure 5.2.

The flow of Figure 5.2 summarizes the sequence of movement from our physical world of material indexing/shearing to eu's and randomization in the experiment (Figure 5.2a), to data presentation (Figure 5.2b and c). From the scatter diagram, we can clearly see dispersion in our length, associated with the randomly selected pressure level treatments.

Location is measured by $\bar{Y}_{i.}$ in the CRD (Figure 5.2c). Dispersion is assessed by observing the scatter in the +'s within and between each treatment. This scatter is measured by computing sample variances s_i^2 for

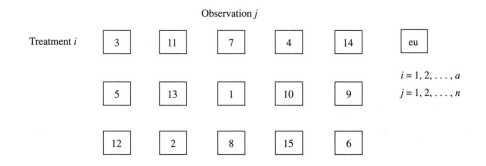

Observation *j*

Treatment *i*

| 3 | 11 | 7 | 4 | 14 | eu |

$i = 1, 2, \ldots, a$
$j = 1, 2, \ldots, n$

| 5 | 13 | 1 | 10 | 9 |

| 12 | 2 | 8 | 15 | 6 |

Each eu has an equal chance of being assigned to any treatment before the fact.
Each eu is assigned to each treatment completely at random.

(a)

Figure 5.2 CRD physical layout, data table, and scatter plot (Roll Pressure Case). (a) Single-factor CRD physical layout.

(continued)

Treatment levels (pressure, N/cm²)	Observations (length, mm)					$Y_{i.}$	$\bar{Y}_{i.}$
100	996.1	1000.3	999.5	1001.2	997.6	4994.7	998.94
110	998.7	1001.5	1000.3	1003.1	999.9	5003.5	1000.70
130	1002.1	1001.4	1003.7	1002.2	1004.3	5013.7	1002.74

$Y_{..} = 15011.90 \qquad \bar{Y}_{..} = 1000.79$

$$Y_{i.} = \sum_{j=1}^{n} Y_{ij} \qquad Y_{..} = \sum_{i=1}^{a}\sum_{j=1}^{n} Y_{ij} \qquad \bar{Y}_{i.} = \frac{Y_{i.}}{n} \qquad \bar{Y}_{..} = \frac{Y_{..}}{an} \qquad s_i^2 = \frac{\sum_{j=1}^{n_i}(Y_{ij}-\bar{Y}_{i.})^2}{n_i-1} \qquad s_{pooled}^2 = \frac{\sum_{i=1}^{a} s_i^2\,(n_i-1)}{\sum_{i=1}^{a}(n_i-1)}$$

(b)

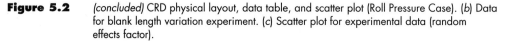

(c)

Figure 5.2 *(concluded)* CRD physical layout, data table, and scatter plot (Roll Pressure Case). (b) Data for blank length variation experiment. (c) Scatter plot for experimental data (random effects factor).

each treatment. The individual treatment variances are pooled (a weighted average taken of the sample variances, weighted by the degree of freedom, $n_i - 1$, for each). This overall measure of dispersion is subsequently labeled "experimental error" in the CRD ANOVA presented in Table 5.4. This statistic is consistent with our earlier generic definition of experimental error; see Table 5.2.

The CRD and its associated analysis, the analysis of variance (ANOVA), allow us to systematically isolate and assess variation due to the treatment (roll pressure) and variation due to all other factors not specifically controlled in our subprocess.

Table 5.4 ANOVA table for meter stick length variation (Roll Pressure Case).

Source	df	SS	MS	F	P Value
Total corrected	$15 - 1 = 14$	70.349			
Treatment (between pressures)	$3 - 1 = 2$	36.165	18.083	6.35	0.0132
Exp. error (within pressures)	$3(4 - 1) = 12$	34.184	2.849		

Sample calculations:
$SS_{tc} = 996.1^2 + 1000.3^2 + \ldots + 1004.3^2 - (15011.9^2/15) = 70.349$
$SS_{trt} = 4994.7^2 + 5003.5^2 + 5013.7^2 - (15011.9^2/15) = 36.165$
$SS_{err} = 70.349 - 36.165 = 34.184$

The numerical results in the CRD are typically displayed in ANOVA tables as shown in Table 5.3. Table 5.3a displays a random effects model. This ANOVA pertains to a balanced experiment with a treatment levels and n observations per treatment.

Using the single-factor CRD ANOVA format shown in Table 5.3a, we develop each ANOVA row. Specific ANOVA results from the meter stick length data are presented in Table 5.4. Here, we interpret the SS_{err} as measuring the failure of each eu, as judged by sheared blank length, to respond the same, within each roller pressure treatment. We interpret the SS_{trt} as measuring the failure of eu's to respond the same, as measured by treatment averages, between all roller pressure treatments.

Our F test results are summarized in Table 5.4. Here, we see $F_{calc} = 6.35$ with 2 and 12 degrees of freedom. From our F tables in Section 8, Table VIII.4, we bracket our F_{calc} between $F_{0.01, 2, 12} = 6.93$ and $F_{0.025, 2, 12} = 5.10$ and state that

$$0.01 < P \text{ value} < 0.025$$

The P value = 0.0132 was obtained by using the SAS computer aid.

Next, we consider our specific random effects hypotheses,

$H_0: \sigma_\tau^2 = 0$ (no variation is due to the roll pressure—length variation is not influenced by roll pressure within the experimental range)

$H_1: \sigma_\tau^2 > 0$ (variation is due to the roll pressure—length variation is influenced by roll pressure within the experimental range)

Here, b.o.e., we reject H_0 at the $\alpha = 0.05$ level of significance, and conclude that roll pressure is influencing our blank length variation. In other words, on the basis of the evidence at hand, interpreting our P value, we would expect to see about a 1.3% chance of obtaining a more extreme F value, given that H_0 is true.

The evidence clearly supports the rejection of H_0; i.e., the P value = $0.0132 < \alpha = 0.05$. Hence, we estimate both σ_ϵ^2 and σ_τ^2. Since

$$EMS_{err} = \sigma_\epsilon^2$$

$$MS_{err} = \hat{\sigma}_\epsilon^2 = 2.849$$

Since

$$EMS_{trt} = \sigma_\epsilon^2 + n\sigma_\tau^2$$

$$\hat{\sigma}_\tau^2 = \frac{MS_{trt} - MS_{err}}{n} = \frac{18.083 - 2.849}{5} = 3.047$$

In order to interpret our results, we state that the variance associated with the population of meter stick blanks, Y_{ij}, is

$$\text{Var } (Y_{ij}) = \sigma_\epsilon^2 + \sigma_\tau^2$$

and its estimate is

$$\hat{\text{Var}} (Y_{ij}) = 2.849 + 3.047 = 5.896$$

In other words, we are looking at a length average estimate of $\bar{Y}.. = 1000.79$ mm and a standard deviation estimate of $(5.896)^{1/2} = 2.428$ mm.

We can also state that our evidence suggests that roughly $3.047/5.896 = 0.517$, or about 52%, of the variation encountered is attributable to roll pressure variation. Our interpretation is that our product's length variation is heavily influenced by the roller pressure, b.o.e. If this product variation is judged to be excessive, a thorough "variance reduction" project concentrating on roller pressure stability is in order. In this case, the product standard deviation seems large. Here, we must set the length target much larger than 1000 mm, to avoid excessive scrap. An exercise at the end of this chapter addresses this issue.

FIXED EFFECTS MODEL

Fixed effects factors are appropriate when we seek to make an inference to a small set of hand-selected factor levels, e.g., chosen in a nonrandom manner. In general, **with fixed effects factors, we produce analyses and interpretations at two levels: (1) at the experiment, ANOVA table, level and (2) at the treatment or factor level.**

Table 5.3b displays the CRD model in its fixed effects form, along with appropriate expected mean squares. The total corrected (for the overall mean) row is broken out into two independent measures, one for the between-treatment measurement and one for the within-treatment measurement. The expected mean squares (EMS) values represent the expected values for the mean squares. Essentially, the EMS terms are derived by summing the mean square terms over all possible values and simplifying. The mean square (MS) terms provide independent estimates of the EMS terms. We will not discuss the derivation of the EMS values for the CRD, but instead refer our readers to step-by-step developments in texts such as Walpole, Myers, and Myers [1], Montgomery [2], and Box, Hunter, and Hunter [3].

At the experiment level, we test for detectable model effects, i.e., terms in the model, τ_i. We think of the ANOVA table as a series of rows, each measuring a certain effect, indicated by each source column label, and associated with model terms. For example, with respect to the fixed effects CRD ANOVA, Table 5.3b, the total corrected row measures the failure of each individual experimental unit to respond in exactly the same way. If all Y_{ij}'s are exactly the same, then $SS_{tc} = 0$.

We break the total corrected row into two independent measurements: (1) treatment effect and (2) experimental error. In the case of the treatment row we are measuring the failure of each treatment response mean to be the same—if the $\bar{Y}_{i.}$'s are all equal, then $SS_{trt} = 0$. The experimental error row measures the failure of each eu, within each treatment, to respond the same. In this case, if each Y_{ij}, within treatment i, is equal to $\bar{Y}_{i.}$ for all i, then $SS_{err} = 0$.

It is critical to associate ANOVA rows with the physical experiment in order to thoroughly understand and interpret the results. Examination of the relationships of Equations (5.1) through (5.4) with the related ANOVA, Table 5.3b, helps us to "visualize" these statistical measurements in a generic sense. The physical contextual meaning is extracted from the physical observations.

Using the EMS terms, we can construct hypotheses to assess the detection of treatment effect τ_i:

H_0: All $\tau_i = 0$

H_1: At least one $\tau_i \neq 0$

If we set up a ratio of EMS_{trt} to EMS_{err}, as in Table 5.3b, we can in essence isolate the τ_i term. Hence, if the null hypothesis is true, the τ_i^2 terms sum to 0 and we are left with a ratio of σ_ϵ^2 over σ_ϵ^2. In other words, under a true H_0, in a perfectly controlled experiment—i.e., one in which we control all of the X_i's and Z_i's—we obtain two independent estimates of the same value, σ_ϵ^2. And, from the one-tailed F test, we would expect to see an F ratio value of unity, $F = 1$. In any given experiment, when H_0 is true or when the treatment effect τ_i is small relative to the magnitude of the experimental error, we expect to see $F \cong 1$ because of the sampling variation present. Hence, in experimental data, if the treatment effects are 0 or relatively small, it is not uncommon to encounter $F < 1$. When this happens, we see no support for rejecting H_0, and attribute this unusual F ratio to sampling variation. On the other hand, when large treatment effects are present, we see F ratios where $F >> 1$. The preceding interpretation is subject to an adequate model "fit," discussed later.

Because of the model form shown in Equation (5.1), the hypothesis test for the treatment effect above is equivalent to

H_0: $\mu_1 = \mu_2 = \ldots = \mu_a$

H_1: At least one μ_i not equal to the others

This result will become more apparent later when we introduce Equations (5.8), (5.9), and (5.10). **A statistically significant result in the treatment row F test supports the rejection of H_0, and we conclude that sufficient evidence is present to suggest that the average response is not the same at all treatment levels.** Plots of response means are very good at helping to elicit and communicate significant fixed effects, or the lack of fixed effects. Even if the ANOVA indicates no treatment effect, additional observations, more highly controlled conditions in the conduct of the physical experiment, or more precise measurement equipment may eventually lead to the detection of a small effect. On the other hand, there actually may be no effect whatsoever.

The validity of the F test in the ANOVA table depends on assumption 3 in Table 5.3b. Here, we assume the residuals ϵ_{ij} are normally distributed, with $\mu = 0$ and variance σ_ϵ^2. This assumption is necessary to support the validity of the fixed effects treatment hypothesis test. However, the graphical analysis and even the ANOVA SS and MS calculations can be made without regard to the normality assumption. In other words, a fixed effects model can be analyzed on a strictly graphical basis. Nevertheless, the ANOVA method adds detail to the analysis, provided the normality assumption can be satisfied.

Here, we use P values or traditional α values to assess the probability of a type I error. We see $\alpha = 0.05$ or $\alpha = 0.01$ typically used as reasonable levels of risk. The smaller the α we choose, the more conservative we are in risk taking, rejecting H_0 when it is true. By using and stating the P value, we provide more information than a simple "do not reject H_0" or "reject H_0" at a stated α level. Most computer aids provide P values for our convenience, which we include in our interpretation. The point is, if we have a P value available, we need not refer to an F table for a critical or tabulated F value. **As is the case with all hypothesis test interpretations, we state that** our conclusions are based on the evidence at hand (b.o.e.). Other or additional evidence may lead to a different interpretation.

PREDICTION AND RESIDUAL CALCULATIONS

We may want to estimate the value of model terms. Here, we consult our basic CRD model, Equations (5.1) and (5.2), and observe that

$$\hat{\mu} = \bar{Y}_{..} \qquad\qquad [5.8]$$

$$\hat{\tau}_i = \bar{Y}_{i.} - \bar{Y}_{..} \qquad\qquad [5.9]$$

$$\hat{\mu}_{ij} = \bar{Y}_{..} + (\bar{Y}_{i.} - \bar{Y}_{..}) = \bar{Y}_{i.} \qquad\qquad [5.10]$$

By definition, the residual e_{ij} for observation j in treatment i is

$$e_{ij} = Y_{ij} - \hat{\mu}_{ij} \qquad\qquad [5.11]$$

It then follows for the single-factor CRD that

$$\hat{\epsilon}_{ij} = e_{ij} = Y_{ij} - \hat{\mu}_{ij} = Y_{ij} - \bar{Y}_{i.} \qquad\qquad [5.12]$$

Example 5.2 | **ASSEMBLY IMPROVEMENT CASE—FIXED EFFECTS MODEL**   High Lift Corporation designs, builds, and installs material handling solutions for their customers; see examples in Chapters 16 and 17 for additional details. As a response to process improvement efforts in the assembly plant, several assembly team members initiated a benchmarking study of an automobile assembly plant known for innovative assembly processes. Benchmarking is described in Chapter 20. Observations at the automobile assembly plant led to questioning High Lift's assembly processes.

Several general concepts were discussed, including a visual-based inventory/assembly flow, depicted in Figure 17.1. Faced with the challenge of reducing assembly costs, a process team responsible for axle/steering assembly began to question their current activities. Current activities included receiving palletized assemblies from stock, breaking out the palletized assemblies, visually inspecting these assemblies, moving the assemblies to the line, positioning the assemblies on the frame, and fastening them in place. Current practices are strictly manual, with the aid of hand tools and a forklift shared with several other assembly subprocess teams. In addition, several team members float up and down the line, adding labor where needed as assemblies move into the line.

The first step of the improvement effort was to review the subprocess purpose and develop a process improvement purpose to guide activities.

Axle/steering subprocess purpose.   To unpack, visually inspect, and install each axle/steering assembly onto the frame, correctly, without any line delay.

Axle/steering subprocess improvement purpose.   To significantly reduce assembly time and costs.

Over the course of several weeks, several concepts were developed and detailed. The choices were not obvious as to how much, if any, the new concepts would improve operations. In order to study the merit of the potential improvements, it was decided to use experimental tools to measure and assess the productivity possibilities. The team established an experimental purpose and defined the response variable and treatments:

Experimental purpose.   To study the advantages/disadvantages of axle/steering assembly concepts. The experiment should establish the feasibility of each concept, and measure each with respect to assembly time, providing a fair comparison and noting any bottlenecks associated with each concept.

Response variable.   The response variable was defined as the total assembly time required, beginning at the axle/steering assembly staging point and ending with a completed assembly operation on the frame.

Three treatments were developed from the concept work/details available:

Treatment 1.   The present method—pallets and the roving forklift, as described earlier, are used in assembly.

Treatment 2.   The standard pallet method of staging, as in the present method, is used along with a dedicated lifting/positioning crane to move the assembly to the line and position it on the frame. Here, the crane replaces the roving forklift.

Treatment 3.   A special-purpose cart, rather than the pallet at the staging area, is used along with the dedicated lifting/positioning crane. Here, the cart is used as a transport from the supplier rather than the standard pallet.

A fixed effects, single-factor CRD, with three treatments, was selected as the experimental design. Four replicates were selected to be run, on line, in the course of actual production operations. All impacted personnel were informed of the experiment and data. A staging cart was designed and built. All necessary equipment was furnished by vendors. All team members were trained in each treatment method. Careful notes were taken during training.

The eu was defined as one axle/steering assembly. The experimental order was randomized and the experiment was executed. Figure 5.3 displays the results. Figure 5.3*a* depicts the randomization scheme for the CRD.

The order of experimentation is shown in Figure 5.3*a*. Here, order number 1 represents the first treatment run, followed by 2, and so forth. It is important to record the order, as will soon be evident. Figure 5.3*b* displays the observed data, measured in minutes, along with several summary statistics. A scatter plot of the data and summary statistics is presented in Figure 5.3*c*. Here, we see both location and dispersion—the geometric counterpart of Equations (5.1) through (5.4).

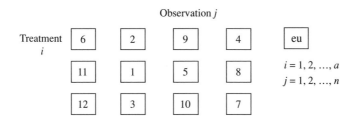

Each eu has an equal chance of being assigned to any treatment before the fact.
Each eu is assigned to each treatment completely at random.

(*a*)

Figure 5.3   CRD physical layout, data table, and scatter plot (Assembly Improvement Case). (*a*) Single-factor CRD physical layout and randomization.

(*continued*)

Treatment levels (assembly methods)	Observations (time, minutes)				$Y_{i.}$	$\bar{Y}_{i.}$
1 (present–pallet and forklift)	64	56	49	73	242	60.50
2 (pallet and lifting device)	26	52	37	69	184	46.00
3 (cart and lifting device)	23	15	41	28	107	26.75
					$Y_{..} = 533$	$\bar{Y}_{..} = 44.417$

$$Y_{i.} = \sum_{j=1}^{n} Y_{ij} \quad Y_{..} = \sum_{i=1}^{a}\sum_{j=1}^{n} Y_{ij} \quad \bar{Y}_{i.} = \frac{Y_{i.}}{n} \quad \bar{Y}_{..} = \frac{Y_{..}}{an} \quad s_i^2 = \frac{\sum_{j=1}^{n_i}(Y_{ij}-\bar{Y}_{i.})^2}{n_i - 1} \quad s_{pooled}^2 = \frac{\sum_{i=1}^{a} s_i^2 (n_i - 1)}{\sum_{i=1}^{a}(n_i - 1)}$$

(b)

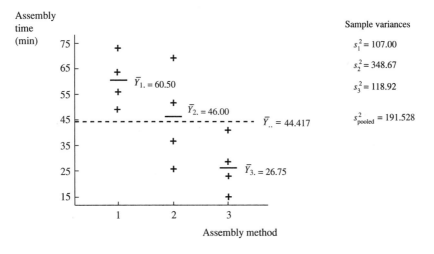

(c)

Figure 5.3   (concluded) CRD physical layout, data table, and scatter plot (Assembly Improvement Case). (b) Data for assembly methods. (c) Scatter plot for experimental data (fixed effects factor).

The CRD fixed effects model is appropriate for this case since we have hand-selected our treatment levels, in a nonrandom fashion. Our CRD model for the fixed effects treatment factor is developed as in Equation (5.1). Our sum of squares is expanded as in Equations (5.3) and (5.4). The CRD fixed effects model details are summarized in Table 5.3b. Using our ANOVA form in Table 5.3b, we have developed an ANOVA table; see Table 5.5. The experimental error row measures the failure of each axle/steering assembly to be installed in the same amount of time, when assembled within each of the three methods. The treatment row measures the failure of each method's average assembly time response to be the same.

Results and sample calculations are shown in Table 5.5. Our overall F test in the ANOVA produces $F = 5.99$ and P value = 0.0222. SAS PROC GLM was used as a computer aid here [4, 5]. If we rely on

Table 5.5 ANOVA table for assembly alternatives (Assembly Improvement Case)

Source	df	SS	MS	F	P Value
Total corrected	$12 - 1 = 11$	4016.917			
Treatment (between assembly methods)	$3 - 1 = 2$	2293.167	1146.583	5.99	0.0222
Experimental error (within assembly methods)	$3(4 - 1) = 9$	1723.750	191.528		

Sample calculations

$SS_{tc} = (64^2 + 56^2 + \ldots + 41^2 + 28^2) - (533^2/12) = 4016.917$

$SS_{trt} = (242^2 + 184^2 + 107^2)/3 - (533^2/12) = 2293.167$

$SS_{err} = 4016.917 - 2293.167 = 1723.750$

manual calculations, using our F tables, Section 8, Table VIII.4, we would say $0.01 < P$ value < 0.025, since our $F_{calc} = 5.99$ falls between tabulated $F_{0.01,2,9} = 8.02$ and $F_{0.025,2,9} = 5.71$. For our fixed effects factor of assembly methods,

H_0: All $\tau_i = 0$ or all μ_i are equal (no assembly effect exists; all assembly methods produce the same mean response time)

H_1: At least one $\tau_i \neq 0$ or not all μ_i are equal (an assembly effect exists; at least one assembly method produces a different mean response time than the others)

On the basis of our evidence, we reject H_0 in favor of H_1. This conclusion is also supported qualitatively by our plot; see Figure 5.3c.

For our single-factor CRD model-related estimates, from Equations (5.8) through (5.12),

$$\hat{\mu} = 44.417 \text{ minutes}$$

$$\hat{\tau}_1 = 60.500 - 44.417 = 16.083 \text{ minutes}$$

$$\hat{\tau}_2 = 46.000 - 44.417 = 1.583 \text{ minutes}$$

$$\hat{\tau}_3 = 26.750 - 44.417 = -17.667 \text{ minutes}$$

$$\hat{\mu}_{i=1} = \bar{Y}_{..} + (\bar{Y}_{1.} - \bar{Y}_{..}) = \bar{Y}_{1.} = 60.500 \text{ minutes}$$

$$\hat{\mu}_{i=2} = \bar{Y}_{..} + (\bar{Y}_{2.} - \bar{Y}_{..}) = \bar{Y}_{2.} = 46.000 \text{ minutes}$$

$$\hat{\mu}_{i=3} = \bar{Y}_{..} + (\bar{Y}_{3.} - \bar{Y}_{..}) = \bar{Y}_{3.} = 26.750 \text{ minutes}$$

$$e_{11} = 64 - 60.5 = 3.50 \text{ minutes}$$

$$e_{12} = 56 - 60.5 = -4.50 \text{ minutes}$$

$$\cdot$$
$$\cdot$$
$$\cdot$$

$$e_{34} = 28 - 26.75 = 1.25 \text{ minutes}$$

At this point, we have a good deal of information. We can clearly state that we see strong evidence to suggest that at least one assembly method is responding differently from the others, with respect to average assembly time, P value $= 0.022$. In other words, at the $\alpha = 0.05$ level of significance, we reject H_0, b.o.e. However, we are still at a loss to recommend an assembly method for our smaller-is-better productivity criteria.

TREATMENT LEVEL INTERVAL ANALYSIS (FIXED EFFECTS FACTORS)

At the treatment level, with fixed effects factors, our objective is usually to distinguish differences between treatment means (bigger and smaller is better) **and/or between a treatment mean and a target value** (nominal is best). In these situations, we use the MS_{err} as an estimate of pooled variation—**from our empirical evidence.**

We can compute confidence intervals on our mean estimates or develop error bars for our estimates. A confidence interval on a mean can be calculated as

$$\hat{\mu}_i \pm t_{\alpha/2,\, df_{err}} \sqrt{\frac{MS_{err}}{n'}} \qquad\qquad \textbf{[5.13]}$$

where n' represents the number of observations in the mean calculation.

Error bars for each individual mean can be established in terms of multiples of standard errors of the mean

$$\hat{\mu}_i \pm b \sqrt{\frac{MS_{err}}{n'}} \qquad\qquad \textbf{[5.14]}$$

where b is the desired multiple, e.g., $b = 1$, $b = 2$. For large samples, $b = 2$ represents approximately a 95% confidence interval. **These intervals prove useful when faced with a nominal is best response.** We leave further development of the nominal-is-best case for end-of-chapter exercises.

Example 5.3 | **ASSEMBLY IMPROVEMENT CASE REVISITED—TREATMENT LEVEL CONFIDENCE INTERVALS**
Previously we developed point estimates for our treatment level means. Now, using our point estimates along with our ANOVA results, we set confidence intervals on our treatment means. For a 95% confidence level, our two-sided confidence interval for each mean is

$$\hat{\mu}_i \pm t_{\alpha/2,\, df_{err}} \sqrt{\frac{MS_{err}}{n'}}$$

where $t_{\alpha/2,\, df_{err}} = t_{0.025,\, 9} = 2.262$.

For treatment 1

$$60.50 \pm 2.262 \sqrt{\frac{191.528}{4}}$$

$$60.50 \pm 15.65 \text{ minutes}$$

For treatment 2

$$46.00 \pm 15.65 \text{ minutes}$$

For treatment 3

$$26.75 \pm 15.65 \text{ minutes}$$

Here, we see reasonably good precision in our mean response estimates. This precision is due to the power of the ANOVA method, in that we are using a pooled, df = 9, measure of variation.

Had we used each treatment data set alone to form 95% confidence intervals on the means, we would have obtained

$$\hat{\mu}_i \pm t_{\alpha/2,3} \sqrt{\frac{s_i^2}{n'}}$$

For treatment 1,

$$60.50 \pm 3.182 \sqrt{\frac{107.000}{4}}$$

$$60.50 \pm 16.46 \text{ minutes}$$

For treatment 2,

$$46.00 \pm 3.182 \sqrt{\frac{348.667}{4}}$$

$$46.00 \pm 29.71 \text{ minutes}$$

For treatment 3,

$$26.75 \pm 3.182 \sqrt{\frac{118.917}{4}}$$

$$26.75 \pm 17.35 \text{ minutes}$$

In comparison to our previous confidence intervals, we observe noticeably wider intervals. Hence, we see an increase in precision of mean estimates when we utilize the ANOVA technique and its pooled error concept, as opposed to using each treatment's standard error separately.

TREATMENT MEAN COMPARISONS (FIXED EFFECTS FACTORS)

Regarding bigger or smaller is better inferences, we use pairwise comparisons of treatment means. Several numerical techniques are available for these comparisons, in addition to visual comparisons from our graphs. We introduce only one technique, the least significant differences (LSD) method, and refer our reader to other sources for additional methods; see Kolarik [6], Montgomery [2], Box, Hunter, and Hunter [3].

The LSD test is developed in the same manner as a two-sample t test of means. Hence, the α significance level applies to each pair of sample means tested. The LSD_α test statistic is calculated as

$$\text{LSD}_\alpha = t_{\alpha/2,\, \text{df}_{\text{err}}} \sqrt{\text{MS}_{\text{err}}\left(\frac{1}{n_1} + \frac{1}{n_2}\right)} = t_{\alpha/2,\, \text{df}_{\text{err}}} \sqrt{\frac{2\text{MS}_{\text{err}}}{n'}} \qquad \textbf{[5.15]}$$

Here, n' represents the number of observations making up each treatment sample mean, e.g., $n_1 = n_2 = n'$ for a balanced, single-factor CRD experiment.

The C_2^a hypotheses for a full set of comparisons are

$$H_0: \mu_i = \mu_j \qquad \text{for all } ij \text{ combinations, } i \neq j \qquad \textbf{[5.16]}$$
$$H_1: \mu_i \neq \mu_j$$

where

$$C_2^a = \frac{a!}{2!(a-2)!}$$

If

$$LSD_\alpha < |\hat{\mu}_i - \hat{\mu}_j|$$

then we reject H_0 at the α significance level, b.o.e.

If we use the LSD with its pairwise α and seek to estimate the overall experiment–wise α, we use the form

$$\alpha_{\text{exp-wise}} \cong 1 - (1 - \alpha_{\text{pairwise}})^{C_2^a} \qquad \textbf{[5.17]}$$

For example, if we use $LSD_{0.05}$ for a set of five treatments,

$$\alpha_{\text{exp-wise}} \cong 1 - (0.95)^{C_2^5} = 1 - (0.95)^{10} = 40.1\%$$

Hence, our overall chance of making a type I error is large, considering we are making 10 individual hypothesis tests, each with $\alpha = 0.05$. In such a case, we would likely reduce our pairwise α to, say, 0.01 in order to control the overall experiment–wise type I error rate.

The two-tailed LSD hypothesis, Equation (5.16), **supports a difference in either direction.** In bigger or smaller is better cases, we typically desire to set one-tailed hypotheses. For example, we desire to structure our H_1 as "less than" or "greater than" rather than use a "not equal to" form. **The LSD can be readily modified to a one-tailed comparison, LSD′,** by replacing $\alpha/2$ with α:

$$LSD'_\alpha = t_{\alpha,\, df_{err}} \sqrt{\frac{2MS_{err}}{n'}} \qquad \textbf{[5.18]}$$

In the one-tail case, we might structure a set of hypotheses such as

$H_0: \mu_i = \mu_j$ where i and j represent specific treatment pairs of interest, $i \neq j$

$H_1: \mu_i > \mu_j$ or $H_1: \mu_i < \mu_j$

For $H_1: \mu_i > \mu_j$, if $LSD'_\alpha < \hat{\mu}_i - \hat{\mu}_j$, then we reject H_0

For $H_1: \mu_i < \mu_j$, if $LSD'_\alpha < \hat{\mu}_j - \hat{\mu}_i$, then we reject H_0

Our LSD comparison can be depicted graphically by mapping the LSD value on our plot of treatment means, marking off the LSD distance to each appropriate side of the treatment mean of interest. For interpretation, if other treatment means on our plot fall within the interval, we cannot reject the respective H_0 at the stated α level, based on evidence and the LSD method.

Example 5.4 | **ASSEMBLY IMPROVEMENT CASE REVISITED—TREATMENT COMPARISONS**   Given the original purpose in the assembly improvement case, a smaller assembly time is clearly superior—all other considerations equal, e.g., physical performance, equipment and operational costs, customer service. Hence, we focus on the one-tailed LSD'_α for our pairwise tests. Specifically, we examine the following sequence of hypotheses:

Test 1:

$$H_0: \mu_1 = \mu_3$$
$$H_1: \mu_1 > \mu_3$$

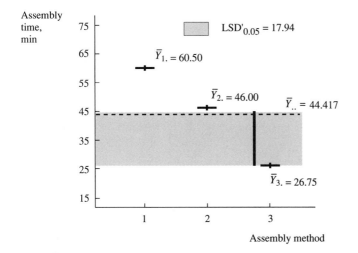

Figure 5.4   Treatment means with one-tailed LSD band superimposed (Assembly Improvement Case).

Test 2:
$$H_0: \mu_2 = \mu_3$$
$$H_1: \mu_2 > \mu_3$$

$$\text{LSD}'_{0.05} = t_{0.05,\,9} \sqrt{\frac{2\text{MS}_{\text{err}}}{n'}} = 1.833 \sqrt{\frac{2(191.528)}{4}} = 17.94 \text{ minutes}$$

Test 1: $\hat{\mu}_1 - \hat{\mu}_3 = 60.50 - 26.75 = 33.75$ minutes

Since $\text{LSD}'_{\alpha} < (\hat{\mu}_1 - \hat{\mu}_3)$, we reject H_0, $\alpha = 0.05$, b.o.e.

Test 2: $\hat{\mu}_2 - \hat{\mu}_3 = 46.00 - 26.75 = 19.25$ minutes

Since $\text{LSD}'_{\alpha} < (\hat{\mu}_2 - \hat{\mu}_3)$, we reject H_0, $\alpha = 0.05$, b.o.e.

Finally, we conclude that treatment 3, the special-purpose cart and dedicated lifting/positioning crane, produces a significantly smaller assembly time than each of the other two treatments, b.o.e., for a one-tailed LSD, $\alpha = 0.05$ criterion.

Our LSD test is presented graphically in Figure 5.4. Here, we have plotted the $\text{LSD}'_{0.05}$ interval of length 17.94 minutes on the upward side of $\hat{\mu}_3$. We can visually observe that the other two treatment means fall above the top of the LSD bar. Thus, our previous conclusion is demonstrated graphically.

5.4   MODEL ADEQUACY

We make a number of assumptions regarding our residuals ϵ_{ij}; see Table 5.3. In order to support our hypothesis testing, we assume our residuals to be independent and normally distributed, $\mu = 0$, $\sigma^2 = \sigma_{\epsilon}^2$. We can assess our assumption in two general ways: (1) with a probability plot or graph or (2) with a statistical goodness-of-fit test. Without computer aids, the first alternative is usually more feasible; with computer aids, both alternatives may be reasonable.

During the course of an experiment, we assume that no systematic change, or "bias," takes place. Good experimental practice demands that we record the order of experimentation, as the experiment is performed. Later we plot our residuals against our order sequence. If we see a systematic pattern, over the physical execution of the experiment—for example, residuals have been mostly negative, but switch to mostly positive at some point—we have evidence that something changed during the course of the experiment and our results may be misleading.

The pooling of variation across treatments provides precision for our estimates in a designed experiment. This concept was demonstrated in Example 5.3 when we developed confidence intervals from the ANOVA MS_{err} and then from the individual treatment variances—the former are narrower than the latter. This pooling effect can also be verified by noting that the pooled variances in both Figures 5.2c and 5.3c are identical to their respective MS_{err} terms in Tables 5.4 and 5.5. In other words, **homogeneity of variance and subsequent pooling across treatments provide the ANOVA technique its power.**

The homogeneity of variance assumption assumes that the variances within each of the treatments are equal and can be "pooled" to form the experimental error. A residuals-predicted values plot helps to assess the validity of the homogeneity of variance assumption. This plot should display no obvious patterns. For example, we should see about equal spreads in the residuals corresponding to each treatment. Bartlett's test provides a numerical alternative to graphical assessment; see Walpole, Myers, and Myers [1].

Computer aids can free us from the drudgery involved in residual analysis. Most computer aids, such as SAS, usually have some provisions to deal with normality, order, and homogeneity of variance analyses [5].

Example 5.5 | **ASSEMBLY IMPROVEMENT CASE REVISITED—MODEL ADEQUACY**   Using the data and results of the assembly methods case (Figure 5.3), first we develop a normal probability plot of residuals and then produce residual-order and residual-predicted value plots.

To assess the normality of residuals assumption, without a computer aid, we develop our $\hat{\mu}_{ij}$ and e_{ij} values as well as our plotting positions, $\hat{F}(t)$'s. We use normal probability plotting paper, provided in Section 8, Table VIII.9, and the traditional mean rank plotting form,

$$\hat{F}(t) = \frac{k - 0.5}{an} \qquad \textbf{[5.19]}$$

where k = rank number and an = number of residuals; each observation produces one residual. Here, the plotting positions, $\hat{F}(t)$'s, represent estimated cumulative mass/density points, relative to the ranks, e.g., 1 through 12, of the residuals.

Table 5.6 lists the ordering of the experiment, from Figure 5.3a, and the observed values Y_{ij}. Since the CRD was used, we then use Equations (5.10), (5.11), and (5.12) to develop our $\hat{\mu}_{ij}$ and e_{ij} values. In Table 5.6, part b, we use the mean rank method, as shown in Equation (5.19), to develop our $\hat{F}(t)$ values, and assign our e_{ij}'s to the mean plotting positions in ascending order by rank.

Once we plot our residual data (Figure 5.5), we see a reasonable pattern of our plotted points about a straight line. The horizontal scale is linear on our normal plotting paper, while the vertical scale is developed relative to the cumulative probability of a normal distribution. Because of the scaling of the axes, plotted points on normal plotting paper will fall roughly on a straight line, if the points represent a sample from a

Table 5.6   Residuals and plotting positions for normal plotting paper.

a.   Observed Data

Observation	Treatment	Y_{ij}	$\hat{\mu}_{ij} = \bar{Y}_{i.}$	$e_{ij} = Y_{ij} - \hat{\mu}_{ij}$
1	2	52	46.00	6.00
2	1	56	60.50	−4.50
3	3	15	26.75	−11.75
4	1	73	60.50	12.50
5	2	37	46.00	−9.00
6	1	64	60.50	3.50
7	3	28	26.75	1.25
8	2	69	46.00	23.00
9	1	49	60.50	−11.50
10	3	41	26.75	14.25
11	2	26	46.00	−20.00
12	3	23	26.75	−3.75

b.   Ranked Residuals (in ascending order)

Rank k	Mean Plotting Position in Percentiles $\hat{F}(t) = (k - 0.5)/an$	e_{ij}
1	0.5/12 = 0.042 4.2%	−20.00
2	1.5/12 = 0.125 12.5%	−11.75
3	2.5/12 = 0.208 20.8%	−11.50
4	3.5/12 = 0.292 29.2%	−9.00
5	4.5/12 = 0.375 37.5%	−4.50
6	5.5/12 = 0.458 45.8%	−3.75
7	6.5/12 = 0.542 54.2%	1.25
8	7.5/12 = 0.625 62.5%	3.50
9	8.5/12 = 0.708 70.8%	6.00
10	9.5/12 = 0.792 79.2%	12.50
11	10.5/12 = 0.875 87.5%	14.25
12	11.5/12 = 0.958 95.8%	23.00

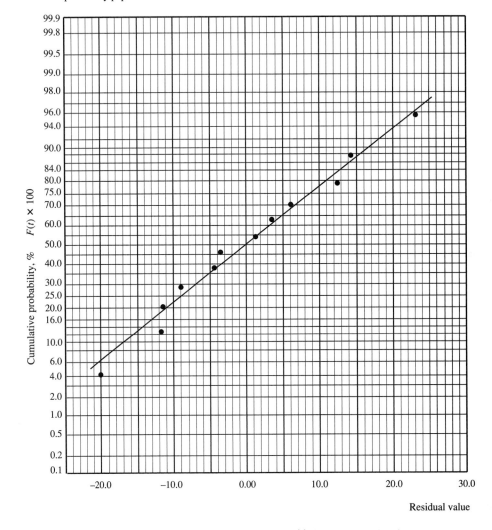

Normal probability paper

Figure 5.5   Residual plot, normal probability paper (Assembly Improvement Case).

normal population. Once we plot our e_{ij}, $\hat{F}(t)$ coordinates, we sketch in a "model" line. If the points fall near the line, we have evidence to suggest that the normal assumption is more or less reasonable, and we can defend the validity of the F test in our ANOVA. If, on the other hand, we are skeptical of the normality plot, we have a dilemma. We can try either a response transformation or nonparametric methods. We refer our readers to Montgomery [2] and Conover [7] for more detailed discussions on transformations and nonparametric methods, respectively.

Our residual-order and residual-predicted value plots are shown in Figure 5.6. To develop the residual-order plot, we simply plot our residuals across time on the horizontal axis similar to a runs chart. We see no pronounced trend or shifts in our e_{ij}'s across time. Therefore, we assume that no serious systematic biases

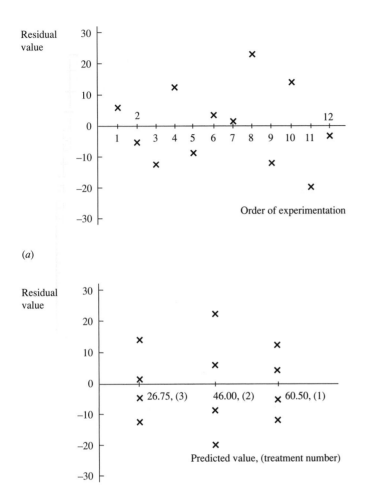

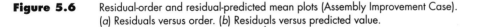

(a)

(b)

Figure 5.6 Residual-order and residual-predicted mean plots (Assembly Improvement Case). (a) Residuals versus order. (b) Residuals versus predicted value.

took place during the course of our experiment. If, on the other hand, we see trends or biases, our results may be misleading and the experiment may have to be rerun. We also plot our residuals against the predicted $\hat{\mu}_{ij}$ values. Here, in Figure 5.6b, the assumption of homogeneity of variance (across treatments) appears valid, as we do not see a great deal of difference in the dispersion from treatment to treatment. In other words, the spread in the plotted residuals within each treatment is about the same. We do see some additional spread in treatment 2, but because of the small sample sizes, it does not merit concern at this point.

5.5   MULTIPLE-FACTOR EXPERIMENTS

Multiple factors can be designed into experiments with what is termed a "factorial arrangement of treatments," (FAT). In a complete, or full, factorial arrangement, all possible treatment combinations (of factor levels) are observed and their responses recorded. Both graphical and numerical analyses are appropriate.

In a FAT, two, three, or more levels of the selected factors are of interest. When two levels of each factor are studied, the arrangement is termed a 2^f FAT, where f represents the number of factors. A full 2^f FAT requires 2^f factor level treatment combinations. When three levels are involved, the arrangement is termed a 3^f, and so on. We may also design factorial experiments with a different number of levels of each factor—a four-factor experiment with $2 \times 3 \times 2 \times 4$ respective levels of the four factors contains 48 factor level combinations.

When we design two-factor experiments, we seek to learn about both factors and their possible interrelationship (interaction) with respect to the response, simultaneously. The single-factor experiment fundamentals apply to multiple-factor experiments in general; however, additional principles are involved. The concept of a treatment now becomes a matter of factor level combinations. For example, if factor A is temperature with a levels, and factor B is time with b levels, we would see ab treatment combinations of these two factors.

With multiple factors, we do not usually develop a treatment row in our ANOVA. We break out the A and B main effects and AB interaction so that we can measure all three. This breakout assists us in process exploration. For example, we can visualize a two-factor experiment by imagining each factor as an adjustable "knob." Here, factors A and B each represent a knob. We are breaking a complicated process with two factors down to its constituent parts for analysis. With our factorial approach we learn about each factor as well as possible factor interactions.

MULTIPLE-FACTOR ANALYSIS OF VARIANCE (ANOVA)

In the ANOVA table we break out rows to quantitatively measure both main effects and interactions. Our design of interest here is a two-factor FAT-CRD. The FAT is developed as combinations of the factor levels, while the CRD is developed by the randomization process where each eu has an equal chance of being assigned to any one of the factor-level treatment combinations. The statistical model is

$$Y_{ijk} = \mu + A_i + B_j + AB_{ij} + \epsilon_{k(ij)} \qquad \textbf{[5.20]}$$

where Y_{ijk} = a response observation
μ = overall mean
A_i = factor A main effect, $i = 1, 2, \ldots, a$
B_j = factor B main effect, $j = 1, 2, \ldots, b$
AB_{ij} = factor A, factor B interaction effect
$\epsilon_{k(ij)}$ = random error, $k = 1, 2, \ldots, n$

Taken together, the A, B, and AB effects constitute what we called a "treatment effect" in our single-factor experiment. The $\epsilon_{k(ij)}$ term represents the experimental error developed from the replicates of the experimental units, which have been treated alike within each treatment combination. The notation is read "k within ij factor level combinations."

Using the dot notation previously introduced in Figure 5.3 and the FAT-CRD model form of Equation (5.20), we express the SS_{tc} equation for our two-factor experiment as

$$\sum_{i=1}^{a} \sum_{j=1}^{b} \sum_{k=1}^{n} (Y_{ijk} - \bar{Y}_{...})^2 = \sum_{i=1}^{a} \sum_{j=1}^{b} \sum_{k=1}^{n} [(\bar{Y}_{i..} - \bar{Y}_{...}) + (\bar{Y}_{.j.} - \bar{Y}_{...})$$
$$+ (\bar{Y}_{ij.} - \bar{Y}_{i..} - \bar{Y}_{.j.} + \bar{Y}_{...}) + (Y_{ijk} - \bar{Y}_{ij.})]^2$$

[5.21]

Expanding and simplifying Equation (5.21), we have

$$\sum_{i=1}^{a} \sum_{j=1}^{b} \sum_{k=1}^{n} (Y_{ijk} - \bar{Y}_{...})^2 = bn \sum_{i=1}^{a} (\bar{Y}_{i..} - \bar{Y}_{...})^2 + an \sum_{j=1}^{b} (\bar{Y}_{.j.} - \bar{Y}_{...})^2$$

[5.22]

$$+ n \sum_{i=1}^{a} \sum_{j=1}^{b} (\bar{Y}_{ij.} - \bar{Y}_{i..} - \bar{Y}_{.j.} + \bar{Y}_{...})^2 + \sum_{i=1}^{a} \sum_{j=1}^{b} \sum_{k=1}^{n} (Y_{ijk} - \bar{Y}_{ij.})^2$$

In Equation (5.22), we have established SS_{tc} on the left-hand side, and partitioned it into SS_A, SS_B, SS_{AB}, and SS_{err}, respectively. Table 5.7 displays the simplified ANOVA row calculations and the EMSs of both fixed and random effects factors. Just as in the single-factor case, **each row of our ANOVA table provides a measurement.** Examining Equation (5.22), we interpret each row in Table 5.7 as

Total corrected. Measures the failure of each eu to respond exactly alike; for $SS_{tc} = 0$, each observation must equal the grand mean.

Main effect A. Measures the failure of the average responses of eu's treated at each level of factor A to be the same, when averaged over factor B and the replications.

Main effect B. Measures the failure of the average responses of eu's treated at each level of factor B to be the same, when averaged over factor A and the replications.

Interaction effect AB. Measures the failure of eu's treated at each A (or B) level to respond the same over all levels of B (or A), when averaged over the replications.

Experimental error. Measures the failure of eu's treated alike to respond the same.

If multiple replications are not made for each treatment combination, the experimental error term cannot be estimated directly. In such cases, higher-order interactions, i.e., three factors or more, are sometimes pooled or combined to some degree and used to estimate the experimental error term. Here, essentially, we would pool the selected mean square values, weighted by their respective degrees of freedom.

Factors in our experiments may be either fixed or random effects related. The former are used to assess the response relative to a hand-selected group of factor levels, just as we did in our single-factor experiments. Random effects factors are useful to estimate components of variation. For example, if we were assessing product variation in a manufacturing process where machines and operators are involved, a two-factor random effects model could be used. Here we could

Table 5.7 Two-factor FAT-CRD ANOVA table with EMS terms.

Model:
$$Y_{ijk} = \mu + A_i + B_j + AB_{ij} + \epsilon_{k(ij)}$$

$i = 1, 2, \ldots, a$
$j = 1, 2, \ldots, b$
$k = 1, 2, \ldots, n$

Source	df	SS	MS	F^*	P	Fixed Effects EMS†	Random Effects EMS†
Total corrected	$abn - 1$	$\sum\limits_{ijk} Y_{ijk}^2 - \dfrac{Y_{\ldots}^2}{abn}$					
A	$a - 1$	$\dfrac{\sum\limits_{i=1}^{a} Y_{i\ldots}^2}{bn} - \dfrac{Y_{\ldots}^2}{abn}$	$\dfrac{SS_A}{df_A}$	$\dfrac{MS_A}{MS_{err}}$	P_A	$\sigma_\epsilon^2 + bn\phi_A$	$\sigma_\epsilon^2 + n\sigma_{AB}^2 + bn\sigma_A^2$
B	$b - 1$	$\dfrac{\sum\limits_{j=1}^{b} Y_{\cdot j\cdot}^2}{an} - \dfrac{Y_{\ldots}^2}{abn}$	$\dfrac{SS_B}{df_B}$	$\dfrac{MS_B}{MS_{err}}$	P_B	$\sigma_\epsilon^2 + an\phi_B$	$\sigma_\epsilon^2 + n\sigma_{AB}^2 + an\sigma_B^2$
AB	$(a-1)(b-1)$	$\dfrac{\sum\limits_{i=1}^{a}\sum\limits_{j=1}^{b} Y_{ij\cdot}^2}{n} - \dfrac{Y_{\ldots}^2}{abn} - SS_A - SS_B$	$\dfrac{SS_{AB}}{df_{AB}}$	$\dfrac{MS_{AB}}{MS_{err}}$	P_{AB}	$\sigma_\epsilon^2 + n\phi_{AB}$	$\sigma_\epsilon^2 + n\sigma_{AB}^2$
Experimental error	$ab(n-1)$	$SS_{tc} - SS_A - SS_B - SS_{AB}$	$\dfrac{SS_{err}}{df_{err}}$			σ_ϵ^2	σ_ϵ^2

NOTE:
$$\phi_A = \frac{\sum A_i^2}{a - 1} \qquad \phi_B = \frac{\sum B_j^2}{b - 1} \qquad \phi_{AB} = \frac{\sum AB_{ij}^2}{(a - 1)(b - 1)}$$

*F tests are structured for fixed effects factors.
†See Table 5.3 for general assumptions and notations format.

determine the proportion of product variation created by each factor; then, we could address the most critical elements through a well-structured variance reduction project.

In both random and fixed effects models, we use computer aids to develop our ANOVAs. For fixed effects factors, we typically use computer aids to develop our graphs as well. We select appropriate information from the ANOVA output and develop or piece together graphs and other analyses, e.g., pairwise tests on the main effect and interaction terms as illustrated in the fixed effects factors case that follows in Example 5.6.

Example 5.6

ENERGY COST/OPERATIONS CASE—TWO FIXED EFFECTS FACTORS As an extension to the Natural Gas Production Case described in the previous chapter (Example 4.3), the plant capacity serves as a process limitation. The gas field extraction process is typically capable of producing up to 120% of the plant capacity for liquefaction. Since the liquefied natural gas (LNG) product is a commodity, once the gas is cleaned and the btu content established, the only way to increase revenues is through volume. One way to increase volume is to lower the chiller temperature and increase the throughput, say to 105% or even 110% of plant capacity. On the other hand, if throughout decreases, say to 95% capacity with low chiller temperatures, plant efficiency is believed to suffer.

At issue is how to best load the plant and how best to target the chiller temperature. Historically, primarily intuitive means were used to run the plant, as detailed cost/energy metrics were limited to global metrics, e.g., monthly energy measures. Further complicating the situation was the fact that field natural gas fuel, largely unmetered, was used in the process along with electricity, which was metered, for powering the facility as a whole.

Recently, improvements in energy metering were completed, whereby energy meters/metrics are available at all major energy consumption points in the process. With this new measurement capability, questions about plant loading to maximize gross profits began to surface. As a first step, a two-factor experiment was proposed to study the relationships between chiller temperature, plant throughput, and energy per unit of product.

Experimental purpose. Determine the relationship between chiller temperature, plant throughput, and energy consumption per unit of product with the intent of finding suitable chiller temperatures to support throughput levels.

Variables. The plant throughput or load, factor L, is defined as a percentage of designed plant capacity. Levels of 95, 100, 105, and 110% are selected for examination. Chiller temperature, factor T, is defined as the chiller operational target level. Levels of $-20, -30$, and $-35°F$ are selected as reasonable. The response is defined as the equivalent energy "cost," in equivalent energy units per unit of output product.

Currently, the temperature target is held near $-35°F$ and plant throughput is pushed up near 110% when possible. The response metric is complicated but now measurable by the newly installed energy metering capabilities. The process energy cost response involves a combination of fuel gas and electricity; hence it is a complex metric. But it is directly convertible to economic cost once the prices of fuel gas and electricity are known.

The experiment is conducted using a 4×3 FAT in a CRD with 12 treatment combinations. One of the 12 load $\times$ chiller combinations is picked at random. Then, the plant is run at that level for 24 hours. Another combination is selected at random and another 24-hour run is made. In all, three replications, or 36 runs, are made. The energy meters are reset and recalibrated before and after each run. The entire plant is used for the experiment, since the treatment combinations are all suitable for producing LNG. Hence, no additional risks of off-specification product are taken.

The CRD model and hypotheses are presented below.

$$Y_{ijk} = \mu + L_i + T_j + LT_{ij} + \epsilon_{k(ij)}$$ **[5.23]**

where
Y_{ijk} = the response in energy cost per product unit output
μ = overall mean
L_i = factor L main effect, $i = 1, 2, 3, 4$
T_j = factor T main effect, $j = 1, 2, 3$
LT_{ij} = factor L, factor T interaction effect
$\epsilon_{k(ij)}$ = random error, $k = 1, 2, 3$ replications

Main effect L:

H_0: All $L_i = 0$ or all μ_{Li} are equal (there is no throughput load effect present on the average energy cost, averaged over temperature and replications)

H_1: At least one $L_i \neq 0$ or at least one μ_{Li} is not equal to the others (there is a significant throughput load effect present on the average energy cost, averaged over temperature and replications)

Main effect T:

H_0: All $T_j = 0$ or all μ_{Tj} are equal (there is no temperature effect present on the average energy cost, averaged over loads and replications)

H_1: At least one $T_j \neq 0$ or at least one μ_{Tj} is not equal to the others (there is a significant temperature effect present on the average energy cost, averaged over loads and replications)

Interaction LT:

H_0: All $LT_{ij} = 0$ or all μ_{LTij} are equal (all load-temperature combinations produce the same average energy cost, averaged over replications)

H_1: At least one $LT_{ij} \neq 0$ or at least one μ_{LTij} is not equal to the others (not all load-temperature combinations produce the same average energy cost, averaged over replications)

Data collected from this smaller-is-better experiment are summarized in Table 5.8. Main effect and interaction plots are presented in Figure 5.7. A computer-generated ANOVA table broken out for the two-factor FAT-CRD is presented in Table 5.9. From the graphical depictions we see what appear to be

Table 5.8　Experimental data and summary (Energy Cost/Operations Case); response is in energy units/production unit.

		Chiller Temperature (j)			
		−20ºF	−30ºF	−35ºF	
Plant loading (i)	95%	60	50	101	
		69　$Y_{11.} = 215$	76　$Y_{12.} = 191$	92　$Y_{13.} = 308$	$Y_{1..} = 714$
		86	65	115	
		$\bar{Y}_{11.} = 71.667$	$\bar{Y}_{12.} = 63.667$	$\bar{Y}_{13.} = 102.667$	$\bar{Y}_{1..} = 79.333$
		$s_{11.} = 13.204$	$s_{12.} = 13.051$	$s_{13.} = 11.590$	
	100%	48	49	70	
		65　$Y_{21.} = 168$	45　$Y_{22.} = 157$	89　$\bar{Y}_{23.} = 235$	$Y_{2..} = 560$
		55	63	76	
		$\bar{Y}_{21.} = 56.000$	$\bar{Y}_{22.} = 52.333$	$\bar{Y}_{23.} = 78.333$	$\bar{Y}_{2..} = 62.222$
		$s_{21.} = 8.544$	$s_{22.} = 9.452$	$s_{23.} = 9.713$	
	105%	59	45	84	
		71　$\bar{Y}_{31.} = 195$	53　$Y_{32.} = 158$	49　$Y_{33.} = 194$	$Y_{3..} = 547$
		65	60	61	
		$\bar{Y}_{31.} = 65.000$	$\bar{Y}_{32.} = 52.667$	$\bar{Y}_{33.} = 64.667$	$\bar{Y}_{3..} = 60.778$
		$s_{31.} = 6.000$	$s_{32.} = 7.506$	$s_{33.} = 17.786$	
	110%	92	68	76	
		76　$Y_{41.} = 266$	85　$Y_{42.} = 222$	64　$Y_{43.} = 223$	$Y_{4..} = 711$
		98	69	83	
		$\bar{Y}_{41.} = 88.667$	$\bar{Y}_{42.} = 74.000$	$\bar{Y}_{43.} = 74.333$	$\bar{Y}_{4..} = 79.000$
		$s_{41.} = 11.372$	$s_{42.} = 9.539$	$s_{43.} = 9.609$	
		$Y_{.1.} = 844$	$Y_{.2.} = 728$	$Y_{.3.} = 960$	$Y_{...} = 2532$
		$\bar{Y}_{.1.} = 70.333$	$\bar{Y}_{.2.} = 60.667$	$\bar{Y}_{.3.} = 80.000$	$\bar{Y} = 70.333$

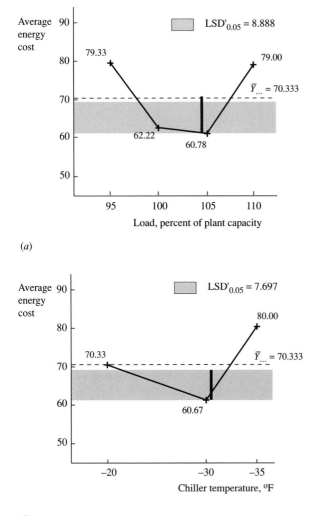

(a)

(b)

Figure 5.7 Two-factor main effects and interaction plots (Energy Cost/Operations Case). (a) Load main effect (L) plot. (b) Temperature main effect (T) plot.

significant main effects and interactions for this smaller-is-better analysis. The ANOVA table confirms these appearances. Working with the graphs and ANOVA table, two interpretations follow—one at $\alpha = 0.05$, the other at a more conservative $\alpha = 0.01$.

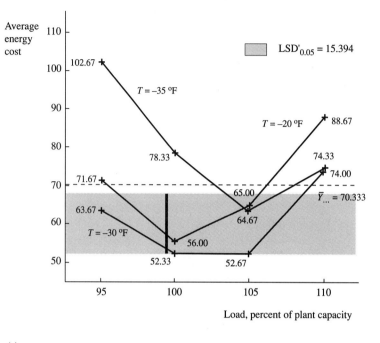

(c)

Figure 5.7 (continued)   Two-factor main effects and interaction plots (Energy Cost/Operations Case). (c) Load-temperature interaction effect (LT) plot.

ANOVA Interpretation Using α = 0.05   In our evidence, we observe graphical indications that interaction is present in the nonparallelness in the temperature profiles, plotted over the loads; see Figure 5.7c. Our P value = 0.0237 for LT interaction, Table 5.9, indicates that we should reject H_0, judging at the α = 0.05 level, and conclude significant interaction is present, b.o.e. In other words, we conclude that the average response is not the same at all load-temperature combinations.

Based on evidence, main effect L is significant at the α = 0.05 level, P value = 0.0009. Hence, we reject the H_0 and conclude that a significant loading effect is present, b.o.e. In other words, we see a failure of the average response to be the same for at least one throughput load level. This conclusion is clearly supported in Figure 5.7a.

Likewise, main effect T is significant at the α = 0.05 level, P value = 0.0011. And we reject H_0, concluding that a significant chiller temperature effect is present, b.o.e. In other words, we see a failure of the average response to be the same for at least one temperature level. Graphical support for this conclusion is presented in Figure 5.7b.

When we encounter a significant interaction, we provide a joint interpretation of the responses—in this case we interpret both L and T effects on the response together. Hence, we acknowledge that both main effects are significant at the α = 0.05 level, but proceed to focus on

Table 5.9 LNG production process FAT-CRD ANOVA (computer generated)

Dependent Variable: COST   ENERGY COST OF OPERATIONS

Source	DF	Sum of Squares	Mean Square	F Value	Pr > F
Model	11	7270.0000	660.9091	5.44	0.0003
Error	24	2914.0000	121.4167		
C Total	35	10184.0000			

R-Square	C.V.	Root MSE	COST Mean
0.7139	15.6667	11.0189	70.3333

Source	DF	Type I SS	Mean Square	F Value	Pr > F
L	3	2818.8889	939.6296	7.74	0.0009
T	2	2242.6667	1121.3333	9.24	0.0011
L*T	6	2208.4444	368.0741	3.03	0.0237

Observation	Observed Value	Predicted Value	Residual
1	60.0000	71.6667	-11.6667
2	69.0000	71.6667	- 2.6667
3	86.0000	71.6667	14.3333
4	50.0000	63.6667	-13.6667
5	76.0000	63.6667	12.3333
6	65.0000	63.6667	1.3333
7	101.0000	102.6667	- 1.6667
8	92.0000	102.6667	-10.6667
9	115.0000	102.6667	12.3333
10	48.0000	56.0000	- 8.0000
11	65.0000	56.0000	9.0000
12	55.0000	56.0000	- 1.0000
13	49.0000	52.3333	- 3.3333
14	45.0000	52.3333	- 7.3333
15	63.0000	52.3333	10.6667
16	70.0000	78.3333	- 8.3333
17	89.0000	78.3333	10.6667
18	76.0000	78.3333	- 2.3333
19	59.0000	65.0000	- 6.0000
20	71.0000	65.0000	6.0000
21	65.0000	65.0000	0.0000
22	45.0000	52.6667	- 7.6667
23	53.0000	52.6667	0.3333
24	60.0000	52.6667	7.3333
25	84.0000	64.6667	19.3333
26	49.0000	64.6667	- 15.667
27	61.0000	64.6667	- 3.6667
28	92.0000	88.6667	3.3333
29	76.0000	88.6667	-12.6667
30	98.0000	88.6667	9.3333
31	68.0000	74.0000	- 6.0000
32	85.0000	74.0000	11.0000
33	69.0000	74.0000	- 5.0000
34	76.0000	74.3333	1.6667
35	64.0000	74.3333	-10.3333
36	83.0000	74.3333	8.6667

the interaction graph, Figure 5.7c. We explain and examine our effects relative to the 12 load × temperature combinations, seeking the best combination—the one that yields the smallest energy cost in this case.

ANOVA Interpretation Using $\alpha = 0.01$ If we use a significance level of $\alpha = 0.01$ for our interpretation, we describe our findings in a different manner. Here, b.o.e., we reject both main effect H_0's with P value $= 0.0009$ for main effect L and P value $= 0.0011$ for main effect T, observing that both P values are less than $\alpha = 0.01$. However, at $\alpha = 0.01$, we do not reject the null hypothesis of no interaction effect when we encounter its P value $= 0.0237$.

   Absolutely no interaction between two variables produces a parallel signature between factor levels in interaction plots. Examination of Figure 5.7c indicates that our temperature profile lines, across the loading levels, are clearly not parallel. However, using $\alpha = 0.01$ for our interpretation, we conclude that the interaction is not extreme enough to believe that the LT_{ij} term in our model, Equation (5.23), is impacting our average response. Following this line of reasoning, **assuming insignificant interaction,** we interpret load and temperature independently of each other. Hence, we focus on the two main effect plots. In other words, **we examine each main effects plot independently of the other.** For example, we examine load, Figure 5.7a, then we examine temperature, Figure 5.7b. We draw conclusions relative to each without consideration of the other factor.

PAIRWISE TREATMENT COMPARISONS (FIXED EFFECTS FACTORS)

After the ANOVA table and its hypotheses and inference have been completed, we proceed to our LSD tests to establish treatment combination differences, b.o.e. Here, **we establish LSD statistics for each main effect as well as the interaction effects**—provided interaction and main effects were deemed significant previously. Using the $\alpha = 0.05$ level of significance, both one- and two-tailed LSD statistics are produced.

   Main effect L:

$$\text{LSD}'_{\alpha = 0.05} = t_{\alpha, \text{df}_{\text{err}}} \sqrt{\frac{2\text{MS}_{\text{err}}}{n'}}$$

$$= 1.711 \sqrt{\frac{2(121.417)}{9}} = 8.888$$

$$\text{LSD}_{\alpha = 0.05} = t_{\alpha/2, \text{df}_{\text{err}}} \sqrt{\frac{2\text{MS}_{\text{err}}}{n'}}$$

$$= 2.064 \sqrt{\frac{2(121.417)}{9}} = 10.721$$

Main effect T:

$$\text{LSD}'_{\alpha = 0.05} = 1.711 \sqrt{\frac{2(121.417)}{12}} = 7.697$$

$$\text{LSD}_{\alpha = 0.05} = 2.064 \sqrt{\frac{2(121.417)}{12}} = 9.285$$

Interaction LT:

$$\text{LSD}'_{\alpha = 0.05} = 1.711 \sqrt{\frac{2(121.417)}{3}} = 15.394$$

$$\text{LSD}_{\alpha = 0.05} = 2.064 \sqrt{\frac{2(121.417)}{3}} = 18.570$$

Figure 5.7 provides a graphical depiction of the summarized data with one-tail LSD graphics superimposed.

Following up on our interpretation, we clearly see significant differences in the average energy costs per unit of product in some combinations, b.o.e., using the LSD criteria. For example, from the main effects plots, the 100 and 105% loads appear to yield lower average costs than the 95 and 110% loads, averaged over temperatures and replications. The −30ºF chiller temperature yields a smaller average cost than those at the −20ºF and −35ºF levels, averaged over loads and replications.

Taking only the main effects into consideration, we would recommend loads of 100 or 105% with a −30ºF chiller temperature. However, when we examine the load × temperature interaction plot, our results are not so clear. If we mark off our one-tailed $\text{LSD}'_{0.05} = 15.394$ from our lowest LT combination response (100% load and −30ºF), we see several other LT combinations within this bar. Namely, combinations 95%, −30ºF; 100%, −20ºF; 105%, −20ºF; 105%, −35ºF; and 105%, −30ºF all fall within this interval, and are considered indistinguishable from the 100%, −30ºF combination average response, b.o.e., $\text{LSD}'_{\alpha=0.05}$.

Hence, our decision, b.o.e., is not entirely clear. But considering higher throughput more attractive, i.e., we have more LNG to sell, and higher temperatures more attractive in terms of maintenance and equipment wear, we would most likely favor operations in the 105% load and −30ºF area, b.o.e. However, we would continue to collect statistics to obtain more precise estimates of energy costs in this operation region to verify our initial inclination.

As a note in passing, we would validate our model and inferences by assessing the normality of our residuals, shown in Table 5.9, verifying the homogeneity of variance assumption by plotting our residuals versus our predicted values, and assessing our residuals as to order of experimentation. These assessments are logical extensions to those detailed in the single-factor case. These tasks are left for an end-of-chapter exercise.

5.6   SUMMARY

The CRD model is only one of many possible experimental design models available. We have overviewed the CRD model in one and two factors. We can readily develop the CRD in three or more factors in full factorials or fractional factorials. These designs plus others such as the randomized complete block and split plot designs are described by Kolarik [6], Montgomery [2], Box, Hunter, and Hunter [3], and other authors.

    In addition to the classical models, Taguchi-based models are available. Here, variation is systematically introduced into the design through the introduction and manipulation of noise variables, using noise matrices. Then, signal-to-noise ratio logarithmic transformations are made regarding the responses. These designs are overviewed by Nair et al. [8] and Kolarik [6] and are described in detail by Taguchi [9], Phadke [10], Ross [11], and Roy [12].

REVIEW AND DISCOVERY EXERCISES

REVIEW

5.1.   Review the Roll Pressure Case (Example 5.1). Develop an analysis of the residuals. Assess the normality, homogeneity of variance, and general validity of the experimental data regarding order sequence. Interpret your results.

5.2.   Revisit the Energy Cost/Operations Case (Example 5.6) and develop a residuals analysis. Assess the normality and homogeneity of variance. Interpret your results.

5.3.   Presented with a two-factor, random effects experiment, develop appropriate F tests and estimation techniques for components of variation. Begin with the information displayed in Table 5.7.

5.4.   For one or more of the following cases, described in Section 7, develop a thorough analysis of the experimental data provided. Include both graphical as well as inferential analyses. Use an $\alpha = 0.05$ for inferences.
Case VII.5:   Big City Waterworks
Case VII.10:   Downtown Bakery—pH Measurement
Case VII.18:   M-Stick Manufacturing
Case VII.19:   Night Hauler Trucking
Case VII.23:   Re-Use—Sensor Precision

DISCOVERY

5.5.   Typically, discussions regarding the analysis of fixed effects factor experiments are centered around a bigger-is-better or smaller-is-better response scenario. In these two cases, the experimental protocol is clear—we use techniques such as the LSD. However, needs in practice many times focus on a nominal-is-best case. Using basic concepts of statistics, namely the confidence interval approach, develop an analytical protocol for a nominal-is-best case. In your development, begin with a single-factor experiment, and then include a two-factor experiment. Assume multiple observations in each treatment/treatment combination–replication.

5.6. For one or more of the following cases, described in Section 7, develop a thorough analysis of the experimental data provided. Include both graphical as well as inferential analyses. Use an $\alpha = 0.05$ for inferences.

Case VII.1: AA Fiberglass

Case VII.12: Hard-Shell Aquaculture

Case VII.15: High-Precision—Collar Measurement

Case VII.16: Link-Lock Chain

5.7. For Case VII:26: Squeaky Clean Laundry, described in Section 7, develop a conceptual argument to support the purpose described in the case.

5.8. For Example 5.1, Roll Pressure Case, assess the practical significance of the results and what they mean relative to possible blank length targets at the shear subprocess. Base your arguments on scrap losses.

REFERENCES

1. R. E. Walpole, R. H. Myers, and S. L. Myers, *Probability and Statistics for Engineers and Scientists*, 6th ed., Upper Saddle River, NJ: Prentice Hall, 1998.

2. D. C. Montgomery, *Design and Analysis of Experiments*, 4th ed., New York: Wiley, 1997.

3. G. E. P. Box, W. G. Hunter, and J. S. Hunter, *Statistics for Experimenters*, New York: Wiley, 1978.

4. *SAS Procedures Guide*, Version 6, Cary, NC: SAS Institute, 1990.

5. *SAS/STAT User's Guide*, Vols. 1 and 2, Version 6, Cary, NC: SAS Institute, 1990.

6. W. J. Kolarik, *Creating Quality: Concepts, Systems, Strategies, and Tools*, New York: McGraw-Hill, 1995.

7. W. J. Conover, *Practical Nonparametric Statistics*, 2nd ed., New York: Wiley, 1980.

8. V. N. Nair et al., "Taguchi's Parameter Design: A Panel Discussion," *Technometrics*, vol. 34, no. 2, pp. 127–161. 1992.

9. G. Taguchi, *Introduction to Quality Engineering: Designing Quality into Products and Processes*, White Plains, NY: Kraus International, UNIPUB (Asian Productivity Organization), 1986.

10. M. S. Phadke, *Quality Engineering Using Robust Design*, Englewood Cliffs, NJ: Prentice Hall, 1989.

11. P. J. Ross, *Taguchi Techniques for Quality Engineering*, New York: McGraw-Hill, 1988.

12. R. K. Roy, *A Primer on the Taguchi Method*, New York: Van Nostrand Reinhold, 1990.

6

PROCESS RESPONSE MODELING

6.0 INQUIRY

1. What is regression?
2. What is a response surface?
3. How are response surfaces developed?
4. How are response surfaces used to understand processes?

6.1 INTRODUCTION

In general, our graphical and ANOVA analyses of Chapter 5 help to isolate design and noise variables that are driving product or process performance. When we are working with quantitative factors in a fixed effects model, we may want to go beyond identifying critical factors. **Response surface models allow us to predict responses at factor levels other than those originally included in our experiments,** e.g., between our chosen factor levels. We develop a model of the form

$$\hat{\mu} = f (\text{design and noise variables}) = f (\mathbf{X}, \mathbf{Z}) \qquad \textbf{[6.1]}$$

This general model helps us understand our process by providing a mathematical representation of our average response as a function of the variables that are driving it. This predictor allows us to search out high performance factor levels faster than we could otherwise.

6.2 LEAST SQUARES ESTIMATION

Once we have collected our data, we are in a position to hypothesize and fit a model. We typically hypothesize some form of a general linear model and use the least squares technique to "fit" our model. In the least squares estimation technique we develop our predictor $\hat{\mu}$ by minimizing the sum of the squares of deviation between the observed values and the predicted values.

LEAST SQUARES ESTIMATORS

The least squares estimation technique is associated with a general linear model of the form

$$Y = \beta_0 + \beta_1 X_1 + \beta_2 X_2 + \dots \beta_m X_m + \epsilon \qquad \textbf{[6.2]}$$

where
Y = a dependent random variable or response variable
$X_1, X_2, \dots, X_m$ = independent or input model variables
$\beta_0, \beta_1, \dots, \beta_m$ = model parameters
ϵ = an error or residual term

In general,

$$Y_i = b_0 + b_1 X_{1i} + b_2 X_{2i} + \dots + b_m X_{mi} + e_i \qquad i = 1, 2, \dots, n \qquad \textbf{[6.3]}$$

which is also expressed as

$$\hat{\mu}_i = b_0 + b_1 X_{1i} + b_2 X_{2i} + \dots + b_m X_{mi} \qquad i = 1, 2, \dots, n \qquad \textbf{[6.4]}$$

Here,
Y_i = ith response observation
$\hat{\mu}_i$ = ith predicted response (the predicted mean of a population of Y values, indexed by a given level of X)
$b_0, b_1, \dots, b_m$ = estimated model parameters, estimated from the data, e.g.,
$\hat{\beta}_0 = b_0, \dots, \hat{\beta}_m = b_m$
e_i = specific error (or residual) value that represents the difference between an observed Y_i and its predicted value $\hat{\mu}_i$
n = number of observations

We use least squares estimators to minimize the sum of the squared deviations:

$$\sum_{i=1}^{n} e_i^2 = \sum_{i=1}^{n} (Y_i - \hat{\mu}_i)^2 = \sum_{i=1}^{n} [Y_i - (b_0 + b_1 X_{1i} + \dots + b_m X_{mi})]^2 \qquad \textbf{[6.5]}$$

NORMAL EQUATIONS

The sum of squared deviations shown above is minimized by setting the partial derivatives with respect to the parameters $b_0, b_1, \dots, b_m$ equal to 0 and solving. **The result is the general set of classical normal equations;** see Walpole, Myers, and Myers [1], Draper and Smith [2], Box and Draper [3] and Neter, Wasserman, and Kutner [4]:

$$\sum_{i=1}^{n} Y_i = nb_0 + b_1 \sum_{i=1}^{n} X_{1i} + b_2 \sum_{i=1}^{n} X_{2i} + \dots + b_m \sum_{i=1}^{n} X_{mi} \qquad \textbf{[6.6]}$$

$$\sum_{i=1}^{n} X_{1i} Y_i = b_0 \sum_{i=1}^{n} X_{1i} + b_1 \sum_{i=1}^{n} X_{1i}^2 + b_2 \sum_{i=1}^{n} X_{1i} X_{2i} + \dots + b_m \sum_{i=1}^{n} X_{1i} X_{mi} \qquad \textbf{[6.7]}$$

$$\sum_{i=1}^{n} X_{2i}Y_i = b_0 \sum_{i=1}^{n} X_{2i} + b_1 \sum_{i=1}^{n} X_{2i}X_{1i} + b_2 \sum_{i=1}^{n} X_{2i}^2 + \ldots + b_m \sum_{i=1}^{n} X_{2i}X_{mi} \qquad \textbf{[6.8]}$$

$$.$$
$$.$$
$$.$$

$$\sum_{i=1}^{n} X_{mi}Y_i = b_0 \sum_{i=1}^{n} X_{mi} + b_1 \sum_{i=1}^{n} X_{mi}X_{1i} + b_2 \sum_{i=1}^{n} X_{mi}X_{2i} + \ldots + b_m \sum_{i=1}^{n} X_{mi}^2 \qquad \textbf{[6.9]}$$

There are a number of fundamental assumptions associated with our general linear regression model:

1. **The X's and Y's are observed in sets,** e.g., pairs or bivariate random samples in simple linear regression.

2. **The X values are measured with little or no error.**

3. **Each X value indexes a distribution of Y values,** and these distributions have a common variance, σ_{ϵ}^2.

4. **The residuals are independent and distributed normally** with $\mu = 0$, $\sigma^2 = \sigma_{\epsilon}^2$.

The predicted responses, $\hat{\mu}_i$, represent the estimated means of the distribution of Y values mentioned in assumption 3. Assumption 4 is necessary to develop our hypothesis tests, e.g., our t or F tests, discussed later.

6.3   REGRESSION ANALYSIS

In this chapter we use basic regression analysis methods and computer aids to fit our models. We emphasize graphical context as far as possible. Our discussions and presentations of models are clearly limited. Texts by authors such as Walpole, Myers, and Myers [1], Draper and Smith [2], Box and Draper [3], and Neter, Wasserman, and Kutner [4] provide more thorough treatments, including confidence intervals on predicted response estimates.

A generic simple linear regression model, a special case of Equation (6.2) **with only one independent variable, X_1, is of the form**

$$Y = \beta_0 + \beta_1 X_1 + \epsilon \qquad \textbf{[6.10]}$$

A generic fitted predictor for the model is of the form

$$\hat{\mu} = b_0 + b_1 X_1 \qquad \textbf{[6.11]}$$

Here, Y = dependent response variable

$\hat{\mu}$ = predicted response (estimated mean value)

β_0 = intercept parameter

b_0 = estimated value for β_0

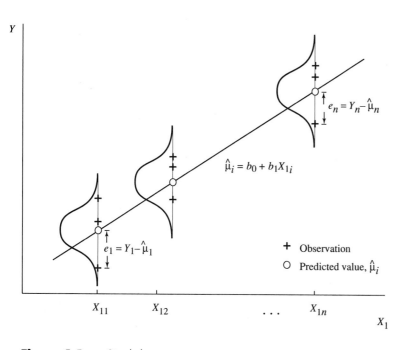

Figure 6.1 Simple linear regression concept.

β_1 = slope parameter
b_1 = estimated value for β_1
X_1 = independent variable
ϵ = error term used to compensate for imperfect fits

Figure 6.1 illustrates the regression concept for the simple linear regression case. Here, we can see the distribution of Y values indexed by each X_1 value. We observe that the model follows a straight line, which in turn locates each $\hat{\mu}$. We also observe that each indexed distribution contains a common variance, σ_ϵ^2.

REGRESSION ANOVAS

Least squares estimators are used heavily in engineering work. In regression analysis and response surface modeling, cross-product and quadratic or higher-order terms are introduced through transformations in the X's. **The regression approach to model fitting and data analysis is general in that it applies to both designed and "undesigned" experiments, whether balanced or unbalanced.** An undesigned experiment consists of data collected from a process, where no specific randomization plan was used. For details on applying the regression approach in general, see Draper and Smith [2], Box and Draper [3], and Neter, Wasserman, and Kutner [4].

Figure 6.2 depicts a simple linear regression predictor model that is compatible with our single-factor CRD experimental results. A group of four replications is shown at X_{1i}. Here X_{1i}

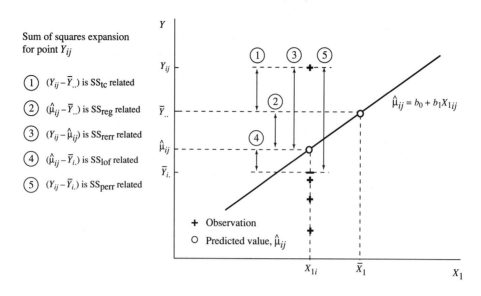

Sum of squares expansion
for point Y_{ij}

① $(Y_{ij} - \bar{Y}_{..})$ is SS_{tc} related

② $(\hat{\mu}_{ij} - \bar{Y}_{..})$ is SS_{reg} related

③ $(Y_{ij} - \hat{\mu}_{ij})$ is SS_{rerr} related

④ $(\hat{\mu}_{ij} - \bar{Y}_{i.})$ is SS_{lof} related

⑤ $(Y_{ij} - \bar{Y}_{i.})$ is SS_{perr} related

Figure 6.2   Sum of squares components, relative to a single observation, Y_{ij}.

represents the ith level of our treatment factor X_1. In order to present a regression discussion consistent with our previous experimental notation, we reintroduce the Y_{ij} response for treatment factor X_{1i} (at treatment level i, $i = 1, 2, \ldots, a$) used in Chapter 5. We have a total of n responses from our experiment.

Labeling an observed value at X_{1i} as Y_{ij}, where the j index represents the jth observation within the ith treatment level, **we develop the regression ANOVA terms.** Expanding,

$$Y_{ij} = \bar{Y}_{..} + (\hat{\mu}_{ij} - \bar{Y}_{..}) + (\bar{Y}_{i.} - \hat{\mu}_{ij}) + (Y_{ij} - \bar{Y}_{i.}) \qquad \textbf{[6.12]}$$

At this point our readers may want to review Equation (5.2) and compare it to Equation (6.12). Now, by rearranging, summing, and squaring both sides, we obtain

$$\sum_{ij} (Y_{ij} - \bar{Y}_{..})^2 = \sum_{ij} (\hat{\mu}_{ij} - \bar{Y}_{..})^2 + \sum_{ij} (\hat{\mu}_{ij} - \bar{Y}_{i.})^2 + \sum_{ij} (Y_{ij} - \bar{Y}_{i.})^2 \qquad \textbf{[6.13]}$$

Or, in other words, **we can represent the terms above as**

$$SS_{tc} = SS_{reg} + SS_{lof} + SS_{perr} \qquad \textbf{[6.14]}$$

An alternative form, which is used when no replication is present, **in our notation is**

$$\sum_{ij} (Y_{ij} - \bar{Y}_{..})^2 = \sum_{ij} (\hat{\mu}_{ij} - \bar{Y}_{..})^2 + \sum_{ij} (Y_{ij} - \hat{\mu}_{ij})^2 \qquad \textbf{[6.15]}$$

$$SS_{tc} = SS_{reg} + SS_{rerr} \qquad \textbf{[6.16]}$$

where SS_{tc} = total corrected sum of squares
SS_{reg} = regression sum of squares
SS_{rerr} = residual error sum of squares
SS_{lof} = lack of fit sum of squares
SS_{perr} = pure error sum of squares

A generic regression ANOVA is displayed in Table 6.1. Here, we see source entries for each of the terms identified above. We typically use computer aids such as SAS [5, 6] to develop these ANOVA tables. But, with enough effort, they can be developed with hand calculations, as indicated in Table 6.1.

To gain further insight into our regression ANOVA, we can relate the geometric relationships in Figure 6.2 with the sum of squares measurements in Equations (6.13) through (6.16) and Table 6.1. These relationships reinforce the fact that each regression ANOVA row provides us with a measurement that helps assess the characteristics of our data and the merits of our model.

Typically, regression ANOVA computer aids, such as PROC REG in SAS, do not develop the SS_{perr} or SS_{lof}. Instead, they usually display a total corrected row (labeled C TOTAL), a regression row (labeled MODEL), and a residual error row (labeled ERROR). If our computer aid is incapable of breaking out pure error and lack-of-fit terms, we can piece together these terms as in Table 6.1. Here, we notice that each X with multiple observations, $n_i > 1$, contributes to df_{perr} and SS_{perr}. The SS_{perr} is the sum of these contributions.

When we use computer aids, our focus shifts from fitting calculations to model structure and interpretations. **We formulate model structures that make sense from both the statistical and physical points of view—and seek out the "best" model.** Best refers to the simplest model (having the fewest terms) that will adequately describe the data.

It is of interest to note that **we may not be able to find a suitable model in all cases.** When our experimental error is relatively large, modeling becomes very challenging. If this "noise," or dispersion "within treatments" (i.e., from a failure of experimental units treated alike to respond the same), becomes too great, we will not be able to find a suitable model under any circumstances, regardless of our statistical brilliance.

RESPONSE SURFACE STRUCTURE

Response surface models assist us in two ways: (1) in the design of a follow-on experiment and (2) in the optimization of a process response. For example, we typically use our model to predict response values at process settings we did not observe in our experiment. In the early stages of discovery, this knowledge aids in configuring our follow-on experiments. In later stages, where we have more confidence in our model, we may use it to establish optimal process response settings.

We present an introduction to response surface modeling. A more detailed treatment is available in Box and Draper [3]. A "full" first-order model in two factors, factor A (e.g., X_1) and factor B (e.g., X_2) is

$$Y = b_0 + b_1(X_1) + b_2(X_2) + e \qquad \textbf{[6.17]}$$

Table 6.1   Generic regression ANOVA format.

Source	df	SS	MS	F	P Value
Total corrected	$n - 1$	$SS_{tc} = \sum_{ij} (Y_{ij} - \bar{Y}_{..})^2$			
Regression	m	$SS_{reg} = \sum_{ij} (\hat{\mu}_{ij} - \bar{Y}_{..})^2$	$\dfrac{SS_{reg}}{df_{reg}}$	$\dfrac{*MS_{reg}}{MS_{rerr}}$	P_{reg}
Residual error	$n - m - 1$	$SS_{rerr} = \sum_{ij} (Y_{ij} - \hat{\mu}_{ij})^2$	$\dfrac{SS_{rerr}}{df_{rerr}}$		
Lack of fit	$df_{rerr} - df_{perr}$	$SS_{rerr} - SS_{perr}$	$\dfrac{SS_{lof}}{df_{lof}}$	$\dfrac{†MS_{lof}}{MS_{perr}}$	P_{lof}
Pure error	$\sum_{i=1}^{a} (n_i - 1)$	$SS_{perr} = \sum_{ij} (Y_{ij} - \bar{Y}_{i.})^2$	$\dfrac{SS_{perr}}{df_{perr}}$		
Pure error components	$n_1 - 1$	$\sum_{j=1}^{n_1} (Y_{1j} - \bar{Y}_{1.})^2$			
	$n_2 - 1$	$\sum_{j=1}^{n_2} (Y_{2j} - \bar{Y}_{2.})^2$			
	$\vdots$	$\vdots$			
	$n_a - 1$	$\sum_{j=1}^{n_a} (Y_{aj} - \bar{Y}_{a.})^2$			

*Overall model test
H_0: All β_{i_*} in the model, $i \neq 0$, are equal to zero (we might as well use $Y_{ij} = \beta_0 + \epsilon_{ij}$ for our model).
H_1: At least one β_i in the model, $i \neq 0$, is not equal to zero.
†Lack of fit test
H_0: The model form is adequate.
H_1: The model form is not adequate (we should seek out an alternate model form).

In general, we can establish first-order models from two-level experiments. A full second-order response surface using the same two factors, *A* and *B,* is

$$Y = b_0 + b_1(X_1) + b_2(X_2) + b_3(X_1 X_2) + b_4(X_1)^2 + b_5(X_2)^2 + e \qquad \textbf{[6.18]}$$

In general, a second-order model requires more than two factor levels.

We take a rather fundamental approach in our response surface modeling in this chapter. Much more sophisticated approaches are available. We start with a full model and then eliminate terms that do not explain a significant amount of the total corrected sum of squares. In practice, we may improvise somewhat by including or excluding a borderline statistically significant term, when doing so allows a more reasonable physical defense of the model. **We justify our model on a physical and a statistical basis. Otherwise we risk advocating what is termed a spurious model:** a good statistical model that fits the data available, but with little, if any, ability to explain the physical process in general.

MODEL SIMPLIFICATION

Our approach here is to develop our response surface model in its full form first. Then we attempt to eliminate terms in the following order:

1. Second-order terms and two-factor interaction terms.

2. First-order terms, provided all associated interaction terms and second-order terms have also been eliminated (i.e., we will not eliminate a first-order term when the factor remains in an interaction term or its second-order term remains in the model).

We also assess the adequacy of our model, assuming a pure error is available for our analysis.

In order to simplify a full model, we use what are called "partial" sums of squares, which are generated by most computer aids. These terms represent the sum of squares corresponding to the situation where each regression model term is fit last—after all other terms have been fit. By examining the fitting of each term last, we can assess its impact in explaining part of the SS_{tc}. For example, we can examine the residual error terms resulting from the full model and reduced models (each missing one term). Once we see the impact of each term fit last, we develop either F or t tests (they are equivalent when we have one numerator degree of freedom); see Draper and Smith [2]. Then, we set a threshold significance level, say $\alpha = 0.15$, and eliminate the term with the highest P value, as long as this value exceeds our threshold P value. In other words, we attempt to eliminate the term that explains the least amount of the SS_{tc} first. We then rerun the model without this term—provided we can justify its elimination—and repeat the process until all terms left are deemed significant. We typically keep the intercept term b_0, regardless of its significance. If we eliminate the intercept term, we force the predictor through the origin.

We assess the significance of our regression model terms quantitatively using t or F tests. Here, we assess each term individually. We set up our hypotheses as

H_0: $\beta_i = 0$ (the ith term in our regression model does not help in explaining the response)

H_1: $\beta_i \neq 0$ (the ith term in our regression model does help in explaining the response)

The calculated t value is

$$t_{calc} = \frac{b_i - 0}{\sqrt{MS_{rerr}c_{ii}}} = \frac{b_i - 0}{s_{bi}}$$ [6.19]

where b_i = least squares estimate of β_i $\quad$ $i = 0, 1, 2, \ldots, m$

c_{ii} = diagonal element of the matrix, $(\mathbf{X}^T\mathbf{X})^{-1}$; see Walpole, Myers, and Myers [1] and Draper and Smith [2]

$$MS_{rerr} = \text{mean square residual error from the model that includes the } i\text{th term}$$
$$s_{bi} = \text{standard error associated with the least squares estimate of } \beta_i$$

If

$$|t_{calc}| > t_{\alpha/2,\, df_{rerr}}$$

then we reject H_0: $\beta_i = 0$ and conclude that the ith term should remain in the model, b.o.e.

If we develop an F test format

$$F_{calc} = \frac{SS(\beta_i | \beta_0, \beta_1, \ldots, \beta_{i-1}, \beta_{i+1}, \ldots, \beta_m)}{MS_{rerr}} \qquad \textbf{[6.20]}$$

where $SS(\beta_i | \beta_0, \beta_1, \ldots, \beta_{i-1}, \beta_{i+1}, \ldots, \beta_m)$ is the sum of squares explained by adding the ith term after all other terms (all terms but the ith term) are already in the model, and

$$SS(\beta_i | \beta_0, \beta_1, \ldots, \beta_{i-1}, \beta_{i+1}, \ldots, \beta_m) = SS(\beta_0, \beta_1, \ldots, \beta_m) - $$
$$SS(\beta_0, \beta_1, \ldots, \beta_{i-1}, \beta_{i+1}, \ldots, \beta_m) \qquad \textbf{[6.21]}$$

Here, MS_{rerr} is the mean square residual error term associated with the model that includes the ith parameter term, β_i.

For this F test,

$$df_{numerator} = 1 \quad \text{and} \quad df_{denominator} = df_{rerr}$$

If

$$F_{calc} > F_{\alpha, 1, df_{rerr}}$$

then we reject H_0: $\beta_i = 0$, and conclude that the ith term should remain in the model, b.o.e. In this case, the t and F tests above are equivalent tests; see Walpole, Myers, and Myers [1] or Draper and Smith [2]. Equivalence is based on the fact that

$$t_{\alpha/2,v}^2 = F_{\alpha,1,v} \qquad \textbf{[6.22]}$$

Obviously, these calculations are difficult to develop manually. But computer aids such as SAS PROC REG provide regression ANOVA rows specifically developed for our convenience in these analyses. When assessing the significance of individual terms in regression models, some computer aids use the t test and some the F test. The F test allows us to test H_0: $\beta_i = 0$, as opposed to H_1: $\beta_i \neq 0$. The t test is a more general test in that we can test for values of $\beta_i = \beta_{i0}$ other than 0. For example, we can test $\beta_{i0} = 0.5$, and so forth, in place of $\beta_{i0} = 0$ in Equation (6.19). In most cases, we use computer aids and their associated tests and their P values to decide which model terms to include or exclude.

The process we have just described, where we start with a full model and simplify by removing terms, is a form of the **backward elimination method.** There are many other techniques used to develop regression models with respect to selecting or rejecting terms. Draper and Smith [2] provide a more detailed discussion of other regression development techniques.

MODEL FIT AND ADEQUACY

The coefficient of multiple determination measure, R^2, is associated with a regression fit. The R^2 value is the proportion of the SS_{tc} "explained" by the regression model. It is calculated as

$$R^2 = \frac{SS_{model}}{SS_{total\ corrected}} = \frac{SS_{reg}}{SS_{tc}} \qquad \textbf{[6.23]}$$

Hence, $0 \leq R^2 \leq 1$, with $R^2 = 1$ representing the "perfect" model fit. The R^2 measure is useful, but can be misleading if used alone in assessing model fit. For n data points, with no replication, an $R^2 = 1$ can be obtained by employing n properly selected coefficients in the model, including β_0, since a model can then be chosen that fits the data exactly; see Draper and Smith [2].

If we use a designed experiment with replication, we can calculate the maximum R^2 possible as

$$R^2_{max} = \frac{SS_{tc} - SS_{perr}}{SS_{tc}} = \frac{SS_{tc} - SS_{err}}{SS_{tc}} \qquad \textbf{[6.24]}$$

where SS_{err} represents the sum of squares experimental error in a designed experiment; see Chapter 5. **No matter what we do in the model, we cannot explain the sum of squares due to the pure error.**

We sometimes develop what is called an adjusted R^2; see Draper and Smith [2]:

$$R^2_{adj} = 1 - \left(\frac{n-1}{n-m+1}\right)(1 - R^2) \qquad \textbf{[6.25]}$$

where R^2 = usual coefficient of multiple determination

n = number of observations in the data set

m = number of parameters (excluding b_0) in the model being fit

We can see that R^2_{adj} considers both the number of observations we have and the number of model terms we are using. As we add (delete) model terms, R^2 will either remain constant or increase (decrease). The R^2_{adj}, on the other hand, may or may not increase (decrease) when we add (delete) terms. We will follow tradition and focus primarily on R^2 rather than R^2_{adj}. Most computer aids provide both measures.

In the case of designed experiments with replications, a pure error term can be developed. If the pure error term is available, we can test for a lack of fit associated with the regression model. We develop an F test where

H_0: The model form is adequate

H_1: The model form is inadequate

$$F_{lof} = \frac{(SS_{rerr} - SS_{perr})\ /\ (df_{rerr} - df_{perr})}{SS_{perr}\ /\ df_{perr}} = \frac{MS_{lof}}{MS_{perr}} \qquad \textbf{[6.26]}$$

If we reject H_0 at the α level of significance, we see strong evidence to suggest that our predicted values do not correspond to the treatment means. This correspondence is measured by the lack-of-fit row in the ANOVA; see Table 6.1 and Equation (6.13). Its significance is then assessed relative to the pure error; see Equation (6.26). In Figure 6.2, item 4 corresponds to lack of fit and item 5 corresponds to pure error.

The pure error term measures the variation within treatments—a function of only the replicates within each treatment. The total corrected row and the pure error row in the ANOVA will not change when we add or delete terms in our regression model. On the other hand, the lack-of-fit term will change because it depends on the terms we include in our regression model—it is developed from the residual error, which depends on the model selected; see Table 6.1.

Example 6.1	**ENERGY COST/OPERATIONS CASE—RESPONSE SURFACE IN ONE FACTOR** Revisiting the Natural Gas Production Case in Chapter 4 and Example 5.6, we develop a response surface model for the energy cost at a chiller temperature of $-30°$F. In this case we demonstrate first- and second-order models. We develop a simple first-order linear regression model of the form

$$\hat{\mu} = b_0 + b_1 X_1$$

Then, we develop a second-order model of the form

$$\hat{\mu} = b_0 + b_1 X_1 + b_2 X_1^2$$

An examination of the response plot, the $-30°$F curve in Figure 5.7c, indicates that exploring a second-order model will probably be necessary to capture the curvature in the plot. But, we will proceed step by step and develop and assess both models.

Restating our data collected at the $-30°$F operating temperature, from Table 5.8, gives

	Treatment Level (Load)			
Replication	**95%**	**100%**	**105%**	**110%**
1	50	49	45	68
2	76	45	53	85
3	65	63	60	69

Developing the terms necessary for our set of classical normal equations gives

i	j	$X_{1i,j}$	Y_{ij}
1	1	95	50
1	2	95	76
1	3	95	65
2	1	100	49
2	2	100	45
2	3	100	63
3	1	105	45
3	2	105	53
3	3	105	60
4	1	110	68
4	2	110	85
4	3	110	69

$$\sum_{ij} X_{1ij} = 1{,}230$$

$$\sum_{ij} X_{1ij}^2 = 126{,}450$$

$$\sum_{ij} X_{1ij}Y_{ij} = 74{,}855$$

$$\sum_{ij} Y_{ij} = Y_{..}\ 728$$

$$n = 12$$

Re-expressing the first normal equation [from the first three terms of Equation (6.6)] gives

$$\Sigma Y_{ij} = nb_0 + b_1 \Sigma X_{1ij}$$

$$728 = 12b_0 + 1{,}230b_1$$

For the second normal equation [likewise from Equation (6.7)],

$$\Sigma X_{1ij}Y_{ij} = b_0 \Sigma X_{1ij} + b_1 \Sigma X_{1ij}^2$$

$$74{,}855 = 1{,}230b_0 + 126{,}450b_1$$

Solving the two equations for b_0 and b_1, we obtain

$$b_0 = -3.567$$
$$b_1 = 0.627$$

Restating our simple linear regression predictor,

$$\hat{\mu} = -3.567 + 0.627(L)$$

where L is plant load in percentage of capacity; $L = X_1$.

Table 6.2 presents a computer-aided solution for the first-order model. Here, we see considerably more analysis than in our hand solution regarding the normal equations. The first-order model is a poor fit, $R^2 = 0.083$, b.o.e. We can construct hypothesis tests to assess the contributions of our parameters as follows:

$H_0: \beta_0 = 0$ (the intercept parameter β_0 is not helping to explain the response in our model)

$H_1: \beta_0 \neq 0$ (the intercept parameter β_0 is helping to explain the response in our model)

Here, we see $b_0 = -3.567$ with a $t_{calc} = (b_0 - 0)/s_{b0} = -3.567/67.624 = -0.053$ and a P value $= 0.959$. We do not reject H_0 and conclude that the intercept term is not helping to explain our response. In general, we tend to ignore this test because we usually do not wish to force the model through the origin. Hence, we typically leave b_0 in our models. Next, we develop

$H_0: \beta_1 = 0$ (the parameter β_1 is not helping to explain the response)

$H_1: \beta_1 \neq 0$ (the parameter β_1 is helping to explain the response)

Here, we see $b_1 = 0.627$ with a $t_{calc} = 0.951$ and a P value $= 0.364$. We cannot reject H_0 here, b.o.e.

Next, we construct an expanded ANOVA table. From our computer-aided results, we see

$$SS_{tc} = 1774.667$$
$$SS_{reg} = 147.267$$
$$SS_{rerr} = 1627.400$$

Table 6.2   Computer-aided results—first-order model (Energy Cost/Operations Case).

Dependent Variable: COST   ENERGY COST OF OPERATIONS

Analysis of Variance

Source	DF	Sum of Squares	Mean Square	F Value	Prob > F
Model	1	147.26667	147.26667	0.905	0.3639
Error	10	1627.40000	162.74000		
C Total	11	1774.66667			

Root MSE	12.75696	R-Square	0.0830	
Dep Mean	60.66667	Adj R-sq	-0.0087	
C.V.	21.02795			

Parameter Estimates

Variable	DF	Parameter Estimate	Standard Error	T for H_0: Parameter=0	Prob > |T|
INTERCEP	1	-3.566667	67.62391589	-0.053	0.9590
L	1	0.626667	0.65876652	0.951	0.3639

Obs	Dep Var COST	Predict Value	Residual
1	50.0000	55.9667	-5.9667
2	76.0000	55.9667	20.0333
3	65.0000	55.9667	9.0333
4	49.0000	59.1000	-10.1000
5	45.0000	59.1000	-14.1000
6	63.0000	59.1000	3.9000
7	45.0000	62.2333	-17.2333
8	53.0000	62.2333	-9.2333
9	60.0000	62.2333	-2.2333
10	68.0000	65.3667	2.6333
11	85.0000	65.3667	19.6333
12	69.0000	65.3667	3.6333

Since we have multiple Y's at our X values, we can compute a pure error row and calculate a lack-of-fit row and analysis. Our ANOVA results are displayed in Table 6.3. Here, we generate our pure error and lack-of-fit rows using hand calculations, taking our information from Table 6.1 and our previous data table.

At this point, we are reasonably sure that we do not have a good predictor, but our ANOVA does not provide clues as to what we should do next—try a new model or give up. Our original plot in Figure 5.7c tends to indicate that a second-order model should be attempted. Here, we expect a second-order model to allow our average response to decrease and then increase as we increase our load. An examination of our residuals also supports this hunch.

Developing our residual terms, where

$$e_{ij} = Y_{ij} - \hat{\mu}_{ij}$$

Table 6.3 Regression ANOVA—first-order model (Energy Cost/Operations Case).

Source	df	SS	MS	F	P Value
Total corrected	11	1774.667			
Regression	1	147.267	147.267	0.905	0.364
Residual error	10	1627.400	162.740		
Lack of fit	$10 - 8 = 2$	813.399	406.700	3.997	$0.05 < P < 0.10$
Pure error	8	814.001	101.750		
For $X_{11} = 95\%$	$3 - 1 = 2$	340.667			
For $X_{12} = 100\%$	$3 - 1 = 2$	178.667			
For $X_{13} = 105\%$	$3 - 1 = 2$	112.667			
For $X_{14} = 110\%$	$3 - 1 = 2$	182.000			

Sample calculations
$SS_{X11=95} = 50^2 + 76^2 + 65^2 - [(50 + 76 + 65)^2/3] = 340.667$
$SS_{X12=100} = 49^2 + 45^2 + 63^2 - [(49 + 45 + 63)^2/3] = 178.667$
$SS_{X13=105} = 45^2 + 53^2 + 60^2 - [(45 + 53 + 60)^2/3] = 112.667$
$SS_{X14=110} = 68^2 + 85^2 + 69^2 - [(68 + 85 + 69)^2/3] = 182.000$

we obtain the following table:

$L = X_1$	Y_{ij}	$\hat{\mu}_{ij}$	e_{ij}
95	50	55.967	−5.967
95	76	55.967	20.033
95	65	55.967	9.033
100	49	59.100	−10.100
100	45	59.100	−14.100
100	63	59.100	3.900
105	45	62.233	−17.233
105	53	62.233	−9.233
105	60	62.233	−2.233
110	68	65.367	2.633
110	85	65.367	19.633
110	69	65.367	3.633

Our residuals from the first-order model are plotted against our X_1's in Figure 6.3. Here, we see an inverted "rainbow" pattern around the zero line. A rainbow or inverted rainbow is an indication that we should add a second-order term to our linear model. Hence, we introduce a second independent variable, called X_2, which we will define as "X_1 squared,"

$$X_{2ij} = X_{1ij}(X_{1ij})$$

In effect, X_2 is a transformation of our original load variable and constitutes a second-order term.

Since we now have three normal equations, from Equations (6.6), (6.7), and (6.8), we present a computer-aided solution in Table 6.4. From Table 6.4 we see $R^2 = 0.534$, along with significant, $\alpha = 0.05$, results for β_0, β_1, and β_2, indicating that each term is helping to explain our response.

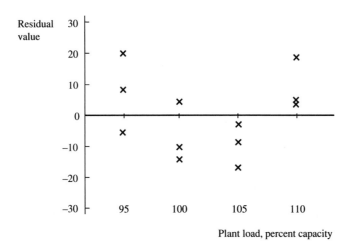

Figure 6.3   Residuals plot—first-order model (Energy Cost/Operations Case).

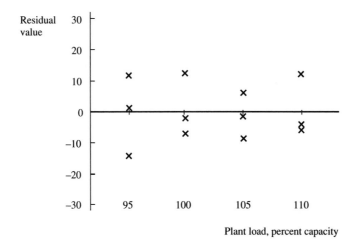

Figure 6.4   Residuals plot—second-order model (Energy Cost/Operations Case).

Restating our second-order regression predictor, we obtain

$$\hat{\mu} = 3418.267 - 66.340(L) + 0.32667(L^2)$$

where L = plant load in percentage of capacity, $L = X_1$, and L^2 = square of plant load, $L^2 = X_2 = X_1(X_1)$.

We have plotted our residuals from Table 6.4 against our X_1's in Figure 6.4. If we see a random pattern of dispersion around our zero line in a plot of residuals, we typically refrain from adding higher-order terms. Our plot in this case appears basically dispersed, so we stop with the second-order model.

Table 6.4 Computer-aided results—second-order model (Energy Cost/Operations Case).

Dependent Variable: COST   ENERGY COST OF OPERATIONS

Analysis of Variance

Source	DF	Sum of Squares	Mean Square	F Value	Prob > F
Model	2	947.60000	473.80000	5.156	0.0322
Error	9	827.06667	91.89630		
C Total	11	1774.66667			

Root MSE	9.58626	R-Square	0.5340	
Dep Mean	60.66667	Adj R-sq	0.4304	
C.V.	15.80152			

Parameter Estimates

Variable	DF	Parameter Estimate	Standard Error	T for H_0: Parameter=0	Prob > |T|
INTERCEP	1	3418.266667	1160.6174041	2.945	0.0163
L	1	-66.340000	22.69737071	-2.923	0.0170
LSQ	1	0.326667	0.11069254	2.951	0.0162

Obs	Dep Var COST	Predict Value	Residual
1	50.0000	64.1333	-14.1333
2	76.0000	64.1333	11.8667
3	65.0000	64.1333	0.8667
4	49.0000	50.9333	-1.9333
5	45.0000	50.9333	-5.9333
6	63.0000	50.9333	12.0667
7	45.0000	54.0667	-9.0667
8	53.0000	54.0667	-1.0667
9	60.0000	54.0667	5.9333
10	68.0000	73.5333	-5.5333
11	85.0000	73.5333	11.4667
12	69.0000	73.5333	-4.5333

We further develop an ANOVA table with a lack-of-fit test for our second-order model in Table 6.5. We note that the total corrected and pure error rows remain the same as with the first-order model. Here, from the lack-of-fit row,

H_0: The second-order model is adequate

H_1: The second-order model is not adequate

We obtain a P value > 0.10. Hence, we conclude, b.o.e., that the second-order model is adequate.

Both the first-order and second-order predictors are superimposed on a scatter plot of the data in Figure 6.5a. Here, we see that the second-order model appears to be much more attractive in that it "explains" the data better. We can also see the dispersion within each treatment level, relative to the treatment means and

Table 6.5   Regression ANOVA—second-order model (Energy Cost/Operations Case).

Source	df	SS	MS	F	P Value
Total corrected	11	1774.667			
Regression	2	947.600	473.800	5.156	0.032
Residual error	9	827.067	91.896		
Lack of fit	1	13.066	13.066	0.128	$P > 0.10$
Pure error	8	814.001	101.750		

the predictors. The dispersion looks reasonably large; hence, we may not be able to develop as good a model as we had hoped. Our R^2_{max} statistic helps us assess the fit, considering the pure error,

$$R^2_{max} = \frac{SS_{tc} - SS_{perr}}{SS_{tc}}$$

$$= \frac{1774.667 - 814.001}{1774.667} = 0.541$$

Here, our second-order model's $R^2 = 0.534$ looks very good, when compared to the R^2_{max}. Figures 6.5*b* and 6.5*c* show pie charts relative to the SS terms. In these pie charts, we can see the proportion of SS_{tc} explained by our models, as well as the residual error, pure error, and lack-of-fit sum of squares components. In summary, we see reasonable evidence to believe

$$\hat{\mu} = 3418.267 - 66.340(L) + 0.32667(L^2)$$

will perform satisfactorily between 95 and 110% of plant capacity load.

In addition to using residual plots to help us structure our models, we use them for assessing our normality assumption. This assessment is done by the procedure developed in Chapter 5. We have left the residual analysis for the assessment of normality as an exercise.

Our previous statement, in which we limit the range of L to the range of the experimental treatments, is critical. We want to be sure that our predictor models are adequate and that they are used within a relevant range. For example, someone using our model for a load of 107% would obtain

$$\hat{\mu}_{L=107} = 3418.267 - 66.340(107) + 0.32667(107^2)$$
$$= 59.93 \text{ energy units/production unit}$$

Loads below 95% and above 110% were not examined in the experiment and not included in the model-building exercise. We simply cannot be sure that our model will provide meaningful prediction below 95% or above 110% loading.

Our previous response surface case developed only one independent variable. **For multifactor experiments, it is likely that we will want to develop a multivariate response surface.** In the multifactor case, we have a number of design or noise variables. We can use the general normal equations, Equations (6.6) through (6.9), to develop our predictor model. However, the hand solution method is rather challenging. Typically, we use a computer aid, such as SAS, to expedite our calculations and develop a regression ANOVA at the same time. We illustrate the multifactor response surface development process with an extended case.

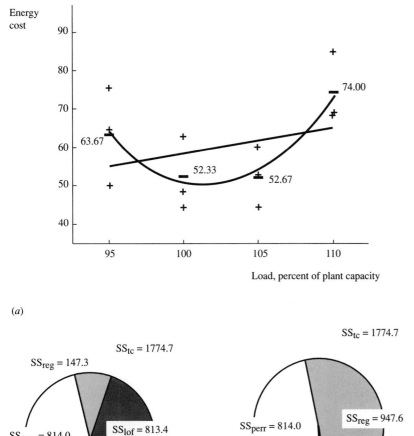

(a)

(b)   (c)

Figure 6.5   First- and second-order models summary (Energy Cost/Operations Case). (a) Scatter plot with first- and second-order models superimposed. (b) Sum of squares for first-order model. (c) Sum of squares for second-order model.

ENERGY COST/OPERATIONS CASE—RESPONSE SURFACE IN TWO FACTORS   Our previous model-fitting exercise focused on a chiller temperature held constant at $-30°F$. However, our experiment from Chapter 5 provided us with data at 12 load × temperature combinations. Now, we use the complete data set (see Table 5.8) and fit a response surface that includes both load and temperature. Our fitting strategy is to fit a full second-order response surface

$$\hat{\mu} = \beta_0 + \beta_1 L + \beta_2 T + \beta_3 LT + \beta_4 L^2 + \beta_5 T^2$$

Example 6.2

Table 6.6　　Computer-aided results—full two-factor model (Energy Cost/Operations Case).

Dependent Variable: COST　　ENERGY COST OF OPERATIONS

Analysis of Variance

Source	DF	Sum of Squares	Mean Square	F Value	Prob > F
Model	5	6625.82381	1325.16476	11.173	0.0001
Error	30	3558.17619	118.60587		
C Total	35	10184.00000			

Root MSE	10.89063	R-Square	0.6506	
Dep Mean	70.33333	Adj R-sq	0.5924	
C.V.	15.48431			

Parameter Estimates

Variable	DF	Parameter Estimate	Standard Error	T for H_0: Parameter=0	Prob > |T|
INTERCEP	1	3438.769048	778.09469253	4.419	0.0001
L	1	-67.112381	14.96031968	-4.486	0.0001
T	1	-2.348413	6.83304845	-0.344	0.7335
LT	1	0.189524	0.05206718	3.640	0.0010
LSQ	1	0.353333	0.07260422	4.867	0.0001
TSQ	1	0.322222	0.07842152	4.109	0.0003

Obs	Dep Var COST	Predict Value	Residual
1	60.0000	67.6881	-7.6881
2	69.0000	67.6881	1.3119
.	.	.	.
.	.	.	.
.	.	.	.
36	83.0000	78.9905	4.0095

and simplify if our t tests on the parameters merit removing and refitting. Our computer-aided solution appears in Table 6.6. Here, we observe that all terms, with the exception of the linear T term, produce significant P values. We also see that the T^2 term is highly significant, $P = 0.0003$. Hence, according to our strategy, detailed earlier in this chapter, we will allow the T term to remain, yielding

$$\hat{\mu} = 3438.77 - 67.11(L) - 2.35(T) + 0.1895(LT) + 0.35333(L^2) + 0.32222(T^2)$$

where L is expressed in plant load capacity percentage (95 to 110%) and T is expressed in chiller degrees Fahrenheit (–20 to –35°F). We leave the lack of fit and residual analyses as exercises at the end of this chapter.

A three-dimensional depiction of the response surface model is shown in Figure 6.6a. An associated contour plot is provided in Figure 6.6b. At this point, the response surface model can be used to estimate mean energy costs for given L and T process inputs, given L varies between 95 and 110% of design load capacity and T varies between –20 and –35°F. The response surface model is also useful in obtaining an L, T combination that provides a minimum cost. Here, we would take partial derivatives, set them equal to zero, and solve for the L^* and T^* that yield a minimum energy cost.

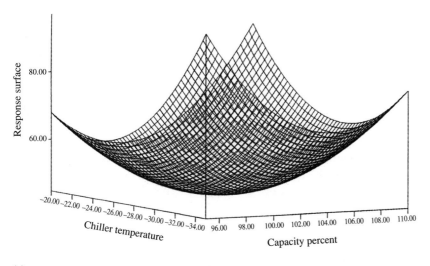

(a)

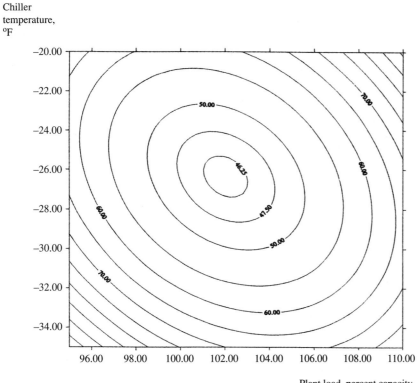

(b)

Figure 6.6 Three-dimensional response surface depiction (Energy Cost/Operations Case). (a) Energy cost response surface. (b) Energy cost contour plot.

6.4   RESPONSE SURFACE DESIGNS

Many process response optimization strategies exist. The objective is to find the optimum response and its associated factor levels as quickly as possible at minimal expense. One strategy is to first develop a relatively broad-based initial experiment that includes the variables we think are driving the response. Next, we run the experiment and fit a response surface from the results. We then use the response surface model to guide us in developing our second experiment, thus moving in closer to the optimal response. This systematic practice tends to work very well when we are dealing with quantitative factors. The usual alternative is a more or less haphazard trial-and-error search over our feasible region or study grid.

Our sequence of experiments may include second-order response surfaces if a first-order response surface is judged to be inadequate. Nevertheless, our objective remains to close in on our best operating setting as quickly and economically as possible. We will briefly discuss and demonstrate two designs that are useful in response surface work: (1) a 2^f FAT-CRD with a center point and (2) a central composite design.

2^f WITH A CENTER POINT

The 2^f design can be used to develop first-order response surfaces. **Using a 2^f design as our base, we can expand our plan to include a center point setting, which we arbitrarily call the "zero" point or "origin,"** for convenience. The center point is, by definition, our best guess at an optimal process setting before the experiment. All other settings can be considered as step-out points to be investigated. Hence, **we can define the step-out as the planned expansion distance from the center point, in each direction, relative to each factor.**

We take one observation at each step-out point and take $n_c \geq 2$ observations at the origin. Using this design we can assess nonlinearity but generally cannot develop a full second-order regression model. For the 2^f with a center point with no replications on the corners or step-outs, we develop the pure error out of the replicated center point observations:

$$\text{SS}_{\text{perr}} = \text{SS}_{\text{err}} = \sum_{i=1}^{n_c} (Y_{ci} - \bar{Y}_c)^2 \qquad \textbf{[6.27]}$$

and

$$\text{df}_{\text{perr}} = \text{df}_{\text{err}} = n_c - 1 \qquad \textbf{[6.28]}$$

Hence

$$\text{MS}_{\text{perr}} = \text{MS}_{\text{err}} = \frac{\text{SS}_{\text{err}}}{\text{df}_{\text{err}}} \qquad \textbf{[6.29]}$$

where $Y_{ci} = i$th center point response

$\bar{Y}_c$ = mean of the n_c center point responses

n_c = number of observations at the design center point

A single degree of freedom curvature row can be added to our ANOVA; see Montgomery [7]. We develop the curvature row sum of squares as

$$SS_{curve} = \frac{n_f n_c (\bar{Y}_f - \bar{Y}_c)^2}{n_f + n_c}$$ [6.30]

with

$$df_{curve} = 1$$ [6.31]

where $\bar{Y}_f$ = average of factorial (corner or step-out) responses and n_f = total number of factorial (corner) observations.

The center point addition to a 2^f design allows us to assess the linearity-nonlinearity characteristics in a two-level design. Otherwise, we require three or more levels of the factors, e.g., a 3^f or possibly a central composite design. It also allows us the option of obtaining a crude pure error analysis, to assess model adequacy, by adding replicates only at the center of the design space.

PACKAGING PROCESS CASE—2^f DESIGN WITH A CENTER   Packaging in food products is a | **Example 6.3** critical process. Robust, airtight seals are required to preserve product "freshness" and shelf life. The sealing subprocess in packaging is critical. One of the leverage points in this subprocess is seal strength. When wrapping materials are joined together, two important leverage variables are sealing temperature and sealing time.

In a process definition effort, see Section 3, requirements for a sealing subprocess for a cookie product were needed. The physical concept for the sealing subprocess is illustrated in Figure 6.7.

Experimental purpose.   Determine "optimal" seal subprocess requirements in order to provide maximum tensile strength in the seal.

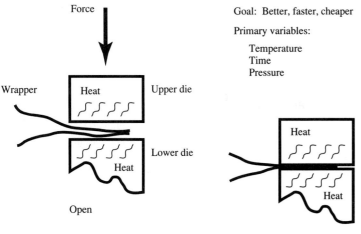

Figure 6.7   Physical sealing subprocess depiction (Packaging Process Case).

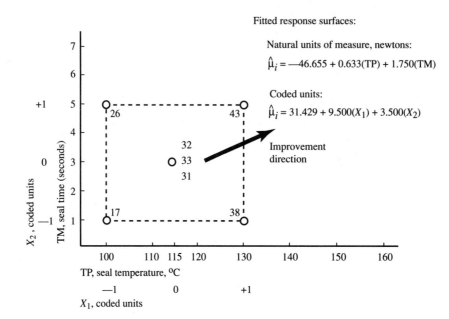

Figure 6.8   Physical boundaries and response data, 2^2 FAT-CRD with a center point design (Packaging Process Case).

Factors and response.   Die temperature (TP) and clamping time (TM) are identified as two design variables. Clamping force is also a critical variable, but it is determined by preset spring tension and is therefore fixed in this experiment. The response selected for this experiment is tensile strength of the seal, as determined by pulling the seal apart.

Design strategy.   Develop a 2^2 with a center CRD experiment and fit a linear response surface. From these results design a follow-on central composite design experiment and build a second-order response surface. Determine "optimum" temperature and time requirements from the results.

Physical experimentation.   Sample coupons, the eu's, are cut from wrapping materials and sealed at the TP, TM combinations, along one edge. Then the loose edges are clamped in a tensile puller and pulled until fracture. The fracture force is recorded in newtons.

Figure 6.8 provides an overview of the 2^f with a center CRD. The factor levels explored, the response data obtained, and the models fit are summarized. Table 6.7 provides an ANOVA table for the data. Here, we see no significant curvature present, P value > 0.10. We see a significant temperature effect, P value < 0.01, and a significant time effect, $0.01 < P$ value < 0.05. Our computer-aided regression results for the natural units of measure appear in Table 6.8*a*.

The coded units as marked in Figure 6.8 are obtained in terms of the natural units

$$X_1 = \frac{TP - 115}{15}$$

$$X_2 = \frac{TM - 3}{2}$$

Table 6.7   ANOVA table—2^2 CRD with a center point (Packaging Process Case).

Source	df	SS	MS	F	P Value
Total corrected	$7 - 1 = 6$	417.71			
TP	$2 - 1 = 1$	361.00	361.00	361.00	$P < 0.01$
TM	$2 - 1 = 1$	49.00	49.00	49.00	$0.01 < P < 0.025$
TP*TM	$(2 - 1)(2 - 1) = 1$	4.00	4.00	4.00	$P > 0.10$
Curvature	1	1.71	1.71	1.71	$P > 0.10$
Experimental error	$3 - 1 = 2$	2.00	1.00		

Sample calculations

$SS_{tc} = (26^2 + 17^2 + 38^2 + 43^2 + 32^2 + 33^2 + 31^2) - (220^2/7) = 417.71$

$*SS_{TP} = [(26 + 17)^2 + (43 + 38)^2]/2 - (124^2/4) = 361.00$

$*SS_{TM} = [(17 + 38)^2 + (26 + 43)^2]/2 - (124^2/4) = 49.00$

$*SS_{TP*TM} = (26^2 + 17^2 + 38^2 + 43^2)/1 - (124^2/4) - 361.00 - 49.00 = 4.00$

$\bar{Y}_f = (26 + 17 + 38 + 43)/4 = 31.00$

$\bar{Y}_c = (32 + 33 + 31)/3 = 32.00$

$SS_{cur} = 4(3)(31 - 32)^2/(4 + 3) = 1.71$

$SS_{err} = (32 - 32)^2 + (33 - 32)^2 + (31 - 32)^2 = 2.00$

*Based only on the four corner responses.

The computer-aided solution in coded units appears in Table 6.8*b*. In both cases, the temperature and time terms are justified in the response surface models, as shown in Figure 6.8. As indicated by the arrow in Figure 6.8, according to the response surface coefficients, both positive, it appears that our optimal operating area is up and to the right of our first experimental factor level boundaries. Hence, for our follow-on experiment we will expand our coverage in the arrow direction.

CENTRAL COMPOSITE DESIGN

FAT designs with at least three factor levels, e.g., 3^f, allow us to develop second-order response surface models. But these models have two primary disadvantages:

1. They are usually inefficient in that they require a great deal of experimentation, especially if we have a relatively large number of factors and want to develop a pure error analysis.

2. They are not rotatable. When the variance of a predicted response $\hat{\mu}$ at some selected set of X's is a function of only the distance from the center of the design space, and is not a function of direction from the center, the form is said to be "rotatable."

The central composite designs are widely used for second-order response surface modeling, both because of their statistical properties (they are rotatable) and because of the practical appeal of their expanded coverage around a center point. Figure 6.9 depicts both a two-factor and a three-factor central composite design layout. By observation, we can see that both layouts can be built up from the 2^f design or the 2^f with a center point design. In other words, we might perform a 2^f with a center point and then add the axial points, if necessary, after an

Table 6.8   Computer-aided results—first-order models (Packaging Process Case).

a. Natural Unit of Measure Regression Results

Dependent Variable: STREN   TENSILE STRENGTH OF SEAL

Analysis of Variance

Source	DF	Sum of Squares	Mean Square	F Value	Prob>F
Model	2	410.00000	205.00000	106.296	0.0003
Error	4	7.71429	1.92857		
C Total	6	417.71429			

Root MSE	1.38873	R-Square	0.9815	
Dep Mean	31.42857	Adj R-sq	0.9723	
C.V.	4.41869			

Parameter Estimates

Variable	DF	Parameter Estimate	Standard Error	T for H_0: Parameter=0	Prob > |T|
INTERCEP	1	−46.654762	5.44973553	−8.561	0.0010
TP	1	0.633333	0.04629100	13.682	0.0002
TM	1	1.750000	0.34718254	5.041	0.0073

b. Coded Unit of Measure Regression Results

Dependent Variable: STREN   TENSILE STRENGTH OF SEAL

Analysis of Variance

Source	DF	Sum of Squares	Mean Square	F Value	Prob>F
Model	2	410.00000	205.00000	106.296	0.0003
Error	4	7.71429	1.92857		
C Total	6	417.71429			

Root MSE	1.38873	R-Square	0.9815	
Dep Mean	31.42857	Adj R-sq	0.9723	
C.V.	4.41869			

Parameter Estimates

Variable	DF	Parameter Estimate	Standard Error	T for H_0: Parameter=0	Prob > |T|
INTERCEP	1	31.428571	0.52489066	59.876	0.0001
X1	1	9.500000	0.69436507	13.682	0.0002
X2	1	3.500000	0.69436507	5.041	0.0073

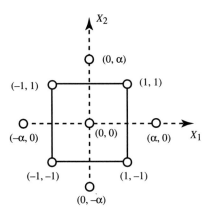

(a)

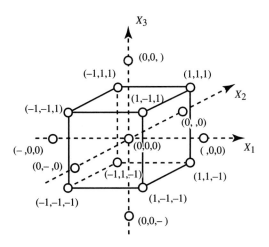

(b)

Figure 6.9   Generic (a) two-factor and (b) three-factor central composite design layouts.

initial analysis. We also see broad coverage of the design space surrounding the center point and extending beyond the 2^f borders on the axes. The α symbol here represents a distance and has no relationship to a type I error probability. Here, we use the α symbol only because it is commonly encountered in the central composite design literature.

Different classes of the central composite design exist. Here, we discuss only the uniform-precision and the orthogonal central composite designs; both are rotatable and allow a pure error analysis. The reader is referred to Cochran and Cox [8] and Box and Draper [3] for further discussions. Tables 6.9 and 6.10 provide the structural requirements for the uniform-precision and the orthogonal central composite designs, respectively.

Table 6.9   Structural requirements for the uniform-precision central composite design.

	Full Factorials				
Number of factors	2	3	4	5	6
Number of vertex points*	4	8	16	32	64
Number of axial points*	4	6	8	10	12
Number of center observations	5	6	7	10	15
Total observations required	13	20	31	52	91
α value	1.414	1.682	2.000	2.378	2.828

*Only one observation is taken at each vertex point and at each axial point.
SOURCE: Adapted, with permission, from G. E. P. Box and J. S. Hunter, "Multifactor Experimental Designs for Exploring Response Surfaces," The Annals of Mathematical Statistics, vol. 28, p. 227, 1957.

Table 6.10   Structural requirements for the orthogonal central composite design.

	Full Factorials				
Number of factors	2	3	4	5	6
Number of vertex points*	4	8	16	32	64
Number of axial points*	4	6	8	10	12
Number of center observations	8	9	12	17	24
Total observations required	16	23	36	59	100
α value	1.414	1.682	2.000	2.378	2.828

*Only one observation is taken at each vertex point and at each axial point.
SOURCE: Adapted, with permission, from G. E. P. Box and J. S. Hunter, "Multifactor Experimental Designs for Exploring Response Surfaces," *The Annals of Mathematical Statistics*, vol. 28, p. 227, 1957.

Example 6.4 | **PACKAGING PROCESS CASE—FOLLOW-ON EXPERIMENT AND MODEL**   On the basis of the results of the linear response surfaces developed in Example 6.3, a follow-on experiment to generate a second-order response surface was designed and executed. Our arrow in Figure 6.8 clearly indicates a shift in our experimental grid up and to the right, i.e., higher temperature and higher time. We set our new experimental grid up in temperature and time with a new center at TP = 140°C and TM = 4 seconds. We use judgment to arrive at this setting.

Here, in the physical process, we desire to reduce our cycle time for productivity enhancement. We believe, on the basis of physical knowledge, we can trade temperature for time, at least up to some point. But we need a model to support our objective. Our follow-on experimental grid is depicted in Figure 6.10.

A central composite design of the uniform-precision class is selected. This design is built to the two-factor design specifications in Table 6.9. We make 13 observations in all. We see the four corners, or vertex points, and the four axial points in Figure 6.10. We use the $\alpha = 1.414$ value from Table 6.9 to construct the axial points. Our 13 experimental coordinates and responses are summarized as follows:

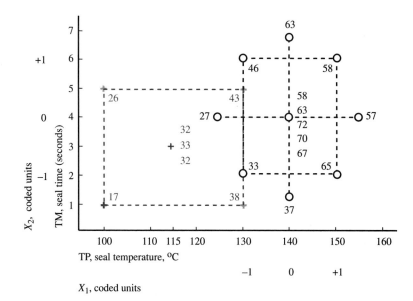

Figure 6.10 Physical boundaries and response data, two-factor uniform precision central composite design (Packaging Process Case).

TP	TM	X_1	X_2	Observed Response
130	2	−1	−1	33
130	6	−1	1	46
150	2	1	−1	65
150	6	1	1	58
140	1.17	0	−1.414	37
140	6.83	0	1.414	63
125.9	4	−1.414	0	27
154.1	4	1.414	0	57
140	4	0	0	58
140	4	0	0	63
140	4	0	0	72
140	4	0	0	70
140	4	0	0	67

Here,

$$X_1 = \frac{TP - 140}{10}$$

$$X_2 = \frac{TM - 4}{2}$$

Table 6.11   Computer-aided results—central composite design (Packaging Process Case).

---

a. Full Second-Order Model in Natural Units of Measure

Dependent Variable: STREN   TENSILE STRENGTH OF SEAL

Analysis of Variance

Source	DF	Sum of Squares	Mean Square	F Value	Prob>F
Model	5	2294.86890	458.97378	11.231	0.0031
Error	7	286.05417	40.86488		
C Total	12	2580.92308			

Root MSE	6.39256	R-Square	0.8892
Dep Mean	55.07692	Adj R-sq	0.8100
C.V.	11.60661		

Parameter Estimates

Variable	DF	Parameter Estimate	Standard Error	T for H_0: Parameter=0	Prob > |T|
INTERCEP	1	-2403.510447	486.20820051	-4.943	0.0017
TP	1	32.652325	6.85064156	4.766	0.0020
TM	1	51.379836	22.91986381	2.242	0.0599
TPTM	1	-0.250000	0.15981411	-1.564	0.1617
TPSQ	1	-0.109180	0.02434641	-4.484	0.0029
TMSQ	1	-1.713368	0.60530253	-2.831	0.0254

(continued)

We use our computer aid to develop our full second-order model. Our full fitted response surface in the observed units for TP (°C) and TM (seconds) is

$$\hat{\mu} = -2403.510 + 32.652(TP) + 51.380(TM) - 0.2500(TP*TM) - 0.10918(TP^2) - 1.71337(TM^2)$$

Results are displayed in Table 6.11. The results in Table 6.11*a* indicate that we can remove the TP × TM cross-product term, *P* value = 0.162. After removal, our results are as shown in Table 6.11*b*. Our final, reduced response surface is

$$\hat{\mu} = -2263.510 + 31.652(TP) + 16.380(TM) - 0.10918(TP^2) - 1.71337(TM^2)$$

A three-dimensional depiction of the strength response surface is provided in Figure 6.11*a*. Likewise, a contour plot is provided in Figure 6.11*b*. Our model is generally applicable over the ranges of 126 ≤ TP ≤ 154°C and 1.2 ≤ TM ≤ 6.8 seconds (the range included in our experiment).

In terms of our coded model, from our computer-aided analysis, our reduced fitted response surface in the X_1 and X_2 units is

$$\hat{\mu}_x = 65.999 + 10.804(X_1) + 5.346(X_2) - 10.8765(X_1^2) - 6.8753(X_2^2)$$

Our model here, in the coded scale, is applicable over the ranges of $-1.414 \leq X_1 \leq 1.414$ and $-1.414 \leq X_2 \leq 1.414$.

Using our previous reduced response surface model in the natural units of measure,

$$\hat{\mu} = -2263.510 + 31.652(TP) + 16.380(TM) - 0.10918(TP^2) - 1.71337(TM^2)$$

Table 6.11 (continued)   Computer-aided results—central composite design (Packaging Process Case).

b. Reduced Second-Order Model in Natural Units of Measure

Dependent Variable: STREN   TENSILE STRENGTH OF SEAL

Analysis of Variance

Source	DF	Sum of Squares	Mean Square	F Value	Prob>F
Model	4	2194.86890	548.71723	11.371	0.0022
Error	8	386.05417	48.25677		
C Total	12	2580.92308			

Root MSE	6.94671	R-Square	0.8504	
Dep Mean	55.07692	Adj R-sq	0.7756	
C.V.	12.61274			

Parameter Estimates

Variable	DF	Parameter Estimate	Standard Error	T for H₀: Parameter=0	Prob > |T|
INTERCEP	1	-2263.510447	519.32771689	-4.359	0.0024
TP	1	31.652325	7.41201327	4.270	0.0027
TM	1	16.379836	5.40350108	3.031	0.0163
TPSQ	1	-0.109180	0.02645690	-4.127	0.0033
TMSQ	1	-1.713368	0.65777369	-2.605	0.0314

Obs	Dep Var STREN	Predict Value	Residual
1	27.0000	29.0338	-2.0338
2	33.0000	32.0586	0.9414
3	46.0000	42.7502	3.2498
4	37.0000	44.7091	-7.7091
5	58.0000	65.9956	-7.9956
6	63.0000	65.9956	-2.9956
7	72.0000	65.9956	6.0044
8	70.0000	65.9956	4.0044
9	67.0000	65.9956	1.0044
10	63.0000	59.8376	3.1624
11	65.0000	53.6980	11.3020
12	58.0000	64.3896	-6.3896
13	57.0000	59.5453	-2.5453

we now solve for our optimal response targets. Taking partial derivatives and setting the resulting equations equal to zero gives

$$\frac{\partial f}{\partial TP} = 31.652 - 2(0.10918)TP = 0$$

$$TP^* = 144.953 \text{ °C}$$

$$\frac{\partial f}{\partial TM} = 16.380 - 2(1.71337)TM = 0$$

$$TM^* = 4.780 \text{ seconds}$$

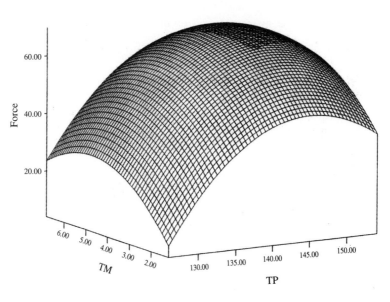

(a)

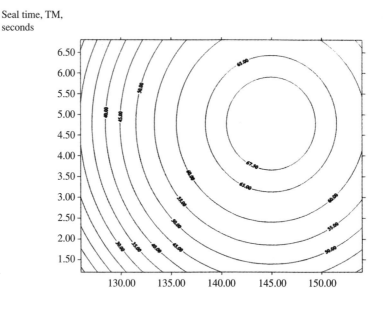

(b)

Figure 6.11 Three-dimensional response surface depiction (Packaging Process Case—Central Composite Design). (a) Seal strength response surface. (b) Seal strength contour plot.

Our estimated maximum strength occurs at TP* = 144.953 °C and TM* = 4.780 seconds. Substituting back into our response surface model, we find our predicted maximum strength is

$$\hat{\mu} = -2263.510 + 31.652(144.953) + 16.380(4.780) - 0.10918(144.953^2) - 1.71337(4.780^2)$$
$$= 69.67 \text{ newtons}$$

At this point, our experimental work is complete. We now use this information to determine the TP, TM combination best suited for our subprocess. Here, we might explore the possibility of reducing TM in order to reduce cycle time. Using our response surface, we can readily follow the surface and perhaps find a suitable TP, TM combination. This quest is left as an end-of-chapter exercise.

The literature is divided on whether to center and code our independent variables or to fit our response surfaces in the natural units of measure. In general, we prefer the units of measure approach. Typically, communication with our customers is enhanced when we can express our results in the units of measure found in the field. In practice, we see both methods in wide use. We must recognize that, given the nature of the calculations involved, our solutions do not match exactly—our predicted values will not totally agree. Draper and Smith [2] and Neter, Wasserman, and Kutner [4] provide more thorough discussions of the data centering/natural unit issue.

REVIEW AND DISCOVERY EXERCISES

REVIEW

6.1. Review the Energy Cost/Operations Case—Response Surface in Two Factors (Example 6.2). Develop a complete ANOVA table and analysis, including lack of fit. Develop an appropriate residuals analysis for the final model. Interpret your analyses.

6.2. Determine the optimal L* and T* values for the Energy Cost/Operations Case—Response Surface in Two Factors (Example 6.2). Is the optimal value reasonable? How does it compare with performance estimates at the current operational values of −35ºF and 110% capacity, as described in Chapter 5?

6.3. Compare and contrast the one- and two-factor energy cost models for the Energy Cost/Operations Case (Examples 6.1 and 6.2). What does the one-factor model tell us? What was gained by the two-factor model? Explain how each model could help us to operate our plant. How much trust should we place in models such as these?

6.4. Review the Packaging Process Cases (Examples 6.3 and 6.4) with respect to the first- and second-order models. Develop complete ANOVA tables and analyses, including lack of fit. Develop an appropriate residuals analysis for the final models. Interpret your analyses.

6.5. Review the Packaging Process Case (Example 6.4). Develop a complete response surface modeling effort in coded units. Determine the optimal operating points in coded units and compare them with their counterparts in natural units. Comment on your results.

6.6. For one or more of the following cases, described in Section 7, develop a thorough response surface analysis using the experimental data provided. Include both graphical as well as inferential analyses. Use an α = 0.05 for inferences.

Case VII.3: Apple Dehydration Exploration

Case VII.5: Big City Waterworks

Case VII.16: Link-Lock Chain

Case VII.19: Night Hauler Trucking

Case VII.30: Tough-Skin—Sheet Metal Welding

DISCOVERY

6.7.   For one or more of the following cases, described in Section 7, develop a thorough response surface analysis using the experimental data provided. Include both graphical as well as inferential analyses. Use an $\alpha = 0.05$ for inferences.

Case VII.2: Apple Core—Dehydration

Case VII.8: Door-to-Door—Pizza Delivery

Case VII.17: LNG—Natural Gas Liquefaction

Case VII.22: Re-Use—Recycling

Case VII.28: Sure-Stick Adhesive

6.8.   Given the reduced model in natural units, from Example 6.4, develop a pragmatic method by which we might study a temperature for time trade-off analysis. Demonstrate your methodology and assess the impact of such a trade-off.

REFERENCES

1.   R. E. Walpole, R. H. Myers, and S. L. Myers, *Probability and Statistics for Engineers and Scientists*, 6th ed. Upper Saddle River, NJ: Prentice Hall, 1998.

2.   N. R. Draper and H. Smith, *Applied Regression Analysis*, 2nd ed., New York, Wiley, 1981.

3.   G. E. P. Box and N. R. Draper, *Empirical Model-Building and Response Surfaces*, New York: Wiley, 1987.

4.   J. Neter, W. Wasserman, and M. H. Kutner, *Applied Linear Statistical Models*, 3rd ed., Homewood, IL: Irwin, 1990.

5.   *SAS Procedures Guide*, Version 6, Cary, NC: SAS Institute, 1990.

6.   *SAS/STAT User's Guide*, Volumes 1 and 2, Version 6, Cary, NC: SAS Institute, 1990.

7.   D. C. Montgomery, *Design and Analysis of Experiments*, 4th ed., New York: Wiley, 1997.

8.   W. G. Cochran and G. M. Cox, *Experimental Designs*, 2nd ed., New York: Wiley, 1957.

3

PROCESS DEFINITION AND REDEFINITION

The purpose of Section 3 is to identify and describe the elements of process definition and redefinition in the context of a production system.

Chapter 7: Process Definition/Redefinition—Output Perspectives

The purpose of Chapter 7 is to overview the elements of process definition/redefinition and to describe the nature of process definition/redefinition with regard to process results.

Chapter 8: Process Definition/Redefinition—Transformation and Input Perspectives

The purpose of Chapter 8 is to describe the nature of process definition/redefinition with regard to process means and creation.

chapter
7

PROCESS DEFINITION/REDEFINITION–
OUTPUT PERSPECTIVES

7.0   INQUIRY

1.   What constitutes process definition/redefinition?
2.   When should a process be redefined?
3.   Why does process definition/redefinition start with results?
4.   How are processes systematically defined/redefined?

7.1   INTRODUCTION

Previously, we noted that processes are fundamental to everything we do—creating, producing, using, and recycling products of any kind. **Process definition and redefinition in general have occurred for as long as we have used processes—since the beginning.** The manner in which process definition or redefinition is thought out and implemented is the central issue here.

Process definition and redefinition are strategic. They occur infrequently, but **their impact is felt throughout the entire organization.** Impacts usually affect the entire production system or at least a major part of it. For example, a redefinition of a process typically impacts its own subprocesses as well as other interacting processes. Redefinition of a subprocess likewise impacts its mother process and other interacting processes. It is not unusual to see significant impacts flowing up to the production system level and affecting external customers in one way or another: physically in function, form, or fit; economically; timelinesswise; and/or customer service–wise.

A generic set of process definition/redefinition elements and subelements is depicted in Figure 7.1. Each element represents a critical part of process definition/redefinition. **Without a full set of process definition/redefinition elements, we take on additional risk of an incomplete address.** That is, **we assume more risk of customer dissatisfaction and business**

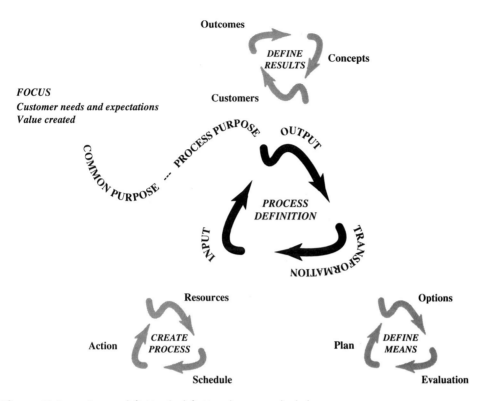

Figure 7.1 Process definition/redefinition elements and subelements.

failure than necessary. In other words, the elements and subelements serve as a process definition/redefinition global positioning device; no matter where we are, we can find our position and establish/verify our direction, relative to our purpose.

The most pronounced characteristic of process definition is its customer-centered and value-created nature. Our elements are driven by competition from both outside and inside our organization, in reactive and proactive fashion. General guidance is provided by common purpose—vision, mission, and core values. Specific guidance is supplied by explicit process purpose.

There are many possible ways to address each process definition/redefinition element and subelement. The manner in which we address these elements and subelements is a direct result of our leadership and creative abilities and behaviors. Typically, we rely on initiatives and tools. **A number of initiatives and tools are available to help generate, express, organize, and communicate process definition/redefinition.** Selected initiatives and tools are listed in Table 7.1, along with a general classification of usefulness. An overview of each initiative and tool is provided in Section 6, Chapters 19 and 20. Several initiatives and tools are used to describe and develop illustrative cases in this section. The critical issue in dealing with initiatives and tools is to remember that the process purpose comes first. **Tools and initiatives exist only to make our pursuit of process purpose more effective and efficient—any pursuit of an initiative or tool for its own sake is misguided.**

Table 7.1   Selected initiatives and tools useful in process definition/redefinition, control, and improvement.*

	Definition/ Redefinition	Control	Improvement	Comments
Initiatives†				
Benchmarking	+++++	+++++	+++++	Helps assess what others are doing
Concurrent engineering	+++++	+	+++	Assures coordination between design and production
Continuous improvement	+++	++	+++++	Addresses worker involvement in improvement
Cycle time reduction	+++	+	+++++	Addresses timeliness of processes
Fifth discipline	+++++	+	+++	Provides guides for organization development
Function value analysis (VA)	++++	++	++++	Focus is usually on product value
ISO 9000	+++++	+++++	+++++	Provides a strategic framework for quality system
Mistakeproofing (poke-yoke)	+++++	+++++	+++++	Seeks to prevent mistakes
Quality awards	+++++	+++++	+++++	A results-based criteria for organization assessment
Quality function deployment (QFD)	+++++	+	++++	Assures customer voice is heard, aids communication/coordination
Re-engineering	+++++	+	++	Stresses totally new process concepts
Robust design	+++++	+++++	+++++	Addresses variance reduction
Six sigma	++++	+++	+++++	Addresses systematic process improvement
Theory of constraints (TOC)	+++++	++	+++++	Systems approach to problem solving
Total quality management (TQM)	+++++	+++++	+++++	Addresses customer focus and participative work environment
Tools†				
Activity/sequence list	+++++	++	+++++	Helpful in implementation scheduling
Break-even analysis	+++++	+	+++	Useful in early estimates of production volume
Capability analysis	++	+++++	++++	Useful in comparing specs to actual production
Cash flow analysis	+++++	+++++	+++++	Appropriate for economic assessment
Cause-effect diagram	+++++	+++++	+++++	Useful in discovering/communicating C-E relationships
Check sheet	+	+++++	+++++	Helpful in early stages of quantification
Control chart	+	+++++	++++	Critical in most control monitoring applications
Correlation/autocorrelation analysis	+++	+++++	+++	Helpful in assessing quantitative relationships
Critical path method (CPM)	+++++	+	++++	Helpful in scheduling and monitoring implementation
Experimental design	+++++	+++++	+++++	Critical in empirical discovery and optimization
Failure mode effects analysis	+++++	+++++	+++++	Critical to discover failure effects/countermeasures
Fault tree analysis	++++	++++	++++	Useful for detailed event/cause analysis
Flowchart	+++++	+++++	+++++	Critical in process depiction and communication
Force field analysis	+++++	+	+++++	Critical in assessing/selling implementation
Gantt chart	+++++	+	+++++	Helpful in implementation scheduling/coordination
Histogram	++	+++++	++++	Helpful in assessing location/dispersion in data
Matrix diagram	+++++	+	++++	Helpful in depicting/communicating relationships
Pareto diagram	+++	+++++	+++++	Helpful in ranking/communicating criticality of losses
Process value chain analysis	+++++	+++++	+++++	Depicts linkage between business and technical cause-effect
Relations diagram	+++++	+++	+++++	Helpful in organizing/communicating basic facts in C-E
Root cause analysis	+++++	+++++	+++++	Helpful in establishing cause-effect relationship
Runs chart	+++	+++++	+++++	Helpful in depicting/communicating results/targets
Scatter diagram	+++	+++++	+++++	Useful in assessing quantitative relationships
Stratification analysis	++++	+++++	+++++	Critical in data analysis

*Range   +++++: very useful to +: may be useful.
†See Chapters 19 and 20 for descriptions.

7.2 TIMING, PERSONNEL, AND EXPOSURE

The need for process definition is obvious; we simply cannot function without processes. The need for redefinition is not always obvious. Process effectiveness and efficiency are critical in the determination of competitive edge; see Chapter 1 for a review. Competitive forces set dynamic benchmarks in both process effectiveness and efficiency, with short-term and long-term organizational prosperity eventually determined, e.g., market leadership, profits, and so on.

We observe several basic philosophies shaping redefinition as to both initiation and scope. Some managers/leaders wait until a competitive disadvantage is obvious in their business metrics to react, e.g., losses are encountered. Others react on negative trends in their business metrics, e.g., a reduction in profits. Others are proactive and base their action on prospects for further effectiveness and efficiency. In the proactive mode, market and technical leverage are pursued through creative process redefinition even though competitive edge is still obvious.

A spectrum of philosophies ranging from problem-related reaction to opportunity-related proaction push and pull us, respectively, toward process redefinition. The critical point is to assess our position and process redefinition options in a systematic manner, and then make timely decisions. The process improvement elements described in Chapter 16 revisit this issue.

Anything accomplished, with regard to process definition or redefinition, is accomplished by people. Here, leadership and creativity determine both the ultimate effectiveness and efficiency of the process.

In order to build the odds of success in our favor in new process endeavors, we generally limit our scope of exposure until we learn about the new process. Hence, we may choose a limited-scale implementation, e.g., one test site. The results that we seek in terms of product-related benefits for our customers may make the site obvious. Or we may have a large organization and seek to use one division or production unit as a test site for a new or redefined process.

The people we choose to define or redefine the process are critical. We know from the principle of participative management that involvement is critical to success.

> People are more likely to modify their behavior when they participate in problem analysis and solution and more likely to carry out decisions they help make (rather than decisions made by someone else).

There are no hard and fast rules governing team selection, but several guidelines apply. **We choose people that qualify in at least one of three respects: (1) people affected by the process** definition or redefinition, **(2) people possessing knowledge and experience** relative to the process definition or redefinition, and **(3) people who will be involved with implementing and operating** the defined or redefined process.

Once we select our people, we move to prepare and empower them for the process definition or redefinition endeavor. We consider personal abilities, knowledge, skills, the expected scope of the definition or redefinition, time demands, and resources required, e.g., meeting facilities, travel, and research materials. Respect for informal time contracts is essential—considering all or at least most personnel have other responsibilities in addition to this new one. Specific topics involving definition and redefinition channels in an organizational structure, as well as channels for process improvement, are discussed in detail in Chapter 18.

7.3   CRITICAL ELEMENTS

As depicted in Figure 3.1, direction supersedes our process definition. **Everything we do in process definition/redefinition is guided by our common purpose.** Our critical production system elements and subelements, Figure 7.1, are expansions of the output, transformation, and input triad.

A clear, razor-sharp process purpose—communicated and coordinated through teamwork—is absolutely necessary. We focus on results. **Results are the logical starting point—** only after we can clearly identify the results sought relative to customers and the production organization itself can we clearly address transformation and means of producing results. Finally, we consider necessary inputs, whereby we actually create a living process through our actions.

Our production system elements and subelements essentially tailor and expand Stages 4, 5, 6, and 7 of the general systems theory methodology presented in Chapter 2; see Figure 2.1. Our expansion provides a higher level of detail, so that high-performing production systems can be brought on line faster and with fewer resources than otherwise possible.

7.4   PRODUCTION SYSTEM LEVEL RESULTS DEFINITION

Our goal in process definition/redefinition is to address and fulfill process purpose in an effective and efficient manner, or at least in a more effective and efficient manner than our competitors. There is no guarantee that either will improve with process redefinition. It is up to us, using leadership and creativity to reduce the risk of failure.

Production systems are unique. Uniqueness applies to both product nature and features, as well as process nature and technologies. For example, products of a commodity nature differ significantly from custom products. Products/processes involving hazardous materials differ significantly from those with more benign materials. And so it goes—**each specific process demands tailoring within our generic definition/redefinition elements.** All process definition/redefinition elements apply, but each is approached in a manner compatible with the physical nature of the product and the processes involved in defining, designing, developing, producing, distributing, marketing, selling, servicing, using, and disposing/recycling the product and byproducts.

Ulrich and Eppinger [1] summarize several attributes associated with five product definition and design efforts. Their summary appears in Table 7.2. We observe that each product has a unique set of attributes. In each case, a unique level of resources and time was expended. In all cases, sequences of processes were required to define, design, develop, and produce the products. While the general nature of these processes was similar, the extent of each varied widely in personal involvement, development time, development cost, and production investment.

While Ulrich and Eppinger focused on products, we feature underlying processes in our discussion. We offer the process definition/redefinition elements and subelements of Figure 7.1 in order to help establish the basic nature of our production system and its constituent processes. **Our process definition/redefinition elements and subelements are applicable regardless of the product or extent of the enterprise.**

External customers use processes to unlock and harvest benefits from their products, just as we use processes to define, design, develop, produce, deliver, market, and sell products. If we understand customer usage processes, then we understand more about our product relative to field

Table 7.2   Product development efforts overview.

	Stanley Tools Jobmaster Screwdriver	Rollerblade Bravoblade In-Line Skates	Hewlett-Packard DeskJet 500 Printer	Chrysler Concorde Automobile	Boeing 777 Airplane
Annual production volume	100,000 units/year	100,000 units/year	1.5 million units/year	250,000 units/year	50 units/year
Sales lifetime	40 years	3 years	3 years	6 years	30 years
Sales price	$3	$200	$365	$19,000	$130 million
Number of unique parts (part numbers)	3 parts	35 parts	200 parts	10,000 parts	130,000 parts
Development time	1 year	2 years	1.5 years	3.5 years	4.5 years
Internal development team (peak size)	3 people	5 people	100 people	850 people	6800 people
External development team (peak size)	3 people	10 people	100 people	1400 people	10,000 people
Development cost	$150,000	$750,000	$50 million	$1 billion	$3 billion
Production investment	$150,000	$1 million	$25 million	$600 million	$3 billion

SOURCE: Reprinted from K. T. Ulrich and S. Eppinger, *Product Design and Development,* New York: McGraw-Hill, 1995, p. 6, with permission.

environments and applications. We help our customers improve their usage processes in order to extract higher levels of benefits while experiencing lower levels of burdens. To this end, we make product modifications, write user's manuals, train users, supply technical support, and so on, all in order to influence our customers' usage processes.

To illustrate the comprehensive nature of process involvement in our production systems, we have developed an extended case sequence for a snack food product. In this case, we encounter a relatively simple product, a cookie, with relatively low definition, design, and development costs. Our production and delivery processes are characterized by high volume and product-process integrity in terms of a perishable product where cleanliness and timeliness are major concerns. Sales and customer service processes are reasonably complex in terms of extensive wholesale and retail marketing channels. Use and disposal processes are relatively simple. Business integration processes are straightforward. The competitive environment is brutal, with competitors ranging from limited production bakeries to national, high-volume integrated food conglomerates.

A summarized process overview, tailored for a bakery production system, is depicted in Figure 7.2. Here, we distinguish between primarily modular processes and primarily integrated processes. **A modular process is characterized by self-contained operations.** If clear input/output requirements are established, we can define a modular process more or less independently of other processes. **An integrated process is characterized by a cross-functional nature—service to many other processes.** Integrated processes typically contain multiple customers, usually widely dispersed, with varying needs. Hence, input/output requirements are usually many and varied, i.e., tailored to individual customers either within, or outside of, the production system. The process

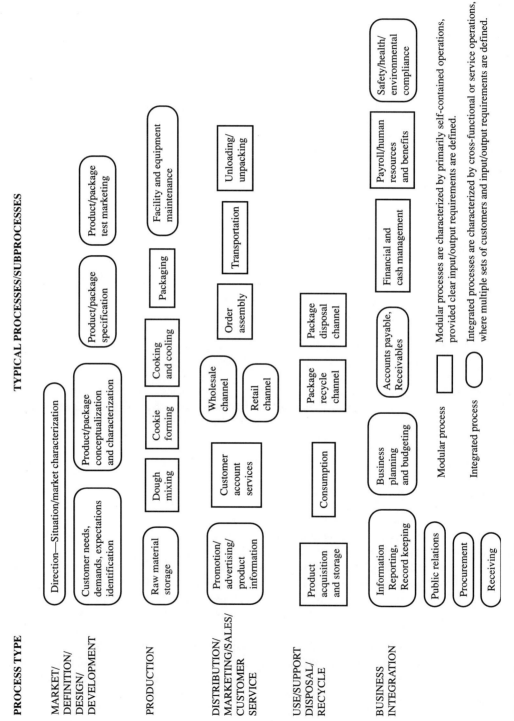

Figure 7.2 Summarized production system process layout overview, tailored for a bakery.

definition/redefinition elements/subelements, as described in Figure 7.1, are applicable in any and all process types.

The processes associated with any production system can be laid out in terms of the seven fundamental process types identified in Chapter 1, just as the bakery-related processes are laid out in Figure 7.2. The complexity of processes varies from one product to another. For example, an automobile production system requires an extensive sequence of definition, design, development, and delivery processes. The sales and customer service processes, e.g., selling, repairing, and maintaining the product, while challenging, are not highly complicated relative to the design and production processes.

Depending on field conditions and applications, automotive use processes are reasonably complex, requiring both knowledge and skill in driving and monitoring the vehicle. Recycling processes in terms of used automobiles and/or parts require reasonably complex processes. Disposal processes of parts, such as tires, present challenges in both disposal technologies as well as upgrading from disposal to recycle technologies, e.g., recycling used tire materials rather than disposing of them. Business integration processes are complex to the extent that the scale of economy in the automotive industry is large, and the products are complex, with facilities distributed throughout the world.

DOWNTOWN BAKERY CASE—DIRECTION AND MARKET   The Downtown Bakery has been   **Example 7.1**
producing a line of specialty breads for about 50 years. Initially, these breads were baked for a local market. About 25 years ago an expansion to regional markets was attempted. Downtown found that as long as they could provide same-day freshness, their regional marketing was successful. Currently, Downtown's bread lines allow for sales of about $5 million per year, derived exclusively from household sales, through local and regional grocery stores. Their products fetch a price premium of about 100 percent over commercial bakeries' premium breads.

Downtown's market characterization over the years points to two competitive edges: (1) taste, texture, and aroma and (2) same-day, dated freshness, with all day-old bread returned and salvaged. Their taste, texture, and aroma edge is attributable to premium ingredients and proprietary recipes. Downtown has developed recipes that also provide unmatched nutritional value. The freshness is due to daily deliveries and a proprietary packaging/material technology. Downtown can avoid artificial preservatives and provide customers with a microwavable/conventional warming package unit that brings out both aroma and flavor in their product in the customer's household environment. In order to gain and maintain these two edges, Downtown's production and delivery costs per product unit currently run about 50 percent higher than those of high-speed commercial bakeries.

Downtown's marketing/production capabilities have grown over the years from a small family business with a handful of employees to about 20 employees. The bakery is characterized by hand kneading and considerable hand work. Basic facilities and equipment are modern and extremely clean. The average employee's tenure is 15 years. Although no formal quality/productivity-related initiatives are present, Downtown employees have a history of providing ideas to management, which has a history of acting on the ideas in a timely fashion.

Downtown is organized on a process basis, with process teams in both marketing/sales and production. Being a small company with long-tenured employees who know each other, trust and communications within the production facility are very good. The company was incorporated several years ago, and as it grew from local to regional markets, employee stock ownership was instituted—and is pointed out as one reason for employee participation in marketing and production improvement.

Recently, at a company recognition ceremony, chocolate chip cookies were served. Several people present commented on the cookies—that they could bake cookies that would taste, look, and smell much better.

The leadership team that had arranged for the refreshments was at first amused at the comments, but later began to think in terms of opportunity. Previously, management had considered expanding bread production but always backed away because of market uncertainties. At this point, a production system definition team was chartered to explore cookie product expansion. The team membership included people with marketing, production, packaging, delivery, sales, purchasing, nutrition, and financial expertise.

The process definition team, along with the team sponsor, Downtown's CEO, first focused on common purpose. Here, they set out to assess their situation and express vision, mission, and core values relative to the possible bakery-cookie enterprise. Realizing they had clear vision, mission, and core values statements for their bread enterprise, they first looked at using or modifying them. A new, additional vision and mission were deemed necessary at this point, since the cookie product differed significantly from their current bread products and likely would be produced in a different facility. However, they recognized the possibility of a combined enterprise, where one global vision, mission, and values statement might be forged out of their present bread-related statements.

The present situation reality was studied carefully, and several documents were produced. Table 7.3 summarizes a chocolate chip cookie market profile. Here, market research documents were obtained from the *Simmons Study of Media and Markets* [2] and summarized. In addition, a creative thinking exercise was initiated, whereby a relations diagram was developed relative to a chocolate chip cookie product and its consumers. This diagram is depicted in Figure 7.3. Here, the situation is described, along with what amounts to an emerging product vision.

From this situation assessment, we conclude that an extensive market for chocolate chip cookies already exists. However, premium-grade cookies are not widely available to the customer. Furthermore, the nutritional value of products currently on the market is marginal. Establishing a premium-grade market segment appears feasible, considering that over half of the current market lies within the 25- to 44-year age range, and about 40% of the households that consume chocolate chip cookies are in the income range of $40,000 per year and above. Retail prices relative to the current market typically range from a high of $5.30 per pound to a low of $2.00 per pound. Direction and common purpose are expressed in the vision, mission, and core values statements growing out of this situation assessment.

Vision.   A Downtown baking enterprise spin-off is envisioned that will be capable of creating and marketing a unique, wholesome, premium-grade chocolate chip cookie that sets new standards in sense appeal—taste, texture, color, and aroma—and is recognized as the cookie of preference and comparison within the young adult market segment.

Mission.   We will simultaneously establish (1) a national/international market niche for a premium-grade chocolate chip cookie product line and (2) a lean process-based production facility and delivery capability that creates and delivers a nutritious product beyond comparison in color, texture, aroma, and taste.

Core Values.   Our production system will offer our customers a uniquely outstanding product at a fair price; offer our employees a safe environment to grow and practice leadership and creative skills, while earning a fair wage; and offer our distributors, retailers, and stakeholders a fair return on their investments.

Once we develop and accept a common purpose, we are in a position to identify our processes and subprocesses. Previously, in Figure 7.2, we depicted several applicable processes. **Each of these processes and subprocesses needs a clear purpose and associated definition.** These definitions require the focus and concentration of people with applicable knowledge and skills, relative to the process or subprocess at hand. The case continuation illustrates the essence of process definition, once we have our general direction and common purpose expressed and accepted.

DOWNTOWN BAKERY CASE—DEFINITION, DESIGN, AND DEVELOPMENT PROCESSES | **Example 7.2**

At this point in the Downtown Bakery Case, we have established interest in a product domain, but we do not have a clear picture of our customers' needs and expectations. Hence, we formulate a purpose for our customer needs, demands, and expectations identification subprocess; see Figure 7.2.

> *Customer needs, demands, and expectations discovery subprocess purpose:*   Identify primary customer needs, demands, and expectations associated with our product, through the eyes of discriminating customers.

We use market research tools such as interviews, focus groups, and surveys to extract details regarding customer needs and expectations. Table 7.4 summarizes our customer needs and expectations as they relate to our product concepts. Critical product characteristics—selling points—are noted in Table 7.4.

At this point, a clearly defined market segment has emerged, and a clear customer/market profile has been identified. A sound basis and reasonable justification for further development are evident.

Table 7.3   Customer/market situation summary (Downtown Bakery Case).

	Soft Chocolate Chip Cookies		Crunchy Chocolate Chip Cookies	
Consumer Category	**Consumers, 1000s**	**Market Share, %**	**Consumers, 1000s**	**Market Share, %**
Age:				
18–24	2,680	9.7	2,526	9.8
25–34	6,786	24.5	7,568	29.2
35–44	7,097	25.5	6,552	25.3
45–54	4,879	17.6	3,347	12.9
55–64	2,732	9.9	2,518	9.7
65 or older	3,537	12.8	3,380	13.1
Total	27,711	100.0	25,891	100.0
Household Income:				
$60,000 or more	5,752	20.8	5,148	19.9
$50,000 or more	7,651	27.6	7,348	28.4
$40,000 or more	10,730	38.7	10,629	41.0
$30,000–$39,000	4,170	15.0	3,848	14.9
$20,000–$29,000	4,515	16.3	4,222	16.3
$10,000–$19,000	4,974	18.0	4,323	16.7
Under $10,000	3,322	12.0	2,869	11.1
Total	27,711	100.0	25,891	100.0
Children in Household:				
2–5 years	4,451	16.1	4,886	18.9
6–11 years	6,070	21.9	6,162	23.8
12–17 years	6,296	22.7	5,147	19.9
Others	17,144	39.3	15,767	37.4
Total	33,961	100.0	31,962	100.0

Current Market	**Price per Pound**
P. F.—soft baked	$5.30
P. F. —chocolate chunk	$5.30
F. A.	$3.20
K.—chip deluxe	$2.00
N. C. A.—chunky	$2.10
N. C. A.—reduced fat	$2.00
Average	$3.32

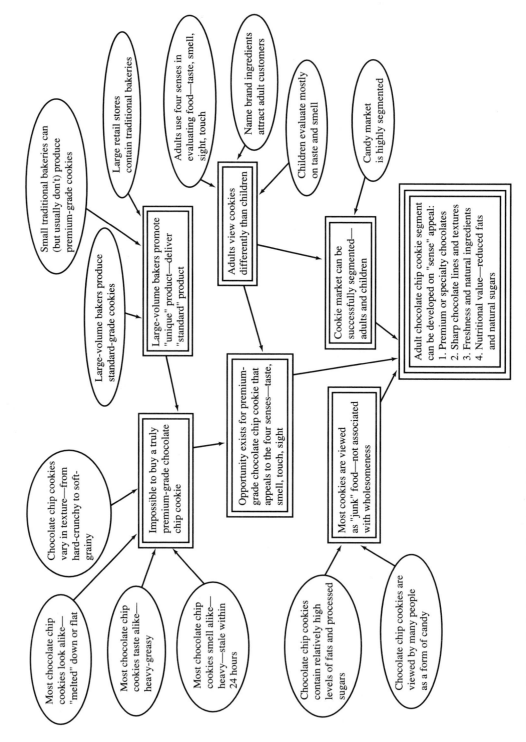

Figure 7.3   Product–market situation–vision relations diagram (Downtown Bakery Case).

Table 7.4   Basic customer needs, demands, and expectations summary.

Issues	Comments
Cookie-related	In focus groups people tended to first observe the cookies presented. They
Taste and texture	had a tendency to choose on the basis of cookie geometry (sharply defined
Rich taste*	chunks of chocolate and midsize cookies) and color (a light brown color was
Tempting smell*	preferred).
Chunks of chocolate	
Chewy*	Although appearance drew the customers to the product, as soon as a
Soft	cookie was selected the customers immediately smelled the product. A rich
No crumbs when eating	"just-baked" aroma was preferred. At this point, they typically tended to bend
Appearance	the cookie, and explained that softness/pliability was associated with freshness.
Look tempting*	
No broken cookies	The next, and obvious, activity observed was that of tasting/eating the cookie.
Right size to eat	Customers in the focus group had a hard time describing taste. A "rich" and
Wholesome*	"chewy" sensation was verbalized.
Package-related	Packaging options were explored. Customers had difficulty opening sealed
Freshness*	plastic packaging and expressed a desire for an easy opening "zip" tab.
Safe	
Choice of package size	
Easy to open	
Easy to close	
Easy to find on shelf	
Environment friendly	
Reasonable price	A consensus for a retail price between $0.25 and $0.75 per cookie (as presented)
	was expressed.

*Critical characteristics—selling points

Next, we introduce the product/package conceptualization and characterization subprocess.

Product/package conceptualization and characterization subprocess purpose.   Define our product in concept, relating both to customer needs and expectations and to competitors.

Starting with the results from the previous subprocesses, we extend our product definition efforts to the technical domain of cookies. Here, we expand from our critical selling points to a technical definition of our product. This subprocess serves as a transition from our marketing information to our technical design. We emerge from this subprocess with a clear design or recipe for our cookie product. Completion of our design/recipe takes place after test marketing. Partial results from this subprocess are summarized in Figure 7.4.

We have chosen the quality function deployment (QFD) format—specifically, a matrix diagram tool—for this subprocess. It allows us to identify our basic technical requirements and competition, and lay them out relative to our critical customer needs, demands, and expectations. The matrix diagram layout allows us to capture a great deal of information in a small space. Our team of technical and business experts was responsible for this development. Here, marketing, financial, product, production, and sales expertise is brought to bear on our common purpose.

Typically, we begin our QFD with the left-hand column: customer needs, demands, and expectations. This information is reproduced from the results of our previous subprocess, i.e., Table 7.4, and arranged in a hierarchical manner. For example, we have identified Cookies, Package, and Reasonable Price as three

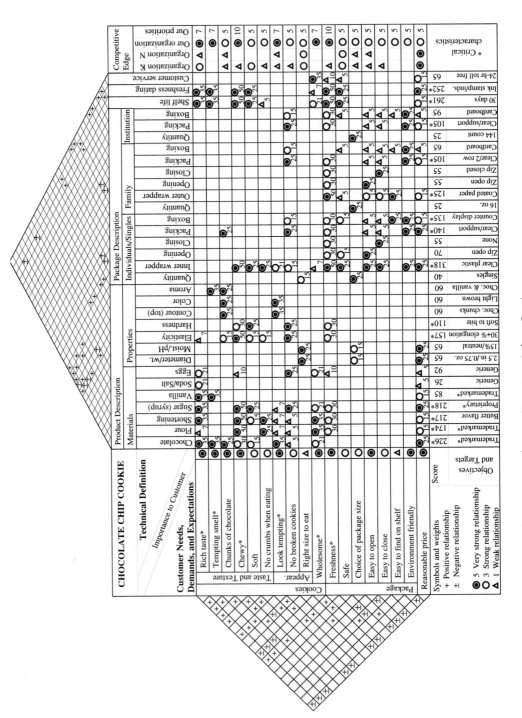

Figure 7.4 Quality function deployment matrix diagram (Downtown Bakery Case).

first-level issues. Then, we have broken Cookies and Package down into second-level issues. Finally, we have broken two second-level cookie issues, Taste and Texture and Appearance, down into third-level issues.

Next, we add the "hat" to the left of this information. The + and − symbols are used to indicate positive and negative relationships between different customer needs, demands, and expectations. The + indicates possible synergistic combinations, while the − indicates possible trade-off combinations. For example, rich taste and tempting smell are related with a +, indicating that they are positively associated and may allow for a synergistic effect in our product. At the intersection of rich taste and reasonable price, we see a −. Because of this negative relationship, we expect to be faced with a trade-off whereby we do our best to accommodate our customers by striking a balance between these two demands. In this case, both demands are very important to the customer; therefore, both have bull's-eyes in the Importance to Customer column. Hence, delivering both rich taste and reasonable price will challenge our creative abilities.

Next, we address the technical definition of our product. We do this by developing the technical definition characteristics across the top of our matrix in Figure 7.4. Here, our results are broken out in three levels. The product and package descriptions are introduced at the top level. Then, secondary and tertiary levels are worked out. Here, we are working toward establishing strength of relationships between the customer needs, demands, and expectations and the technical definition characteristics. First, we establish a qualitative relationship in terms of strength; a bull's-eye represents the strongest relationship. As a reality check, we should see each customer need, demand, and expectation related to at least one technical characteristic; otherwise, our work is not complete. The objective here is to systematically assess and define the product in terms of our customer needs, demands, and expectations.

The quantitative relationship strengths are obtained by setting our priorities in terms of numbers on the far right-hand side. Here, a 10 is used to represent the highest possible priority, and a 5 is used to represent the lowest possible priority. We obtain the relationship strength in the body of the matrix by multiplying our priority strength by our relationship strength; e.g., a bull's-eye is set at a strength of 5, as documented in the lower left-hand box. The score row is developed by summing the relationship scores down each column. In this case, an arbitrary criticality score of 100 was used. We note that the priority strength numbers as well as the relationship strength numbers are optional and arbitrary; i.e., we use our judgment as to defining the magnitude of these numbers. In the end, these scores help us to prioritize our technical characteristics.

Once we have our technical characteristics defined in a verbal sense, we are ready to attempt to define baseline targets. These targets may be qualitative or quantitative. At first, they are typically qualitative, and then, when it is reasonable to do so, they are refined to quantitative values. Here, most of our targets are qualitative. As more and more details of the product-process definition emerge, we move our technical characteristics and targets toward product/process specifications, i.e., an exact recipe for ingredients, cooking temperature, cooking time, and so on.

Our ability to assess and establish competitive edge early in our definition process is critical. At this point, we have introduced two competitors and assessed their product relative to what we perceive as critical customer needs-demands-expectations. This competitive edge assessment, on the far right-hand side of Figure 7.4, helps us put our intentions into perspective with our potential competitors. Here, we can work out our product-related competitive edges.

Once we have a baseline product defined, we think in terms of baseline business results. Here, we can begin to address product cost targets, marketing cost targets, and pricing targets, as well as potential production and sales volume and market share. Although these targets are rather intuitive, they are necessary to bring a business perspective to what has been primarily a technical customer- and product-focused process activity. An abbreviated illustration of baseline business targets appears in Table 7.5.

As an extension to our basic business results parameters and targets in Table 7.5, we can perform a simple break-even analysis. This analysis helps us grasp cost, volume, and profit relationships early in our production system definition/design process. Here, we use basic figures from Table 7.5—we assume variable costs per pound at $2.50 and fixed costs of $2.5 million, along with a wholesale price of $6.50 per pound. Setting our total cost equal to total sales and solving, we obtain an estimated break-even volume of 625,000

Table 7.5   Selected baseline business results parameters and targets (Downtown Bakery Case).

Parameter	Target
Product:	
Production cost	Less than $2.50 per pound
Production volume	More than 300,000 pounds per year
Market:	
Retail price	Less than $8.00 per pound
Wholesale price	Less than $6.50 per pound
Business:	
Capital investment	Less than $2.5 million
Time to market	Less than 6 months
Gross profit margin	More than 33 percent
Return on investment	More than 20 percent
Flexibility:	
Maximum production volume	About 300,000 pounds per year (one shift, 100% capacity)
Additional products	Sugar cookies, oatmeal-raisin cookies

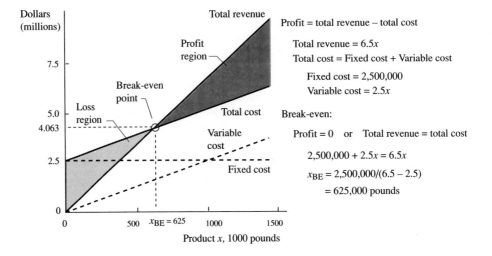

Profit = total revenue – total cost

Total revenue = $6.5x$

Total cost = Fixed cost + Variable cost

Fixed cost = $2,500,000$

Variable cost = $2.5x$

Break-even:

Profit = 0   or   Total revenue = total cost

$$2,500,000 + 2.5x = 6.5x$$

$$x_{BE} = 2,500,000/(6.5 - 2.5)$$

$$= 625,000 \text{ pounds}$$

Figure 7.5   Break-even analysis graph and numeric solution (Downtown Bakery Case).

pounds. Figure 7.5 provides a graphical depiction of our initial cost/volume/profit estimate and the solution to our break-even equation.

    With a baseline product-business definition, we are now ready to enter the design/development product and package specifications and test marketing subprocesses. As a reality check, to protect our assets, we must identify either a product competitive edge or a process competitive edge, or both, before proceeding—otherwise we stop. We note that our illustrations are partial and simple, but systematic. Further details of these subprocesses are omitted for brevity.

7.5 PROCESS LEVEL RESULTS DEFINITION

Initially, our production system focus, as a whole, is on customer needs and expectations as well as value-created transformations viewed from an organizationwide perspective, primarily aligned with external customers. **When we tighten our focus down to specific processes, we view results considering both internal customers as well as external customers.**

CUSTOMERS

Regardless of customer classification we must understand our customers—their needs, demands, and expectations. They may be halfway around the world or located at the next process station. Our customers may be external, product buyers/users, or internal, product builders/support providers. In either case, **we know that the customer's experience of quality is human-need-based.**

The individual quality experience (IQE) model, as described in Chapter 1, Figure 1.3, relates the basic human needs structure to attitudes and behaviors. In the IQE cycle, **customer experience involves:**

1. *Observing.* Sensing and perceiving (within our physical and social environments).
2. *Assessment and interpretation.* Thinking in both the cognitive (logical) and affective (emotional) dimensions.
3. *Attitudes and behaviors.* Decisions and initiatives (forming attitudes, deciding, and taking action or remaining inactive).

The IQE model helps us associate the customer with our product and production system. For example, **customers experience/perceive physical, economic, timeliness, and customer service performance, and then shape their attitudes and behaviors accordingly.**

Understanding our customers and making sure our suppliers understand us, as their customers, is critical. Details regarding the understanding of process customers/suppliers are presented in Table 7.6. **We ultimately wrap our product definition around our customer's needs and expectations.** Our goal is clear, but the ways and means of address are varied and many. For example, we can work from an unfulfilled or underfulfilled customer need toward a product concept. Or, we can work from a new basic technology/product concept toward an unfulfilled or underfulfilled customer need. **The objective is to maximize customer benefits, minimize customer burdens, and reap the associated business rewards.**

We formulate, express, and communicate a razor-sharp purpose, in a customer context, regarding results, i.e., outcomes, likely to be created by our process or subprocess as it will ultimately be defined or redefined. We express this purpose in one sentence. This sentence is critical to establishing our specific direction, as well as to keeping us on course as events unfold. **Our process purpose serves to keep us focused on whole-process solutions,** so as to avoid diverging toward partial-process solutions. Even though our expression is brief, our effort to reach this directional expression of purpose may be considerable.

We identify process boundaries so that we can focus our efforts. We define boundaries as clear-cut interfaces with our suppliers on one side, i.e., input, and customers on the other side, i.e.,

Table 7.6   Details related to process customers/suppliers.

Action	Description
Set direction	Develop and express process purpose as to what product-based outcomes (quality experience) we intend to provide for our customers.
Identify process boundaries, suppliers, and customers	Identify the boundaries associated with the process. Identify process suppliers and customers. Describe the customer profile in general and then uniquely identify customers as a group.
Discover true needs/demands/expectations	Get to know customers/suppliers as a group and discover and relate needs/demands/expectations. Define products to needs/demands/expectations (not the converse) and visualize customer benefits and burdens.
Coalesce advantage points	Harvest available data and information to describe customers and significant advantages our product offers. Clearly verbalize what we, as customers, want/need/demand from our suppliers.

output. **People at the output and input boundaries provide clues as to what constitutes an acceptable process result/outcome.** Although somewhat arbitrary, these boundaries usually encompass logical segments or blocks of physical activities.

In most cases customers are able to verbalize their needs and expectations; in other cases verbalization is difficult. Usually, internal customers can adequately express their needs and expectations. Regarding commercial products, external customers can usually describe their needs, demands, and expectations. In the case of redefinition, we usually have a pre-existing customer base, and perhaps a rich historical basis to examine and expand. In all cases, we rely on direct experience, similarities, and perhaps test-case exposures, e.g., surveys, interviews, focus groups, and test marketing.

When entirely new products are considered, customers may be unable to verbalize their needs and expectations, or these verbalizations may be misleading. For example, the needs for automobiles, wireless phones, and microcomputers were not immediately clear at the time of invention. The development of associated infrastructures, such as highways, signal carriers, and software, was generally necessary before customers could visualize benefits from these new inventions and technologies. At the time of invention, before the infrastructures were developed, potential customers would likely state that they didn't see benefits and hence didn't need the product.

In cases where there is doubt as to the accuracy and precision of customer perception of needs and expectations, we may choose to develop visual extensions/models of the product for our potential customers and/or resort to sophisticated market research tools. In all cases, we engage our customers, potential customers, and former customers as best we can.

We visualize customer needs and expectations in the context of outcomes or results— customer benefits and burdens that can be offered and realized through processes. We are essentially simulating or preplaying the customer's experience of quality in a proactive manner, i.e., through the individual quality experience model. We are using nontechnical terms, or customer language, for the most part in this discovery exercise.

DOWNTOWN BAKERY CASE—PRODUCTION PROCESS BOUNDARIES, CUSTOMERS, AND | **Example 7.3**
SUPPLIERS   In this case extension, we focus exclusively on the production process definition regarding the Downtown Bakery's cookie spin-off enterprise. Following up on the production system vision, mission, and core values expressed previously, we define our production process purpose and illustrate the "define results" triad depicted in Figure 7.1.

> *Production process purpose.*   Given product and packaging specifications, produce and package a product that meets all specifications at all times, and actively participate/lead in product, package, and production improvements so as to assure customer satisfaction.

Once our process purpose is clear, we are ready to define our production process boundaries, suppliers, and customers. The depiction in Figure 7.6 initiates our definition. Here, we place the production process in the center, and then identify our boundaries at both the input and output sides. At this point, we are not primarily concerned with the details within our boundaries. Rather, we focus on the boundaries themselves in order to describe our customers and their demands on the output side, and our production process demands, as customers, of our suppliers on the input side.

From Figure 7.6, we can see a variety of supplies inbound, and products and by-products outbound. We have identified basic needs and expectations for the identified suppliers as well as the identified customers. On the customer's side, we have identified a chain of customers starting with our own distribution/ marketing/sales/service process. Here, we start with the next subprocess to touch the packaged product, i.e., the distribution subprocess. Then, we identify the channel by which our product flows to the ultimate consumer, i.e., the people who make the decision to buy our product and consume our cookies—wholesalers, food services, retailers, buyers, consumers.

We see that the impact of the production process is far-reaching. For example, wholesaler and retailer demands, as well as purchaser and consumer demands, must be discovered and considered in the production process. This interaction appears rather obvious from our systems perspective, but it may be obscured by the day-to-day production pressures within a production process. We want to make sure that this obscurity is removed in process definition and never appears within the actual production process once it is implemented. We want every member of the production team to always be aware of the consequences of their decisions and actions. Eventually, the process value chain, Figure 3.5, will convey this message.

OUTCOMES

Customer outcomes are critical—we define baseline technical and business requirements to link customer needs and expectations to our products and business. Here, we define relationships between our customer's world, as well as our supplier's world, and our technical world, so that we can eventually produce maximum customer benefits with minimum customer burdens and positive customer outcomes. We define our basic business requirements in order to put the technical issues in perspective. Table 7.7 summarizes the actions necessary to define our baseline requirements regarding our substitute product characteristics and business expectations with respect to our customer needs and selling points.

In order to develop a comprehensive description of product and customer service outcomes, we rely on systematic methods. Here, we use initiatives and tools from Table 7.1 or develop and build our own, as appropriate for our purpose. We choose a finished product, or a specific intermediate product basis for description, depending on our purpose. Weakness at this point ripples through all subsequent activities depicted in Figure 7.1 and yields a process that falls short in effectiveness (quality) and efficiency (productivity).

Production process purpose: Given product and packaging specifications, produce and package a product that meets all specifications at all times, and actively participate/lead in product, package, and production improvements so as to assure customer satisfaction.

Primary supplies

Physical:
Ingredient materials
Packaging materials
Energy/water
Cleaning supplies

Informational:
Product recipe
Process control targets
Product orders
Process performance reports

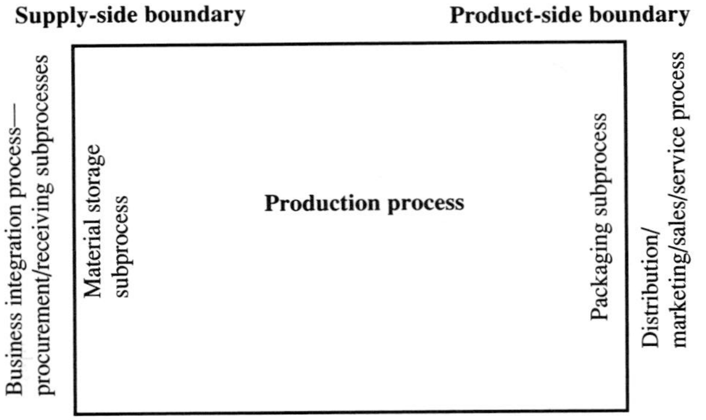

Supply-side boundary

Business integration process—
procurement/receiving subprocesses

Material storage
subprocess

Production process

Packaging subprocess

Distribution/
marketing/sales/service process

Product-side boundary

Primary products/by-products

Packaged cookies
Scrap cookies
Wastes (solid/liquid)

Primary Suppliers
[upstream suppliers]

Physical:
Procurement/
receiving subprocesses

[Ingredient vendors
Packaging material vendors
Utility company
Cleaning supply vendors]

Informational:
Definition/design/
development process
Distribution/marketing/sales/
customer service process
Business integration process

Supplier-Side Demands
(summarized)

The right ingredients/materials/
supplies/uninterrupted flow

The right ingredients/amounts/on time
Right materials/amounts/on time
Right utilities/uninterrupted
Right supplies/amount/on time

Clear recipe/specifications/instructions

Concise product orders

Informative and timely reports

Primary Customers
[downstream customers]

Physical:
Distribution/marketing/sales/
customer service process

[Wholesalers
Food services
Retailers
Buyers
Consumers]

Informational:

Business integration process

Customer-Side Outcomes
(summarized)

Uniform product/meeting all specifications
On-time production/steady flow

Fresh/long shelf life/high profit
margin/attractive display/attractive
package/easy to open/tempting
appearance/aroma/texture/taste/
nutritional value/reasonable price

Low production costs

Figure 7.6   Production process customer-side outcomes and supplier-side demands [Downtown Bakery Case].

Table 7.7   Details related to process outcomes.

Actions	Description
Review and verify customer needs, expectations, and selling points	These true quality characteristics have been established previously.
Define technical essence of product (produced by the process at hand)	Define baseline substitute product categories and characteristics to match true quality characteristics.

Relate customer-based needs and expectations to technical product characteristics
    Customer-side outcomes
    Supplier-side demands
	Take the perspective of process customers, include customer-side needs/demands/expectations and supplier-side needs/demands/expectations.

Address general targets regarding products/supplies and business results	Tentatively identify critical substitute quality characteristics, parameters, and targets. Develop a fundamental set of technical/market/business expectations relative to process customers and suppliers.

DOWNTOWN BAKERY CASE—PRODUCTION PROCESS OUTCOMES   At this point in our **Example 7.4** development of the Downtown Bakery's production process definition, we follow up on customer needs and expectations identified in Figure 7.6. Here, we begin with the identified needs and expectations as expressed by our process customers, i.e., distribution/marketing/sales/service and business integration. We take these expressions as they are and translate them into technical language. This translation is necessary in order to express these general needs in a form that we can address and measure within our production process.

Likewise, we seek out business-related outcomes that we expect from our production process. Both the technical and business characteristics eventually become a part of our process value chain (PVC) model, Figure 3.5. In order to integrate our technical and business characteristics into our PVC, we develop related parameters, i.e., metrics. With these metrics at hand, we set targets, i.e., best values for these parameters, to the extent possible. Here, our targets are of three basic types: smaller is better, larger is better, and nominal is best.

Both technical and business outcome characteristics are summarized in Table 7.8. Here, we see a wide variety of desired customer outcomes and supplier demands, viewing ourselves as our suppliers' customers. The information in Table 7.8 is not complete in the sense of specific metrics and targets, but does provide direction as to baseline characteristics and parameters that must ultimately be addressed. In the target column, we note the primary sources that are involved with the targeting, in conjunction with the customer and/or supplier.

Eventually, as our process is further defined, we add detail to these characteristics, parameters, and targets. But, for now, we settle for baseline descriptions. Ultimately, when we complete our process definition plan, as described in Chapter 8, we have a detailed description/development of our baseline characteristics, parameters, and targets. These details then lead into our process control triad, as described in Section 4, Chapters 9 to 15.

CONCEPTS

Once we describe our desired outcomes with respect to our customers, suppliers, and our business, we explore process concepts. Detailed actions associated with concept exploration and impact assessment are provided in Table 7.9. We first review our baseline product technology

Table 7.8   Production process outcomes/demands summary (Downtown Bakery Case).

Production Process Interface	Customer Side—Outcomes Supplier Side—Demands	Parameter/Target Description
Primary Customers		
Distribution/marketing/sales/ customer service process	Meet all product specifications	All specifications essential to making product
	Meet all packaging specifications	All specifications essential to packaging product
	Meet production schedule	Timeliness/product flow related
Business integration process	Control production costs	Unit cost Throughput Scrap/rework Downtime Process cost to plan
Primary Suppliers		
Definition/design/ development process	Recipe specifications Packaging specifications	All specifications essential to making product
		All specifications essential to packaging product
Distribution/marketing/sales/ customer service process	Product orders	Wholesale/retail Institutional
Business integration process (procurement/receiving subprocess)	Acquisition of ingredients, packaging materials, energy, water, cleaning supplies	Meet all technical specifications Meet all amount and timeliness specifications

Table 7.9   Details related to concepts.

Actions	Description
Review outcomes and demands	These outcomes/demands have previously been expressed in terms of customer-side outcomes/parameters/targets and supplier-side demands/parameters/targets.
Identify all relevant product/process concepts	Examine the outcomes/demands and break out the whole process in meaningful subprocesses using block diagrams. In the case of process redefinition, this hierarchy already exists, but may need modification.
Assess quality and productivity impacts of possible process concepts	Review the process concepts in light of customer needs, demands, and expectations, as well as supplier demands and business results. Identify advantages and disadvantages associated with each concept.
Identify product/process bottlenecks and limitations	Recycle and replay the product/process concepts and identify bottlenecks and limitations, along with the general degree of difficulty associated with resolution.
Summarize feasible process concepts/eliminate infeasible concepts	Considering customer needs and expectations and product requirements, list feasible core concepts with advantages, disadvantages, and bottlenecks.

requirements. This review helps us to keep our priorities straight, i.e., enhance customer benefits and reduce burdens, which in turn yields business results.

Our goal is to identify all relevant process concepts that can in some way help us to fulfill our process purpose and produce our desired outcomes. First, we study our baseline product/by-product characteristics or requirements as a whole. We carefully define our product as a physical entity. For example, we might describe product assemblies or components in enough detail to begin to match core process concepts to them.

At this point, we identify process concepts that are appropriate or compatible with our product/by-product requirements and form a product-process flow. For example, we superimpose our process options on our product and vice versa. This product/process hierarchy is expressed qualitatively in a technical sense in words and graphics such as product/process sketches, cause-effect diagrams, and block diagrams, e.g., flowcharts.

By simultaneously considering product and process concepts, we typically modify product features, e.g., materials, to accommodate processes in order to yield economic or timeliness gains, while holding or even enhancing physical and/or customer service performance. For example, we might substitute plastic for metal in order to gain corrosion resistance (physical product performance) and fabrication efficiency in equipment, energy, and cycle time savings. **Our aim is to create synergy, in quality and productivity results, as we identify our core process technology concepts.**

Initially, we do not critically assess the process concepts generated. Rather, we stress creativity in our thinking and conceptualization. The more complete our process concepts, the more likely we are to eventually develop a competitive process definition solution.

As a final step in our concept descriptions, we focus on advantages/disadvantages associated with our concepts. Here, we think in terms of both quality and productivity. Our advantages/disadvantages thinking and list help us transition into the next leg of our triad—transformation—as depicted in Figure 7.1.

DOWNTOWN BAKERY CASE—PRODUCTION PROCESS CONCEPTS Resuming our discussion **Example 7.5** and development of the Downtown Bakery cookie spin-off enterprise, we remain focused on the production process definition, specifically addressing process concepts. Here, we consider two basic concepts: (1) private label/coproducer subcontracting and (2) in-house production. These two concepts differ a great deal in their impact on our production system. However, both concepts appear to offer our customers virtually identical products and packages. Basic overviews of each concept appear in Figure 7.7.

The private label concept in Figure 7.7a identifies the boundaries between Downtown's processes and our coproducer's production process. Here, in subconcept 1, we supply our coproducer with product orders and product and package specifications, including our recipe, and expect them to perform material procurement, production, and packaging. We then pick up our packaged product at their plant and move it to our warehouse, where we assemble our orders and link up with our own distribution subprocess.

In subconcept 2, our strategy is to protect our recipe; hence, we supply all materials regarding our cookie product to our coproducer. In this subconcept, the Downtown Bakery procures, measures, and mixes all ingredients at Downtown's facilities, and then delivers them to the coproducer. Subconcept 2 is identical to subconcept 1, after the mixed ingredients are received by the coproducer.

In the private label concept, we do not actually develop a production process of our own, but utilize one that already exists, owned by our coproducer. One concern here is quality control. We have introduced a coproducer relations subprocess in order to certify and audit our contractor's production process. This added

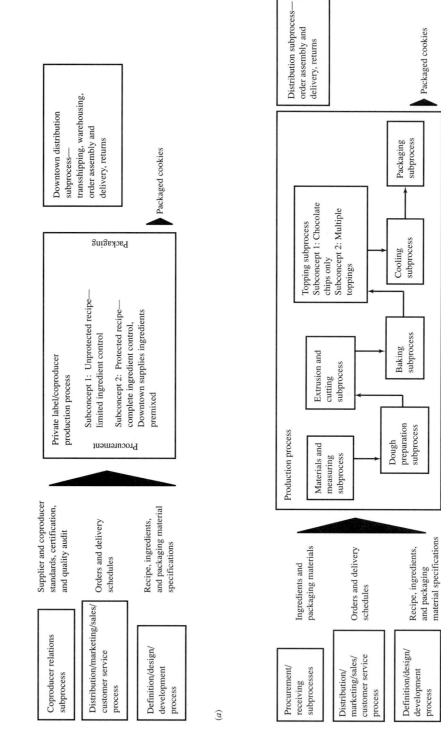

Figure 7.7 Production process concepts (Downtown Bakery Case). (a) Concept 1—private label/coproducer production process. (b) Concept 2—Downtown/in-house production process.

Table 7.10   Primary private-label concept impact distinctions—notes (Downtown Bakery Case).

Concept Issue	Impact
Advantages	
Production facilities/equipment personnel furnished by coproducer.	Capital investment is avoided; costs/risks are shifted to coproducer. Personnel requirements are reduced; cost/benefits responsibilities are shifted to coproducer.
Scrap, nonconforming product, is responsibility of coproducer.	Risk and costs of nonconformance are shifted to coproducer.
Records and bookkeeping, production-process-related, are furnished by the coproducer.	Personnel requirements are reduced; personnel/costs/benefits responsibilities are shifted to coproducer.
Materials acquisition furnished by the coproducer in subconcept 1.	Personnel requirements are reduced; personnel/costs/benefits responsibilities are shifted to coproducer.
Materials acquisition and mixing done by Downtown in subconcept 2.	Keeping the recipe secret partially assures product integrity, partially assuring a unique product/market niche—and maybe protecting our premium price.
Disadvantages	
Downtown acts more like a broker/distributor than a producer.	Gross profit in the 20% range is associated with broker/distributor, whereas the coproducer expects up to 35% gross profit.
Coproducer may ask for a new contract.	Coproducer may squeeze Downtown with contract escalation.
Coproducer may have limited capacity for growth and share production facilities with other customers.	Coproducer may limit market expansion.
Downtown has reduced control over product quality and production schedules.	Premium product image may be compromised; demand may not be satisfied.
Downtown recipe may be lost in subconcept 1. Coproducer or competitor may obtain product recipe and successfully copy/swipe product.	Marketing image and market share may be lost.

subprocess on our part is deemed necessary so that we can assure the integrity of "their" product/output in a proactive manner.

Our in-house production concept provides for our own production process, made up of several subprocesses, all of which we define, design, develop, implement, and operate. Within this concept, we have two basic subconcepts. One allows for a production process for only chocolate chip cookies. The second allows for a generic cookie production process. For example, we might change our recipe and produce sugar cookies in addition to chocolate chip cookies. This second subconcept allows us flexibility in terms of building new market niches. When we compare these two subconcepts, we see the same basic subprocesses, with subtle differences in our raw material holding facilities, dough preparation, and topping subprocesses.

When we compare the private label concept with the in-house production process concept, we see substantial differences. With the exception of the product, the nature of the production function looks very different through the eyes of the Downtown Bakery. Several of the primary private label concept distinctions are summarized in Table 7.10. Here, we see significant advantages in avoiding facilities and equipment capital expenses, as well as avoiding higher payrolls and benefits. However, we see definite compromises in our product integrity and reputation. Subconcept 2 allows some protection for our recipe, but limitations in production volume and perceived product integrity still exist. Business and marketing considerations lead to

Table 7.11 Primary in-house production concept impact distinctions—notes (Downtown Bakery Case).

Concept Issue	Impact
Advantages	
Product integrity, including nutritional value, can be carefully controlled and improved rapidly if necessary. Product integrity can be used as a promotional feature.	Market niche can be established/enhanced/protected.
Downtown controls marketing channels and production resources.	Gross profit margins can be increased as high as the 55% range.
Production capacity can be increased in both the near and long terms.	Near-term and long-term expansion is possible.
Subconcept 1 allows for focus on one product, i.e., chocolate chip cookies.	Line changeovers are eliminated—increased efficiency.
Product expansion, e.g., to sugar cookies, is possible with subconcept 2. Diversified cookie product lines can be established.	New markets can be explored, profits enhanced, and business risks limited.
Disadvantages	
High capital equipment and facilities requirements.	Initial capital/start-up costs are increased due to need for production facilities and equipment.
More employees needed to operate production process. More training necessary.	Higher personnel-related costs.
Extensive production safety/hygiene requirements.	Higher cost of compliance, e.g., OSHA, FDA, ... Higher exposure to safety/liability issues.
More extensive records and bookkeeping, production-process-related.	Higher personnel/management costs.
Subconcept 1 limits product line to chocolate chip cookie product.	Option for product line growth is limited.
Subconcept 2 requires higher level of process equipment and training.	Additional equipment and training costs are encountered in changeover capability.

major bottleneck concerns in gross profit margin and market niche establishment for a premium-priced product.

Downtown's in-house concept distinctions are summarized in Table 7.11. Here, we see an autonomous production facility with advantages in assuring product integrity and preserving our proprietary recipe. However, we see higher levels of capital required for facilities and equipment. We see more extensive requirements for personnel, training, and related issues. Subconcept 2 is attractive in the sense that expansion in other cookie lines is supported from the beginning. However, it comes at an increased cost in capital equipment and training. From a business standpoint, the increased gross profit margins are very attractive. Marketing advantages related to in-house production, and associating the premium-grade product's uniqueness to exclusive production processes, are appealing. The major bottlenecks in the in-house concept are raising sufficient capital for financing the spin-off enterprise and controlling the business risk inherent with the launch of a new product in an essentially untested market niche.

In a head-to-head comparison, the most pronounced difference is in quality control. This difference might be minor in a basic commodity product. However, in a premium product, e.g., one priced at nearly 2 times commodity competitors, this difference is significant both in the nature of the final product and in the reputation of the product. These differences fall in both the physical and customer service domains. Outside of the quality perspective, we see differences in economics and timeliness. These differences are substantial and fall within the productivity domain.

In the end, in order to move to the next step in our production process definition, we evaluate our concepts and narrow our possibilities. Here, all things considered, we intend to pursue the in-house production process concept so that we can more effectively address our customers' needs and expectations. Production process control and product association with the goodwill of the Downtown Bakery name are the overriding factors in our decision to pursue the in-house concept and drop the private label concept. The prospect for a higher gross profit margin is also a major impact issue. At this point, we have not decided as to a single product, i.e., strictly chocolate chip cookies, or a more general production process. We address this issue in Chapter 8.

Once we have explored product-process concepts, we replay each concept through our results triad; see Figure 7.1. During this replay exercise we consider customer outcomes in the context of our product/process concepts, **looking for potential bottlenecks created by limitations** to our ability to establish product-based or process-based competitive edges. Successfully addressing these limitations produces early and relatively inexpensive gains before we commit additional resources to our production system. Here, we are in the definition phase—$1 category—see Example 1.4. Major changes are readily made at this point with the stroke of a pen or mouse.

REVIEW AND DISCOVERY EXERCISES

REVIEW

7.1. Explain the difference between a process definition and a redefinition. What characteristics/challenges does each present?

7.2. What relationship exists between process definition/redefinition and initiatives and tools? Explain.

7.3. Explain the difference between proactive and reactive process redefinition.

7.4. Who should be involved with process definition/redefinition efforts? Why?

7.5. What is the goal of process definition/redefinition? Explain.

7.6. What is the difference between a modular process and an integrated process? Explain. Provide an example of a modular process and an integrated process.

7.7. Why are customers so fundamental to process definition/redefinition? Explain.

7.8. Why is a razor-sharp purpose necessary in process definition/redefinition? Explain.

7.9. What are process boundaries and why are they necessary in process definition/redefinition? Explain.

7.10. What constitute true needs, demands, and expectations on the customer's part? How are they identified and expressed? Explain.

7.11. How do we go about defining the technical essence of a product? Explain.

7.12. How do we associate and map true customer needs, demands, and expectations to substitute quality characteristics? Explain.

7.13.   How do we set targets for products/supplies and business results? Explain.

7.14.   What constitutes a basic product/process concept? Explain.

DISCOVERY

7.15. All organizations do not focus an equal amount of attention/resources on each of the seven fundamental processes. Why?

7.16. After examining the processes identified in Figure 7.3, explain why we have broken the seven fundamental processes into only five groups. How do we know which processes can be grouped? Explain.

7.17. Why do we focus on process results/outcomes first, before we focus on process conversions and/or resources? Explain.

7.18. One of the objectives in process definition/redefinition is to maximize creativity. Standardization has long been recognized as retarding creativity. How is it that we can suggest a definition/redefinition format, as depicted and expanded in the results triad from Figure 7.1, and expect maximum creativity? Explain.

The following project challenges are best worked as team projects. In academic exercises with college students, we have found small teams of from four to seven students work best. For practicing professionals, teams sized according to the project scale are best. For example, an extensive project that requires a variety of expertise requires a larger team. As an outcome, we expect to see a well-functioning team produce a logical and meaningful process definition/redefinition. Each team is responsible for both effective and efficient oral presentations as well as written presentations consisting of both graphical and prose communication elements. Teamwork fundamentals are discussed in Chapter 18.

SECTION 3, PROJECT 1

Select a need or market (e.g., purposeful theme) of your choice and develop a process definition sequence analogous to that of the Downtown Bakery Case, tracing the development from the market through the product definition, design/development, production, distribution/marketing/sales/ customer service, and business integration. Feel free to extend your development to the use/support and recycle/disposal processes. Use the development in Chapters 7 and 8 as a guide, but feel free to expand beyond the scope and detail presented in the text.

SECTION 3, PROJECT 2

Identify an organization of your choice, e.g., service, hardgood, or basic material producer. Identify a fundamental process or subprocess and develop a process redefinition. Here, we start with an existing process description and redefine the process, using our process definition/redefinition elements, as expanded and discussed in Chapters 7 and 8. Our expanded materials and the Downtown Bakery Case inserts provide a guide. Feel free to expand and enhance the level of detail in your work beyond that specifically illustrated in the Downtown Bakery Case examples.

SECTION 3, PROJECT 3

Identify a fundamental process or subprocess of your choice in a service, hardgood, or basic material production system. Using a "clean sheet of paper," apply our process definition elements, as expanded and discussed in Chapters 7 and 8. Our expanded materials and the Downtown Bakery Case inserts provide a guide. Feel free to expand and enhance the level of detail in your work beyond that specifically illustrated in the Downtown Bakery Case examples.

REFERENCES

1. Ulrich, K. T., and S. D. Eppinger, *Product Design and Development*, New York: McGraw-Hill, 1995.

2. *Simmons Study of Media and Markets: Chewing Gum, Candy, Cookies and Snacks*, p. 19, 30th ed., New York: 1993.

PROCESS DEFINITION/REDEFINITION— TRANSFORMATION AND INPUT PERSPECTIVES

8.0 INQUIRY

1. How are process means discovered? Evaluated?
2. What is a process leverage point? What is a process limitation?
3. How are processes created?
4. How does process implementation differ from process planning?
5. How does process definition/redefinition interface with process control and improvement?

8.1 INTRODUCTION

Up to this point, we have focused on process output and results. **When we clearly know what we want to accomplish, we focus on process means and process creation;** see Figure 7.1. Our discussion of process means focuses on options, evaluation, and detailed plans. Follow-through in process creation, bringing a new or redefined process to life, focuses on implementation. Here specific resources are identified, schedules are developed, and action is taken.

8.2 PROCESS MEANS

There are usually several ways in which we can transform a process concept into a living process. However, these means are not typically obvious, nor are they necessarily easy to discover and refine. We begin with a disciplined, systematic approach that starts by constructing basic

Table 8.1   Details related to options.

Action	Description
Express/re-express process and subprocess purposes	Consider and express physical, economic, timeliness, and customer service performance dimensions in process and subprocess purposes. Check for compatibility with production system vision, mission, and values.
Formulate basic process layouts	From the existing process concepts, develop high-level process flows. Focus on basic process technologies/functions. Identify basic subprocesses, i.e., clusters within processes. Include suppliers and customers.
Formulate subprocess layouts	Using the basic process layouts defined, supply details to the subprocess level. Focus on basic process/subprocess technologies/functions. Identify basic subprocesses, i.e., clusters within processes. Include suppliers and customers.
Superimpose process teams on layouts	Reexamine the process and subprocess flows and cluster them in a logical manner for teamwork. A semidetailed process layout for each process definition or redefinition alternative is created.
Identify subprocess boundaries and interfaces	Position the process definition or redefinition in the overall context of the production system (i.e., customers and suppliers) and describe interfaces with other processes.

process options. Then, we proceed to define and refine details to the point of a feasible plan. This focus demands choice, i.e., decisions, after the basic alternatives, facts, and estimates are established. Commitment to defining details in technical process requirements down to subprocess levels follows. For example, we describe and select a feasible process option—then we develop the plan to make it work effectively and efficiently.

OPTIONS

In order to formulate feasible process alternatives, we add detail to our process concepts. The sequence of actions described in Table 8.1 provides a systematic approach. First, we revisit process purpose and extend to subprocess purposes. We test each purpose relative to our production system vision, mission, and core values. Each process and subprocess, as expressed through its purpose, should add or create value for our customers. Otherwise, it is subject to elimination or realignment. **We expect that every process and its subprocesses should create potential value in our product**—to produce customer benefits and/or reduce customer burdens, and ultimately yield value to both external and internal customers and stakeholders.

We develop basic process layouts using flowcharting tools. Typically, a simple block diagram will provide a starting point. **We develop a new block diagram for each process alternative.** In a similar manner, **we break out the subprocesses within each process as subprocess blocks and eventually cascade the level of detail down to the point of a feasible plan through subsubprocesses.**

We look for natural clusters of process/subprocess functions/activities that lend themselves to operational teaming. Then, we superimpose teams on our layouts, clustering subprocesses when necessary to form effective and efficient team domains. **For these clusters, we identify specific process and subprocess inputs and outputs and their respective suppliers and customers.** Then, we lay out and view our processes and subprocesses together in more detail, in order to identify clear-cut boundaries and describe interfaces.

Example 8.1 | **DOWNTOWN BAKERY CASE—PRODUCTION PROCESS OPTIONS**   We now return to the Downtown Bakery production process concepts, as depicted in Figure 7.7. On the basis of previous arguments and analysis, we select the in-house production process concept for development in this case extension. Our production process purpose appears in Figure 7.6. At this point, we note our production process boundaries on one side as the procurement/receiving subprocesses (contained in the business integration process) and on the other side as the distribution/marketing/sales/service process.

We expand our concept using our flowchart tool; tool details are provided in Chapter 20. Here, we define five subprocesses: (1) materials, (2) dough, (3) extruding/rolling/cutting/baking, (4) topping/cooling, and (5) packaging. Each of these subprocesses is expanded and depicted in Figure 8.1, along with representative subprocess purposes. Each subprocess is made up of sub-subprocesses, as depicted by the individual, labeled boxes in Figure 8.1. We use our knowledge of baking to identify these subprocesses and their sub-subprocesses.

In addition to our process flow, we identify logical concerns. In this case, we mark several places where waste/scrap might appear. These notes of concern help in both structuring and evaluating our alternatives, and provide a means for the value-created focus of the process basis.

In order to place people in our process picture, we group our subprocesses into team clusters. These clusters are depicted in Figure 8.2. Here, we see logical clusters that contain whole subprocesses. This natural clustering is devised in order to allow our people to address the natural flow of our product. This address is based on both responsibility for subprocesses and their sub-subprocesses as well as the authority to effectively and efficiently produce our product. Team purposes are made up from subprocess purposes, just as team domains are made up of subprocesses.

We continue to add detail to our process definition by focusing on individual team/subprocess units. Here, we emphasize customer outcomes and the demands we make on our suppliers. A summarized example regarding the cookie/topping team is depicted in Figure 8.3. Here, we define and describe our desired products/by-products and supplies in terms of outcomes and demands at the boundaries of our team/subprocess combination. Within each team/sub-subprocess combination, we likewise define products and supplies, and respective outcomes and demands.

The reasoning for our systematic process definition is to follow a logical progression, adding detail to our process definition as necessary. The physical nature of our product, as expressed through physical transformation of inputs to outputs, drives this development of detail. **We concentrate on what needs to be done without a tight focus on how, exactly, we will accomplish our outcomes.** The "how" details will follow when we address our alternatives and their subsequent evaluation.

Our process layouts, although reasonably simple—we add no more complexity than absolutely necessary—serve several purposes later. They aid us in making process choices, as well as serve as a basis for the subsequent development of detailed process requirements. Furthermore, they serve as starting points for process control and process improvement activities, described in Sections 4 and 5, respectively.

Production process purpose: Given product and packaging specifications, produce and package a product that meets all specifications at all times, and actively participate/lead in product, package, and production improvements so as to assure customer satisfaction.

Materials subprocess purpose: Store and move the right materials, in the right quantities, at the right time, to serve the needs of the dough subprocess.

• • •

Packaging subprocess purpose: Produce a wrapped and packaged product that meets all specifications, on time, every time.

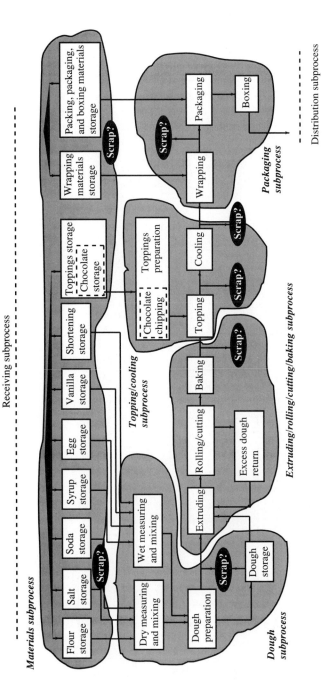

Figure 8.1 Production process flowchart with subprocesses and potential scrap locations identified (Downtown Bakery Case).

191

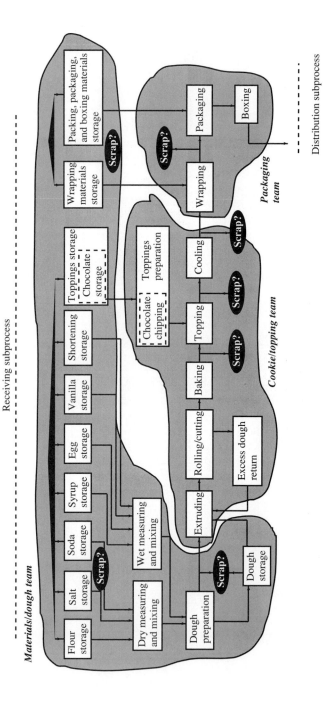

Figure 8.2   Production process flowchart with teams identified (Downtown Bakery Case).

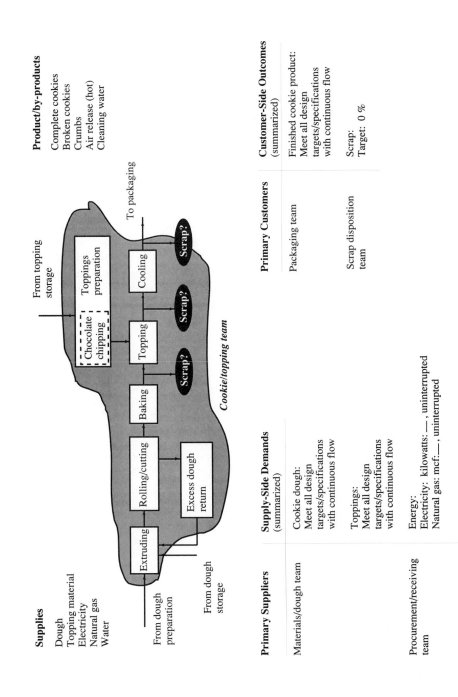

Figure 8.3   Cookie/topping team customer-side outcomes and supplier-side demands (Downtown Bakery Case).

EVALUATION

Process choice is established through evaluation and decision. To set up decision formats, several things must happen. Table 8.2 enumerates our basic sequence. First, we describe our alternatives. At this point, we begin with our process and subprocess purposes, process and subprocess flowcharts, inputs, and outputs.

We add detail to our basic options, re-expressing them as alternatives. Each alternative is described in terms of characteristics relevant for decision making. Each and every process definition or redefinition deserves to be handled on an exceptional basis. However, relevant aspects of impact include (1) physical performance, (2) economic performance and scale (cost/revenue, volume), (3) timeliness performance, (4) customer service performance, and (5) process flexibility. Descriptions of alternatives include detailed qualitative and quantitative estimates regarding critical impact in both the short term and the long term.

A critical aspect of process definition or redefinition is to improve the production system as a whole. Care is taken to visualize or possibly simulate the impact that the process definition or redefinition will have beyond its own boundaries—impacts on other processes.

Once we have generated, described, and estimated our alternatives in all decision-relevant aspects, we structure comparisons. Here, **for relative comparisons, we assess only parameters of significant difference; for relative decision purposes, parameters that do not differ between alternatives need not be assessed.** Provided we can express our alternatives in

Table 8.2   Details related to evaluation.

Action	Description
Develop and describe feasible alternatives	From process-subprocess layouts, structure feasible alternatives, and position each in the context of the process being defined or redefined.

Assess outputs offered and associated risk:
    Physical (function, form, fit)
    Economics and scale
    Timeliness
    Customer service
    Flexibility
    Other
	For each alternative listed describe significant output features, develop outcome estimates (some will be quantitative, some qualitative), and identify output concerns.

Assess inputs required:
    People
    Machines
    Facilities
    Materials/supplies
    Other
	For each alternative listed, estimate resource levels necessary for addressing each outcome, and identify input concerns.

Express output-input balances, differences, comparisons:
    Quantitative parameters
    Qualitative parameters
	Set up a logical comparison format that will consider all relevant output and input parameters of difference.

Choose process alternative	From the above analysis, and any other relevant information, choose the most appropriate alternative.

meaningful physical, economic, timeliness, customer service, and flexibility parameters, both qualitative and quantitative methods exist to perform the evaluations. Generally accepted rules for economic analysis exist, e.g., discounted cash-flow techniques and tools. If we can express our alternatives with reasonably accurate and precise estimates, then we can make reasonably sound economic comparisons. However, when significant factors other than economic factors surface, our decisions also hinge on how we weigh these factors.

In the end, **our decision process contains some measure of subjectivity due to perceived risk in both markets and variables** that we cannot accurately and precisely quantify. Here, **we define risk in two basic dimensions:** (1) **probability of outcome occurrence** and (2) **magnitude of outcome occurrence.** Sometimes higher levels of risk, e.g., lower probability of success and higher payoff, are assumed, and both our technology and people are challenged. At other times low levels of risk, e.g., modest payoff and high probability of success, are pursued. Limited-scale implementation is used to test, tune, and tame alternatives. Hence, our process definition and re-definition protocol is robust but not totally immune to risk of failure.

DOWNTOWN BAKERY CASE—EVALUATION OF PROCESS OPTIONS   In this case extension, | **Example 8.2**

we focus on the detailed decisions necessary to develop a production system plan. In particular, we focus on the baking sub-subprocess. Within the baking sub-subprocess, two primary technologies provide alternatives: (1) conventional forced-air cooking technology or (2) infrared radiation cooking technology. In order to set up our evaluation, we identify a list of relevant, critical decision parameters. This list appears in Figure 8.4. Here, we focus on costs associated with equipment, facilities, energy, scrap, labor, and service/maintenance. In addition, product uniformity, throughput time, and process reliability are identified as critical decision parameters.

Each alternative is depicted as an in-out/to-from graphic in Figure 8.4. These graphics are helpful to communicate the physical nature of our alternatives. However, they are not sufficient to support a decision. For decision support, we detail our critical decision parameter list. First, we construct a cash-flow model using only parameters of difference. Here, a parameter of difference is defined as a cost parameter that differs with respect to the alternatives at hand. For example, initial costs are considerably different for these two alternatives, while their utilities hookup costs are not significantly different. Hence, we list equipment cost in our model and do not list utilities hookup costs. By using only the parameters of difference, we simplify our model and still obtain a fair relative comparison in terms of choosing between these two alternatives. However, we are unable to estimate total costs; i.e., our simplification allows us to model only on the basis of differences.

Comparative economic model results are provided in Tables 8.3 and 8.4 for the conventional and the infrared alternatives, respectively. The economic parameters of difference are summarized on a cost basis. This basis does not allow a classical payout period analysis, as there are no revenues associated with our economic parameters of difference. Instead, we have used a net present value/cost method whereby we have discounted our cost back to the "present time," i.e., in year zero dollars. This discounting method is used to reflect and adjust for the time value of money. In this case, we have used a minimum attractive rate of return, MARR, of 15 percent. This rate was chosen as it represents a reasonable expectation for a rate of return on company assets. For example, if we used the money we are considering spending for our oven otherwise, we could earn about 15% on this money. We therefore expect that an oven investment should yield at least 15% or more to be worthwhile.

In this case, both oven alternatives are capable of meeting our volume needs. The lower net present value/cost (NPV) of $(283,629) for the infrared oven provides it with a significant economic advantage over the conventional alternative, NPV of $(459,732)—even though the conventional alternative has a lower

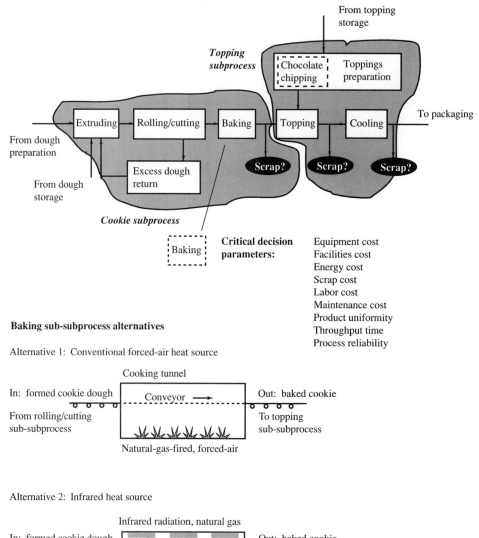

Baking sub-subprocess alternatives

Alternative 1: Conventional forced-air heat source

Alternative 2: Infrared heat source

Figure 8.4   Baking sub-subprocess alternatives (Downtown Bakery Case).

Table 8.3   Cash-flow model—baking sub-subprocess alternative 1, conventional oven (Downtown Bakery Case).

Parameters of Difference

Time Frame—End of Year	0	1	2	3	4	5	6
Facilities	$ (30,000)	0	0	0	(5,000)	0	0
Equipment (leased)	(15,000)	(10,000)	(10,000)	(25,000)	(10,000)	(10,000)	(10,000)
Service/maintenance	0	0	(2,000)	(5,000)	0	(2,000)	(5,000)
Labor	0	(22,500)	(22,500)	(22,500)	(22,500)	(22,500)	(22,500)
Energy:							
Natural gas	0	(10,000)	(18,000)	(25,000)	(32,000)	(40,000)	(50,000)
Electricity	0	(800)	(1,000)	(1,300)	(1,700)	(2,100)	(2,500)
Scrap loss	0	(6,500)	(15,000)	(22,000)	(30,000)	(40,000)	(50,000)
Source/sorting inspection	0	(15,000)	(18,000)	(21,000)	(22,500)	(25,000)	(26,000)
Total	$ (45,000)	$(64,800)	$(86,500)	$(121,800)	$(123,700)	$(141,600)	$(166,000)
Present value (15%)	$ (45,000)	$(56,348)	$(65,406)	$ (80,085)	$ (70,726)	$ (70,400)	$ (71,766)
Net present value (Year 0)	$(459,732)						
MARR discount rate	0.15						

Other Critical Parameters

Product uniformity	Wide variation in color and texture from side to side in baking tunnel
Throughput time	12 minutes
Process reliability	90% uptime

NOTES: 1. For the conventional oven alternative, the equipment life is estimated at 3 years, with a replacement oven installed at the beginning of year 4, with facilities installation charges at the end of year 4.

2. The alternative is characterized by low equipment costs, high energy costs, high labor costs, high scrap costs/high product variation, long throughput time, and low process reliability.

initial lease/cost. The infrared alternative provides significant gains in energy and scrap costs. Hence, over the 6-year time horizon, the infrared alternative looks very attractive. The net present value is calculated as

$$PV_n = \text{total cash flow entry } [(1 + i)^{-n}]$$

$$NPV = \sum_{\text{all } n} PV_n \qquad\qquad \textbf{[8.1]}$$

where PV_n = present value associated with the cash flow from time period n
       NPV = net present value at the present time, sometimes called year 0
        i = minimum attractive rate of return (MARR)
       n = time period index

Further discussion of discounted cash-flow tools is provided in Chapter 20.

Tables 8.3 and 8.4 also summarize findings regarding other relevant critical decision parameters. Here, a lack of product uniformity with respect to cooking effects, i.e., over- and undercooked product at the edges and center, respectively, is a distinguishing characteristic of the conventional oven. This leads to the high scrap costs for the conventional oven. The infrared oven produces a more uniform cooking effect that manifests itself in a low scrap rate. The infrared oven also produces a shorter cycle or throughput time, i.e., 8 minutes versus 12 minutes. This shorter time allows more product per shift to be produced. In addition, the infrared alternative has a slight edge in process reliability, i.e., estimates of 95% compared to 90%. Here, we define process reliability in general terms, as a ratio of uptime to total time (uptime plus downtime).

Table 8.4   Cash-flow model—baking sub-subprocess alternative 2, infrared oven (Downtown Bakery Case).

Parameters of Difference

Time Frame—End of Year	0	1	2	3	4	5	6
Facilities	$ (35,000)	0	0	0	0	0	0
Equipment (leased)	(20,000)	(20,000)	(20,000)	(20,000)	(20,000)	(20,000)	(20,000)
Service/maintenance	0	0	(800)	(1,200)	(5,000)	(1,500)	(2,500)
Labor	0	(22,500)	(22,500)	(22,500)	(22,500)	(22,500)	(22,500)
Energy:							
Natural gas	0	(8,000)	(10,500)	(13,500)	(17,500)	(21,000)	(25,000)
Electricity	0	(300)	(400)	(550)	(800)	(1,100)	(1,500)
Scrap loss	0	(400)	(550)	(650)	(900)	(1,200)	(1,500)
Source/sorting inspection	0	(200)	(250)	(300)	(350)	(400)	(450)
Total	$ (55,000)	$(51,400)	$(55,000)	$(58,700)	$(67,050)	$(67,700)	$(73,450)
Present value (15%)	$ (55,000)	$(44,696)	$(41,588)	$(38,596)	$(38,336)	$(33,659)	$(31,754)
Net present value (Year 0)	$(283,629)						
MARR discount rate	0.15						

Other Critical Parameters

Product uniformity	Highly uniform product in color and texture
Throughput time	8 minutes
Process reliability	95% uptime

NOTES: 1. For the infrared oven alternative, the equipment life is estimated at 6 years.

2. The alternative is characterized by high equipment costs, low energy costs, low scrap costs/low product variation, low inspection costs, low throughput time, and high process reliability.

Our decision as to an oven/cooking tunnel alternative is straightforward. The infrared alternative is a clear winner. In interpreting analyses such as this one, we keep in mind that we make choices only with respect to the alternatives we were able to capture in our analyses. If a better alternative really does exist, and we fail to identify it in our alternative set, we may eventually place ourselves at a competitive disadvantage when our competition does discover and implement the overlooked alternative.

PLAN

Once we select a process alternative, we commit ourselves to further development in the form of defining technical details necessary to make the process alternative work within our production system. Table 8.5 addresses process planning relative to process definition/redefinition.

Our plans begin with the process and subprocess flow layouts previously generated. We verify process and subprocess inputs and outputs, suppliers and customers. Then, we set targets and specific requirements for our inputs and outputs. In some cases, off-line experiments and response surface models, as described in Chapters 5 and 6, are used to determine targets. In other cases, targets and specifications may be more obvious.

We define critical process and subprocess metrics in order to track and eventually control our process and subprocess results. Our metrics include relevant parameters associated with

Table 8.5   Details related to plan.

Action	Description
Identify process customers and suppliers and define detailed input and output requirements	Using the process flow for the chosen process, review customers and suppliers and respective inputs and outputs at the boundaries. Then, determine operating targets and specifications associated with these inputs and outputs.
Identify subprocess customers and suppliers and define input and output requirements	Using the subprocess flow for the chosen process alternative, identify customers and suppliers and inputs and outputs at the boundaries. Then, determine operating targets and specifications associated with these inputs and outputs.
Identify critical metrics	From both process and subprocess flows, identify metrics that will calibrate critical aspects of physical performance, economics, timeliness, and customer service.
Preidentify critical control points and process variable targets: Physical Economic Timeliness Customer service	Isolate cause-based variables within the process/subprocess that can be controlled. Then determine "best" target levels, using off-line experiments if necessary.
Develop PVC and related models	Identify linkages between business metrics and physical variables that influence them.

physical, economic, timeliness, and customer service performance. **Along with the more obvious product-output process-related metrics, we examine basic cause-effect relationships between the output metrics and fundamental upstream variables within the process boundaries.** For example, chiller load and temperature may impact production costs of natural gas. But the precise relationships may not be clear. A designed experiment and response surface model can be produced to quantify this relationship and help set process targets; for example, see the Energy Cost/Operations Cases developed in Chapters 5 and 6, Examples 5.6, 6.1, and 6.2. Once our targets are set, we resort to process control methods to stabilize and hold to the targets.

Finally, **we construct the process value chain (PVC) diagram to connect our process results to our process variables. The PVC structure is used to document and understand our process, as well as to communicate and reinforce the importance of process vigilance and control.** For example, if upstream variables drift off target, then process results will eventually drift off target as well. A complete process value chain structure connects high-level business metrics, such as return on investment, to process metrics, to subprocess metrics, and finally to process variables. Additionally, we construct cash-flow models that help us project business results into the future.

DOWNTOWN BAKERY CASE—PRODUCTION PROCESS PLAN   Revisiting our Downtown | **Example 8.3**
Bakery production process, we add more detail, eventually to the point of producing a workable process plan. Here, we focus on each sub-subprocess in a physical sense, describing in-out/from-to characteristics as well as internal physical characteristics that we can use to provide further process descriptions as well as link with process control.

A broad-based process cause-effect analysis is helpful in identifying basic variables that potentially impact our process results. Figure 8.5 depicts a C-E diagram broken out by basic processes/subprocesses.

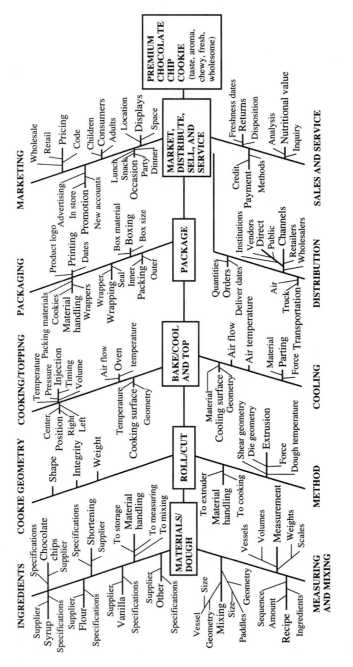

Figure 8.5 Downtown Bakery process/subprocess cause-effect diagram (Downtown Bakery Case).

Such a diagram provides a starting point for more detailed cause-effect analyses focused on each subprocess/sub-subprocess. We cascade our level of detail down to the point of locating meaningful process control variables.

For illustrative purposes, we have focused on two critical sub-subprocesses, baking and topping. Figure 8.6 depicts the flow and physical nature of each. Although these depictions are highly summarized, we eventually define details regarding quality, business, operations, and maintenance requirements, and any other requirements that are essential for effective and efficient sub-subprocess operations. We also examine inter-relationships between the sub-subprocesses so that they become complementary rather than competitive. For example, we want to prevent a gain in one sub-subprocess from creating a loss in another one.

Our in-out/from-to analyses are purpose-driven so that we can satisfy our process needs and our process customers. Finally, we focus within our sub-subprocesses and identify control issues useful in defining a process control scheme. Figure 8.7 depicts several issues that are critical to the topping sub-subprocess. Here, we see position, timing, temperature, weight, and product appeal as driving issues. We seek to control these issues by locating leverage points and levers in our physical process and applying process control strategies to each. These strategies as well as details on locating and exploiting process leverage points are described in Section 4.

At this point, we identify our comprehensive business metrics, e.g., return on investment, and begin to construct our process value chain, PVC. C-E analysis serves to establish the left-hand side of the PVC. Figure 8.8 depicts a partial PVC structure. This chain stretches from return on investment back to basic process variables. When fully developed, PVCs are extensive, branching back through each process to basic variables. A PVC is essentially a process-based model of our organization; using this model, we can locate control and improvement points. The PVC allows all personnel to see how what they do impacts our business as well as our customers in terms of cause, effect, relative to targets and benchmarks.

As an extension of our detailed analysis format, we develop cash-flow models for entire processes, as well as for entire production systems. With regard to the Downtown Bakery enterprise, we develop a production system cash-flow model, shown in Table 8.6. Here, we identify our parameters associated with cash flow and map them against a time frame of 5 years. The first year is broken into quarters, i.e., 3 months each. The production schedule at the top of Table 8.6 provides the volume basis for our production process. Here, we see a gradual ramp-up in production to 500,000 pounds per year in year 5. We also see and compensate for scrap/waste, which when deducted leaves us with an estimate of salable product.

In this case, we include all economic parameters, not just those of interest or difference, because we want to develop an enterprise economic model. In general, our cash-flows include both costs and revenues. However, in the production process we see only costs. When we develop a similar model for our distribution/marketing/sales/customer service process, we see both costs and revenues, i.e., positive and negative cash flows. We can develop submodels for each row entry in Table 8.6. For example, we could, but did not, break out the equipment row and include entries for each piece of equipment in the production process. Using a spreadsheet, we can link all of these entries together and provide a model that we can use to assess the economic impact of different production schemes on our production process as a whole.

In this particular production process model, we include a unit cost analysis and a net present value analysis. The unit cost analysis is crude, as it simply divides our total 5-year cash flow by anticipated product units. But it is critical in order to summarize our projected production costs, i.e., $1.89 per pound, so that we can benchmark against competitors. When we subtract our estimated unit production cost from the price of the product to our distributors, i.e., $6.00 per pound, we can obtain a gross profit figure of $4.11 per pound.

The net present value for the production process is useful to provide a benchmark figure for comparing production process alternatives as a whole. In other words, we can compare entire process alternatives, provided we have counterpart process net present values, and provided we use the same time horizon for all comparable processes. If our time horizons differ, we can develop what are termed *equivalent annual values,* whereby we take our net present value and spread it over our time horizon, using our MARR value. Cash-flow modeling is explained in more detail in Chapter 20.

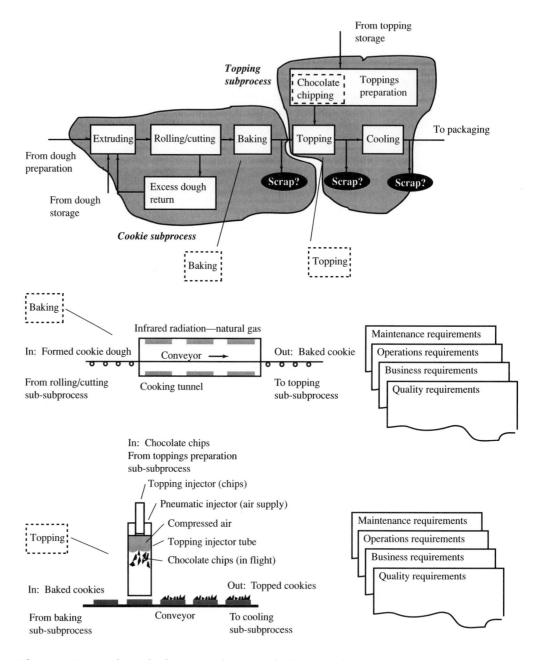

Figure 8.6   Baking sub-subprocess and topping sub-subprocess plans (Downtown Bakery Case).

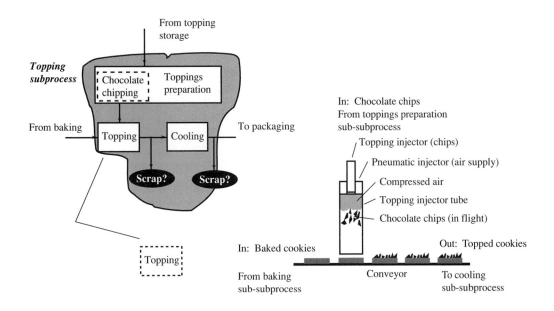

Process control—Topping sub-subprocess

Issue	Focus	Leverage	Strategy
Position	Conveyor Cookie base Injector tube	Alignment—chips dispersed on cookie	APC
Timing	Chips Air shot	Force—chips embedded in cookie	APC
Temperature	Chips Cookie base	Penetration—chips attached to cookie	APC
Package label weight	Chips Cookie base	Combined weight— product weight as labeled	APC and SPC
Appeal	Cookie	Appearance— color/surface texture	SPC

APC: Automatic process control
SPC: Statistical process control

Figure 8.7 Topping sub-subprocess control plans (summarized) (Downtown Bakery Case).

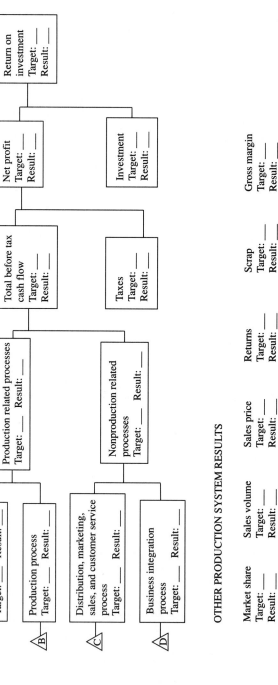

Figure 8.8 Downtown Bakery process value chain diagram (Downtown Bakery Case).

(continued)

CONTROLLABLE AND UNCONTROLLABLE VARIABLES

SUBPROCESS BUSINESS
AND TECHNICAL RESULTS

PROCESS BUSINESS
AND TECHNICAL RESULTS

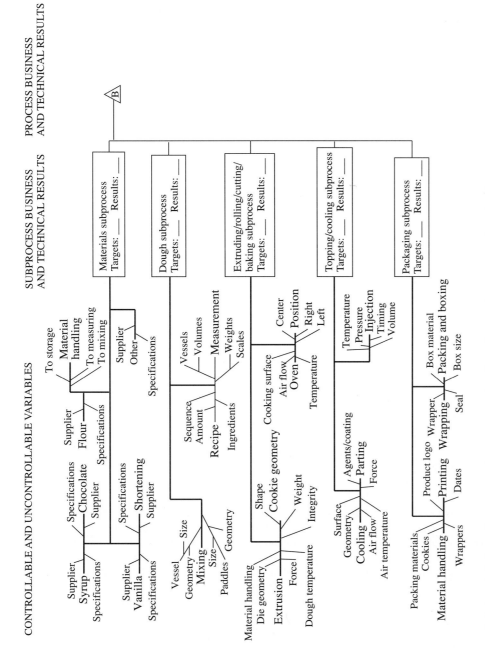

Figure 8.8 (concluded)   Downtown Bakery process value chain diagram (Downtown Bakery Case).

Table 8.6   Cash-flow model—production process, 5-year projection (Downtown Bakery Case).

Time Frame—End of Period	Q1 (Yr. 0.25)	Q2 (Yr. 0.50)	Q3 (Yr. 0.75)	Q4 (Yr. 1.0)	Year 2	Year 3	Year 4	Year 5
Production								
Production units, pounds	0	0	60,000	70,000	280,000	325,000	400,000	500,000
Scrap, %	0	0	2	1	0.5	0.5	0.5	0.5
Net salable production, pounds	0	0	58,800	69,300	278,600	323,375	398,000	497,500
Production Cash Flows								
Facilities	$ (141,000)	$ (33,000)	$ (15,000)	$ (15,000)	$ (15,000)	$ (15,000)	$ (15,000)	$ (15,000)
Equipment	(166,500)	(27,000)	(21,200)	(22,000)	(85,200)	(85,200)	(96,500)	(96,500)
Service/maintenance	0	0	(800)	(1,000)	(4,100)	(4,800)	(5,900)	(7,400)
Materials	(750)	(1,500)	(17,000)	(20,000)	(83,000)	(96,000)	(118,000)	(148,000)
Supplies	(500)	(1,000)	(2,000)	(2,500)	(10,000)	(11,500)	(14,000)	(18,000)
Personnel:								
Operations	(56,000)	(56,000)	(56,000)	(56,000)	(225,000)	(225,000)	(225,000)	(225,000)
Management	(11,500)	(11,500)	(11,500)	(11,500)	(47,000)	(47,000)	(47,000)	(47,000)
Energy/utilities	(500)	(750)	(4,000)	(5,000)	(18,000)	(21,000)	(31,000)	(48,000)
Wastes	(200)	(200)	(1,100)	(1,500)	(5,700)	(6,700)	(8,300)	(10,500)
Compliance (property taxes/ licenses, etc.)	(27,800)	(1,900)	(750)	(750)	(6,500)	(6,500)	(6,500)	(6,500)
Other	(25,500)	(2,000)	(2,000)	(2,000)	(8,000)	(8,000)	(8,000)	(8,000)
Total	$ (429,750)	$(134,850)	$(131,350)	$(137,250)	$(507,500)	$(526,700)	$(575,200)	$(629,900)
Net Present Value Analysis:								
Present value of total (15%)	$ (414,994)	$(125,748)	$(118,279)	$(119,348)	$(383,743)	$(346,314)	$(328,872)	$(313,172)
Net present value	$(2,150,469)							
MARR discount rate	0.15							

Unit Cost Analysis (Total Cash Flow/Salable Production in Pounds)

Total production costs	$(3,072,500)
Total salable production, pounds	1,625,575
Ave. unit production cost, $/lb	$(1.890)

If we choose, we can build an entire economic model for a production system by combining all process models. Table 8.7 illustrates such a model. We have omitted the usage and disposal processes in this model, as we desire to focus only on the production system processes for which we have direct control. At this level, we perform a payback analysis as well as a net present value analysis. The *payback analysis* simply cumulates the total cash flows. When this cumulated value changes from negative to positive, we define this change point as the payback period. In this case, we see a payback period of about 3 years.

The net present value, $1,034,779, reflects an estimated discounted cash value relative to the entire production system over the 5-year time horizon. Here, we are looking for a positive value to support a decision to proceed with our production system plans. A negative value implies that we are not meeting our MARR regarding our entire production system. If we encounter a negative NPV, the overall implication is that we could better invest our money in other alternatives.

Table 8.7 Cash-flow model—Production system, 5-year projection (Downtown Bakery Case).

Time Frame—End of Period	Q1 (Yr. 0.25)	Q2 (Yr. 0.50)	Q3 (Yr. 0.75)	Q4 (Yr. 1.0)	Year 2	Year 3	Year 4	Year 5
Production								
Production units, pounds	0	0	60,000	70,000	280,000	325,000	400,000	500,000
Scrap, %	0	0	2.00	1.00	0.50	0.50	0.50	0.50
Returns, %	0	0	8.00	4.00	1.50	1.50	1.50	1.50
Production sold, pounds	0	0	54,096	66,528	274,421	318,524	392,030	490,038
Process Cash Flows								
Definition/design/ development	$ (50,000)	$ (15,000)	$ (5,000)	$ (5,000)	$ (20,000)	$ (22,000)	$ (25,000)	$ (27,500)
Production	(429,750)	(134,850)	(131,350)	(137,250)	(507,500)	(526,700)	(575,200)	(629,900)
Distribution	(35,800)	(16,000)	(29,000)	(31,200)	(116,450)	(129,550)	(143,700)	(180,150)
Marketing/sales/customer service	(91,500)	(53,500)	266,955	332,529	1,359,408	1,606,696	2,006,011	2,533,764
Business integration	(158,900)	(66,900)	(70,500)	(73,400)	(293,000)	(312,500)	(323,500)	(333,000)
Total cash flow—production system	$ (765,950)	$ (286,250)	$ 31,105	$ 85,679	$ 422,458	$ 615,946	$ 938,611	$1,363,214
Payback Analysis								
Cumulative cash flow	$ (765,950)	$(1,052,200)	$(1,021,095)	$(935,416)	$(512,959)	$ 102,988	$1,041,599	$2,404,813
Payback period	3 years							
Present Value Analysis								
Present value of cash flow (15%)	$ (739,649)	$ (266,930)	$ 28,010	$ 74,503	$319,439	$ 404,995	$ 536,654	$ 677,758
Net present value	$1,034,779							
MARR discount rate	0.15							

Positive NPVs indicate that the production system, as modeled, produces more than the MARR. High positives, in general, indicate a higher rate of return. Short paybacks indicate a profitable production system, provided that the cash flows do not reverse their trends over time, turning back to negative cumulated totals. This reversal is unusual, but could and does take place, e.g., due to high periodic maintenance or some other large cash-flow reversal from litigation issues brought about by customers or governmental bodies. Examples include pollution problems, personal injury/death accidents, and so on. If these external possibilities appear to be significant, we can include the related usage and disposal/recycle processes in our model, thus adding a level of comprehensiveness to our analysis.

An inadequate NPV and/or payback period does not automatically imply that we should abandon our planned production system. But it does imply that our current definition is not satisfactory, and that we should explore and improve our process definition so that we reduce our risk of business failure.

In order to build and support our production system process entries in Table 8.7, we develop subschedules by process. Table 8.8 depicts several supporting process subschedules. First, the distribution subprocess is broken out into major cash-flow categories. Then, marketing/sales/customer service is broken out by cost and revenue categories. In the marketing/sales/customer service breakout schedule, we see both a cost portion and a revenue portion. The revenue portion is driven directly from the production volumes, considering

Table 8.8   Cash flow supporting schedules—Production system, 5-year projection (Downtown Bakery Case).

Time Frame—End of Period	Q1 (Yr. 0.25)	Q2 (Yr. 0.50)	Q3 (Yr. 0.75)	Q4 (Yr. 1.0)	Year 2	Year 3	Year 4	Year 5
Distribution Subprocess Cash Flows								
Facilities	$ (20,000)	$ (1,000)	$ (1,000)	$ (1,000)	$ (5,000)	$ (5,000)	$ (5,000)	$ (5,000)
Equipment (leased)	(5,400)	(5,400)	(5,400)	(5,400)	(21,600)	(21,600)	(21,600)	(32,400)
Service/maintenance	(1,500)	0	(500)	(500)	(2,500)	(2,500)	(3,000)	(3,000)
Materials (cases, etc.)	(500)	(500)	(1,000)	(1,000)	(3,000)	(3,000)	(4,000)	(5,000)
Supplies (fuel, parts, etc.)	(500)	(500)	(5,500)	(6,000)	(25,000)	(30,000)	(35,000)	(45,000)
Personnel:								
Operations	(4,000)	(5,000)	(7,000)	(8,000)	(35,000)	(40,000)	(45,000)	(55,000)
Management	(2,500)	(500)	(1,500)	(2,000)	(7,000)	(8,000)	(9,000)	(10,000)
Utilities	(250)	(250)	(450)	(650)	(2,500)	(3,500)	(4,500)	(5,500)
Wastes	0	0	(150)	(150)	(350)	(450)	(600)	(750)
Compliance (property taxes/ licenses, etc.)	(850)	(850)	(4,500)	(4,500)	(6,500)	(7,500)	(8,000)	(10,500)
Other	(2,000)	(2,000)	(2,000)	(2,000)	(8,000)	(8,000)	(8,000)	(8,000)
Total—distribution process	$ (35,800)	$(16,000)	$(29,000)	$(31,200)	$(116,450)	$(129,550)	$(143,700)	$(180,150)
Business Integration Subprocesses Cash Flows								
Information/record keeping	$ (35,000)	$ (7,000)	$ (7,000)	$ (7,000)	$ (28,000)	$ (35,000)	$ (40,000)	$ (42,000)
Business planning	(11,600)	(11,600)	(11,600)	(11,600)	(46,000)	(46,000)	(46,000)	(46,000)
Accounts payable/receivable	(16,300)	(6,300)	(6,300)	(6,300)	(25,000)	(25,000)	(25,000)	(25,000)
Financial and cash management	(15,000)	(15,000)	(15,000)	(15,000)	(60,000)	(60,000)	(60,000)	(60,000)
Payroll/human resources/ benefits	(8,500)	(12,000)	(6,000)	(6,000)	(23,000)	(23,000)	(23,000)	(23,000)
Safety/health/environmental	(12,000)	(2,000)	(5,000)	(6,500)	(25,000)	(30,000)	(30,000)	(30,000)
Public relations	(4,500)	(1,500)	(4,500)	(5,000)	(22,000)	(24,000)	(24,000)	(24,000)
Procurement	(15,000)	(4,000)	(6,300)	(6,500)	(26,000)	(29,000)	(33,000)	(38,000)
Receiving	(16,000)	(2,500)	(3,800)	(4,500)	(18,000)	(20,500)	(22,500)	(25,000)
Other	(25,000)	(5,000)	(5,000)	(5,000)	(20,000)	(20,000)	(20,000)	(20,000)
Total—business integration process	$ (158,900)	$(66,900)	$(70,500)	$(73,400)	$(293,000)	$(312,500)	$(323,500)	$ (333,000)

(continued)

returns and scrap. Here, we see rather large positive cash flows starting in the third quarter. However, from the production system schedule, Table 8.7, we pick up the costs from all other processes and see a slower cash-flow growth for the production system as a whole. Finally, the business integration process is broken out by subprocesses. Each subprocess is typically built up from other more detailed models, all linked together in the spreadsheet.

Cases up to this point have illustrated the concept of moving toward more and more detail. The partial economic, cash-flow models developed for alternative selection provide details but not a comprehensive look at the processes as a whole. **Our process and production system enterprise economic models provide a comprehensive view of possible economic outcomes.** Typically, we include federal/state taxes in our cash-flow models, but this detail has been omitted in order to hold model complexity down to a minimum.

Table 8.8 (concluded)   Cash flow supporting schedules—Production system, 5-year projection (Downtown Bakery Case).

Time Frame—End of Period	Q1 (Yr. 0.25)	Q2 (Yr. 0.50)	Q3 (Yr. 0.75)	Q4 (Yr. 1.0)	Year 2	Year 3	Year 4	Year 5
Marketing/Sales/Customer Service Subprocess Cash Flows								
Costs								
Facilities	$ (10,000)	$ (3,000)	$ (3,000)	$ (3,000)	$ (10,000)	$ (12,000)	$ (15,000)	$ (20,000)
Equipment	(25,000)	(500)	(800)	(1,000)	(5,000)	(7,000)	(9,000)	(11,500)
Service/maintenance	(1,500)	(200)	(300)	(400)	(2,000)	(2,500)	(3,000)	(3,500)
Supplies	(10,000)	(800)	(1,500)	(2,500)	(7,500)	(8,500)	(9,500)	(10,500)
Personnel:								
Operations	(15,000)	(20,000)	(20,000)	(20,000)	(85,000)	(90,000)	(100,000)	(120,000)
Management	(5,000)	(5,000)	(7,500)	(7,500)	(25,000)	(30,000)	(35,000)	(40,000)
Commissions (5% of salables)	0	0	(16,229)	(19,958)	(82,326)	(95,557)	(117,609)	(147,011)
Advertising/promotion	(20,000)	(20,000)	(15,000)	(15,000)	(60,000)	(50,000)	(50,000)	(50,000)
Energy/utilities	(1,000)	(1,000)	(1,000)	(1,000)	(8,500)	(8,500)	(9,000)	(9,000)
Compliance (property taxes/ licenses, etc.)	(500)	(500)	(500)	(500)	(2,000)	(2,000)	(2,000)	(2,000)
Other	(3,500)	(2,500)	(1,500)	(1,500)	(8,500)	(8,500)	(8,500)	(8,500)
Total costs—marketing/sales/ service	$(91,500)	$(53,500)	$(67,329)	$(72,358)	$(295,826)	$(314,557)	$(358,609)	$(422,011)
*Revenues**								
Product sales	0	0	$324,576	$399,168	$1,646,526	$1,911,146	$2,352,180	$2,940,225
Returns sales	0	0	9,408	5,544	8,358	9,701	11,940	14,925
Scrap sales	0	0	300	175	350	406	500	625
Other	0	0	0	0	0	0	0	0
Total revenues—marketing/sales/ service	$0	$0	$334,284	$404,887	$1,655,234	$1,921,254	$2,364,620	$2,955,775
Total—marketing/sales/ service process	$(91,500)	$(53,500)	$266,955	$332,529	$1,359,408	$1,606,696	$2,006,011	$2,533,764

*Sales price per pound: product $6.00, returns $2.00, scrap $0.25.

Within our cash-flow models, we have two basic choices for schedule breakouts: (1) by cost/revenue categories and (2) by subprocesses. Typically, for complex or extensive processes, we expect to see a breakout by subprocesses and sub-subprocesses and then break these subprocesses and sub-subprocesses out by cost category. The critical point is to include all relevant costs/revenues attributable to each process. Completeness is essential so that we provide a comprehensive estimate of process/production system cash flows. Depending on what we see in our estimated cash flows, we may need to back up and reconsider our alternative, with cost reductions in mind.

The composite package of annotated process/subprocess flow layouts, cause-effect documents, critical process point locations, targets and specifications, supplier and customer descriptions, the PVC structure, and cash-flow models makes up our process plan/model. Our process plan allows us to logically express and communicate our intentions and to move into the process creation subelement in Figure 7.1 **with the confidence that we know what to do and how to do it.**

8.3 PROCESS CREATION

Process creation issues are very different from planning-based issues—change is initiated and risk of failure is introduced. People (suppliers, operators, engineers, managers, external customers, stakeholders) are affected, equipment is affected, facilities are affected, materials and supplies are affected. No matter how brilliant the new process plan, missteps in implementation render it ineffective and inefficient. In other words, **plans describe our process; we create a reality through implementation.**

RESOURCES

Processes are not created in isolation; they interface with other processes—their people, facilities, equipment, and so on. This fact is reflected in process implementation and operations. Once we review and preplay our plan, locating and resolving possible bottlenecks and ambiguities in our previous work, we are ready to detail the process input resources necessary for implementation and operations; see Table 8.9.

The identification of resources and their sources may prove to be a simple matter in limited-scale processes, or it may be a very challenging matter in the case of large-scale processes. **First, we describe the exact nature of, and source for, our resources,** ranging from personnel-related knowledge and skills, to training, to equipment, to tooling, to facilities, to raw materials, to energy, to purchased parts and supplies, to information, to software. **Then, we obtain bids and quotations and select sources/suppliers for our required resources.**

Some resources are more difficult to obtain and take longer to obtain than others. These resources are critical components. Special attention is necessary to assure that critical resources are in place when needed. Hence, critical resources are clearly identified so that they can be scheduled properly and expedited accordingly in the schedule and action subelements, respectively.

SCHEDULE

At this point, the process plan is completed, and technical process requirements and resources are detailed. **Scheduling is the subelement we use to organize and coordinate our details, so that**

Table 8.9   Details related to input resources.

Actions	Description
Review/preplay the process definition/ redefinition plan	Working with the process plan, carefully review/preview the process and subprocesses in a whole production system context. Supply any missing details to complete the plan and resolve identified process bottlenecks.
Identify resources and sources	Identify exact resources and amounts needed to fulfill the process plan. Obtain bids and quotations.
Identify critical components	Locate critical process components that are likely to need special attention as to source and/or timeliness.

Table 8.10   Details related to schedule.

Actions	Description
Identify activities and milestones	Given the process plan and resource list, identify/list all implementation activities that are necessary to put the process in place, e.g., tear-outs, procurements, installations, and so on. Identify critical implementation activities and project milestones.
Time-sequence activities	Sequence the implementation activities identified above in terms of order of execution, with estimated time to completion for each activity.
Time-sequence resources to activities	Associate/align resources with activities, so that resource inputs are time-sequenced with activities.
Construct the implementation schedule	Piece together the activities, timing, and resource inputs to form a total implementation plan on a calendar time basis.

Identify critical implementation metrics:
  Physical
  Economic
  Timeliness
  Customer service
	Set up meaningful implementation execution metrics in the appropriate categories to monitor the implementation phase.

we can effectively and efficiently implement our plan—and create a living process. Table 8.10 addresses the scheduling subelement.

First, we identify activities and events. These activities and events correspond to the process plan. Usually, activity lists are used to summarize activities and events. **An activity is an action of some type that requires a time duration for accomplishment. An event happens at a point in time,** e.g., the beginning or ending point of an activity. **Our critical events represent milestones—points at which we reassess our progress.**

Next, we time-sequence our activities and events, noting our milestones in this sequence. This sequencing effort includes developing time estimates for each activity. Given a start date, we can build a schedule in calendar time. Depending on the complexity of the process implementation, this schedule may utilize one or more of several tools, e.g., a Gantt chart and/or the critical path method (CPM), both described in Chapter 20.

Activity sequences occur in three ways. First, activities can be **sequential,** whereby one activity must be finished before the next activity can begin. Second, activities can be **parallel,** whereby two or more activities can be ongoing at the same time and do not depend on each other. Third, activities can be **coupled,** whereby one or more activities are linked so that their progress as individual activities is mutually dependent in one way or another.

Considering the physical nature of each activity as well as the sequential nature of all activities together, we develop an activity list structure that labels each activity in the physical context and displays activity sequences. In addition, we estimate a duration for each activity for scheduling purposes. Given a start date, we can build a schedule for implementation. Depending on the complexity of the process, this schedule may vary from a simple Gantt chart to a complex computer-generated and -monitored CPM plan. Ultimately we produce a complete schedule that includes activities, durations, events, resources, and resource order points.

Finally, **we identify critical implementation-based metrics to measure our effectiveness and efficiency in the action subelement.** Here, our aim is to meet or better our scheduled plan. In evaluating our implementation we consider physical, economic, timeliness, and customer service performance parameters associated with implementation activities.

Example 8.4 | **DOWNTOWN BAKERY CASE—PRODUCTION PROCESS IMPLEMENTATION**   As a result of foresight in pursuing an expansion initiative, Downtown Bakery has received an offer to locate the new production facility near the present facility, in an economic enterprise zone. Furthermore, Downtown has been offered an abandoned building in this zone. This building once housed a small manufacturing facility. The building structure is well suited to the bakery expansion; however, the roof and interior spaces are not suited for this expansion as they currently stand. On the basis of an attractive incentive and tax abatement package, management has decided to pursue this offer.

Working though the production system plan, facilities and equipment plans have been developed. Contractors and vendors have furnished bids corresponding to the breakout of the production system plan. A summarized activity list appears in Table 8.11. This list contains activity descriptions, symbols, sequences, and estimated durations for each activity. Activities include tear-outs, facilities modifications, equipment installation, equipment integration, limited-scale production, and full-scale production. The activity symbols are necessary in order to simplify our scheduling layouts. The sequencing is necessary so that we can plan our implementation and activities in a logical order, helping our contractors and vendors to be at the right place, at the right time, with the right stuff. We have marked our milestone activities with a * in our activity list in order to expedite project control. Our objective is to have our production process in a full-scale production mode as soon as possible.

In order to schedule our implementation, we use the critical path method (CPM) scheduling tool. Our CPM diagram is depicted in Figure 8.9. Here, we have taken the symbol, sequence, and duration information from Table 8.11 and constructed a network of activities. The network flows from left to right in a time sequence. Each activity is represented by a node, i.e., a circle. Within each circle we have listed the activity's symbol and its estimated time duration. Other information shown includes earliest start time (ES), earliest completion time (EC), latest start time (LS), and latest completion time (LC). These estimates are provided for each node/activity on the network. The critical path is shown by bold arrows. The critical path is defined as that path which determines the minimum completion time for the entire implementation project. If a delay occurs on any activity on the critical path, then the project duration will be increased. Hence, we monitor the activities on the critical path very carefully. In addition, milestone activity/event points have been highlighted.

Table 8.12 has been constructed in order both to facilitate our network development and to summarize our results. Here, we have repeated our activity descriptions, symbols, and durations. We have listed our ES, EC, LS, and LC estimates, which match those in Figure 8.9. Additionally, we have included total slack (TS) and free slack (FS) estimates. Details regarding CPM network development and calculations are provided in Chapter 20.

Once we have an implementation plan and schedule developed, we use it to help us guide and support our implementation. For example, we can use updated CPM graphics and tables to update our plan as activities are completed. Additionally, we can project changes in subsequent activity estimates. Here, we use the same basic rules that we used to develop the initial CPM network, but begin at the end of the completed event. Hence, we produce updated ES, EC, LS, and LC estimates for the remaining activities, in addition to redeveloping our slack estimates. We can also determine if our critical path has changed as a result of our activity changes.

Table 8.11   Production process activities, sequences, and durations (Downtown Bakery Case).

Activity Description	Activity Symbol	Predecessor	Duration, Days
Interior tear-out	A	—	15
Roofing tear-off	B	—	5
Floor tear-out*	C*	A	21
Roofing installation	D	B	10
Footings installation—dig/form/pour/cure	E	C	12
Underfloor utilities/drains installation	F	E	8
Floor installation—form/pour/cure	G	F, D	4
Docks installation—form/pour/cure	H	E	7
Overhead utilities installation*	I*	G	14
Outside doors installation	J	G, H	3
Boiler/pumps/air compressor installation	K	G	5
Refrigeration units installation	L	G	3
Air-conditioning system installation	M	G	12
Storage bins/loading elevators/conveyors installation	N	I	4
Measuring/mixing drums/vessels/conveyors installation	O	N	3
Extruder/rolling/cutting equipment installation	P	I	2
Oven/cooking tunnel/conveyor installation	Q	I	4
Topping preparation/injection equipment/conveyors installation	R	N	3
Cooling tunnel/hookups installation	S	I	5
Wrapping/packaging equipment/conveyors installation	T	I	4
Materials handling system/integration/test installation*	U*	O, P, Q, R, S, T	5
Inside doors/curtains installation	V	U	3
Quality assurance laboratory equipment installation	W	V	5
Floor surface/markings installation	X	U	7
Office areas finish-out	Y	I	14
Limited-scale production/start-up*	Z*	V, W, X	5
Test/tune/mistakeproof production system	AA	Z	10
Full-scale production/start-up*	BB*	AA	6

| *Indicates milestone activities; milestone events occur at the end of the marked activity.

   Milestone events allow us to focus on critical points in our implementation activities. At these points in time we reflect on our previous activities in terms of physical, timeliness, economic, and customer service performance. For example, hiring and training activities may be linked to physical production system implementation activities. Any differences in actual and planned dates may require adjustments in hiring and training to allow a smooth production system start-up.

ACTION

Action requires a variety of activities; see Table 8.13. We follow the process implementation schedule, updating when necessary. Our plan most likely includes activities that test parts of the process, and perhaps interface with other processes, as they are installed. Some redesign and rework is usually necessary. Eventually, the whole process is ready to check out and test.

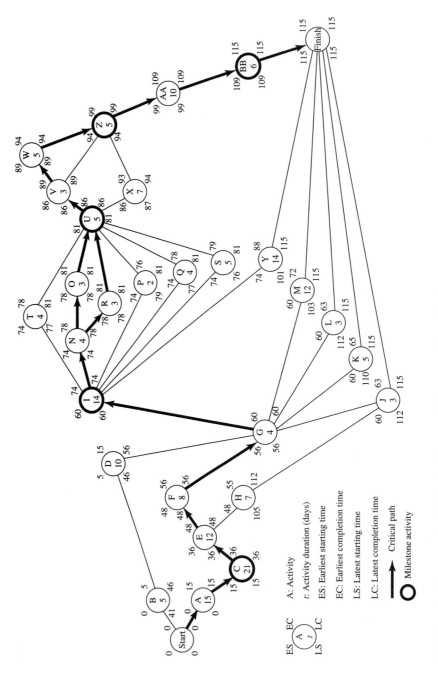

Figure 8.9 Production process CPM schedule network (Downtown Bakery Case).

Table 8.12 Production process scheduling details (Downtown Bakery Case).

Activity Description	Activity Symbol	Duration, Days	ES	EC	LS	LC	TS	FS	Critical Path?
Interior tear-out	A	15	0	15	0	15	0	0	Yes
Roofing tear-off	B	5	0	5	41	46	41	41	—
Floor tearout*	C*	21	15	36	15	36	0	0	Yes
Roofing installation	D	10	5	15	46	56	41	41	—
Footings installation— dig/form/pour/cure	E	12	36	48	36	48	0	0	Yes
Underfloor utilities/drain installation	F	8	48	56	48	56	0	0	Yes
Floor install—form/pour/cure	G	4	56	60	56	60	0	0	Yes
Docks install—form/pour/cure	H	7	48	55	105	112	57	5	—
Overhead utilities installation*	I*	14	60	74	60	74	0	0	Yes
Outside doors installation	J	3	60	63	112	115	52	52	—
Boiler/pumps/air compressor installation	K	5	60	65	110	115	50	50	—
Refrigeration units installation	L	3	60	63	112	115	52	52	—
Air-conditioning system installation	M	12	60	72	103	115	43	43	—
Storage bins/loading elevators/ conveyors installation	N	4	74	78	74	78	0	0	Yes
Measuring/mixing drums/vessels/ conveyors installation	O	3	78	81	78	81	0	0	Yes
Extruder/rolling/cutting equipment installation	P	2	74	76	79	81	5	5	—
Oven/cooking tunnel/conveyor installation	Q	4	74	78	77	81	3	3	—
Topping preparation/injection equipment/conveyors installation	R	3	78	81	78	81	0	0	Yes
Cooling tunnel/hookups installation	S	5	74	79	76	81	2	2	—
Wrapping/packaging equipment/ conveyors installation	T	4	74	78	77	81	3	3	—
Materials handling system/ integration/test installation*	U*	5	81	86	81	86	0	0	Yes
Inside doors/curtains installation	V	3	86	89	86	89	0	0	Yes
Quality assurance laboratory equipment installation	W	5	89	94	89	94	0	0	Yes
Floor surface/markings installation	X	7	86	93	87	94	1	1	—
Office areas finish-out	Y	14	74	88	101	115	27	27	—
Limited-scale production/start-up*	Z*	5	94	99	94	99	0	0	Yes
Test/tune/mistakeproof system	AA	10	99	109	99	109	0	0	Yes
Full-scale production/startup*	BB*	6	109	115	109	115	0	0	Yes

*Indicates milestone activities; milestone occurs at the end of the marked activity.

Table 8.13   Details related to action—limited-scale.

Action	Description
Execute scheduled plan	Install the process in accordance with the implementation schedule.
Test	Check out subprocesses piece by piece, as they are installed.
Commence limited-scale operations	Begin operations on a limited scale and observe closely for bottlenecks and related limitations.
Assess process performance: Physical Economics Timeliness Customer service	Collect and display data per critical metrics and critical process variables, including designed on-line experiments aimed at optimizing process results.
Mistakeproof and tune process	Address actual and potential process bottlenecks and limitations that appear in limited-scale operations.

Once sufficient implementation is completed, we pursue limited-scale process operations. Limited-scale operations allow us a trial period to test, mistakeproof, and tune the process before full-scale operations begin. We examine operations and results with respect to our targets and requirements. We assess physical, economic, timeliness, and customer service performance aspects of the new process. We are looking for bottlenecks and the limitations that create them—limitations create bottlenecks and limit process outputs.

Mistakeproofing and bottleneck resolution are approached in a variety of ways. Approaches may be reactive in nature: we isolate a process limitation, trace it to a root cause, and resolve it. Or they may be proactive in nature: we identify a process leverage point, isolate the process lever, and exploit it. In either case, **we look for opportunities to make process changes, such that mistakes in full-scale process operations are difficult, if not impossible, to make.** Initiatives and tools such as mistakeproofing, failure mode and effects analysis, fault tree analysis, and cause-effect analysis are helpful here; see Chapters 19 and 20 for details.

Finally, **we fine-tune the process.** Here, we make any necessary adjustments to process targets/operations regarding equipment settings or process variables before we commence full-scale operations. We may use designed experiments and resulting process response surface models to help tune or optimize the process. The Packaging Process Case in Chapter 6 (Examples 6.3 and 6.4) provides examples of process exploration and modeling regarding process targeting in start-up operations.

By the time we commence full-scale operations (see Table 8.14) our process and product metrics and their targets are identified. Additionally, sensors and data collection systems are in place so that we can assess on-line results. Tools such as runs charts help us to characterize our critical metrics relative to our targets. Refer to Chapter 4 for details regarding process characterization. Our process control leverage points and their sensors provide us with data useful to assess process stability. Here, **we use process control technology** (see Section 4, Chapters 9 to 15) **to target and stabilize our process, i.e., monitor and adjust critical process variables.**

Table 8.14   Details related to action—full-scale.

Action	Description
Commence full-scale operations	Move tuned and mistakeproofed process on-line.
Assess results: 　　Physical 　　Economics 　　Timeliness 　　Customer service	Collect, analyze, and display data per critical process metrics.
Stabilize process/subprocess	Apply process control technology (see Section 4).
Improve process	Apply process improvement technology (see Section 5).
Redefine process	Redefine process as new opportunities and technologies emerge (revisit Section 3).

As process experience is gained, we identify opportunities for process improvement, within the current process definition. To pursue these opportunities, we use process improvement technology; see Section 5, Chapters 16 and 17. **Eventually, we rethink our process by repeating the protocols described in this section—we redefine our process.**

REVIEW AND DISCOVERY EXERCISES

REVIEW

8.1. Why do we constantly focus on process and subprocess purposes? Explain.

8.2. Why/how do we cluster subprocesses into team domains? Explain.

8.3. What constitutes a natural boundary for a process/subprocess? Explain.

8.4. On what basis do we choose one alternative over another? Explain.

8.5. What is the minimum attractive rate of return, and how is it used in the net present worth method of analysis? Explain.

8.6. Why is it absolutely essential that we use the same time horizon for all alternatives assessed with the net present value method? Explain.

8.7. How can we handle both quarters and annual periods in the same process cash-flow model, regarding net present value calculations? Explain.

8.8. What constitutes a critical metric in the context of a process or subprocess? Explain.

8.9. What is a process control point? How are they identified? Explain.

8.10. How do we construct a PVC structure? What basic tools are indispensable in PVC construction? Explain.

8.11. What purpose does a process-based cash-flow model serve? How are process-based cash-flow models constructed? What parameters do they contain? Explain.

8.12. How does implementation differ from planning? Explain.

8.13. What is an activity? What is an event? How are they related? Explain.

8.14. What is a milestone in the context of an implementation plan? Explain.

8.15. What is an activity/sequence list? How is it constructed? See Chapter 20 for details.

8.16. What is a critical path? Explain.

8.17. Explain what is accomplished on (1) a forward pass in a CPM network and (2) a backward pass. See Chapter 20 for details.

8.18. How do we know when to switch from limited-scale to full-scale action in process definition/redefinition? Explain.

DISCOVERY

8.19. Does every critical metric need/deserve a target? Explain.

8.20. Define risk. What are its components? How does it affect process decisions? A brief review of risk is provided in the failure mode and effects tool section of Chapter 20.

8.21. Why does the equivalent annual value method work for projects with different time horizons? Explain.

8.22. Seamless project integration is a term that refers to a smooth continuum of activity/action involving planning and implementation. How can/should seamless integration be approached in process definition/redefinition? Explain.

8.23. In the Downtown Bakery's CPM network, shown in Figure 8.9, what would be the result if activity P was extended by 4 days? By 5 days? By 6 days? Address both project duration and critical path.

8.24. How is a CPM network a useful tool for in-progress monitoring of implementation? Explain.

See Chapter 7 Review and Discovery Exercises for project ideas.

PROCESS CONTROL

The purpose of Section 4 is to identify and describe the elements of process control in the context of a production system.

Chapter 9: Process Control—Concepts and Options

The purpose of Chapter 9 is to overview the elements of process control and introduce process control options and strategies.

Chapter 10: Process Monitoring—Variables Control Charts for Grouped Measurements

The purpose of Chapter 10 is to present fundamental modeling concepts and techniques for variables data that are collected in groups.

Chapter 11: Process Monitoring—Variables Control Charts for Individual Measurements and Related Topics

The purpose of Chapter 11 is to present fundamental modeling concepts and techniques for variables data collected as individual observations and to present concepts and models for process capability.

Chapter 12: Process Monitoring—Attributes Control Charts for Classification Measurements

The purpose of Chapter 12 is to present fundamental attributes modeling concepts and techniques for data collected in terms of defective items and number of defects.

Chapter 13: Process Monitoring—Nontraditional SPC Concepts and Models

The purpose of Chapter 13 is to discuss applications and misapplications of control charts as well as to introduce fundamental multivariate statistical process control models.

Chapter 14: Process Adjustment—Introduction to Automatic Process Control, Conventional Models

The purpose of Chapter 14 is to introduce fundamentals of discrete and continuous automatic process control (APC) modeling concepts and techniques.

Chapter 15: Process Adjustment—Introduction to Automatic Process Control, Unconventional Models

The purpose of Chapter 15 is to introduce nontraditional APC models such as neural networks and expert systems as well as combined APC/SPC models.

9

PROCESS CONTROL—CONCEPTS AND OPTIONS

9.0 INQUIRY

1. What is process control?
2. Why is process control essential?
3. What process control options exist?
4. How do process monitoring for special causes and process adjustment differ?
5. How is process control accomplished?

9.1 INTRODUCTION

A production system is a process hierarchy, consisting of basic processes and their respective sub-processes and sub-subprocesses. **Process control is a critical part of operations. Process control is a complex combination of measurement, comparison, and correction.** Box, Coleman, and Baxley [1] and Box and Luceno [2] cite two techniques for dealing with process control issues: techniques of process monitoring and techniques of process adjustment. As an extension of their discussion, **we consider process control strategies that address two broad issues:**

1. *Process monitoring* strategies, focused on process disruptions/special cause elimination—the detection, isolation, and removal of influences over and above common cause or natural variation that enters a process by virtue of controllable or uncontrollable variables.

2. *Process adjustment* strategies, focused on process regulation/adjustment—the manipulation of identified controllable input/transformation variables so as to influence the value of an output variable.

We design our process controls to effectively and efficiently work with both issues. These two issues overlap to some degree, and our process control strategies tend to reflect this overlap. **The process control focus is on understanding and exploiting cause-effect relationships between controllable, uncontrollable, and response variables.**

Three basic process control strategies exist: (1) **subjective-based**—where a human senses/measures, compares, and corrects intuitively, with limited objective/quantitative measurement and/or comparison; (2) **objective-based**—where a human, with the aid of instruments and mathematical/statistical analytical models/tools, senses/measures, compares, and corrects; and (3) **equipment-based**—where mechanical, electromechanical, and/or electronic equipment are utilized to perform the entire sequence of sensing/measuring, comparing, and correcting. These basic strategies in the context of open- and closed-loop processes were introduced in Chapter 3. These strategies and related models are further addressed in this section.

9.2   CRITICAL ELEMENTS OF PROCESS CONTROL

The process control triad depicted in Figure 3.1—measurement, comparison, and correction—is expanded in the form of basic elements and subelements in Figure 9.1. Direction for process control is provided through process purpose, goals, objectives, and targets. **We always start with a**

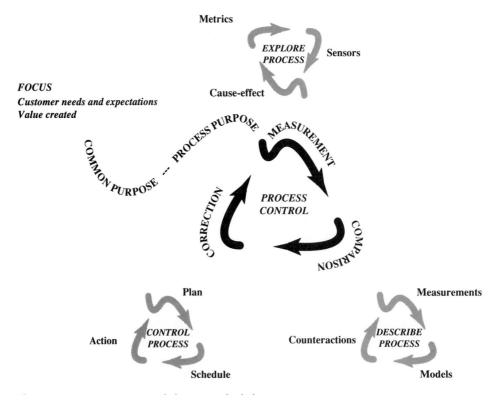

Figure 9.1   Process control elements and subelements.

clear process control purpose, relating the nature of our process and our goals, objectives, and targets. We seek to control from process leverage points, measuring, comparing, and correcting high-leverage variables.

Measurement issues in terms of process characterization and exploration have been addressed previously in Section 2. Relevant questions addressed in Chapter 4 include: Why do we measure? What do we measure? How do we measure? Here, we apply and expand our discussion: Where do we measure? When do we measure? How do we use the data collected to help control the process? In summary, based on cause-effect understanding, we identify our critical process leverage variables, define a scale of measure, and develop appropriate sensors, relevant to our process control purpose.

At the comparison element we describe our process. We characterize our process both physically and statistically, exploring relevant cause-effect issues. We focus on control-based issues regarding the need for counteraction. We determine an appropriate control strategy and model, perhaps monitoring a process using a process control chart associated with closed-loop manual controls, or perhaps adjusting a process input to keep an output near a setpoint with closed-loop automatic process controls. In the case of process monitoring, once we are satisfied that our model is valid, we link our model to special cause identification if possible, or at least an indication that signals the need for some sort of corrective action. The need for corrective action is signaled by the model, but corrective action is taken in the physical world—in the real process, not the process model—in order to remove the effect of the special cause. In the adjustment case, we typically track our process output, and when deviations occur, as judged against our setpoint/target, we adjust an appropriate input variable.

Next, we formulate our process control plan in the context of the whole process and schedule its implementation. In terms of action, we launch a limited-scale control scheme. We observe carefully, testing and tuning our control scheme. Once we are reasonably satisfied with its effectiveness and efficiency, we are ready for full-scale implementation. In full-scale operations, we focus on holding to target and stabilizing our process results.

Our goal in general, regardless of our process control strategy—subjective, objective, or equipment—is to consistently operate as close as possible to our target over time, so that all results meet specifications. We seek on-target processes that are stable. A stable process is in control and performing at its best, considering the current process definition.

Process stability, given the process is set to target, implies the very best operation possible, given the present state of our process—incoming materials, machines, and operator skills. Once a process has been targeted and stabilized, more effective or efficient process performance can occur only through system changes—the responsibility of management and empowered employees—through process improvement as described in Section 5, or process redefinition as described in Section 3.

Instability is created when a disruptive special cause or disturbance is present. Special causes shift either the process location or the process dispersion, or both. These shifts may be small, large, abrupt, gradual, or any combination. They are always of a physical nature in origin, introduced by natural forces, human errors, and so on. Shifts are judged relative to the process location and dispersion under stable conditions. **An unstable or disrupted process is not operating up to its potential.** Hence, more effective or efficient process performance can be obtained by taking effective counteraction to remove or offset the effect of the special cause and bring the process to a stable condition—approaching our target or "best" operating point, with minimal variation.

Process adjustment focuses on cause-effect identification linkages between a process output and a process input variable(s), whereby we manipulate a known input(s) to move/keep the output to target. Adjustments may be introduced manually or automatically. For example, to adjust the temperature within a process, we may adjust a steam flow valve. The adjustment might be based on a feedback signal reflecting the magnitude of error between our process output measurement and a target/setpoint.

A stable, on-target process does not necessarily meet all product/process specifications. Targets and specifications result from customer needs and demands, while process stability results from process operations. The concept of process capability relates these two different worlds. **Process capability is a critical performance measure that addresses process results with respect to product specifications.** Hence, process capability measures are widely used in practice to measure our own ability or a supplier's ability to meet technical specifications. Process capability and process capability indices are discussed in detail in Chapter 11.

A number of initiatives and tools have been developed to address process control. Selected initiatives and tools are listed in Table 7.1 and are related to the measurement, comparison, and corrective elements. Process control initiatives and tools range from highly qualitative, e.g., C-E diagrams to highly quantitative, e.g., statistical process control (SPC) charts. In general, we use our initiatives and tools to help us describe and understand the physical nature of our process, establish fundamental cause-effect relationships between variables, measure and model these relationships quantitatively, detect process disruptions, and develop criteria for initiating process counteractions, and adjustments.

9.3 PROCESS CONTROL OPTIONS AND GROWTH

Process control plays a critical role in extracting process results. **For any given process, we have several fundamental control options.** Table 9.1 overviews our basic choices. Essentially, **we choose some form of either open-loop or closed-loop process structure.** The **open-loop** option lacks active intervention during the process cycle; during the process cycle the process output has no influence or impact on process activities. Here, process instructions are preset for the entire process cycle. However, between process cycles we have the opportunity to make adjustments.

In **closed-loop** processes the process actions are affected by or tied to process surroundings, inputs, and outputs. In **feedback control,** output response is sensed and compared to a target/ setpoint value, with adjustment made during the process cycle accordingly. In **feedforward control,** potential or anticipated process disturbances associated with environmental, input, or process transformation variables are sensed and weighed against predicted response outputs, and adjustments are made accordingly during the process cycle. We may choose to pursue human-assisted or automatic closed-loop options in feedback and/or feedforward control configurations, all containing active corrective capabilities during the process cycle.

Both the nature of the process and our level of process understanding dictate our process control strategy. As we gain process understanding, we grow in our ability to provide effective and efficient process control. Figure 9.2 provides insight into process control growth regarding our ability to measure or compare process characteristics and knowledge or ability to take counteraction. When we encounter a process we cannot adequately characterize and measure, and

Table 9.1 Fundamental process/process control choices.

Process Control Option	Description	Example
Open loop	The process cycle is preset (no active intervention during each cycle).	Toaster, dishwasher, conventional computer numerical control (CNC) operations
Closed loop (human-assisted)	Human efforts are involved in measurement, comparison, and/or correction activities during the process cycle.	Manual, mechanized operations —includes involvement of driver, operator in process activities
Closed loop (automatic)	Hardware and/or software are used in measurement, comparison, and correction activities during the process cycle.	Heating/cooling system, flexible manufacturing cell
Feedback (in closed-loop processes)	Output is measured and compared —reactive correction to controllable variables.	Response to too hot or too cold in heat/cooling system
Feedforward (in closed-loop processes)	Inputs are measured and compared —proactive correction to controllable variables.	Avoiding an animal in the roadway while driving a car or truck

lack understanding as to definitive counteraction, we are typically in an open-loop, trial-and-error position—in the lower left quadrant of Figure 9.2. As we gain both measurement and counteraction knowledge or ability, we move upward and to the right, respectively, in Figure 9.2.

In general, we encounter three basic control strategies: (1) **subjective-based control**—a form of manual control, (2) **objective-based, statistical process control (SPC)**—usually another form of manual control, (3) **equipment-based, automatic process control (APC)**—machine control. Any one of these strategies/options may be appropriate, depending on our ability to measure and understand our process. Our ability to move from lower left upward and/or to the right in our grid in Figure 9.2 is dictated by two fundamentally different process-related aspects. First, we must be able to "read" our process in terms of where our output variables Ys and input variables Xs and Zs are operating and identify disruptions, drifts, shifts, and so on. Second, we must be able to determine appropriate counteraction and take counteraction when necessary.

The first fundamental deals with measurement and comparing. What do we measure? How often do we measure? What do we use as the basis for declaring a significant disruption, drift, or shift? The second fundamental deals with physical proaction or reaction in terms of counteraction. What should we do and how fast can it be done, relative to the process cycle? What or who will take counteraction? When should it be initiated?

Regarding the first fundamental, all the while knowing that the physical world is probabilistic, we may choose to treat our variables as deterministic. In the deterministic case, we ignore common cause variation. We compare our measure to our target—typically a deterministic value—and render our decision for the necessity of counteraction. Many conventional automatic process control systems, such as thermostats, use deterministic models to trigger adjustment counteraction to offset process output changes.

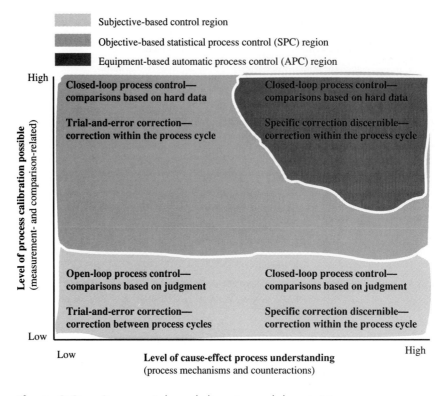

Figure 9.2   Process control growth dimensions and characteristics.

On the other hand, we can treat or model our variables in a probabilistic manner. Here, we estimate a common cause variation level in our process measurements and use either process result-generated or target-based statistical comparison criteria as a basis to render our decision for the necessity of counteraction. Human-based closed-loop control systems using SPC utilize probabilistic models.

We should recognize that **process measurement, comparison, and correction are impacted by process control strategy and technology, as well as our ability to understand our process**. Hence, a competitive position in process control assumes we aggressively pursue both process knowledge and process control technology. It is up to us to assess and balance out physical, economic, timeliness, and customer service performance, with respect to process control, relative to customer needs and expectations, considering the positions of our competitors.

Example 9.1 | **AUTOMATIC CLOSED-LOOP FEEDBACK CONTROL CASE**   Within our process control structures, we realize that all variables are actually probabilistic. However, we choose to treat some process input, transformation, and output variables as deterministic in our process models. For example, in a typical heating/cooling closed-loop feedback control system, as depicted in Figure 3.4c, we measure temperature and treat the data as deterministic. Here, we choose to ignore natural variation in the temperature and variation introduced by the sensor, with respect to the metric presented.

We can justify this treatment on the basis of relatively accurate and precise sensors and strong process inertia. Process inertia in this case results from the fact that heat transfer in any given medium is not instantaneous. For example, the temperature 1 minute hence is dependent on, and can be predicted reasonably well from, the present temperature. Our comparison is made from sensed temperature data, compared to our target/setpoint, without regard for variation. Here, we simply "compare" two numbers with each other and make a decision to take corrective action: turn on or off our equipment. For example, if our sensed temperature drops below our setpoint, the heating system turns on. When the temperature increases above the setpoint, the heating unit turns off. This treatment simplifies our model and works well, provided bias and imprecision are small in relation to the sensed temperature measurement.

In other words, our ability to sense and measure our output's impact on our surroundings, express a clear target/setpoint, compare our measurements with our target, and identify appropriate counteraction is high. This combination allows us to utilize effective and efficient automatic closed-loop process control, allowing for both effectiveness and efficiency, relative to other process control technologies. This case falls in the upper right quadrant of Figure 9.2 and is clearly an example of a process adjustment issue.

9.4 PROCESS CONTROL MODELS OVERVIEW

Given our three basic process control strategies/options—subjective-based, objective-based SPC, and equipment-based APC—a number of models exist. Figure 9.3 depicts these three options and related models.

Subjective-based process control has existed as long as humankind has existed. Subjective-based process control relies on personal experience and judgment to sense, compare, and correct process activities. **A wide variety of intuitive human strategies and "models" exist. To one degree or another these mental models correspond to SPC and APC models.** For example, in all human activities, we intuitively judge when to seek out special causes as we perceive a disturbance, and we intuitively adjust input variables when we perceive a shift or drift in process outputs.

Subjective-based process control was significantly improved with the advent of statistical methods. Sampling methods and probabilistic process models were introduced by Walter Shewhart and his peers in the 1920s. **The term *statistical process control* is used to describe sampling-based models applicable for monitoring on-line process activities. Although SPC is a valuable aid in justifying the need for counteraction, it typically does not identify appropriate counteraction.** Hence, the need for human intervention and reasoning aimed at effective and efficient counteraction is apparent.

Since its introduction, SPC-based knowledge has grown considerably. Our applications of SPC focus on leverage variables monitored or observed in the presence of natural variation. An upset or disturbance as a result of special cause variation manifests itself in such a way that a shift or change is produced in the data. SPC provides a quantitative basis for special cause detection by helping to isolate unusual points or patterns in the data. Once a special cause is detected, we then seek to identify it and its source in the physical process. Then, its effect is eliminated to produce process operations reflecting only common cause variation. The cycle repeats.

SPC technology can be divided several ways. A primary division between conventional SPC and unconventional SPC depends on the level of process inertia in the process metric or variable. Process inertia is related to physical elements that occur naturally. Process inertia

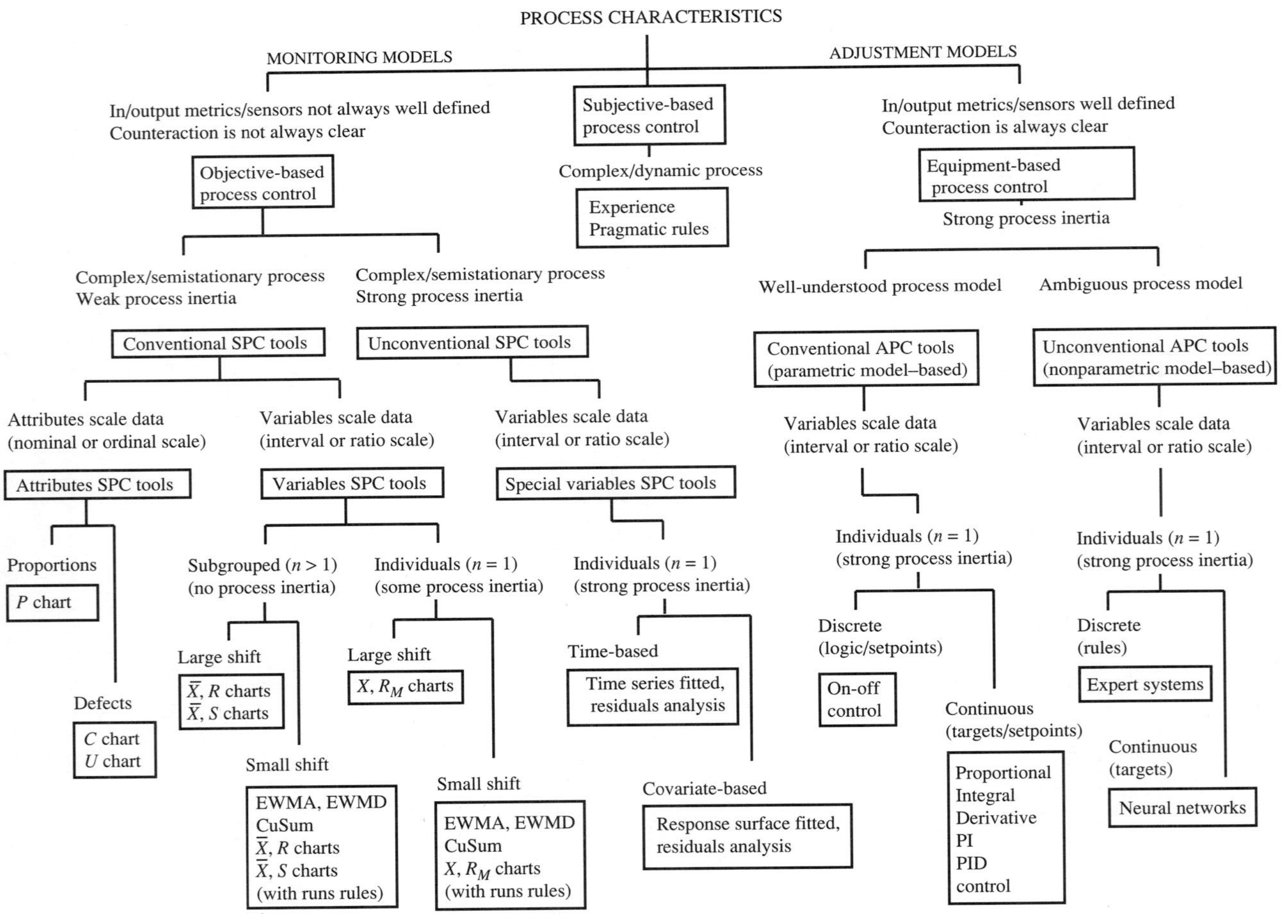

Figure 9.3 Process control model options.

results from physical principles and mechanisms, such as heat transfer, fluid flow, and so on, whereby subsequent performance is a direct result of present performance. Process inertia is observed as a matter of degree in the physical world. **In general, the more predictable a future response is relative to a current response, the higher the process inertia level.** Continuous process flows typically manifest strong process inertia characteristics. Process inertia is calibrated in a mathematical sense by time or order serial correlation in empirical data. Here, we typically see a sequence-related correlation within a data stream, termed *autocorrelation;* please refer to Chapter 4 for more details.

Referring to Figure 9.3, **divisions in conventional SPC are drawn between attributes-based models and variables-based models.** Other subdivisions are drawn between defectives and defects, within attributes models, and rational subgrouped and individual measures within variables models.

Unconventional SPC tools tend to mathematically compensate for the portion of natural variation attributable to time and/or order of production related physical effects. These methods are more complicated than their conventional counterparts, since the dynamic nature of the metrics they model is typically complicated. Nevertheless, they are useful in process control and are applied when conventional methods prove inadequate.

Modern automatic process control (APC) technology has evolved from general control theory. The body of knowledge in this area is extensive. Applications range from simple on/off applications such as thermostats to telemetric guidance control systems for space exploration. **In general, APC seeks to hold a response/output to a target value/setpoint by manipulating an associated input variable.**

Figure 9.3 depicts two branches of automatic process control based on the nature of the process model used to execute the control. **One sub-branch of conventional APC consists of discontinuous control action**, which includes on/off switching technology. **A second sub-branch of conventional APC includes continuous control action such as proportional, integral, and/or derivative control. Unconventional APC offers several options.** Rule-based **expert systems** offer logical arguments in response to control action or process challenges. **Neural networks** offer "black box" numerical options in process modeling—useful to relate inputs to outputs.

9.5 INTRODUCTION TO SPC MODELS

Most SPC process control models have their roots in statistical inference but have been adapted to a charting or graphical form for ease of use by engineers and process operators. The graphical format, as practiced—in order to simplify the mathematics to one degree or another—compromises statistical theory. But the result is useful for everyday applications on production floors. Probabilistic SPC models are utilized by humans or encased in automatic closed-loop process control models.

Control charting technology utilizes in-process, or on-line, sampling techniques to help us monitor a process. **The purpose of control charting is to indicate when the process is functioning as intended and when corrective action to offset special cause disruption is necessary,** i.e., to provide early warning of process upsets. Control charting can be considered proactive or preventive if warnings allow us to avoid significant amounts of off-specification product; otherwise it is reactive.

Table 9.2   Statistical process control indications and errors.

Sample Indication from SPC (Sample-Based View of the Process)	Process Reality (God's Eye View of the Process)	
	Stable (In Control)	**Unstable (Out of Control)**
Stable (in control)	Correct indication	Failure to detect (type II error†)
Unstable (out of control)	False alarm (type I error*)	Correct indication

*Type I error: On the basis of our sampled information, we declare a process unstable (out of statistical control), when the process is really stable (in statistical control).

†Type II error: On the basis of our sampled information, we declare a process stable (in statistical control), when the process is really unstable (out of statistical control).

Control charts are "watchdog" tools that provide us with indications of in-control or out-of-control status. It is important to note that an in-control process is considered stable; see Deming [3]. An out-of-control process is said to be unstable. Process stability implies the very best operation possible, given the present state of our production system definition—incoming materials, machines, and operator skills. **Enhanced performance in an on-target stable system can occur only through system changes involving improvement or redefinition work; both are the responsibility of management and empowered employees.**

Instability is created when a special cause, or disturbance, is present. Special causes shift either the process metric's location or dispersion, or both. The shift is judged in a statistical sense, relative to the process location and dispersion under stable conditions. An unstable or shifted process is not operating up to its potential.

Since process control charts rely on sampling, they are subject to statistical errors. For example, **a false alarm** (when a process shift is indicated, and in fact, there is none) or **a failure to detect** a process shift (when a process shift really has occurred and we did not detect the shift from our sample information) may be encountered. Table 9.2 relates these two error classifications to type I and type II statistical errors, respectively. Once an indication of a process shift is detected, it is up to the operators, engineers, and other technical people to locate the special cause(s) and take corrective action.

Once a quality characteristic is defined, a target set, and specifications stated, we have a benchmark for judging a process or product. For example, a meter stick might be described as follows:

Critical quality characteristic: overall length

Target: 1000 mm

Specification: LSL = 999 mm, USL = 1001 mm

Each meter stick is perfect if it is exactly 1000 mm long; it is acceptable if 999 mm ≤ measured length ≤ 1001 mm. Otherwise, it does not conform to this two-sided specification.

A simplified production process for a meter stick was described in Figure 4.4. Here, the finished product length is critical, as are several other characteristics. In assessing process control strategies for this process, we look upstream to the point at which the length is determined. At the shear and deburr subprocesses, we see a leverage point. We monitor the sheared length after deburr with a variables SPC tool. Assuming we were producing our meter sticks in a reasonably large quantity, the depictions and descriptions in Figure 9.4 illustrate the nature of both variables and attributes data regarding sheared length.

The meter stick results in Figure 9.4*a*, *b*, and *c* portray the single, measurable product/process characteristic of overall length. Here, we might apply SPC at the upstream deburring subprocess point. The normal, or Gaussian, probability mass function is typically used to model quality characteristics when variables measures are used. Such a model is represented by the smooth curves in Figure 9.4*a, b,* and *c*. Variables process control models are addressed in Chapters 10, 11, and 13.

Binomial probability mass functions are typically used to model quality characteristics when two possible outcomes are considered; see Figure 9.4*d* and *e*. In this figure, the stack heights of product represent proportions conforming and nonconforming. Results are reported as a fraction defective or a fraction acceptable. In this situation we might apply a *P*-chart model at the deburr subprocess. With such a model, we could cover one characteristic, e.g., overall length, or several, e.g., length, width, and straightness. The *P*-chart model and other attribute-based models are addressed in Chapter 12.

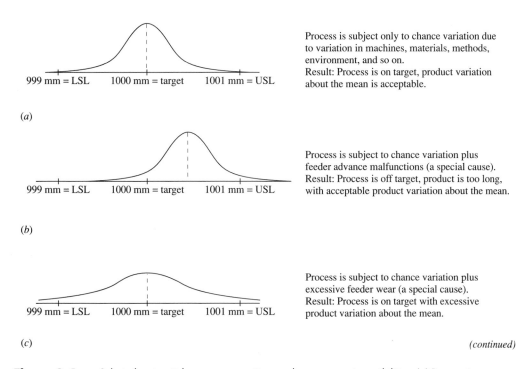

(*a*)

Process is subject only to chance variation due to variation in machines, materials, methods, environment, and so on.
Result: Process is on target, product variation about the mean is acceptable.

(*b*)

Process is subject to chance variation plus feeder advance malfunctions (a special cause).
Result: Process is off target, product is too long, with acceptable product variation about the mean.

(*c*)

Process is subject to chance variation plus excessive feeder wear (a special cause).
Result: Process is on target with excessive product variation about the mean.

(*continued*)

Figure 9.4 Selected meter stick process operations and measurement possibilities. (*a*) Process is operating as it was intended (stable or in control). (*b*) Process is experiencing a location shift due to feeder advance malfunction (unstable or out of control). (*c*) Process variation increases due to excessive feeder wear (unstable or out of control).

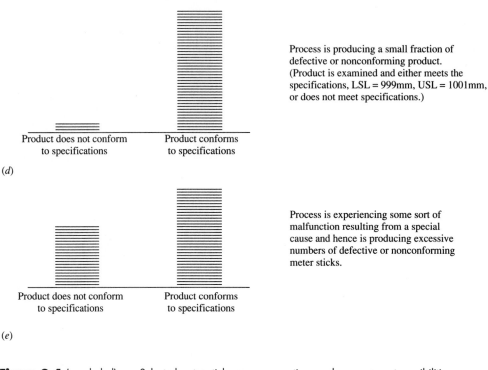

Process is producing a small fraction of defective or nonconforming product. (Product is examined and either meets the specifications, LSL = 999mm, USL = 1001mm, or does not meet specifications.)

Product does not conform to specifications

Product conforms to specifications

(*d*)

Process is experiencing some sort of malfunction resulting from a special cause and hence is producing excessive numbers of defective or nonconforming meter sticks.

Product does not conform to specifications

Product conforms to specifications

(*e*)

Figure 9.4 (concluded)   Selected meter stick process operations and measurement possibilities. (*d*) Process is operating as it was intended (stable or in control). (*e*) Process is not operating as it was intended (unstable or out of control).

REVIEW AND DISCOVERY EXERCISES

REVIEW

9.1.   What is the focus or purpose of process control? Explain.

9.2.   Define and explain the concept of process stability.

9.3.   Define and explain the concept of process capability.

9.4.   Using Figure 9.3, identify the basic process control strategies. What general conditions must be met for each to be justifiable? Explain.

9.5.   Explain how and why SPC indications can be in error. Relate these errors to statistical errors of types I and II.

9.6.   What is the major distinction between SPC on a variables data basis and SPC on an attributes basis? Explain.

DISCOVERY

9.7. Why is process control necessary? Explain.

9.8. What common elements run through process control, regardless of the option taken, e.g., subjective-based, objective-based SPC, or equpment-based APC? Explain.

9.9. An on-target, stable process may not be capable, and a capable process may not be on-target, stable. Explain.

9.10 The ideal situation in process control is for a zero defects outcome. Which concept, feed-back control or feedforward control, is more conducive to zero defects? Why? Here, zero defects is related to in-process correction that totally prevents rework and scrap.

9.11. What would be the likely result of using an automatic process control strategy on a low-inertia process? What would be the likely result of using a SPC process control strategy on a high-inertia process? Explain.

9.12. Select a basic closed-loop process in your personal life, e.g., driving an automobile, wash-ing your hands, writing a note, putting on your shoes; describe the process purpose, and briefly characterize the process (using flowcharts and/or cause-effect diagrams, etc., as described in Chapter 4). Identify and describe appropriate control points and the respective measurement, comparison, and correction that you take to control your identified process. Which elements of control are feedback in nature? Feedforward in nature?

REFERENCES

1. Box, G. E. P., D. E. Coleman, and R. V. Baxley, Jr., "A Comparison of Statistical Process Control and Engineering Process Control," *Journal of Quality Technology*, vol. 29, no. 2, pp. 128–130, 1997.

2. Box, G. E. P., and A. Luceno, *Statistical Control by Monitoring Feedback Adjustment*, New York: Wiley, 1997.

3. Deming, W. E., *Out of the Crisis*, Cambridge, MA: MIT Center for Advanced Engineering Studies, 1986.

Process Monitoring—Variables Control Charts for Grouped Measurements

10.0 Inquiry

1. How do we detect special cause influences?
2. When can we subgroup data? What advantages does subgrouping offer?
3. How are subgrouped control charts constructed? Interpreted?
4. Why are small shifts difficult to detect? How can we detect small shifts?

10.1 Introduction

In Chapter 10 we introduce statistical process control (SPC) charts that use subgrouped data. First, classical Shewhart charts—the X-bar chart, the range, R, chart, and the standard deviation, S, chart—are introduced. **We describe the concepts underlying these charts, their mechanics, and their interpretation.** To deal with sustained small shifts, we introduce runs rules for these charts. We introduce the exponentially weighted moving average (EWMA) and deviation (EWMD) and the cumulative sum (CuSum) models as alternative SPC tools for grouped measurements. Finally, we explore the statistical theory behind control charts.

10.2 SPC Model Rationale for Variables Data

Variables data can be summarized and analyzed in a number of ways, including traditional data analysis tools such as runs charts, histograms, and summary statistics. However, using these traditional tools to do so is somewhat ineffective and inefficient. We describe a number of specialized graphical techniques that allow us to monitor both location and dispersion characteristics of our process metric. In our discussions, we develop a number of SPC charts based on the arithmetic mean, the most commonly used measure of location. Two measures of dispersion that are extremely useful in statistical quality control are the range and the standard deviation.

In dealing with variables quality characteristics, it is customary to use separate charts for monitoring location and dispersion measures. Since our processes operate over time or order of production, we track location and dispersion over time or order of production and continuously assess process stability, monitoring for the presence of special cause disruptions across time or order of production.

SPC CONCEPT

In SPC, it is customary to assume that a process is "in control," or stable, until sufficient evidence is discovered to the contrary. The assumption that the process is in control in general implies that both the location parameter and the dispersion or variation parameter are not influenced by a special cause—they are essentially "on target." In an approximate sense, we conceptualize a set of statistical hypotheses for each of these parameters. Here, H_0 refers to the null hypothesis and H_1 refers to the alternative hypothesis.

Hypothesis for the location parameter:

H_0: The process population mean is on target—the process is operating under common cause variation, without special cause influence present.

H_1: The process population mean is not on target—the process is operating under common cause variation, with an additional special cause influence present.

Hypothesis for the dispersion parameter:

H_0: The process population variation about the mean is on target—the process is operating under common cause variation without special cause influence present.

H_1: The process population variation about the mean is not on target—the process is operating under common cause variation, with special cause influence present.

We typically use one control chart to monitor location and another chart, of the set, to monitor dispersion. **In a crude sense, we are testing each of these hypothesis sets each time we plot a point on our variables control charts.** Hence, we are asking ourselves whether or not to assume that the process is stable with respect to location and likewise for dispersion. If a special cause is present and the process is upset or disturbed, and therefore unstable, we seek to identify the physical cause and remove its influence on the production process. **Ideally, when a special cause disruption enters our process, we detect its presence through our monitor; essentially we reject one or the other or both of the null hypotheses above.** However, in practice, we must contend with the presence of *failure-to-detect (type II) errors* and *false alarm (type I) errors,* as described in Table 9.3.

Variables control charts provide three basic benefits: (1) **they display the average and/or target level of a process variable**—location or central tendency, (2) **they display the variation in a process variable**—dispersion, and (3) **they display the consistency in our process over**

time or order of production. Provided we have a target for our control chart metric, we can also display this target on the chart. Later, we will discuss the possibilities of developing our SPC chart parameters solely on the basis of targets, rather than on the basis of past, observed measurements.

Example 10.1

PROCESS MONITORING FUNDAMENTALS   Before proceeding further with a detailed description of the different types of variables control charts, we illustrate the use of an X-bar chart for monitoring the location parameter and an R, or range, chart for monitoring the dispersion parameter. In this example, we revisit the meter stick production process previously described in Chapters 4 and 9. As before, our process lever here is the length of the blanks after deburring. The characteristic of critical importance is the length of the meter stick. Our target is 1000 mm. The tolerance limits (the explicit quality requirement applicable to each and every meter stick) for the length of the meter stick are specified as 1000 ± 1 mm.

These substitute product characteristics result from our customer's true quality characteristics of demanding a household measuring device that is the right length. Even though customers desire a meter stick that is exactly 1 m long, we know that we cannot make each meter stick exactly 1 m long due to the nature of our materials, manufacturing processes, and so forth—common cause variation. Hence, we have agreed to the substitute characteristics above.

Approximately once every hour the lengths of five meter stick blanks that have just been sheared and deburred were measured. Each sample of five meter sticks is called a subgroup. Averages and ranges for each subgroup have been calculated. A range is the difference between the largest measured value and the smallest measured value in each subgroup. The data obtained are shown in Table 10.1. These data were plotted in a scatter plot, across order of production, by subgroup, as shown in Figure 10.1a.

Table 10.1   Length measurements (in millimeters) (Example 10.1).

Subgroup Number	Length Measurements x_{ij}					Average $\bar{x}_j$	Range R_j
1	999.90	999.80	999.60	999.70	999.50	999.70	0.40
2	999.40	999.40	999.70	999.50	999.30	999.46	0.40
3	999.30	999.30	999.50	999.30	999.50	999.38	0.20
4	999.50	999.60	999.60	999.50	999.80	999.60	0.30
5	999.50	1000.00	1000.00	1000.00	999.80	999.86	0.50
6	999.50	999.50	999.40	999.60	999.60	999.52	0.20
7	999.60	999.60	999.90	999.50	999.40	999.60	0.50
8	998.80	999.60	999.80	999.80	999.80	999.56	1.00
9	1000.60	999.90	999.80	998.70	999.40	999.68	1.90
10	999.90	999.80	999.90	1000.40	1000.40	1,000.08	0.60
11	999.70	1000.10	999.80	999.70	1000.10	999.88	0.40
12	999.90	1000.10	1000.10	1000.20	1000.30	1000.12	0.40
13	999.90	1000.30	999.80	998.90	999.60	999.70	1.40
14	999.90	999.80	1000.00	999.70	999.60	999.80	0.40
15	999.30	1000.00	999.60	999.70	999.80	999.68	0.70
16	999.10	999.40	999.60	999.60	999.60	999.46	0.50
17	999.60	999.30	999.70	999.60	999.80	999.60	0.50
18	998.80	999.10	999.30	999.00	999.40	999.12	0.60
19	999.80	999.90	999.20	999.00	999.50	999.48	0.90
20	999.60	999.80	999.80	1000.20	999.90	999.86	0.60
Totals						19,993.14	12.40

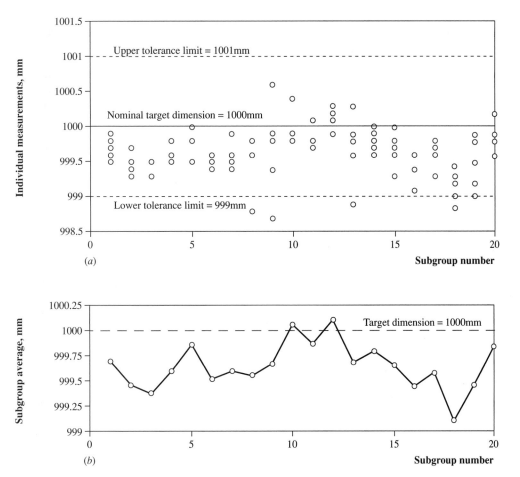

Figure 10.1 Two charts that are *not* control charts but depict control chart information (Example 10.1). (a) Individual measurements of meter sticks. (b) Average of samples of five.

Figure 10.1a shows individual meter stick measurements plotted for each subgroup, as well as the nominal-is-best target, and upper and lower tolerance limits. Here, we can judge each meter stick relative to the product specifications. All the meter sticks examined met the product design specifications, with the exception of four.

Figure 10.1b shows the averages of these subgroups. The chart in Figure 10.1b shows trends more clearly than the chart shown in Figure 10.1a. However, it is developed on the basis of subgroup means of samples of size $n = 5$, rather than on the basis of individual meter stick lengths. The target can be displayed on a plot of subgroup means, because the mean of a population of average values is equal to the mean of a population of individual values. However, we cannot judge the product subgroup means using the product design specifications. Such a judgment leads to a serious error.

It is the individual meter stick, not the subgroup average, that has to meet the tolerances. Averages of a subgroup often fall within tolerance limits even though some of the individual articles in the subgroup are

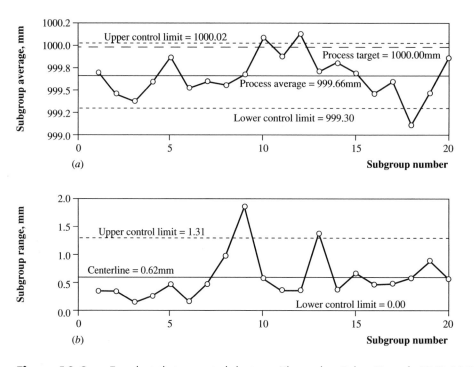

Figure 10.2   Two charts that *are* control charts, an *X*-bar and an *R* chart (Example 10.1). (*a*) Control chart for averages, *X*-bars. (*b*) Control chart for ranges, *R*'s.

outside the limits. We observed this in samples 8, 9, 13, and 18, in which the averages were between 999 and 1001 mm even though one meter stick in each group measured less than 999 mm.

Developing plots of subgroup means appears counterintuitive from a physical perspective. However, it makes a great deal of sense from a statistical perspective since we gain precision in our estimates as we increase sample size. This concept applied to our discussion of experimental design in Chapter 5, and it applies here as well. Our result is the *X*-bar control chart. Figure 10.2*a* shows an *X*-bar control chart for the averages, i.e., *X*-bars. Our readers should note that this is Figure 10.1*b* with the additional of control limits. Figure 10.2*b* shows an *R* control chart, for the ranges.

Before choosing the *X*-bar and *R* chart models, we consider the degree of process inertia. Here, we justify our subgrouping on the basis of both physical and statistical arguments that little, if any, process inertia exists in our shearing and deburring subprocesses. Because of the nature of the shearing process characterized in Figure 4.4, we shear one blank at a time. We have no reason to believe that the length of a subsequent blank will be affected by the shearing of a previous blank, so long as we are anticipating only common cause variation. In order to assess the level of process inertia in more detail, we resort to the analyses described in Chapter 4. Figure 4.5 provides both a runs plot and a correlogram for our meter stick data stream, suggesting low inertia from a statistical perspective. Hence, we are justified in our SPC tool choice.

Each of these control charts is drawn with a solid line to indicate the average value of the statistic that is plotted. The grand average $\bar{\bar{x}}$ (the average of the subgroup averages) is 999.66 mm. This number is calculated as

$$\overline{\overline{x}} = \frac{\sum\limits_{j=1}^{m} \overline{x}_j}{m} = \frac{19{,}993.14}{20} = 999.66 \text{ mm}$$

Likewise, the average of the ranges is 0.62 mm. This number is calculated as

$$\overline{R} = \frac{\sum\limits_{j=1}^{m} R_j}{m} = \frac{12.40}{20} = 0.62 \text{ mm}$$

The charts show dashed lines marked "upper control limit" and "lower control limit." The distances of the control limits from the line showing the average value depend on the distribution of the statistics of interest and the subgroup size. The methods for calculating control chart limits are explained later in this chapter. The limits shown in Figure 10.2 are called **trial limits.** Before projecting them into the future (past sample 20) to control future production, they need to be modified because points outside of the limits indicate instability.

As the charts now stand, the process shows a lack of control—instability. Three points (subgroups 10, 12, and 18) are outside the control limits on the X-bar chart. Two points (subgroups 9 and 13) are outside the control limits on the R chart. These points indicate the presence of special causes in the shearing and/or deburring subprocesses; the factors contributing to these upsets in product length should be identified and addressed. These outlying points in our control chart were found in retrospect, as the trial control limits were not established until the end of subgroup 20. We search for physical causes and remove these influences as best we can, on the basis of our process logs.

The control charts in Figure 10.2 provide evidence that there is an opportunity to improve our process by using process control technology, without explicit process redefinition or improvement initiatives. Once the process is centered at (near) the target value of 1000 mm and stable, only process improvement or redefinition initiatives will produce further gains. When a process is operating to target and stable, that's all we can ask in the way of process control of our current process—operators, materials, machines, etc.

The dividends from control charting come in the application of the control limits to future production. Timely indications of instability, from the charts, aid in hunting down and eliminating the effects of special causes before the process is allowed to operate in an unstable condition for any length of time, as well as in helping to prevent recurrence of the process upset.

It is critical to understand that SPC charts are incomplete with respect to comprehensive process control in two regards: (1) they do not tell us whether or not we are meeting tolerance specifications consistently, and (2) **they neither explicitly identify nor remove special causes.** Once the control charts indicate that a process is in control with respect to both location and dispersion, we may obtain a false feeling of confidence that the product meets specifications. The purpose of process control charts is to address process stability, not to tell us if we are meeting tolerance specifications. Furthermore, process control charts do not identify special causes, and they do not tell us how to eliminate effects of special causes. Control charts basically serve us as statistical warning devices for process disruptions. **It is up to us to develop meaningful process logs to account for and document physical characteristics, actions taken, and results obtained.**

SUBGROUPING RATIONALE

When our data are independent, across time or order of production, we can use subgrouping effectively. The level of physical inertia—physical influence across time or order of production—in our process metric essentially determines independence or a lack thereof. We can justify subgrouping of data when we can establish that our observations are independent—where little or no process inertia exists in our process measurements. We make this determination through process characterization, as described in Chapter 4.

   Logical subgrouping allows us to obtain enhanced statistical performance in our control charts. The enhancement we seek is a reduced chance of failing to detect special cause—a reduced type II error probability. **We can gain this performance edge in two ways. First, as we increase our sample/subgroup size, we increase the statistical precision of our location and dispersion estimates. Second, if we sample within each subgroup, close together in time/ production order, we minimize the chance of a special cause occurring within each subgroup.** And, hence, our dispersion statistics are less likely to be influenced (increased) by special causes when we develop our control chart limits. These gains argue for subgrouping when process inertia is not a factor in the physical world. If process inertia is a major factor, our statistical performance is distorted. These distortions are addressed in Chapter 13.

10.3   NOTATION FOR SUBGROUPED SPC MODELS

Our conventional SPC tools parallel traditional statistical inference tools, but do so in a graphical, and approximate, format. The graphical format offers the advantage of displaying our data across time or order of production, offering us a visual record of process activity as well as simple rules of interpretation. Underlying the graphical content are a host of calculations. We introduce a set of notations to allow us to describe the mechanics of the tools.

GENERAL CONTROL CHART SYMBOLS

n: The number of observations or measurements in a subgroup, i.e., subgroup size

m: The number of subgroups in the preliminary data set used to set up a control chart

b: The control limit spread in number of standard deviations (chart sigmas) of the plotted quantity

$\mu, \hat{\mu}$: The population mean, estimated population mean for the population of individual measurements or items

$\sigma, \hat{\sigma}$: The population standard deviation, estimated population standard deviation for a population of individual measurements or items

$\sigma_{\bar{x}}, \hat{\sigma}_{\bar{x}}$: The population standard deviation, estimated population standard deviation for a population of means

R, S, X-BAR CHART SYMBOLS

R_j: A subgroup sample range value for the *j*th subgroup or sample

$\bar{R}$: The mean of a group of subgroup sample range values

LCL_R UCL_R: The lower, upper control limits for the *R* chart

s_j: A subgroup sample standard deviation for the *j*th subgroup

$\bar{s}$: The mean of a group of subgroup sample standard deviations

σ_0: A target or assumed population standard deviation (given rather than calculated)

LCL_S, UCL_S: The lower, upper control limits for the *S* chart

x_i: The *i*th individual measurement within a subgroup

$\bar{x}_j$: A subgroup sample mean for the *j*th subgroup or sample

$\bar{\bar{x}}$: The mean of a group of subgroup sample means

$\mu_0, \bar{x}_0$: A target or assumed mean value (given rather than calculated)

$LCL_{\bar{x}}$, $UCL_{\bar{x}}$: The lower, upper control limits for the *X*-bar chart

EXPONENTIAL WEIGHTED MOVING AVERAGE AND DEVIATION (EWMA AND EWMD) CHART SYMBOLS

r: A weighting factor that can assume values between 0 and 1

W_{j-k}: Weight associated with the observation $\bar{x}_{j-k}$, where $\bar{x}_j$ is the most recent observation

A_j: The exponential weighted moving average (EWMA) for observation *j*

V_j: The exponential weighted moving deviation (EWMD) for observation *j*

D_j: The absolute deviation at observation *j*, equal to $|\bar{x}_j - A_{j-1}|$

LCL_A, UCL_A: The lower, upper control limits for the EWMA chart

LCL_V, UCL_V: The lower, upper control limits for the EWMD chart

CUMULATIVE SUM (CUSUM) CHART SYMBOLS

y_c: A user-selected CuSum chart scaling factor (vertical over horizontal units), selected so that the CuSum chart horizontal-to-vertical scale and the V-mask angle will produce an attractive and legible chart

θ_c: A V-mask angle for a CuSum chart

D_c: The process shift magnitude a CuSum chart should detect with very high probability

L_j: A CuSum value for the *j*th sample, e.g., subgroup

L_j^+ and L_j^-: Tabular CuSum chart plotting points

k and h: Constants that determine the statistical performance of the tabular CuSum

10.4 SHEWHART X-BAR, R, AND S CONTROL CHART CONCEPTS AND MECHANICS

The X-bar, R, and S charts are sometimes referred to as Shewhart charts because of their heritage. **The development of X-bar, R, and S chart mechanics is based on the process/product metric being independent and normally distributed.** However, the X-bar chart itself (but not the R and S charts) is robust to deviations from normality, as covered by the Control Limit Theorem (CLT).

NORMAL MODEL

In general, the variables SPC models developed here are based on the normal, or Gaussian, probability mass function. The normal distribution has two parameters, μ and σ, where μ is the mean and σ is the standard deviation. If we define the random variable X as a measurable product/process characteristic with mean μ and standard deviation σ, then the probability mass and the cumulative mass functions are

$$f(x; \mu, \sigma^2) = \frac{1}{\sigma\sqrt{2\pi}} \exp\left[-\frac{(x - \mu)^2}{2\sigma^2}\right] \qquad -\infty < x < \infty$$

and

$$F(x; \mu, \sigma^2) = \Phi\left[\frac{x - \mu}{\sigma}\right] \qquad -\infty < x < \infty$$

where Φ represents the standard normal density function cumulated from left to right. A table of the cumulated standard normal distribution, or mass function, appears in Table VIII.1, Section 8.

X-BAR CONTROL CHART MECHANICS

Suppose that we select subgroup samples of size n, taken at regular and frequent intervals from a process. **Each sample indicates how the process is behaving at one "point" in time in terms of location and dispersion,** judged through the subgroup mean and its range or standard deviation. **We seek to determine if the process is stable or unstable.** If the process is subject only to chance or common causes, the $\bar{x}$'s and the R's or the s's should be randomly distributed within certain probabilistic limits, called *control chart limits* or simply *control limits*.

The general form to construct the X-bar chart control limits, assuming all subgroups are the same size n and that the process mean μ and the process standard deviation σ are known, is

$$\left.\begin{array}{l} \text{UCL}_{\bar{X}} \\[2ex] \text{LCL}_{\bar{X}} \end{array}\right\} = \mu \pm b\sigma_{\bar{x}} \qquad \textbf{[10.1]}$$

$$\text{Centerline} = \mu$$

It is common practice in the United States to set $b = 3$. In this case, the type I error probability can be found in our standard normal table, $\alpha = 2P(Z < -3) = 0.0026$, assuming normality and one sample point beyond the control limits. We typically estimate μ and σ from $\bar{\bar{x}}$ and $\bar{R}$, respectively. Hence, **the ± 3-sigma control limits for our X-bar control chart are**

$$\left.\begin{array}{l} \text{UCL}_{\bar{X}} \\ \\ \text{LCL}_{\bar{X}} \end{array}\right\} = \bar{\bar{x}} \pm \frac{A\bar{R}}{d_2} = \bar{\bar{x}} \pm \frac{3\bar{R}}{d_2\sqrt{n}} = \bar{\bar{x}} \pm A_2\bar{R} \qquad \textbf{[10.2]}$$

$$\text{Centerline} = \bar{\bar{x}}$$

where $\quad \bar{x}_j = \dfrac{\sum\limits_{i=1}^{n} x_i}{n} \quad$ for each subgroup

and $\quad \bar{\bar{x}} = \hat{\mu} = \dfrac{\sum\limits_{j=1}^{m} \bar{x}_j}{m}$

Also, $\quad R_j = (X_{\max} - X_{\min}) \quad$ in each subgroup

and $\quad \bar{R} = \dfrac{\sum\limits_{j=1}^{m} R_j}{m} \quad$ where j is used as a subgroup index.

For a 3-sigma X-bar ($b = 3$) chart where μ and σ are estimated from $\bar{\bar{x}}$ and $\bar{s}$, respectively,

$$\left.\begin{array}{l} \text{UCL}_{\bar{X}} \\ \\ \text{LCL}_{\bar{X}} \end{array}\right\} = \bar{\bar{x}} \pm \frac{3\bar{s}}{c_4\sqrt{n}} = \bar{\bar{x}} \pm A_3\bar{s} \qquad \textbf{[10.3]}$$

$$\text{Centerline} = \bar{\bar{x}}$$

where $\quad s_j = \sqrt{\dfrac{\sum\limits_{i=1}^{n}(x_i - \bar{x}_j)^2}{n-1}} \quad$ for each subgroup

and $\quad \bar{s} = \dfrac{\sum\limits_{j=1}^{m} s_j}{m}$

The control chart constants d_2, c_4, A, A_2, and A_3 are as listed in Table VIII.5, Section 8. The constants A, A_2, and A_3 were developed for convenience in calculating the 3-sigma limits, whereas the constants d_2 and c_4 are related to the range and standard deviation distributions, respectively.

Once the X-bar chart is set up using the computed control limits and centerline, and the process is judged to be stable, we plot the $\bar{x}_j$ values as they become available. We connect the points together with a solid line and use the chart to monitor the process.

The fact that all subgroup means $\bar{x}_j$'s fall within the 3-sigma limits on the X-bar chart and exhibit a random pattern indicates a stable process. However, departures from expected process behavior may not always manifest themselves on the control chart immediately, or for that matter at all. In other words, if the mean has shifted, then the null hypothesis is now false, but the data may not show this shift in a manner we can detect on our chart. This failure to detect a shift results in a type II statistical error. Although narrower limits, e.g., 2-sigma limits, $b = 2$, would allow faster detection of special causes, such limits would also increase the chance of false alarms, the type I statistical error. That is, when the process is actually in statistical control, but a sample mean $\bar{x}_j$ falls outside the control limits, we falsely conclude that the process is out of control and begin looking for a nonexistent special cause.

The selection of the appropriate placement of the UCL and LCL, that is, the selection of the type I error probability, is an economic and timeliness issue. **The intent is to design the control limits and subgroup size in such a way as to balance the economic consequences of (1) failing to detect a special cause when it is present** and (2) **falsely detecting a special cause when one is not present.** Practice over the years has led to the use of 3-sigma limits as a good choice for balancing these two undesirable consequences; see DeVor et al. [1]. In the appendix to this chapter, we develop operating characteristic (OC) curves for the X-bar chart to quantitatively illustrate these concepts. In Chapter 13 we simulate several cases of stability and instability and demonstrate these concepts empirically.

R AND *S* CONTROL CHART MECHANICS

So far, we have confined ourselves to a discussion of the detection of the process mean shifts. However, **special cause disturbances may cause changes in the dispersion.** Another hypothesis-testing argument, similar to that of the X-bar chart, can be put forward for the behavior of a series of ranges and standard deviations of the same samples. **The sampling distributions of the range and the standard deviation provide the bases for the establishment of dispersion chart control limits.**

The *R* chart is a very popular control chart used to monitor the dispersion associated with a process variable. Its simplicity of construction and maintenance makes it very popular; however, the sample range provides an abstract measure of dispersion. For example, as the sample size increases, we expect the range to increase, but the amount of increase is difficult to visualize. With commonplace use of computer aids, e.g., spreadsheets, the simplicity of calculation that the *R* chart once held over the *S* chart has pretty much disappeared.

The range is a good measure of variation for small subgroup sizes, e.g., $2 \leq n \leq 5$. As n increases, the utility of the range measure, as a measure of dispersion, falls off, since it only reflects the information in the two extreme points in the subgroup; see Montgomery [2]. Once $n > 5$ or so, the standard deviation measure is preferred, since it utilizes all subgroup points. Hence, one rule of thumb is to use *R* charts for small subgroup sizes and *S* charts for larger subgroup sizes. The $n = 5$ or $n = 6$ dividing line is arbitrary, but once $n \geq 10$ the *S* chart is clearly the better chart from the statistical point of view.

To establish a control chart for the range, we must be specific about the quality characteristic's underlying distribution. This is so because the distribution of ranges does not exhibit the same robust behavior as that of the distribution of averages—the CLT does not protect us here, as it does in the X-bar chart case. If we can assume that the individual measurements are normally and independently distributed, we can establish a relationship between the standard deviation of the ranges

σ_R and the standard deviation of measurements σ. We refer our readers to Duncan [3] for details on range distributions.

Statistical theory has established the relationship between the standard deviation of the range and the standard deviation of the normally distributed quality characteristic measurement, random variable X, as

$$\sigma_R = d_3 \sigma$$

where d_3 is a known function of n, the subgroup size. Given the relationship between R and σ, we have

$$\hat{\sigma}_R = \frac{d_3}{d_2} \bar{R}$$

where d_2 and d_3 are tabulated in Table VIII.5.

The general form for the R chart, assuming all subgroups are the same size n and the range mean μ_R is known, is

$$\left. \begin{array}{c} \text{UCL}_R \\[1em] \text{LCL}_R \end{array} \right\} = \mu_R \pm b\, \sigma_R \qquad \textbf{[10.4]}$$

$$\text{Centerline} = \mu_R$$

For a 3-sigma R chart where σ_R is estimated from $\bar{R}$,

$$\text{UCL}_R = \bar{R} + 3\,\frac{d_3}{d_2}\,\bar{R} = D_4\bar{R} \qquad \textbf{[10.5]}$$

$$\text{LCL}_R = \bar{R} - 3\,\frac{d_3}{d_2}\,\bar{R} = D_3\bar{R} \qquad \textbf{[10.6]}$$

$$\text{Centerline} = \bar{R}$$

where $R_j = (x_{\max} - x_{\min})$ in each subgroup and

$$\bar{R} = \frac{\sum\limits_{j=1}^{m} R_j}{m}$$

Here j is used as a subgroup index. If the LCL calculated is less than 0, no LCL exists, or we set the $\text{LCL}_R = 0$. The constants d_2, d_3, D_3, and D_4 are as listed in Table VIII.5.

Once the R chart is set up with the computed control limits and centerline, and the process is judged to be stable, we plot the R_j values as they become available. We connect the points together with a solid line and use the chart to monitor the process.

As explained above, the S chart is preferred as a control chart to monitor dispersion when the subgroup sample sizes are relatively large. It tracks the subgroup standard deviations; hence, it is statistically straightforward in its interpretation. As is the case with the R chart, we must assume that the individual measurements are normally distributed in order to justify use of the S chart.

Assuming that random variable X is normally distributed, statistical theory has established the relationship between the standard deviation of the subgroup and the standard deviation of the quality characteristic measurement X.

For a 3-sigma S chart where σ_s is estimated from $\bar{s}$, the control limits are

$$\text{UCL}_S = \bar{s} + 3\ \frac{\bar{s}\sqrt{1 - c_4^2}}{c_4} = B_4\bar{s} \qquad \textbf{[10.7]}$$

$$\text{LCL}_S = \bar{s} - 3\ \frac{\bar{s}\sqrt{1 - c_4^2}}{c_4} = B_3\bar{s} \qquad \textbf{[10.8]}$$

$$\text{Centerline} = \bar{s}$$

If the lower control limit calculated is less than 0, no LCL exists or we set it to 0. The constants c_4, B_3, and B_4 are as listed in Table VIII.5.

Here,

where

$$s_j = \sqrt{\frac{\sum\limits_{i=1}^{n}(x_i - \bar{x}_j)^2}{n - 1}}$$

and

$$\bar{x}_j = \frac{\sum\limits_{i=1}^{n} x_i}{n} \qquad \text{for each subgroup}$$

$$\bar{s} = \frac{\sum\limits_{j=1}^{m} s_j}{m}$$

Here j is used as a subgroup index.

Once the S chart is set up with the computed control limits and centerline, and stability is indicated, we plot the s_j values on the S chart as they become available. We connect the points together with a solid line and use the chart to monitor the process.

Since the process standard deviation σ plays a key role in describing our process dispersion in SPC, a comment is in order. In statistics, we typically develop estimates of μ and σ^2 using

$$\hat{\mu} = \bar{x} = \frac{\sum\limits_{i=1}^{n} x_i}{n}$$

$$\hat{\sigma}^2 = s^2 = \frac{\sum\limits_{i=1}^{n}(x_i - \bar{x})^2}{n - 1}$$

where i represents a sample point index. Here, s^2 can be shown to be an unbiased estimate of σ^2; see Walpole et al. [4]. But $s = \sqrt{s^2}$ is biased as an estimate of σ, and thus, so also is $\bar{s}$. The c_4

factor is used to adjust out this bias. Scanning the c_4 column in Table VIII.5, we can see that this bias is very small as we approach, say, $n = 30$. But, for small n's, the usual case in SPC, the bias is large. Hence, the c_4 adjustment in Equation (10.9a) is critical.

$$\hat{\sigma} = \frac{\bar{s}}{c_4} \qquad \text{[10.9a]}$$

We can also estimate $\hat{\sigma}$ from our R chart,

$$\hat{\sigma} = \frac{\bar{R}}{d_2} \qquad \text{[10.9b]}$$

Here, again, c_4 and d_2 values are listed in Table VIII.5. For more details on how to develop the factors we have previously introduced (d_2, d_3, c_4), we refer our readers to Duncan [3].

X-BAR AND R CHART MECHANICS   We revisit the meter stick production process discussed in | **Example 10.2**
Example 10.1 and construct the 3-sigma trial limits for the X-bar and R charts. What we have are 20 samples ($m = 20$) of size 5 ($n = 5$) meter sticks. From Table 10.1 it is evident that

$$\bar{\bar{x}} = \frac{19{,}993.14}{20} = 999.66 \text{ mm}$$

$$\bar{R} = \frac{12.40}{20} = 0.62 \text{ mm}$$

From Equation (10.2), the 3-sigma control limits for the X-bar chart are

$$\text{UCL}_{\bar{X}} = \bar{\bar{x}} + A_2\bar{R} = 999.66 + 0.58(0.62) = 1000.02 \text{ mm}$$
$$\text{LCL}_{\bar{X}} = \bar{\bar{x}} - A_2\bar{R} = 999.66 - 0.58(0.62) = 999.30 \text{ mm}$$
$$\text{Centerline} = \bar{\bar{x}} = 999.66 \text{ mm}$$

From Equations (10.5) and (10.6), the 3-sigma control limits for the R chart are

$$\text{UCL}_R = D_4\bar{R} = 2.11(0.62) = 1.31 \text{ mm}$$
$$\text{LCL}_R = D_3\bar{R} = 0(0.62) = 0 \text{ mm}$$
$$\text{Centerline} = \bar{R} = 0.62 \text{ mm}$$

These control chart limits for the X-bar and R charts were shown in Figure 10.2. It was previously mentioned that in order for the limits to be meaningful, the process under consideration must be in statistical control. Figure 10.2 makes it clear that the meter stick production process is not in statistical control. In fact, both the X-bar chart and the R chart show points outside the trial control limits, which would be very unlikely if the process were actually stable.

Since the process appears to be unstable, the variation in the data collected is very likely due to both the variation created by chance or common causes and that created by special causes. **In order to arrive at control chart limits that represent a stable process, the special cause effects must be eliminated.** At this point in our example, we assume these causes were identified and their effects eliminated. Hence, we eliminate subgroups 10, 12, and 18 on the X-bar chart and subgroups 9 and 13 on the R chart and recompute our trial control limits, excluding these subgroups.

We are now left with 15 samples ($m = 15$). The revised grand average, $\bar{\bar{x}} = 999.63$ mm, is calculated as

$$\bar{\bar{x}} = \frac{\displaystyle\sum_{j=1}^{m} \bar{x}_j}{m} = \frac{14,994.44}{15} = 999.63 \text{ mm}$$

The average of the ranges is 0.50 mm. This number is calculated as

$$\bar{R} = \frac{\displaystyle\sum_{j=1}^{m} R_j}{m} = \frac{7.50}{15} = 0.50 \text{ mm}$$

The *revised control limits* for both the X-bar chart and the R chart are calculated as follows: From Equation (10.2), the 3-sigma control limits for the revised X-bar chart are

$$\text{UCL}_{\bar{X}} = \bar{\bar{x}} + A_2\bar{R} = 999.63 + 0.58(0.50) = 999.92 \text{ mm}$$
$$\text{LCL}_{\bar{X}} = \bar{\bar{x}} - A_2\bar{R} = 999.63 - 0.58(0.50) = 999.34 \text{ mm}$$
$$\text{Centerline} = \bar{\bar{x}} = 999.63 \text{ mm}$$

From Equations (10.5) and (10.6), the revised 3-sigma control limits for the R chart are

$$\text{UCL}_R = D_4\bar{R} = 2.11(0.50) = 1.06 \text{ mm}$$
$$\text{LCL}_R = D_3\bar{R} = 0(0.50) = 0.00 \text{ mm}$$
$$\text{Centerline} = \bar{R} = 0.50 \text{ mm}$$

These limits, called revised control limits, are shown in Figure 10.3. All 15 subgroups are now within the revised control limits on both charts. This being the case, it is safe to extend the control limits into the future to monitor future production. Now, to estimate σ, the population standard deviation,

$$\hat{\sigma} = \frac{\bar{R}}{d_2} = \frac{0.50}{2.326} = 0.215 \text{ mm}$$

We want the control charts to reflect our current, stable process; any future process changes demand that we reevaluate our control charts to ensure that they reflect current operations. Should the physical process change, care should be exercised in recomputing new, appropriate control limits. It is of interest to note here that our target of 1000 mm is higher than our process average. Although the process exhibits statistical stability, we typically are concerned about this deviation from target in the sense that we may have problems meeting individual product specifications. This topic, one of process capability, is discussed at length in Chapter 11.

10.5   INTERPRETATION OF SHEWHART CONTROL CHARTS

In working with the Shewhart control charts, the information from each sample (a rational subgroup with sample size greater than 1) is judged to determine whether or not it indicates the presence of a special cause disturbance. **Unless the evidence is significant, in favor of the**

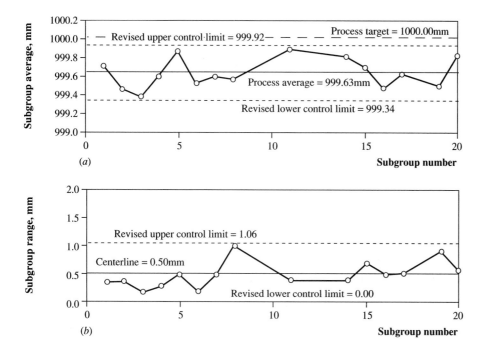

Figure 10.3 Revised control charts for meter stick production process (Example 10.2). (a) Control chart for averages, X-bars, with revised control limits. (b) Control chart for ranges, R's, with revised control limits.

occurrence of a special cause disturbance, we declare that the process is "in control," or stable, and subject only to common cause variation.

The *X*-bar, *R*, or *S* chart combinations possess two desirable properties when little or no process inertia is present and the process metric is normally distributed. **First, each subgroup is independent of other subgroups,** so that we see new independent information in each chart point. This feature helps us focus on current process operations, as opposed to mixing old data or information with data from each new sample or subgroup. **Second, we can establish sampling schemes relatively close in time within subgroups and relatively far in time between subgroups, whereby we can usually capture only common cause variation within subgroups and thereby isolate special cause variation between subgroups.** This feature helps us to isolate and estimate common cause variation more precisely. If we mix common cause and special cause variation, we will typically overestimate common cause variation, which will push our control limits out further than they should be. Hence, our chart's statistical properties will be affected.

BASIC INTERPRETATION

There are two process characteristics of general interest in the SPC scenario for variables data: the location and the dispersion. **We are particularly interested in determining any point in time at which either location or dispersion changes.** From a statistical point of view, we set up crude hypothesis tests each time we sample and plot subgroup statistics. Here, type I (false alarm) and type II (failure to detect a process disturbance) errors are applicable; refer to Table 9.3.

Figure 10.4 graphically depicts the hypothesis-testing concept. Figure 10.4a depicts a normal population of individual process characteristic measurements. In Figure 10.4b, we see two sample results obtained from the population in Figure 10.4a, denoted as $\bar{x}_1$ and $\bar{x}_2$. The critical regions (for H_0 rejection) are shaded. They are established on the basis of the sample/subgroup size n, the process standard deviation σ, and the type I error probability (α risk) we are willing to accept. For the sample result $\bar{x}_1$ plotted, we would reject the null hypothesis, while the result $\bar{x}_2$ leads to the conclusion that we cannot reject H_0. Here, $\bar{x}_1$ fell in the lower critical region while $\bar{x}_2$ did not fall in a critical region.

Figure 10.4c shows the two sample/subgroup means $\bar{x}_1$, $\bar{x}_2$ plotted on an X-bar chart. To the right of these two points we have drawn an "in-control" sequence, or signal (center position). An "out-of-control" sequence (far right) is also shown. **Under the assumption that the sampling results arise from a constant system of chance causes, common cause, the plotted values should evolve randomly over time.** The center sequence in Figure 10.4c illustrates this case. On the other hand, **if the sampling results show evidence of a nonrandom pattern,** such as the sequence to the right, **a special cause disturbance may be present and investigation is warranted.** From comparing Figures 10.4b and c, we see that the distribution of means is rotated 90° to form a control chart and that the critical regions are defined by the control chart limits.

Our use of subgroup sizes $n > 1$ brings the central limit theorem (CLT) to bear on the population of subgroup means, displayed by our X-bar chart. See Walpole et al. [4] for a more detailed discussion of the CLT. According to the CLT, the distribution of sample means will approach a normal distribution (Gaussian distribution) for reasonably large subgroup sizes, even if the population sampled is not normally distributed. If the population is normally distributed, then the CLT tells us that the distribution of means will also be normal. Grant and Leavenworth [5] maintain that samples of size 3 to 5 offer reasonable protection for X-bar charts, even if the underlying population is not normally distributed. Walpole et al. [4] use a sample size of 30 as a guide to assume normality in distributions of means.

In practice, if the population of individual measurements sampled is normally distributed or near normally distributed to begin with, the subgroup size issue becomes a matter of sampling economics and statistical error probabilities. Dealing with larger subgroups is more expensive and time-consuming, but probabilities of failures to detect are lower. This topic is discussed in detail in the appendix to this chapter.

The ± 3-sigma control limits are widely used in practice. Therefore, the control chart mechanics previously described in this chapter are based on 3-sigma limits, i.e., ±3 standard deviations associated with the control chart statistic. **In some cases, it may be beneficial to work with control limits other than 3-sigma limits,** e.g., 2.5-sigma limits. In these situations, the usual practice is to (1) calculate 3-sigma limits based on the mechanics presented in this chapter; (2) calculate the 3-sigma distance, the UCL value minus the centerline value; (3) divide by 3 to obtain an estimate for 1 chart-sigma; and (4) calculate the desired UCL and LCL up and down from the centerline in the desired multiple of the chart-sigma. Remember that 1 chart-sigma on the X-bar chart is equal to 1 standard deviation of the mean, i.e., $\hat{\sigma}_{\bar{x}} = \hat{\sigma}/\sqrt{n}$.

The placement of the control chart limits impacts the type I error characteristics. Wider limits yield lower type I error probabilities. In general, the larger the shift in the process statistic, the faster our chart will signal a shift. Small shifts and trends present problems for our Shewhart

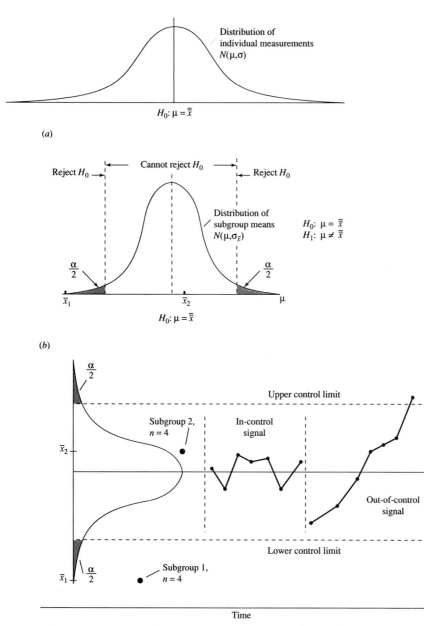

Figure 10.4 Hypothesis testing formulation extended to establish the Shewhart control chart model. (a) Population of individual items sampled, $N(\mu, \sigma)$. (b) Hypothesis testing formulation, subgroup means, $N(\mu, \sigma_{\bar{x}} = \sigma/\sqrt{n})$. (c) Extended formulation in X-bar chart form.

charts. **A small shift, when our signal criterion is defined by one point outside our control limits, is difficult to detect. Two methods are used to address this problem:** (1) **pragmatic runs rules,** described below, and (2) **other models/charts** such as the EWMA or CuSum chart, described in subsequent sections of this chapter.

PRAGMATIC INTERPRETATION

In practice, **the null hypothesis of a stable process is rejected and it is likely that a special cause is present, when any of the following is observed:**

1. **An extreme point outside of the control limits** (usually ± 3-sigma limits).

2. **An unusual pattern.**

3. **A run or trend** (upward, downward, or level).

The most widely used pragmatic chart control indicators are based on the AT&T runs rules; see AT&T Technologies [6]. These runs rules are usually applied to a 3-sigma control chart after it has been divided into zones. Zone A is the innermost zone, which is ± 1 sigma ($b = 1$ in the chart mechanics) from the centerline. Zone B extends beyond zone A by an additional standard deviation. Hence, the outer bounds of zone B are at ± 2 sigma ($b = 2$) and the inner bounds of zone B are at ± 1 sigma. The third zone, zone C, is represented by outer bounds at ± 3 sigma ($b = 3$) and inner bounds at ± 2 sigma ($b = 2$). Figure 10.5 depicts the zones for an X-bar chart. The probabilities shown are based on a normal distribution. The band probabilities will be different for charts other than the X-bar chart (e.g., the R and S charts). In practice, very little interest is expressed in exact probabilities for these other charts, which are based on distributions other than the normal distribution. It is not uncommon to see runs rules similar to those used on X-bar charts applied to other types of charts in order to help isolate small shifts and identify patterns or trends.

Some widely used runs rules are listed below. These runs rules are based on the AT&T [6] and Hoskins et al. [7] (Motorola) runs rules. They take a pragmatic approach to control chart applications. Rule 1, below, is considered universal for control charts. **Actual practice in any given facility may entail use of all of these rules, some of these rules, or even different runs rules. A process is declared out of statistical control**—action should be taken to determine special cause—**when**

1. **A point falls beyond the control limits,** usually set at ± 3 sigma.

2. **Two out of three consecutive points fall beyond the same ± 2-sigma band** (beyond the same B zone).

3. **Four out of five consecutive points fall beyond the same ± 1-sigma band** (beyond the same A zone).

4. **A run of seven consecutive points falls:**

    a. **Above the center line.**

    b. **Below the center line.**

    c. **In a continuous upward pattern.**

    d. **In a continuous downward pattern.**

5. **Nine out of 10 points fall within the ± 1-sigma band** (both A zones).

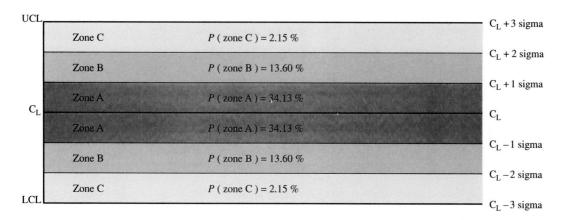

Figure 10.5   Control chart zones and normal distribution–based probabilities for a 3-sigma X-bar chart.

Figure 10.6 (next page) displays selected examples of the runs rules patterns described above. An X-bar chart from a stable process will display a great deal of activity, or "bounce," up and down. As shown in Figure 10.5, about 68% of the points will fall within the A zones, 27% in the B zones, and 4% in the C zones, with a great deal of movement from zone to zone.

10.6   SHEWHART CONTROL CHART OC CURVES AND AVERAGE RUN LENGTHS

An operating characteristic (OC) curve can be very helpful in assessing type I and type II statistical error probabilities. A generic two-tailed OC curve is shown in Figure 10.7 on page 255. The vertical axis displays P_a, the probability of accepting H_0 as true. All values of P_a, except one, represent β's, or type II error probabilities. The one particular P_a where H_0 is true represents $1 - \alpha$. All points other than the point where H_0 is true represent cases where some H_1 is true. When the parameter value, i.e., μ, shifts, $1 - P_a$ represents the probability of correctly detecting the shift on the first sample or subgroup taken after the shift occurs. **OC curves can be useful in control chart design by helping to identify the probability of a false alarm, α, and the probability of a failure to detect a shift, β.** Generic control chart strategies and their related false alarm and failure to detect characteristics are provided in Table 10.2 on page 255.

In addition to OC curves, **quality control chart performance is measured by average run length (ARL).** The ARL is the average or expected number of samples or subgroups before the control chart provides an out-of-control signal. If the null hypothesis, H_0, is true, the ARL_0 indicates the expected number of correct indications or sample points before a type I error. In this case a large ARL_0 is desirable.

If, on the other hand, a shift has occurred and some H_1 is true, then ARL_1 indicates the expected number of "incorrect" indications (failures to detect) before we encounter a "correct"

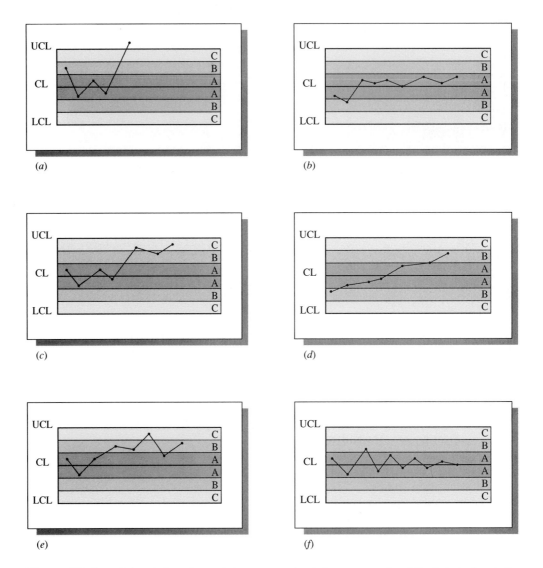

Figure 10.6   Runs rule–based control chart patterns that indicate process instability (a lack of control).
(a) One point outside control limits. (b) Seven consecutive points above the centerline.
(c) Two out of three points in zone C or beyond. (d) Seven consecutive points in an upward
trend. (e) Four out of five points in zone B or beyond. (f) Nine out of 10 consecutive points
fall within plus or minus one sigma from the centerline. The runs in b through f are
meaningful to only those control charts that assume independence between successive
sample points (they are not applicable to CuSum, EWMA, EWMD, and R_M control charts).

Table 10.2   Generic control chart design strategies.

Strategy To Detect a Process Shift of a Given Magnitude	P (False Alarm), α	P (Failure To Detect Shift), β
Increase sample size *n*; same *b* (sigma limit) multiple	No change	↓
Decreasing *b*, in order to pull the control limits in	↑	↓
Use of runs rules in addition to points outside control limits	↑	↓

indication of a process shift. Hence, a small ARL_1 is desirable. Here, the minimum ARL_1 is 1. In other words, we must encounter at least one sample or subgroup to detect a shift.

OC curves and ARLs are described quantitatively in the appendix to this chapter. Simulation studies are used to study ARL performance in complex situations where runs rules are used. Chapter 13 provides several cases of SPC performance assessment using both normal and non-normal data, runs rules, and process inertia.

10.7   PROBABILITY LIMITS FOR SHEWHART CONTROL CHARTS

When we use probability limits on control charts, we refer to the probability of the distribution that lies beyond or beneath the control limits. Whereas sigma limits refer to the distance from the centerline out to the control limits. **Control chart probability limits relate directly to the type I error probability α.**

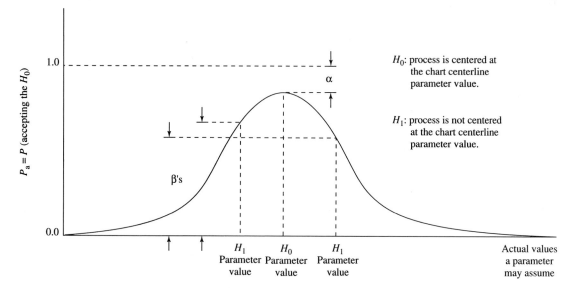

Figure 10.7   Generic two-tailed OC curve.

The probability limit concept is compared with the sigma limit concept in Figure 10.8. It is possible to develop probability limits for any control chart in which the underlying model distribution is quantifiable. We will briefly discuss probability limits for the X-bar, R, and S charts and then introduce the S^2 chart. See Oakland [8], Grant and Leavenworth [5], and Duncan [3] for more details.

X-bar chart probability limits are straightforward in that a standard normal distribution is applied, once the "tail" areas are associated with the desired probability levels. For example, a $UCL_{0.975}$ and an $LCL_{0.025}$ are associated with $b = \pm 1.96$. It is more difficult to deal with the R and S charts. In the example below, we will see that the R and S chart probability limits are not symmetrically placed about the centerline, as is the case in general for sigma limits. We also notice that LCL_R and LCL_S are not set at 0 for small subgroup sizes, as they were in the previous $\pm b$-sigma examples.

From a statistical perspective, probability limits make sense, but in the United States, sigma limits see much wider use because of their familiarity and the ease in calculation. Of course, for the X-bar chart it does not make any difference; we can simply adjust the b value. For convenience, special $A_{2,0.xxx}$ and $A_{3,0.xxx}$ probability limit factors have been calculated for use with X-bar, R and X-bar, S probability limit control chart sets. These values are shown in Table VIII.6 in Section 8. Counterpart $D_{0.xxx}$ and $B_{0.xxx}$ constants for R and S charts, respectively, are also presented.

| Example 10.3 | **PROBABILITY CONTROL LIMITS**   Using the meter stick production process data from Example 10.2, calculate the probability control limits for the X-bar, R, and S control charts with type I error probability equal to 0.01. In other words, determine the $UCL_{0.995}$ and $LCL_{0.005}$ for the three control charts. |

Solution

We can determine from the reduced data set, after eliminating subgroups 9, 10, 12, 13, and 18, that the subgroup size $n = 5$, the grand average $\bar{\bar{x}} = 999.63$ mm, the mean of subgroup ranges $\bar{R} = 0.50$ mm, and the mean of subgroup sample standard deviations $\bar{s} = 0.206$ mm.

For the X-bar control chart based on $\bar{R}$, the probability limits are

$$UCL_{\bar{X},0.995} = \bar{\bar{x}} + A_{2,0.995}\,\bar{R} = 999.63 + 0.50(0.50) = 999.88 \text{ mm}$$

$$LCL_{\bar{X},0.005} = \bar{\bar{x}} - A_{2,0.005}\,\bar{R} = 999.63 - 0.50(0.50) = 999.38 \text{ mm}$$

$$\text{Centerline} = \bar{\bar{x}} = 999.63 \text{ mm}$$

For the R control chart,

$$UCL_{R,0.995} = D_{0.995}\,\bar{R} = 2.10(0.50) = 1.05 \text{ mm}$$

$$LCL_{R,0.005} = D_{0.005}\,\bar{R} = 0.24(0.50) = 0.12 \text{ mm}$$

$$\text{Centerline} = \bar{R} = 0.50 \text{ mm}$$

For the S control chart,

$$UCL_{S,0.995} = B_{0.995}\,\bar{s} = 2.04(0.206) = 0.420 \text{ mm}$$

$$LCL_{S,0.005} = B_{0.005}\,\bar{s} = 0.23(0.206) = 0.047 \text{ mm}$$

$$\text{Centerline} = \bar{s} = 0.206 \text{ mm}$$

All constants are obtained from Table VIII.6.

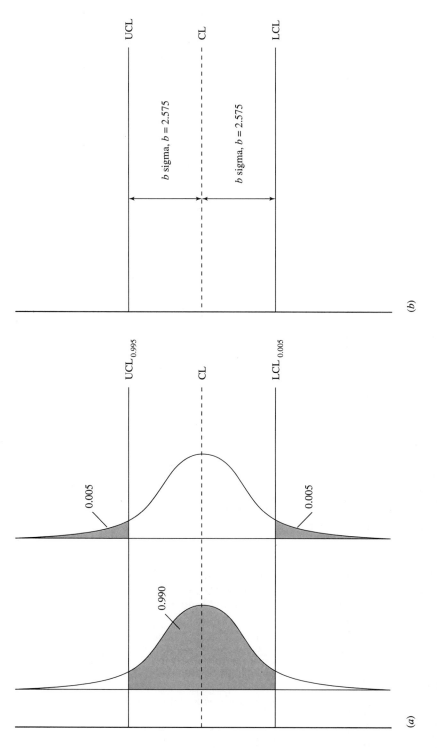

Figure 10.8　Probability and sigma limit concepts for the $\bar{X}$-bar control chart. (*a*) Probability limit concept. (*b*) Sigma limit concept.

An S^2 **chart, rather than an S chart, can be used for subgrouped data. This chart lends itself naturally to probability limits.** Here, we sample just as in the case of the S chart; however, we track S^2 rather than S for each subgroup of size n.

and

$$s_j^2 = \frac{\sum\limits_{i=1}^{n}(x_i - \overline{x}_j)^2}{n - 1} \qquad \textbf{[10.10]}$$

$$\overline{s}^2 = \frac{\sum\limits_{j=1}^{m} s_j^2}{m}$$

Here,

$$UCL_S^2 = \frac{\overline{s}^2}{n - 1} \, \chi^2_{\alpha/2,\, n-1} \qquad \textbf{[10.11]}$$

$$LCL_S^2 = \frac{\overline{s}^2}{n - 1} \, \chi^2_{1-(\alpha/2),\, n-1} \qquad \textbf{[10.12]}$$

$$\text{Centerline} = \overline{s}^2$$

where $\chi^2_{\alpha/2,n-1}$ and $\chi^2_{1-(\alpha/2),n-1}$ are the chi-squared upper and lower critical values, respectively, given $n - 1$ degrees of freedom; see Section 8, Table VIII.3.

Chart interpretation follows conventional S chart rules. However, we must recognize that the units of measure on the S^2 chart are expressed in units squared, thus rendering the S^2 chart somewhat more abstract than the S chart.

10.8   **EWMA AND EWMD CONTROL CHARTS**

The X-bar and R or S charts are not extremely effective in detecting shifts when the shift is small. The addition of "runs" rules helps solve this problem to some extent. But, other charts can be used to address this issue as well. In this section, we will introduce the use of exponentially weighted moving averages (EWMAs) to monitor location and exponentially weighted moving deviations (EWMDs) for tracking process dispersion, adapted from Sweet [9]. Here, we describe the EWMA and EWMD control charts for subgrouped data. In the next chapter, we will describe the same control charts for individual data. **The EWMA and EWMD charts are considered effective for detecting small sustained shifts.**

An EWMA is a moving average of past data where each data point is assigned a weight. These weights decrease in an exponentially decaying fashion from the present into the remote past. The moving average tends to be a reflection of the more recent process performance, as greater weight is allocated to the most recent data. The amount of decrease in the weights over time is an exponential function of the weighting factor r which can assume values between 0 and

1. **By choosing the weighting factor r we can essentially tailor our EWMA and EWMD charts for timeliness or time sensitivity.** This feature, plus their ability to signal small sustained shifts, makes the EWMA and EWMD charts attractive SPC model alternatives.

The weighting factor associated with observation $\bar{x}_{j-k}$ is

$$W_{j-k} = r(1 - r)^k$$

When a small value of r is used, the moving average at sample point j carries with it a great amount of inertia from the past. Hence, it is relatively insensitive to short-lived changes in the process. **For control chart applications where fast response to process shifts is desired, a relatively large weighting factor, say $r = 0.2$ or $r = 0.5$, is used;** see DeVor et al. [1]. Figure 10.9, based on the weighting relationship above, shows the decaying behavior for some commonly used values of r.

If a shift in the mean occurs, the EWMAs will gradually, depending on r, move to the new mean of the process, while the exponentially weighted moving deviations, EWMDs, will remain unchanged. On the other hand, **if there is a shift in the process dispersion, the EWMDs will gradually move to the new level,** while the EWMAs still vary about the same process mean. In order to construct the EWMA and EWMD control charts, we have two options: (1) use target values such as μ_0 and σ_0 or (2) use past data to estimate μ and σ. The first option is straightforward, provided we have reasonable targets established. The second option requires us to collect data from the process. In the second option, we calculate estimates of the process mean and standard deviation by using a reasonably large number of subgroups.

Here, $\displaystyle \bar{x}_j = \frac{\sum_{i=1}^{n} x_i}{n}$ for each subgroup

$$\hat{\mu} = \bar{\bar{x}} = \frac{\sum_{j=1}^{m} \bar{x}_j}{m}$$

and $\displaystyle s_j = \sqrt{\frac{\sum_{i=1}^{n} (x_i - \bar{x}_j)^2}{n - 1}}$ for each subgroup

$$\bar{s} = \frac{\sum_{j=1}^{m} s_j}{m}$$

and $\displaystyle \hat{\sigma} = \frac{\bar{s}}{c_4}$

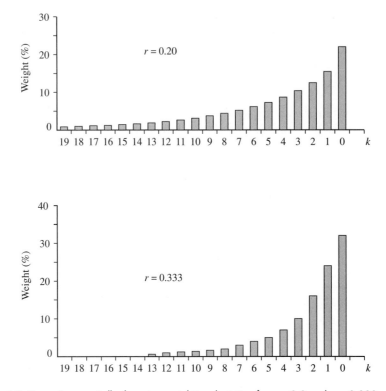

Figure 10.9　Exponentially decaying weighting depiction for $r = 0.2$ and $r = 0.333$.

Here i and j are used as the within subgroup index and the subgroup index, respectively. Then,

$$\hat{\sigma}_{\bar{x}} = \frac{\hat{\sigma}}{\sqrt{n}}$$

We compute EWMAs, A_j, and EWMDs, V_j, as follows:

$$A_j = r\bar{x}_j + (1 - r) A_{j-1} \qquad \text{where } A_0 = \hat{\mu} \text{ or } \mu_0 \tag{10.13}$$

$$V_j = rD_j + (1 - r) V_{j-1} \qquad \text{where } D_j = |\bar{x}_j - A_{j-1}|, V_0 = \hat{\sigma}_{\bar{x}} \text{ or } \sigma_{\bar{x}_0} \tag{10.14}$$

and $\quad \sigma_{\bar{x}_0} = \dfrac{\sigma_0}{\sqrt{n}}$

The control limits and the A_j and V_j centerlines are calculated as

$$\left. \begin{array}{l} \text{UCL}_A \\ \\ \text{LCL}_A \end{array} \right\} = \hat{\mu} \quad \text{or} \quad \mu_0 \pm A^*\hat{\sigma}_{\bar{x}} \quad \text{or} \quad A^*\sigma_{\bar{x}_0} \tag{10.15}$$

Centerline $= \hat{\mu}$ or μ_0

$$\text{UCL}_V = D_2^* \hat{\sigma}_{\bar{x}} \text{ or } D_2^* \sigma_{\bar{x}_0} \qquad \textbf{[10.16]}$$

$$\text{LCL}_V = D_1^* \hat{\sigma}_{\bar{x}} \text{ or } D_1^* \sigma_{\bar{x}_0} \qquad \textbf{[10.17]}$$

$$\text{Centerline} = d_2^* \hat{\sigma}_{\bar{x}} \text{ or } d_2^* \hat{\sigma}_{\bar{x}_0}$$

where all A^*, d_2^*, D_1^*, and D_2^* constants are as listed in Table VIII.7, Section 8.

We plot all of the m, A_j values on the A chart and the m, V_j values on the V chart and interpret the charts, with respect to points falling above/below the control limits. Once stability is established, we extend our charts to monitor future process operations.

SUBGROUPED EWMA AND EWMD   A printed circuit board (PCB) manufacturer builds PCBs from | **Example 10.4**
layers of insulating board material and circuits sandwiched in between. Once the basic PCB is cured, toward the end of PCB fabrication various electronic components, such as resistors and chips, are mounted on the board's surface. This mounting subprocess involves several steps. One of the first steps is the placement of solder pads at appropriate locations on the board. Components are then placed on these pads, and the solder pads are melted to form a bond between the circuit contacts on the board and the leads of the component.

Solder pads must be placed accurately and precisely over the appropriate circuit contact. This placement is done using a pneumatic arm that extends across the board (the Y direction), as the board travels (the X direction) via conveyor perpendicular to the travel of the pneumatic arm. One process challenge is to stop the board in the X direction in position so that the solder pad is located over the circuit contact. This alignment situation is shown in Figure 10.10.

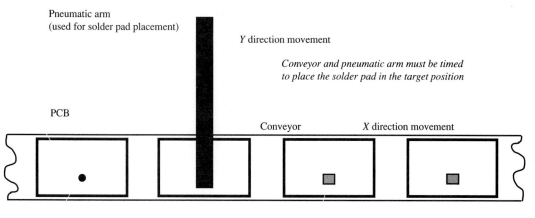

Figure 10.10   Printed circuit board conveyor with PCBs and pneumatic arm, simplified depiction (Example 10.4).

Table 10.3   Printed circuit board location data (Example 10.4)

Sub-group	Board Location Deviation, x_{ij}			Subgroup Average $\bar{x}_j$	EWMA and EWMD Statistics		
	1	2	3		A_j	D_j	V_j
1	−0.00402	−0.00309	−0.02042	−0.00918	−0.0046	0.0092	0.0190
2	0.05408	0.08795	−0.00896	0.04436	0.0199	0.0489	0.0340
3	0.04588	0.03835	−0.05926	0.00832	0.0141	0.0116	0.0228
4	0.01839	0.01865	0.02223	0.01975	0.0169	0.0056	0.0142
5	−0.13447	0.09491	0.04377	0.00140	0.0092	0.0155	0.0149
6	−0.06491	0.01824	0.04246	−0.00140	0.0039	0.0106	0.0127
7	−0.01469	−0.04438	−0.02768	−0.02892	−0.0125	0.0328	0.0228
8	−0.02617	0.05391	0.00690	0.01155	−0.0005	0.0241	0.0234
9	−0.05081	0.05422	0.03294	0.01212	0.0058	0.0126	0.0180
10	0.01934	0.04390	0.01097	0.02474	0.0153	0.0189	0.0185
11	−0.04829	−0.00438	0.00543	−0.01575	−0.0002	0.0310	0.0247
12	−0.04173	0.05669	−0.05752	−0.01419	−0.0072	0.0140	0.0193
13	0.03423	0.04487	0.09676	0.05862	0.0257	0.0658	0.0426
14	−0.03304	0.01334	−0.02725	−0.01565	0.0050	0.0414	0.0420
15	−0.05986	0.07018	−0.00151	0.00294	0.0040	0.0021	0.0220
16	0.05760	0.04773	−0.02392	0.02714	0.0156	0.0232	0.0226
17	−0.00045	−0.00758	−0.03315	−0.01373	0.0009	0.0293	0.0259
18	0.04517	−0.04534	−0.02498	−0.00838	−0.0037	0.0093	0.0176
19	−0.05799	−0.02861	0.02806	−0.01951	−0.0116	0.0158	0.0167
20	0.04891	−0.05253	0.00635	0.00091	−0.0054	0.0125	0.0146
21	0.01624	−0.06697	0.02469	−0.00868	−0.0070	0.0033	0.0090
22	0.01866	−0.00615	0.06397	0.02549	0.0092	0.0325	0.0207
23	0.03635	0.01638	−0.07690	−0.00806	0.0006	0.0173	0.0190
24	−0.07829	−0.07536	−0.05502	−0.06956	−0.0345	0.0701	0.0446
25	0.04474	0.04218	−0.01587	0.02368	−0.0054	0.0582	0.0514

In this process, many critical quality characteristics exist and are monitored. One particularly critical quality characteristic is the deviation from the center of the circuit contact to the center of the solder pad in the X direction, once the pad is set in place. The obvious target for this metric is zero. The data displayed in Table 10.3 on the left side were collected from this process. Deviations were measured in millimeters, and a subgroup size of $n = 3$ boards was used. Targets were set at $\mu_0 = 0$ mm and $\sigma_0 = 0.05$ mm.

Small sustained shifts are expected in this process, so that the EWMA and EWMD charts are useful. Using $n = 3$, $\mu_0 = 0$ mm, and $\sigma_0 = 0.05$ mm, and $r = 0.5$, develop the appropriate subgrouped EWMA and EWMD control charts.

Solution

Calculations regarding A_j, D_j, and V_j were made according to Equations (10.13) and (10.14). These results are displayed on the right side of Table 10.3. The control limits are calculated below.

$$\text{UCL}_A = \mu_0 + A^*\sigma_{\bar{x}_0} = 0 + 1.732(0.0500/\sqrt{3}) = 0.0500$$

$$\text{LCL}_A = \mu_0 - A^*\sigma_{\bar{x}_0} = 0 - 1.732(0.0500/\sqrt{3}) = -0.0500$$

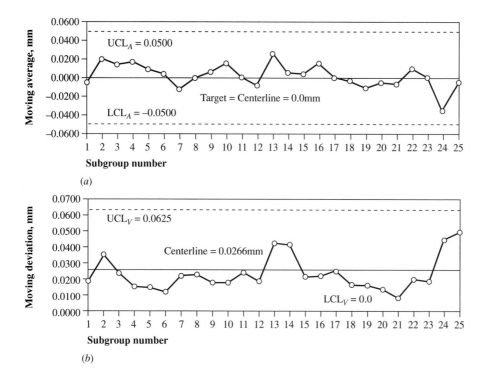

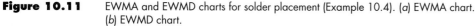

Figure 10.11 EWMA and EWMD charts for solder placement (Example 10.4). (a) EWMA chart. (b) EWMD chart.

$$\text{Centerline} = \mu_0 = 0$$

$$\text{UCL}_V = D_2^* \, \sigma_{\bar{x}_0} = 2.164(0.0500/\sqrt{3}) = 0.0625$$

$$\text{LCL}_V = D_1^* \, \sigma_{\bar{x}_0} = 0(0.0500/\sqrt{3}) = 0$$

$$\text{Centerline} = d_2^* \, \sigma_{\bar{x}_0} = 0.921(0.0500/\sqrt{3}) = 0.0266$$

The charts obtained are displayed in Figure 10.11. Here, the process appears to be on target and stable. In this case, target values were used since $\mu_0 = 0$ mm is the ideal location, and $\sigma_0 = 0.05$ mm is reasonable, on the basis of a great deal of experience on this line, with this pneumatic arm.

10.9 CuSum Control Charts

In this section, we introduce the CuSum chart for subgrouped data as an SPC tool recognized for its ability to perform well in the presence of small sustained shifts. At this point, we should note that subgrouping may not be the best way to apply the CuSum charting tool. Several authors, including Hawkins [10] and Montgomery [2], argue that CuSum technology (or EWMA technology) is more cost-effective when used with individual observations. The exception is when there are appreciable economics associated with subgrouped sampling. Individual data EWMA and CuSum charts are described in Chapter 11.

The CuSum charts have their origin in sequential sampling theory, i.e., sequential sampling schemes; see Page [11, 12] and Barnard [13]. This sequential analysis origin is different from the Shewhart chart origins, even though the two technologies serve essentially the same application—that of on-line monitoring/shift detection associated with a process variable.

We describe the classical V-mask CuSum chart as well as a contemporary tabular version of the CuSum control chart. The V mask dates back to the 1950s when data analysis capabilities were limited. Today, with computer aids, such as spreadsheets, the V mask has lost favor. Montgomery [2] summarizes criticism of the classical V mask from both practical and statistical perspectives.

The CuSum control chart was designed to identify slight but sustained shifts in a process. The statistic to be accumulated and charted may be an actual observation or the deviation of an observation from a desired target. The CuSum chart, as its name implies, accumulates information from one subgroup to the next. Hence, it does not enjoy the independence property (between plotted points) that the X-bar chart enjoys. It carries momentum across all previous observations. **The CuSum applies an equal weighting across all past observations, whereas the X-bar chart reflects information in only the current subgroup, and the EWMA weights the most recent subgroup more heavily.** Hence, the CuSum chart is not "memoryless" in relation to past observations.

CuSum charts work on a "trend" principle and can be designed for both location and dispersion parameters. If the CuSum chart tool is to be used with subgrouped data, we suggest using an S or R chart along with the CuSum location chart. The CuSum dispersion charts are awkward to deal with; see Grant and Leavenworth [5]. We will discuss the mechanics for a CuSum process location chart in order to introduce our readers to the CuSum concept. Grant and Leavenworth [5], Duncan [3], Montgomery [2], and DeVor et al. [1] provide additional discussion of CuSum charts.

We use the symbol L_j to represent a CuSum value for the jth subgroup ($n > 1$) from the beginning of the CuSum chart,

$$L_j = \sum_{k=1}^{j} (\bar{x}_k - \mu_0) \qquad\qquad \textbf{[10.18]}$$

where

$$\bar{x}_k = \frac{\sum_{i=1}^{n} x_i}{n}$$

represents the mean of the kth subgroup. Here, μ_0 is a target mean value that corresponds to the target (best) level of our process characteristic. As in the EWMA case, an empirical estimate, $\hat{\mu}$, can be used to replace the target value.

It is common to plot L_j across time or order of production, looking for trends (up or down) moving away from the target, μ_0, value. When we see several points in an upward or downward trend, we sense that a shift may have occurred. The question now becomes one of calibration. How pronounced does the trend have to be before we seek out special causes and take counteractions in the process?

The V mask is sometimes used to represent control limits. Every time a statistic is plotted, the V mask is moved horizontally across the chart to the most recently plotted statistic. If any of the plotted statistics, L_j's, fall outside the mask limits, the process is declared to be out of statistical control. Since a trend or slope principle is involved, the vertical-to-horizontal scaling is critical in the construction of the V mask and related CuSum plot. The general form of the V mask and its dimensions are illustrated in Figure 10.12.

We calculate the V-mask parameters as

$$d_c = 2 \left[\frac{\sigma/\sqrt{n}}{D_c} \right]^2 \ln \frac{1 - \beta}{\alpha} \qquad \text{[10.19]}$$

$$\theta = \tan^{-1} \frac{D_c}{2y_c} \qquad \text{[10.20]}$$

where
- α = probability of a type I (false alarm) error
- β = probability of a type II (failure to detect) error, relative to a shift of D_c
- σ = the standard deviation of the population of individual values; typically we use $\hat{\sigma}$, e.g., $\hat{\sigma} = \bar{R}/d_2$ or $\bar{s}/c_4$ from the R or S chart, or a target value, σ_0
- n = subgroup size
- y_c = graphical scale factor = length of vertical unit/length of horizontal unit
- D_c = process shift magnitude that the chart should detect with very high probability

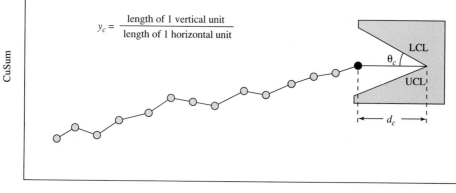

Figure 10.12 CuSum chart and V mask construction.

Using the V mask, we scale our axes according to the y_c value we selected. A y_c between $\sigma_{\bar{x}}$ and $2\sigma_{\bar{x}}$ is recommended in order to develop a presentable CuSum chart. Here, we move the V-mask axis horizontally over the most recently plotted L_j point, locating the marking point directly over L_j. If we observe an L_j point above the LCL arm of the V mask or an L_j point below the UCL arm of the V mask, then we have sufficient evidence to declare that the process is unstable or out of statistical control and seek a special cause. Otherwise, we conclude the process is stable or in statistical control.

Provided we conclude the process is unstable at sample $L_{j=m}$, we may want to estimate the magnitude of the shift Δ. If we judge, looking at the plot, that the shift began back at sample $L_{j=l}$ (i.e., near where the plotted CuSum broke out of the control limit), then

$$\Delta = \frac{L_m - L_l}{m - l}, \, m > l \qquad\qquad \textbf{[10.21]}$$

Once we detect a special cause and address its effect, we restart the physical process. Then, we redevelop our CuSum chart. For example, we simply start again at L_j, where $j = 0$ represents our current reset point in time. Assuming our dispersion did not shift, we use $\hat{\sigma}$ or σ_0 as before as an estimate of σ.

The tabular CuSum chart procedure is considered superior to the classical V-mask limit procedure. Remember, in both cases, the CuSum L_j remains the same. However, the tabular chart tracks L_j^+ and L_j^- rather than L_j directly. Montgomery [2] cites several limitations of the V-mask procedure, concerning issues of construction, interpretation, and flexibility.

Here, **for a two-sided CuSum for subgrouped data, we calculate two CuSum-related statistics, L_j^+ and L_j^-, in a tabular manner where**

$$L_j^+ = \max\,[0, \bar{x}_j - (\mu_0 + k\sigma_{\bar{x}_0}) + L_{j-1}^+] \qquad\qquad \textbf{[10.22]}$$

$$L_j^- = \min\,[0, \bar{x}_j - (\mu_0 - k\sigma_{\bar{x}_0}) + L_{j-1}^-] \qquad\qquad \textbf{[10.23]}$$

where $L_{j=0}^+ = L_{j=0}^- = 0$ and $\sigma_{\bar{x}_0} = \sigma_0/\sqrt{n}$

We can see here that $L_j^+ \geq 0$ and that $L_j^- \leq 0$. Hence, a sustained shift in $\bar{x}_j$ will tend to drift L_j^+ upward. **When L_j^+ drifts up above a limiting value $h\sigma_{\bar{x}_0}$, we declare the process unstable or out of control.** Conversely, a downward sustained shift in $\bar{x}_j$ will tend to drift L_j^- downward. **When L_j^- drifts down below a limiting value $-h\sigma_{\bar{x}_0}$, we declare the process unstable or out of control.** In other words, when

$$L_j^+ > h\sigma_{\bar{x}_0}$$

or

$$L_j^- < -h\sigma_{\bar{x}_0}$$

we declare our process metric to be unstable or out of control, assuming a special cause has created a shift in the process mean.

Here, it is apparent that values selected for k and h will determine the statistical performance of the tabular CuSum chart. Hawkins [14] cites $k = 0.5$ and $h = 4$ or 5 as commonly used values. He states that the choice of k "fine tunes" the CuSum chart to be particularly sensitive to shifts of a given magnitude. The value k is typically set so that the reference value is midway between the in-control mean μ_0 and a selected out-of-control mean μ_1 that we are interested in detecting. For example,

Table 10.4   Hawkins's CuSum k and h combinations for equivalent ±3-sigma performance.

k	0.25	0.50	0.75	1.00	1.25	1.50
h	8.01	4.77	3.34	2.52	1.99	1.61

$$k = \frac{1}{2}\frac{|\mu_0 - \mu_1|}{\sigma_{\bar{x}_0}} \qquad [10.24]$$

Once k is selected, h is selected to address the false alarm rate. In order to match the false alarm rate of the ±3-sigma X-bar chart, Hawkins [10, 14] provides the k, h combinations shown in Table 10.4.

When we encounter a special cause (instability) signal, we investigate and take appropriate counteraction to remove its effect. Then, we restart our tabulations, reinitializing to $j = 0$, $L^+_{j=0} = L^-_{j=0} = 0$, and continue. If σ has shifted or a new target σ_0 is relevant, we reflect this shift in our new L^+_j and L^-_j calculations.

Once a shift has been signaled on the tabular CuSum chart, we may want to determine the approximate magnitude of the shift. The shift timing is determined by locating the sample that initiated the specific drift or trend (that broke out of the limit). For example, if $L^+_{j=m}$ broke through the $h\sigma_{\bar{x}_0}$ limit, then we would declare that the shift occurred between the last sample that yielded $L^+_j = 0$, say sample l, and the subsequent sample that yielded $L^+_j > 0$, sample $l + 1$.

In general, the shift magnitude Δ is estimated by

$$\Delta = \frac{L^+_{j=m}}{m - l} + k\sigma_{\bar{x}_0} \qquad \text{for } L^+_{j=m} > h\sigma_{\bar{x}_0} \qquad [10.25a]$$

$$\Delta = \frac{L^-_{j=m}}{m - l} - k\sigma_{\bar{x}_0} \qquad \text{for } L^-_{j=m} < -h\sigma_{\bar{x}_0} \qquad [10.25b]$$

where l represents the sample index number just prior to the beginning of the trend that eventually pushed through the control limit, $\pm h\sigma_{\bar{x}_0}$.

In Example 10.5 below, we illustrate the tabular CuSum for subgrouped data. We will revisit and expand our CuSum chart technology in Chapter 11, when we discuss SPC charting for individuals data.

TABULAR CUSUM LOCATION CHART FOR PCB DATA   Using the PCB product/process data and targets from Example 10.4, develop a tabular CuSum location chart.

Example 10.5

Solution

The tabular CuSum is tabulated in Table 10.5. Here, we are using $k = 0.5$ and $h = 4.77$, selected from Table 10.4. We make use of Equations (10.22) and (10.23) and a spreadsheet in order to calculate the L^+_j and L^-_j values. Figure 10.13 depicts our resulting tabular CuSum chart. Here, again, as with our EWMA, our process appears to be stable, i.e., all plotted values are within the control limits. Here again, we should note that the tabular CuSum chart does not plot the actual CuSum values.

$$\text{UCL} = h\sigma_{\bar{x}_0} = 4.77(0.05/\sqrt{3}) = 0.1377$$
$$\text{LCL} = -h\sigma_{\bar{x}_0} = -4.77(0.05/\sqrt{3}) = -0.1377$$
$$\text{Centerline} = 0$$

Table 10.5   Printed circuit board location data (Example 10.5)

Sub-group	Board Location Deviation, x_{ij}			Subgroup Average $\bar{x}_j$	CuSum L_j	Tabular CuSum Statistics	
	1	2	3			L_j^+	L_j^-
1	−0.00402	−0.00309	−0.02042	−0.00918	−0.009	0.000	0.000
2	0.05408	0.08795	−0.00896	0.04436	0.035	0.030	0.000
3	0.04588	0.03835	−0.05926	0.00832	0.044	0.024	0.000
4	0.01839	0.01865	0.02223	0.01975	0.063	0.029	0.000
5	−0.13447	0.09491	0.04377	0.00140	0.065	0.016	0.000
6	−0.06491	0.01824	0.04246	−0.00140	0.063	0.000	0.000
7	−0.01469	−0.04438	−0.02768	−0.02892	0.034	0.000	−0.014
8	−0.02617	0.05391	0.00690	0.01155	0.046	0.000	0.000
9	−0.05081	0.05422	0.03294	0.01212	0.058	0.000	0.000
10	0.01934	0.04390	0.01097	0.02474	0.083	0.010	0.000
11	−0.04829	−0.00438	0.00543	−0.01575	0.067	0.000	−0.001
12	−0.04173	0.05669	−0.05752	−0.01419	0.053	0.000	−0.001
13	0.03423	0.04487	0.09676	0.05862	0.111	0.044	0.000
14	−0.03304	0.01334	−0.02725	−0.01565	0.096	0.014	−0.001
15	−0.05986	0.07018	−0.00151	0.00294	0.099	0.003	0.000
16	0.05760	0.04773	−0.02392	0.02714	0.126	0.015	0.000
17	−0.00045	−0.00758	−0.03315	−0.01373	0.112	0.000	0.000
18	0.04517	−0.04534	−0.02498	−0.00838	0.104	0.000	0.000
19	−0.05799	−0.02861	0.02806	−0.01951	0.084	0.000	−0.005
20	0.04891	−0.05253	0.00635	0.00091	0.085	0.000	0.000
21	0.01624	−0.06697	0.02469	−0.00868	0.076	0.000	0.000
22	0.01866	−0.00615	0.06397	0.02549	0.102	0.011	0.000
23	0.03635	0.01638	−0.07690	−0.00806	0.094	0.000	0.000
24	−0.07829	−0.07536	−0.05502	−0.06956	0.024	0.000	−0.055
25	0.04474	0.04218	−0.01587	0.02368	0.048	0.009	−0.017

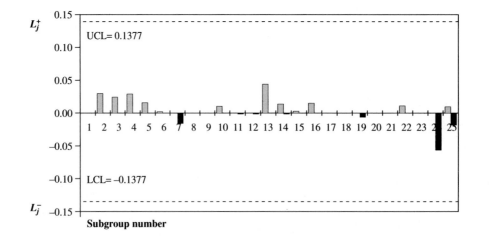

Figure 10.13   Tabular CuSum chart for solder placement (Example 10.5).

10.10 LIMITED DURATION PROCESS RUNS

Classical SPC tools require a significant process history in order to establish chart limits, and assume that we continue high-volume production into the future. This mass production paradigm is valid in many cases but is invalid in others, e.g., in just-in-time production and short-run, custom production. In these cases, process control is critical, just as it is in mass production.

Our traditional SPC tools are limited in dealing with short process runs. However, **several modifications can be made to our classical practices in order to accommodate limited production runs.** We describe two relatively simple modifications. First, we describe deviation from target charts. Second, we describe standardized charts.

DEVIATION FROM TARGET CHARTS

Given a process such as a just-in-time process or a custom production process with a lot size of one, using SPC is difficult. But under some circumstances it can be accomplished. **When we have a product/process characteristic common to different products/processes,** e.g., a length, weight, etc. of different, but similar products, **we can monitor deviations from target values.** Our chart is calibrated in deviations from target. **For our variation chart, we assume that the variation across all different products/processes we are charting is inherently the same.** This assumption may be reasonable if we use the same basic equipment and tooling configurations for the different products/processes.

Here, we simply collect actual measurements and difference them from their target values. As usual, we require 20 to 30 subgroups to establish valid charts. We use our previous control chart technology and mechanics in constructing and interpreting our charts. Sometimes we mark divisions between different parts with dotted, vertical lines on our charts.

STANDARDIZED CHARTS

Given a product/process characteristic where the process cannot be expected to produce a constant variation from short run to short run, we can standardize our data, and then develop standardized control charts based on the classical control chart mechanics. For example, we could use X-bar and R or S chart mechanics where we track standardized values

$$R_{kl}^s = \frac{R_{kl}}{\overline{R}_k}$$

$$s_{kl}^s = \frac{s_{kl}}{\overline{s}_k}$$

$$\overline{x}_{kl}^s = \frac{\overline{x}_{kl} - \mu_{0k}}{\overline{R}_k} \text{ or } \overline{x}_{kl}^s = \frac{\overline{x}_{kl} - \mu_{0k}}{\overline{s}_k}$$

where superscript S indicates a standardized statistic, k represents a specific part or process index, and l represents a subgroup index.

In general, we would obtain $\overline{R}_k$ or $\overline{s}_k$ from historical data with similar parts. Now, we develop X-bar and R or S charts using our standard values. For 3-sigma charts the UCL and LCL for the X-bar chart will be $+A_2$ and $-A_2$, respectively, with a center at zero. For the R or S charts the UCL $= D_4$ or B_4 and LCL $= D_3$ or B_3, respectively, with both centers set at 1. Here again, we need 20 to 30 subgroups to establish our charts.

LIMITED DURATION PROCESS RUN SUMMARY

When we have solid process targets, μ_0 and σ_0, and/or solid process history to draw from, we can address short runs/custom production with traditional SPC chart tools, provided we focus on deviation from target measures. However, such process monitoring must be done along with careful physical and statistical characterization to ensure model validity and integrity. **Process control model validity and integrity are critical—invalid models can lead us in directions that are counterproductive, destroying model credibility in general, in addition to introducing process confusion and losses.**

REVIEW AND DISCOVERY EXERCISES

REVIEW

10.1. Why is it incorrect (misleading) to plot product/process tolerance (specification) limits on subgrouped SPC charts? Explain. Use the meter stick example of Figure 10.1 as a basis for your response.

10.2. Why is it correct (not misleading) to plot product/process targets (nominal is best) on subgrouped SPC charts? Explain. Use the meter stick example of Figure 10.1 as a basis for your response.

10.3. What constitutes instability (a lack of statistical control) in a process? What does it indicate? Explain.

10.4. How complete are SPC charts in the entire effort of process control? What purpose do they serve in process control? Explain.

10.5. How can subgrouping be justified in SPC charting? Explain.

10.6. What advantages does subgrouping offer in SPC charting? Explain.

10.7. Given an X-bar control chart with 3-sigma limits, what is the probability of a single point falling outside the limits? If 2.5-sigma limits are used, what is the probability?

10.8. Why is the S chart preferred over the R chart when n > 5 or 6? Explain.

10.9. Explain the role of the Central Limit Theorem in variables control charting. What does it do for us? How does it impact our assumptions? Explain.

10.10. Why do we construct trial control limits before we finalize our SPC charts for future production? Explain.

10.11. How does special cause manifest itself in a process? How is it manifested in a process data stream? Explain how SPC models attempt to isolate it.

10.12. What is an operating characteristics (OC) curve? What does it measure? How is it interpreted? Explain.

10.13. What is a run length? An average run length (ARL)? Explain.

10.14. How do probability limits differ from sigma limits on SPC charts? Explain.

10.15. How do standardized SPC charts differ from the traditional SPC charts? Why would we desire to use a standardized chart? Explain.

CASE EXERCISES

The following questions pertain to the case descriptions in Section 7. Case selection list:

Case VII.6: Big Dog—Dog Food Packaging

Case VII.7: Bushings International—Machining

Case VII.9: Downtown Bakery—Bread Dough

Case VII.21: Punch-Out—Sheet Metal Fabrication

Case VII.29: TexRosa—Salsa

10.16. Select one or more of the above cases. Develop an X-bar, R chart set from the data provided in the case description. Use 3-sigma limits. Comment on your findings.

10.17. Select one or more of the above cases. Develop an X-bar, S chart set from the data provided in the case description. Use 3-sigma limits. Comment on your findings.

10.18. Select one or more of the above cases. Develop a subgrouped EWMA and EWMD chart set from the data provided in the case description. Select an appropriate r value. Use appropriate target values to calculate the control limits. Comment on your findings.

10.19. Select one or more of the listed cases. Develop a subgrouped CuSum chart from the data provided in the case description. Select appropriate k and h values from Table 10.4. Use appropriate target values to establish the chart. Comment on your findings.

10.20. For the case(s) selected in Problems 10.16 and 10.17, develop probability limits for your charts. Compare the results of your 3-sigma limit case(s) with the probability limit case(s).

10.21. For the case(s) selected in the listed problems, generate an extended, shifted, data set using the method described in Section 7, Equation (VII.1), and plot the extended data on your existing process control charts. Look for shift points and patterns. Comment on your findings.

DISCOVERY

10.22. How do SPC charts relate to hypothesis testing? Explain.

10.23. What assumptions underlie subgrouped SPC charting models? How do we assess whether or not these assumptions are met? Explain.

10.24. What happens if we use an X-bar and R or S chart on process data that carry a high degree of process inertia? Explain.

10.25. How do average run lengths (ARLs) relate to type I and type II statistical errors? Explain.

10.26. How is the ARL curve related to the OC curve? Explain.

10.27. Why do limited production runs pose challenges to SPC charting? Explain.

10.28. Compare and contrast standard unit of measurement SPC—plotting in the observed units of measure—with deviation from target SPC metrics. What is the difference in charting? In interpretation?

10.29. How does the weighting factor r impact the performance of the EWMA/EWMD charts? Explain.

10.30. Why do we prefer to use target values, rather than empirical values, in setting up EWMA/EWMD charts and CuSum charts? Explain.

MORE CHALLENGING CASE EXERCISES

The following questions pertain to the case descriptions in Section 7. *Warning:* These cases contain characteristics that create assumption violations in SPC. Use your results from Chapter 4 to help identify the problem cases and issues. Look for creative solutions to help unlock these cases. Case selection list:

Case VII.4: Back-of-the-Moon—Mining

Case VII.14: High-Precision—Collar Machining

10.31. Select one of the above cases. Develop an X-bar, R chart set from the data provided in the case description. Use 3-sigma limits. Comment on your findings.

10.32. Select one of the above cases. Develop an X-bar, S chart set from the data provided in the case description. Use 3-sigma limits. Comment on your findings.

10.33. For Case VII.14 develop a subgrouped EWMA and EWMD chart set from the data provided in the case description. Select an appropriate r value. Use appropriate target values to calculate the control limits. Comment on your findings.

10.34. For Case VII.14 develop a subgrouped CuSum chart from the data provided in the case description. Select appropriate k and h values from Table 10.4. Use appropriate target values to establish the chart. Comment on your findings.

The following problems pertain to materials in the appendix to this chapter.

10.35. Revisit the case(s) you developed in problems 10.16 and/or 10.17. Using the basic 3-sigma and 2-sigma OC curve information in Table 10.6 as a guide, develop specific location chart OC curves for your cases; i.e., calibrate the shift axis in the units of measure. Interpret your OC curves.

10.36. Revisit the case(s) you developed in problems 10.16 and/or 10.17. Using the basic 3-sigma and 2-sigma ARL curve information in Table 10.7 as a guide, develop specific location chart ARL curves for your cases; i.e., calibrate the shift axis in the units of measure. Interpret your ARL curves.

APPENDIX: OC CURVE CONSTRUCTION AND ARL CALCULATIONS

Operating characteristic (OC) curve construction requires a good deal of effort. The OC curves developed here are highly simplified and do not consider multiple runs rules—only a single point beyond a control limit is considered in detecting the shift. OC curves can provide insight as to how control charts work and don't work. On the following page, Figure 10.14 illustrates the general OC curve construction procedure we use. Notice how the shaded area, P_a, changes with the magnitude of the shift. Larger shifts produce smaller P_a's.

Several OC curves will be constructed to demonstrate the procedure. Computer aids—such as most computer-assisted statistical analysis packages and some spreadsheets—allow us to readily calculate "tail area" probabilities and can help us to develop OC curves. Otherwise, we must rely on tabulated values, which many times must be interpolated.

X-BAR CHART OC CURVES

OC curves can be used to help design X-bar and R charts. Our objective is to quantify our chances of a false alarm and failure to detect, for a true null hypotheses and a variety of true alternative hypotheses, respectively. We develop an OC curve for an X-bar chart in detail. R chart OC curves are more difficult to develop; see Duncan [3] for the development of OC curves for an R chart.

Since the X-bar chart is based on the normal distribution, its OC curve is straightforward. The X-bar chart OC curve will typically be two-tailed, since $LCL_{\bar{x}}$ is usually not backed up against a hard limit, such as the case often is with the 3-sigma R chart, where we have no LCL. We usually develop the right-hand tail of the X-bar chart OC curve and then use the property of symmetry to trace the left-hand tail. This procedure is possible since the normal probability mass function is symmetrical.

We demonstrate the development of a generic X-bar chart OC curve based on process shifts in multiples of σ—the standard deviation of a population of individual items. We assume that the process σ is known, and the process variance does not shift, even when we shift the process mean.

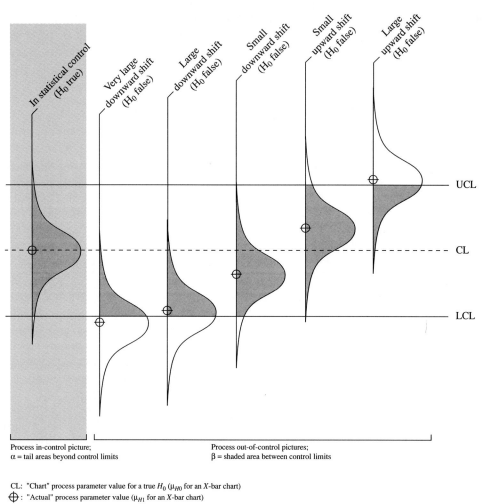

Process in-control picture;
α = tail areas beyond control limits

Process out-of-control pictures;
β = shaded area between control limits

CL: "Chart" process parameter value for a true H_0 (μ_{H0} for an X-bar chart)

$\oplus$: "Actual" process parameter value (μ_{H1} for an X-bar chart)

▨: $P_a = P$ (accepting the H_0 of "in statistical control")

Figure 10.14   Generic X-bar chart OC curve depiction (no runs rules).

We use the standard normal distribution table (see Section 8, Table VIII.1) or computer aids to develop specific probabilities associated with P_a, the probability of accepting H_0. Here,

H_0: The process mean is as the chart states, $\mu = \mu_0$

H_1: The process mean has shifted, $\mu \neq \mu_0$

where μ is the true process mean (unknown to us) and μ_0 is the chart centerline. In addition,

$P(\text{false alarm}) = P(\text{type I error}) = \alpha = P(\text{rejecting } H_0 \mid H_0 \text{ is true})$

$P(\text{failure to detect}) = P(\text{type II error}) = \beta = P(\text{not rejecting } H_0 \mid H_1 \text{ is true})$

Using the standard normal relationship below, and the general approach shown in Figure 10.14, we can reduce the P_a calculation to three factors:

- b: The sigma level, e.g., 3 sigmas
- n: The subgroup size, e.g., 1, 2, 3
- k: The shift in the process mean from the X-bar chart centerline in multiples of σ; e.g., if $k = 0.50$, the process mean shifts by $0.50\ \sigma$

$$P_a = \Phi\left[\frac{UCL - (\mu_{centerline} + k\sigma)}{\sigma/\sqrt{n}}\right] - \Phi\left[\frac{LCL - (\mu_{centerline} + k\sigma)}{\sigma/\sqrt{n}}\right] \qquad [10.26]$$

$$P_a = \Phi[b - k\sqrt{n}] - \Phi[-b - k\sqrt{n}] \qquad [10.27]$$

where Φ is the tail area under the standard normal curve, with the area accumulated from left to right.

For example, to obtain P_a for a shift of 1.5σ on a 3-sigma X-bar chart, with $n = 3$,

$$P_a = \Phi[3 - 1.5\sqrt{3}\,] - \Phi[-3 - 1.5\sqrt{3}\,] = P(Z \le 0.402) - P(Z \le -5.598) \cong 0.656$$

We have used Equation (10.27) and the SAS functions to obtain the OC curve data shown in Table 10.6.

The boxed portions, below the stepped line, in Table 10.6 emphasize the effectiveness of large subgroup sizes in detecting shifts. The top row, $k = 0.00$, provides $P_a = 1 - \alpha$ values. We can see that the β error drops off rapidly as k increases for the larger subgroup sizes, e.g., $n = 5$, $n = 10$, $n = 20$. An OC curve is plotted in Figure 10.15 (page 276) for a 3-sigma X-bar chart. Here again, we can see the dramatic effect of large subgroup sizes on enhancing our probability of shift detection. We should also notice that when we hold the sigma limits constant, the α holds constant. If we want to change α, we must change the control limit sigma level. Hence, in practice, if we want to tailor a chart to a given α and β we must work with the sigma multiple to establish our

Table 10.6 X-bar chart OC curve P_a for $k\sigma$ shifts in the process mean (computer-calculated).

Shift ($k\sigma$)	3-Sigma Chart Control Limits ($b = 3$)					2-Sigma Chart Control Limits ($b = 2$)				
k	$n = 1$	$n = 3$	$n = 5$	$n = 10$	$n = 20$	$n = 1$	$n = 3$	$n = 5$	$n = 10$	$n = 20$
0.00	0.9973	0.9973	0.9973	0.9973	0.9973	0.9545	0.9545	0.9545	0.9545	0.9545
0.25	0.9964	0.9946	0.9925	0.9863	0.9701	0.9477	0.9340	0.9200	0.8840	0.8102
0.50	0.9936	0.9835	0.9701	0.9220	0.7776	0.9270	0.8695	0.8102	0.6622	0.4067
1.00	0.9772	0.8976	0.7776	0.4357	0.0705	0.8400	0.6056	0.4067	0.1226	0.0067
1.50	0.9331	0.6562	0.3617	0.0407	0.0001	0.6912	0.2749	0.0879	0.0030	0.0000
2.00	0.8413	0.3213	0.0705	0.0004	0.0000	0.5000	0.0716	0.0067	0.0000	0.0000
2.50	0.6915	0.0918	0.0048	0.0000	0.0000	0.3085	0.0099	0.0002	0.0000	0.0000
3.00	0.5000	0.0140	0.0000	0.0000	0.0000	0.1587	0.0007	0.0000	0.0000	0.0000
3.50	0.3085	0.0011	0.0000	0.0000	0.0000	0.0668	0.0000	0.0000	0.0000	0.0000
4.00	0.1587	0.0000	0.0000	0.0000	0.0000	0.0228	0.0000	0.0000	0.0000	0.0000
4.50	0.0668	0.0000	0.0000	0.0000	0.0000	0.0062	0.0000	0.0000	0.0000	0.0000
5.00	0.0228	0.0000	0.0000	0.0000	0.0000	0.0013	0.0000	0.0000	0.0000	0.0000

SOURCE: SAS function PROBNORM(z); normal distribution tail-area estimates.

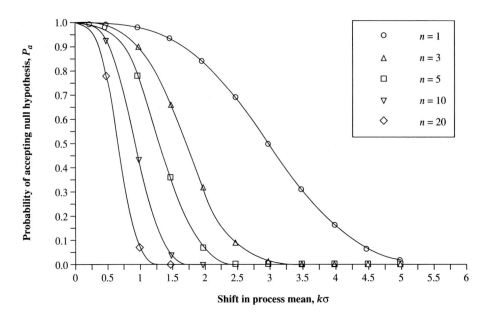

Figure 10.15   X-bar chart OC curves, 3-sigma limits.

desired α and then work with the subgroup size to establish the desired β at a given shift value, e.g., 1σ, 1.5σ.

Example 10.6 | **SHIFT DETECTION PROBABILITIES**   Given a 2-sigma X-bar chart and a process where $\bar{\bar{x}} = 150$, $\sigma = 5$, and $n = 3$:

1. What is the probability that a shift will not be detected after sampling one subgroup (after the shift), if the process mean has shifted to $\mu_1 = 155$?
2. What is the probability of a false alarm when $\mu_0 = 150$?
3. How many subgroups would be required in order to have at least a 95 percent chance of detecting a shift from $\bar{\bar{x}} = 150$ to $\bar{\bar{x}} = 155$? Assume only one point beyond a control limit is used to signal detection.

Solution 1

For a shift to $\mu_1 = 155$,

$$k = \frac{155 - 150}{5} = 1, b = 2, n = 3$$

$$P_a = \Phi\,[2 - 1\sqrt{3}\,] - \Phi\,[-2 - 1\sqrt{3}\,]$$

$$= P(Z \le 0.268) - P(Z \le -3.732) \cong 0.6056 \cong \beta$$

This answer is also available in Table 10.6.

Solution 2

When $\mu_0 = 150$,

$$k = \frac{150 - 150}{5} = 0, b = 2, n = 3$$

$$P_a = \Phi \, [2 - 0] - \Phi \, [-2 - 0]$$

$$= P(Z \leq 2) - P(Z \leq -2) = 0.9545 = 1 - \alpha$$

$$\alpha = 1 - 0.9545 = 0.0455$$

Solution 3

From I, $\beta \cong 0.6056$. Since P(detecting a shift) $= 1 - P$(not detecting a shift),

$$P\text{(not detecting shift after } m \text{ subgroups)} < 1 - 0.95 = 0.05$$

Therefore,

$$(\beta)^m < 0.05$$

$$(0.6056)^m = 0.05$$

$$m = \frac{\ln 0.05}{\ln 0.6056} = \frac{-2.996}{-0.5015} = 5.97$$

Hence, we would need at least six subgroups.

ARL CALCULATIONS

Our ARL calculations in this discussion are based on the geometric distribution. The geometric distribution is defined as a sequence or geometric progression of trials, with the success-failure probability held constant from trial to trial; see Walpole et al. [4]. For the random variable X, we define the geometric distribution probability mass function as

$$f(x; p) = p(1 - p)^{x-1} \qquad x = 1, 2, 3, \ldots$$

where $p = $ the probability of our control chart signaling an out-of-control process, and $(1 - p) = $ the probability of our control chart signaling an in-control process. The expected value, mean, of the geometric distribution is

$$E[X] = \mu = \frac{1}{p} \qquad\qquad \textbf{[10.28]}$$

If we substitute $1 - P_a$ from our OC curve calculations into Equation (10.28) for p, we can develop the ARL.

$$ARL = \frac{1}{1 - P_a} \qquad\qquad \textbf{[10.29]}$$

At this point, we can clearly see the ARL-OC curve relationship. We can develop ARL curves from OC curves to provide guidance in assessing the effectiveness of our SPC charts. Table 10.7 (on the following page) provides ARLs for the 2- and 3-sigma X-bar chart OC curves developed in the previous section; see Table 10.6. The 3- sigma ARLs are plotted in Figure 10.16. At this point, we should compare Figure 10.15 with Figure 10.16 (on the following page). We can clearly

observe that the ARLs and the OC curve P_a's both provide meaningful performance measures for assessing the sensitivity of SPC charts to process shift detection and to false alarms.

In cases where we cannot readily develop OC curves for our charts, we can use simulation techniques to gain insight into ARLs (as well as type I and type II error probabilities). These empirical means can be applied to both simple cases consisting of one point outside of control

Table 10.7   X-bar ARLs for $k\sigma$ shifts in the process mean (computer-calculated)

Shift ($k\sigma$)	3-Sigma Chart Control Limits ($b = 3$)					2-Sigma Chart Control Limits ($b = 2$)				
k	$n = 1$	$n = 3$	$n = 5$	$n = 10$	$n = 20$	$n = 1$	$n = 3$	$n = 5$	$n = 10$	$n = 20$
0.00	370.370	370.370	370.370	370.370	370.370	21.978	21.978	21.978	21.978	21.978
0.25	277.778	185.185	133.333	72.993	33.445	19.120	15.152	12.500	8.621	5.269
0.50	156.250	60.606	33.445	12.821	4.496	13.699	7.663	5.269	2.960	1.685
1.00	43.860	9.766	4.496	1.772	1.076	6.250	2.535	1.685	1.140	1.007
1.50	14.948	2.909	1.567	1.042	1.000	3.238	1.379	1.096	1.003	1.000
2.00	6.301	1.473	1.076	1.000	1.000	2.000	1.077	1.007	1.000	1.000
2.50	3.241	1.101	1.005	1.000	1.000	1.446	1.010	1.000	1.000	1.000
3.00	2.000	1.014	1.000	1.000	1.000	1.189	1.001	1.000	1.000	1.000
3.50	1.446	1.001	1.000	1.000	1.000	1.072	1.000	1.000	1.000	1.000
4.00	1.189	1.000	1.000	1.000	1.000	1.023	1.000	1.000	1.000	1.000
4.50	1.072	1.000	1.000	1.000	1.000	1.006	1.000	1.000	1.000	1.000
5.00	1.023	1.000	1.000	1.000	1.000	1.001	1.000	1.000	1.000	1.000

$\text{ARL} = 1/(1 - P_a)$

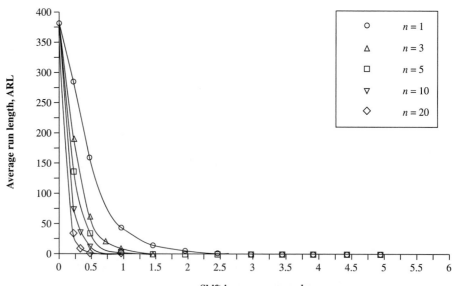

Figure 10.16   X-bar chart ARL curves, 3-sigma limits.

limits, as well as complicated cases including runs rules, nonnormal populations, and so on. Several cases are presented in Chapter 13.

The process we follow is to develop a simulated process data stream, and then apply our SPC discipline on the data stream by sampling and assessing in- and out-of-control status. Here, we count the number of in-control indications we observe before we encounter an out-of-control indication. We keep score of these encounters generating run lengths. Then, we average our run lengths for empirical ARLs.

REFERENCES

1. R. E. DeVor, T. Chang, and J. W. Sutherland, *Statistical Quality Design and Control: Contemporary Concepts and Methods*, New York: Macmillan, 1992.

2. D. C. Montgomery, *Introduction to Statistical Quality Control*, 3rd ed., New York: Wiley, 1996.

3. A. J. Duncan, *Quality Control and Industrial Statistics*, 5th ed., Homewood, IL: Irwin, 1986.

4. R. E. Walpole, R. H. Myers, and S. L. Myers, *Probability and Statistics for Engineers and Scientists*, 6th ed., Upper Saddle River, NJ: Prentice Hall, 1998.

5. E. L. Grant and R. S. Leavenworth, *Statistical Quality Control*, 7th ed., New York: McGraw-Hill, 1996.

6. AT&T Technologies, *Statistical Quality Control Handbook*, 11th printing, Indianapolis: AT&T Technologies, 1985.

7. J. Hoskins, B. Stuart, and J. Taylor, *Statistical Process Control,* Phoenix: Motorola, BR392/D.

8. J. S. Oakland, *Statistical Process Control*, New York: Wiley, 1986.

9. A. L. Sweet, "Control Charts Using Coupled Exponentially Weighted Moving Averages," *Transactions of IIE*, vol. 18, no. 1, pp. 26–33, 1986.

10. D. M. Hawkins, "Cumulative Sum Control Charting: An Underutilized SPC Tool," *Quality Engineering*, vol. 5, no. 3, pp. 463–477, 1993.

11. E. S. Page, "Continuous Inspection Schemes," *Biometrics*, vol. 41, 1954.

12. E. S. Page, "Cumulative Sum Control Charts," *Technometrics*, vol. 3, 1961.

13. G. A. Barnard, "Control Charts and Stochastic Processes," *Journal of the Royal Statistical Society*, vol. 21, no. 8, 1959.

14. D. M. Hawkins, "A Fast Accurate Approximation for Average Run Lengths of CuSum Control Charts," *Journal of Quality Technology*, vol. 24, no. 1, pp. 37–43, 1992.

chapter 1

Process Monitoring—Variables Control Charts for Individual Measurements and Related Topics

---

11.0 Inquiry

1. When are individual measurements appropriate?
2. How are individuals charts constructed? Interpreted?
3. Where should control points appear in process flows?
4. What is a capable process? How is process capability measured?
5. How are gauge accuracy and precision measured?

11.1 Introduction

When subgrouping is feasible, the Shewhart charts described in Chapter 10 provide advantages in statistical precision over charts that do not use subgrouping. However, **subgrouping is not always feasible because of inherent physical, economic, or timeliness constraints.** Forcing subgrouping in circumstances where it is not appropriate will create ineffectiveness and inefficiency in process control activities. **One recourse is to resort to individuals, rather than subgrouped, models and tools.**

This chapter focuses on the individuals branch of the classical variables SPC models; see Figure 9.3. Here, we introduce the individuals (X), moving average ($\bar{X}_M$), and moving range (R_M) charts, followed by a reintroduction of the exponentially weighted moving average (EWMA) and exponentially weighted moving deviation (EWMD) charts. Then, we reintroduce the cumulative sum, CuSum, control chart for individuals data.

In addition to SPC models, several related topics are included in this chapter. We describe process sampling and chart control point placement. Then, we introduce process capability and related topics. Finally, gauge studies are presented.

11.2 SPC MODEL RATIONALE FOR INDIVIDUALS DATA

Our physical processes are driven by the world around us. These processes neither know nor care about producing independent measurable characteristics for us to measure and model. **We fit our process control schemes to processes—processes cannot be expected to conform to our modeling assumptions.** Many processes are characterized by process inertia to one degree or another. Fundamentals of process inertia and data characterization to assess the nature and degree of process inertia were presented in Chapter 4.

Process inertia manifests itself in several ways. For example, temperatures and pressures move as a result of physical conditions resulting from heat and mass transfer. We see common process conditions in many processes—in multifaceted tooling (e.g., multimold cavities), in certain process characteristics (e.g., multiple features cut from the same reference in CNC work, continuous production such as rolling or extruding), in inspection procedures (e.g., multiple probe readings from the same basic product feature, frequent samples from a continuous flow process stream), and so on.

Process inertia in the physical world creates data streams that lack independence. When we subgroup data, we assume independence between measurements in order to estimate our within-subgroup variation, which in turn drives both our location and dispersion charts and process control decisions. **High levels of process inertia create problems with classical Shewhart SPC tools in several ways.** For example, rapid sampling (as the time between samples approaches zero) from a continuous-flow process yields within subgroup variation estimates that reflect only the imprecision in the measuring instrument (rather than common cause variation in the process). Other analogous examples result from multiple-probe readings from sensors concerning dimensions.

We have several means of dealing with process inertia, depending on the nature of the process and its degree of inertia. Provided our process is stationary, with limited inertia, we can resort to individuals (a subgroup size of $n = 1$) SPC models and tools directly. **Our strategy in this chapter, with traditional individuals SPC tools, is to modify our sampling scheme so as to capture and utilize common cause variation in our charts as best we can using individual data points.** However, we know that at best we lose some degree of statistical precision compared to the case where we have independence and pursue a subgrouping strategy. And, at worst, we have to abandon traditional SPC tools. Nevertheless, when we encounter process inertia, i.e., nonindependence between sample points, SPC is challenging. **Typically, we try to strike a balance between process needs and statistical performance.** The chiller case in Chapter 4 (Example 4.3) illustrated a case where traditional SPC just would not perform as a result of a high level of process inertia and the influence of fluctuating covariables, e.g., the ambient temperature. Later, in Chapter 13, we explore several nontraditional options to address this issue.

As in the case of the Shewhart subgrouped charts, **we can associate our individuals charts with hypothesis testing across time or order of production.** We assume the process is in control or stable until we see credible evidence to the contrary. We consider both location and dispersion. Our hypotheses are identical to those stated in Chapter 10; however, our models are based on subgroup size $n = 1$.

11.3 NOTATION FOR INDIVIDUALS SPC MODELS

GENERAL CONTROL CHART SYMBOLS

m: The number of samples (of size $n = 1$) in the preliminary data set used to set up a control chart

b: The control limit spread in number of standard deviations (chart-sigmas) of the plotted quantity

$\mu, \hat{\mu}$: The population mean, estimated population mean for the population of individual measurements or items

$\sigma, \hat{\sigma}$: The population standard deviation, estimated population standard deviation for a population of individual measurements or items

$\sigma_{\bar{x}}, \hat{\sigma}_{\bar{x}}$: The population standard deviation, estimated population standard deviation for a population of means

X, $\bar{X}_M$, AND R_M CHART SYMBOLS

n_a: The artificial subgroup size used for the $\bar{X}_M$ or R_M chart

σ_0: A target or assumed population standard deviation (given rather than calculated)

x_j: The jth individual measurement or sample value

$\bar{x}_{M_j}$: A moving subgroup sample average value for the jth artificial subgroup or sample

$\bar{x}$: The mean for the m samples

$\bar{\bar{x}}_M$: The mean of a group of moving artificial subgroup sample means

$\mu_0, \bar{x}_0$: A target or assumed mean value (given rather than calculated)

$\text{LCL}_X, \text{UCL}_X$: The lower, upper control limits for the X chart

$\text{LCL}_{\bar{X}_M}, \text{UCL}_{\bar{X}_M}$ The lower, upper control limits for the $\bar{X}_M$ chart

R_{M_j}: A moving subgroup sample range value for the jth artificial subgroup or sample

$\bar{R}_M$: The mean of a group of moving artificial subgroup sample range values

$\text{LCL}_{R_M}, \text{UCL}_{R_M}$: The lower, upper control limits for the moving range chart

EXPONENTIAL WEIGHTED MOVING AVERAGE AND DEVIATION (EWMA AND EWMD) CHART SYMBOLS

r: A weighting factor that can assume values between 0 and 1

W_{j-k}: Weight associated with the observation x_{j-k}, where x_j is the most recent observation

A_j: The exponential weighted moving average (EWMA) for observation j

V_j: The exponential weighted moving deviation (EWMD) for observation j

D_j: The absolute deviation at observation j, equal to $|x_j - A_{j-1}|$

LCL_A, UCL_A: The lower, upper control limits for the EWMA chart

LCL_V, UCL_V: The lower, upper control limits for the EWMD chart

CUMULATIVE SUM (CuSum) CHART SYMBOLS

L_j: A CuSum value for the jth sample

L_j^+ and L_j^-: Tabular CuSum chart plotting points

k and h: Constants that determine the statistical performance of the tabular CuSum

11.4 X, $\bar{X}_M$, AND R_M CONTROL CHART CONCEPTS AND MECHANICS

The X chart (sometimes called a "runs chart" when no control limits are applied) and R_M charts are relatively simple to construct. **The construction of X and R_M charts is similar to that of X-bar and R charts, except that now samples consist of individual measurements and R_M is the range of a group of consecutive individual measurements combined "artificially" to form an artificial subgroup of size n_a.**

The moving range R_M is calculated primarily for the purpose of estimating common cause variability in the process. The artificial samples formed from successive measurements should be small in size and taken reasonably close together in time or order of production in order to minimize the chance of including, in the artificial subgroup, any data arising from unstable conditions. Special cause occurrences within the subgroup typically will inflate the common cause variability estimate and erode the chart's sensitivity. However, in the presence of limited process inertia, as measured by autocorrelation, see Chapter 4, we sometimes spread our samples to some degree in order to help estimate the common cause variation. This practice of spreading is pragmatic and not supported by theory, but when we encounter processes that do not entirely measure up to our independence assumption, we typically rely on pragmatism in order to attempt a workable, but not ideal, solution.

Applications of X and R_M control charts receive mixed reviews. Grant and Leavenworth [1] tend to discourage their use. DeVor et al. [2] state that the X and R_M charts are perhaps the most misused (often abused) of all of the charts in common use today. The DeVor et al. reasoning focuses on two points: (1) the charts often have inflated within-subgroup variability because too much time has transpired from one measurement to another, and (2) these charts are typically constructed with a small number of data points and, hence, are subject to large sampling errors. These arguments are generally valid when individual measurements are taken and used in cases where natural subgrouping can and should be used instead. But, in other applications, the X and R_M charts are feasible alternatives. Nevertheless, discretion is necessary on the engineer's part when process inertia is present.

If we assume that random variable X, the product/process characteristic measurement of interest, is normally and independently distributed, we can set up probabilistic control limits for the X chart that can help us in monitoring the process location parameter. The general form to construct X chart control limits, assuming that X is normally distributed and that the process mean μ and process standard deviation σ are known, is

$$\left. \begin{array}{c} \text{UCL}_X \\ \text{LCL}_X \end{array} \right\} = \mu \pm b\sigma \qquad \textbf{[11.1]}$$
$$\text{Center line} = \mu$$

For the X chart we use a subgroup size of $n = 1$. For its companion R_M control chart we select a small artificial subgroup size ($n_a = 2$ or $n_a = 3$) to minimize opportunity for special causes to arise within the artificial samples, or moving subgroups. Larger artificial sample sizes will tend to inflate our estimate of variation, since they are collected over a longer time period. We should note here that we are forced to use $n_a \geq 2$ for the R_M chart, since single points will not, by themselves, allow us to deal with dispersion.

Assuming that the individual observations from our process are at least approximately independent and normally distributed, it is possible to construct the control limits by estimating the process parameters μ and σ from $\bar{x}$ and $\bar{R}_M$, respectively. To do so, it is essential that the number of individual measurements m made for constructing the control limits be at least 25.

We calculate the sample mean as

$$\bar{x} = \hat{\mu} = \frac{\sum\limits_{j=1}^{m} x_j}{m}$$

This sample mean serves as the centerline for the X chart. We next compute sample moving ranges for the artificial samples of size n_a starting with R_{M1}, which is the difference between the largest and the smallest values in the first artificial sample $(x_1, \ldots, x_{n_a})$. Each time we obtain a new sample, x_j value, from the process, we repeat this computation for each succeeding moving sample of artificial size n_a by adding the latest x_j value and dropping the oldest x_j value. We calculate an average moving range $\bar{R}_M$ from the $(m - n_a + 1)$ sample moving ranges,

$$\bar{R}_M = \frac{\sum\limits_{j=n_a}^{m} R_{Mj}}{m - n_a + 1}$$

This average serves as the centerline for the R_M chart.

The 3-sigma control limits for the X chart are

$$\left.\begin{array}{l} \text{UCL}_X \\[2em] \text{LCL}_X \end{array}\right\} = \bar{x} \pm 3\,\frac{\bar{R}_M}{d_2} \qquad\qquad \textbf{[11.2]}$$

$$\text{Centerline} = \bar{x}$$

and **the 3-sigma control limits for the R_M chart are**

$$\text{UCL}_{RM} = D_4 \bar{R}_M \qquad\qquad \textbf{[11.3]}$$

$$\text{LCL}_{RM} = D_3 \bar{R}_M \qquad \text{or 0, whichever is larger} \qquad \textbf{[11.4]}$$

$$\text{Centerline} = \bar{R}_M$$

All constants, d_2, D_3, and D_4, are as listed in Table VIII.5.

Once the X chart and R_M chart are set up using the computed control limits and centerlines, we plot the x_j values on the X chart and the R_{Mj} values on the R_M chart. For each chart, we connect the points together with a solid line. Provided the trial charts indicate stability, we use the chart parameters to monitor the process.

The R_M chart provides a plotting point for each individual sample taken, once the first n_a measurements are obtained. For a moving subgroup size of n_a, the effect of any one given sample will be contained in n_a consecutive plotted points. Hence, we see a relatively slow response, depending on sample frequency, in picking up a shift. In examining X and R_M control charts, given that the artificial subgroups overlap, the interpretations of the X and R_M charts are not independent. Sometimes only the X chart is analyzed for out-of-control signals, while the R_M chart is evaluated primarily to assure proper construction of the X chart.

The moving average control chart ($\bar{X}_M$ chart) is an alternative to the X chart and pairs up with the R_M chart. The $\bar{X}_M$ chart tends to smooth the location chart by including more than one x_j value in each plotted $\bar{x}_{Mj}$ point. We develop moving averages for our x_j values using artificial subgroups of size n_a. Then, **we develop our $\bar{X}_M$ chart parameters in a similar manner to** those for the X-bar chart. For the 3-sigma $\bar{X}_M$ chart,

$$\left.\begin{array}{l} \text{UCL}_{\bar{X}_M} \\[2em] \text{UCL}_{\bar{X}_M} \end{array}\right\} = \bar{\bar{x}}_M \pm A_2 \bar{R}_M \qquad\qquad \textbf{[11.5]}$$

$$\text{Centerline} = \bar{\bar{x}}_M$$

where

$$\bar{\bar{x}}_M = \frac{\sum\limits_{j=n_a}^{m} \bar{x}_{Mj}}{m - n_a + 1}$$

Example 11.1 | **STEEL PRODUCTION CASE**   Flatt Sheet Metal Company is a major supplier of mild steel sheet metal to a leading automobile manufacturer. Facing tighter automobile quality requirements, the design engineers from the automobile plant have tightened their supplier quality requirements. The sheet-metal company is experiencing a situation in which the rolling mills and furnaces require frequent attention to maintain the desired tolerances.

To identify potential areas of quality and productivity improvement, it was decided to monitor the production process using X and R_M control charts. Here, rational subgrouping with $n > 1$ is not meaningful because of strong process inertia—intrasheet variation is negligible in this process as a result of the material passing through the rolls. A schematic illustration of the flat-rolling process is shown in Figure 11.1. Basically, the process includes feeding cast metal slabs at elevated temperatures iteratively between pressurized cylindrical rollers, which flatten the slabs to sheets. At the end of the process, the sheets are subject to cold rolling to achieve the desired thickness dimension and surface finish. The produced sheets are sampled for their thickness, uniformity, and roughness by dial comparators and profilometers.

The quality characteristic selected for study in this case is the thickness of the sheet. The product tolerance set by the purchaser is 2.2500 ± 0.0075 mm. The sheet thicknesses (in millimeters) of 32 sheets produced during a 4-hour period are displayed in Table 11.1.

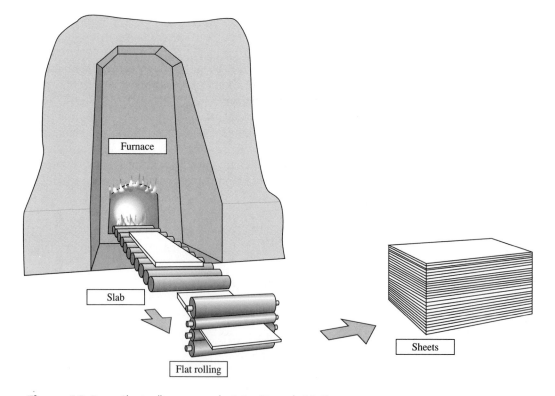

Figure 11.1   Sheet rolling process depiction (Example 11.1).

Table 11.1 Thickness measurements of 32 sheets (Example 11.1).

Sample Number	x_j	R_{Mj}	R_M Calculations ($n_a = 2$)
1	2.2331		
2	2.2403	0.0072	$R_{M2} = \lvert 2.2403 - 2.2331 \rvert = 0.0072$
3	2.2473	0.0070	$R_{M3} = \lvert 2.2473 - 2.2403 \rvert = 0.0070$
4	2.2525	0.0052	$R_{M4} = \lvert 2.2525 - 2.2473 \rvert = 0.0052$
5	2.2497	0.0028	
6	2.2511	0.0014	
7	2.2521	0.0010	.
8	2.2507	0.0014	.
9	2.2342	0.0165	.
10	2.2395	0.0053	
11	2.2510	0.0115	
12	2.2459	0.0051	
13	2.2518	0.0059	
14	2.2547	0.0029	
15	2.2550	0.0003	
16	2.2502	0.0048	
17	2.2341	0.0161	
18	2.2423	0.0082	
19	2.2500	0.0077	
20	2.2487	0.0013	
21	2.2488	0.0001	
22	2.2522	0.0034	
23	2.2548	0.0026	
24	2.2477	0.0071	
25	2.2316	0.0161	
26	2.2453	0.0137	
27	2.2511	0.0058	
28	2.2445	0.0066	
29	2.2531	0.0086	
30	2.2511	0.0020	
31	2.2523	0.0012	
32	2.2509	0.0014	$R_{M32} = \lvert 2.2509 - 2.2523 \rvert = 0.0014$
Totals	71.9176	0.1802	

To determine the R_M values, an artificial sample size of $n_a = 2$ was selected. The R_{Mj} values are therefore given by

$$R_{Mj} = \lvert x_j - x_{j-1} \rvert \qquad \text{for } j = 2, \ldots, m$$

Table 11.1 displays the 31 calculated moving range, R_{Mj}, values that were extracted from the data by using the artificial subgroup, $n_a = 2$. From Table 11.1 it is evident that

$$\bar{x} = \frac{71.9176}{32} = 2.2474 \text{ mm}$$

$$\bar{R}_M = \frac{0.1802}{31} = 0.0058 \text{ mm}$$

From Equation (11.2), the 3-sigma control limits for the X chart are

$$UCL_X = \bar{x} + 3\,\frac{\bar{R}_M}{d_2} = 2.2474 + 3\,\frac{0.0058}{1.128} = 2.2628 \text{ mm}$$

$$LCL_X = \bar{x} - 3\,\frac{\bar{R}_M}{d_2} = 2.2474 - 3\,\frac{0.0058}{1.128} = 2.2320 \text{ mm}$$

$$\text{Centerline} = \bar{x} = 2.2474 \text{ mm}$$

From Equations (11.3) and (11.4), the 3-sigma control limits for the R_M chart are

$$UCL_{R_M} = D_4\bar{R}_M = 3.27\,(0.0058) = 0.0190 \text{ mm}$$

$$LCL_{R_M} = D_3\bar{R}_M = 0\,(0.0058) = 0 \text{ mm}$$

$$\text{Centerline} = \bar{R}_M = 0.0058 \text{ mm}$$

These trial control limits for the X and R_M charts, along with tolerance limits for sheet thickness, are shown in Figure 11.2.

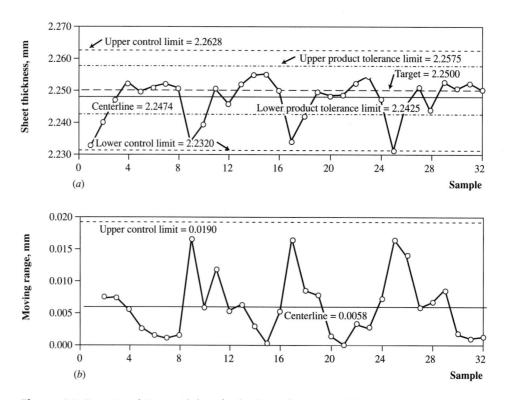

Figure 11.2 X and R_M control charts for the sheet rolling process (Example 11.1). (a) X control chart with 3-sigma control limits and tolerance limits. (b) R_M control chart with 3-sigma control limits.

In examining the X and R_M control charts in Figure 11.2, we see that only one point, sample 25, is beyond the limits. Also, close observation of the charts shows nonrandom sequences. As was mentioned earlier, the 32 examples shown in the charts represent 4 hours of production. It can be seen from the X control chart that the first two samples from every hour of production (eight samples represent 1 hour) are low in thickness in comparison to the rest of the samples. It can also be seen from Figure 11.2a that these first samples are also outside the tolerance limits and will end up as scrap or unacceptable sheet. Here, we should note that product tolerance limits as well as the target can be placed on the X chart, since we are working with subgroups of $n = 1$ sheet.

The production team investigated possible causes for these nonrandom patterns and found a special cause. They determined that, because the scrap collected during every production hour goes into the furnace at the beginning of the next hour of production (within 4 or 5 min), the steel obtained during scrap reprocessing is more ductile and weaker in grain strength.

In order to eliminate the effect of this special cause, scrap material will be fed into the furnace at a uniform rate throughout the production period. After implementation of this process change, 32 more samples were collected from the production process (tabulated in Table 11.2). The new calculated control limits for the X and R_M charts are shown in Figure 11.3, along with the tolerance limits. The new control chart indicates a stable process.

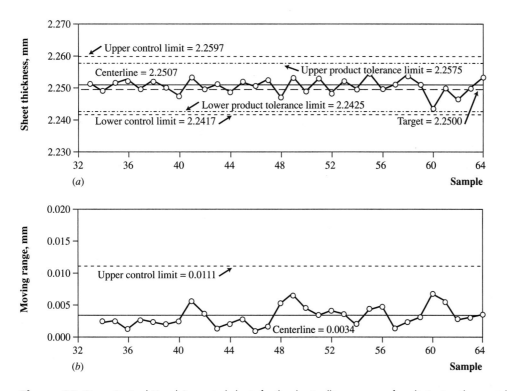

Figure 11.3 Revised X and R_M control charts for the sheet rolling process after eliminating the special cause (Example 11.1). (a) X control chart with 3-sigma control limits and tolerance limits. (b) R_M control chart with 3-sigma limits.

Table 11.2   Thickness measurements of 32 sheets after eliminating the special cause (Example 11.1).

Sample Number	x_j	R_{Mj}	R_M Calculations ($n_a = 2$)
33	2.2511		
34	2.2490	0.0021	$R_{M34} = \|2.2490 - 2.2511\| = 0.0021$
35	2.2513	0.0023	$R_{M35} = \|2.2513 - 2.2490\| = 0.0023$
36	2.2525	0.0012	$R_{M36} = \|2.2525 - 2.2513\| = 0.0012$
37	2.2497	0.0028	
38	2.2521	0.0024	
39	2.2501	0.0020	
40	2.2477	0.0024	.
41	2.2532	0.0055	
42	2.2495	0.0037	.
43	2.2510	0.0015	
44	2.2489	0.0021	
45	2.2518	0.0029	
46	2.2507	0.0011	
47	2.2525	0.0018	
48	2.2472	0.0053	
49	2.2541	0.0069	
50	2.2493	0.0048	
51	2.2531	0.0038	
52	2.2487	0.0044	
53	2.2524	0.0037	
54	2.2502	0.0022	
55	2.2548	0.0046	
56	2.2497	0.0051	
57	2.2516	0.0019	
58	2.2543	0.0027	
59	2.2511	0.0032	
60	2.2445	0.0066	
61	2.2501	0.0056	
62	2.2471	0.0030	
63	2.2503	0.0032	
64	2.2539	0.0036	$R_{M64} = \|2.2539 - 2.2503\| = 0.0036$
Totals	72.0235	0.1044	

A visual examination of the X chart in Figure 11.3 points up the fact that the product tolerance limits are inside the 3-sigma control limits. Intuitively, this occurrence is not appealing. However, at this point, we do not have a quantitative means to compare these two metrics (the target and tolerance metrics come from the customer world, while the center line and control limits are from the production world). Later in this chapter we describe the concept of a capability index so that we can bring these two worlds together.

11.5 EWMA AND EWMD CONTROL CHARTS

The X and R_M **charts are not extremely effective in detecting shifts in the process mean or process variance, particularly when the shift is small.** The addition of "runs" rules, as discussed in Chapter 10, helps solve this problem to some extent. But other charts can be used to address this issue as well. In this section, we reintroduce the use of exponentially weighted moving averages (EWMAs) to monitor location and exponentially weighted moving deviations (EWMDs) for tracking process dispersion. **The EWMA and EWMD charts work well with individuals data and are considered effective for detecting small sustained shifts.** Here again, we are assuming independent, normally distributed data.

As previously stated in Chapter 10, an EWMA is a moving average of past data where each data point is assigned a weight. These weights decrease in an exponentially decaying fashion from the present into the remote past; review Figure 10.10. Thus, the moving average tends to be a reflection of the more recent process performance, as greater weight is allocated to the most recent data. The amount of decrease in the weights over time is an exponential function of the weighting factor r, which can assume values between 0 and 1. **By choosing the weighting factor r, we can essentially tailor our EWMA and EWMD charts for timeliness or time sensitivity.** This feature, plus their ability to signal small sustained shifts, make the EWMA and EWMD charts very attractive SPC model alternatives.

The weighting factor associated with observation x_{j-k} is

$$W_{j-k} = r(1-r)^k$$

When a small value of r is used, the moving average at sample point j carries with it a great amount of inertia from the past. Hence, it is relatively insensitive to short-lived changes in the process. **For control chart applications where fast response to process shifts is desired, a relatively large weighting factor,** say $r = 0.2$ or $r = 0.5$, **is used.**

In selecting r, the following relationship between the weighting factor r and the sample size n for Shewhart control charts is often used (Sweet [3]):

$$r = \frac{2}{n+1}$$

We can observe that if $r = 1$, for the EWMA, all the weight is given to the current single observation, which is equivalent to the X chart.

If a shift in the mean occurs, the EWMAs will gradually, depending on r, move to the new mean of the process, while the exponentially weighted moving deviations (EWMDs) will remain unchanged. On the other hand, **if there is a shift in the process variability, the EWMDs will gradually move to the new level,** while the EWMAs still will vary about the same process mean. Just as in the subgrouped case, in order to construct the EWMA and EWMD control charts for individual data values, we have two options: (1) use target values for μ_0 and σ_0 or (2) use past data to estimate μ and σ. If we choose the first option, then we should take care in selecting appropriate targets. The location target is usually obvious, but the standard deviation target is not. The standard deviation target should reflect the common cause process dispersion; over- or understatement will result in poor statistical performance. If we choose the second option, we should take care to develop our estimates with reasonably large samples, under stable process conditions, so as to obtain precise estimates of location and common cause variation. Otherwise, chart performance will be compromised. If we choose to estimate μ and σ from our data,

$$\bar{x} = \hat{\mu} = \frac{\sum\limits_{j=1}^{m} x_j}{m}$$

and

$$s = \hat{\sigma} = \sqrt{\frac{\sum\limits_{j=1}^{m}(x_j - \bar{x})^2}{m - 1}}$$

We compute EWMAs (A_j) and EWMDs (V_j) as follows:

$$A_j = rx_j + (1 - r)A_{j-1} \qquad \text{where } A_0 = \hat{\mu} \text{ or } \mu_0 \tag{11.6}$$

$$V_j = rD_j + (1 - r)V_{j-1} \qquad \text{where } D_j = |x_j - A_{j-1}| \text{ and } V_0 = \hat{\sigma} \text{ or } \sigma_0 \tag{11.7}$$

The control limits and the A_j and V_j centerlines are calculated as

$$\left.\begin{array}{c} \text{UCL}_A \\ \\ \text{LCL}_A \end{array}\right\} = \hat{\mu} \text{ or } \mu_0 \pm A^*\hat{\sigma} \text{ or } A^*\sigma_0 \tag{11.8}$$

$$\text{Centerline} = \hat{\mu} \text{ or } \mu_0$$

$$\text{UCL}_V = D_2^*\hat{\sigma} \text{ or } D_2^*\sigma_0 \tag{11.9}$$

$$\text{LCL}_V = D_1^*\hat{\sigma} \text{ or } D_1^*\sigma_0 \tag{11.10}$$

$$\text{Centerline} = d_2^*\hat{\sigma} \text{ or } d_2^*\sigma_0$$

where all A^*, d_2^*, D_1^*, and D_2^* constants are as listed in Table VIII.7, Section 8.

We plot all of the m, A_j values on the A chart and the m, V_j values on the V chart and interpret the charts to determine if the process is stable in terms of both process location and dispersion. Once stability is indicated, we extend the charts to monitor future process operations.

Example 11.2 | **INDIVIDUALS EWMA AND EWMD CHARTS**   In this example, we revisit the PCB manufacturing process control situation introduced and described in Example 10.5. We approach the X-direction circuit contact location deviation control issue using individuals, rather than subgrouped data. Our objective is to develop EWMA and EWMD charts.

Solution

Here, our process data are displayed on the left side in Table 11.3. We have selected $r = 0.5$ so that we emphasize the most recent data points. Our A_j, D_j, and V_j calculations have been performed by using Equations (11.6) and (11.7) and target values of $\mu_0 = 0$ mm and $\sigma_0 = 0.05$ mm. Our EWMA and EWMD control limits and centers were calculated by using Equations (11.8), (11.9), and (11.10).

Table 11.3 Printed circuit board location data (Example 11.2)

Sample Number	Board Location Deviation	EWMA and EWMD Statistics		
	x_j	A_j	D_j	V_j
1	0.01760	0.0088	0.0176	0.0338
2	−0.01222	−0.0017	0.0210	0.0274
3	0.02591	0.0121	0.0276	0.0275
4	−0.04518	−0.0165	0.0573	0.0424
5	0.01495	−0.0008	0.0315	0.0369
6	0.02429	0.0117	0.0251	0.0310
7	−0.04695	−0.0176	0.0587	0.0449
8	−0.00873	−0.0132	0.0089	0.0269
9	0.02318	0.0050	0.0363	0.0316
10	0.01314	0.0091	0.0081	0.0199
11	−0.03755	−0.0142	0.0466	0.0332
12	0.07726	0.0315	0.0915	0.0624
13	−0.03241	−0.0004	0.0639	0.0631
14	−0.01475	−0.0076	0.0143	0.0387
15	0.02370	0.0081	0.0313	0.0350
16	0.03615	0.0221	0.0281	0.0316
17	0.00409	0.0131	0.0180	0.0248
18	−0.05661	−0.0218	0.0697	0.0472
19	0.00282	−0.0095	0.0246	0.0359
20	0.02587	0.0082	0.0353	0.0356
21	20.03634	20.0141	0.0445	0.0401
22	20.01946	20.0168	0.0054	0.0227
23	20.02748	20.0221	0.0107	0.0167
24	0.02120	20.0005	0.0433	0.0300
25	0.00218	0.0009	0.0026	0.0163

$$\text{UCL}_A = \mu_0 + A^*\sigma_0 = 0 + 1.732(0.05) = 0.0866$$

$$\text{LCL}_A = \mu_0 - A^*\sigma_0 = 0 - 1.732(0.05) = -0.0866$$

$$\text{Centerline} = \mu_0 = 0$$

$$\text{UCL}_V = D_2^*\sigma_0 = 2.164(0.05) = 0.1082$$

$$\text{LCL}_V = D_1^*\sigma_0 = 0$$

$$\text{Centerline} = d_2^*\sigma_0 = 0.921(0.05) = 0.0461$$

The resulting charts are displayed in Figure 11.4. Here, we observe process stability. By comparing these charts to their counterparts in Figure 10.11, we observe the wider limits in the individuals data example.

From the data, we can calculate that $\bar{x} = -0.0010$ mm and $\hat{\sigma} = 0.0318$ mm. Obviously, we would obtain different charts using these empirical values. In Figure 11.4, the data are located in the center of the charts, with considerable space within the control limits. This spacing indicates that our target mean is close to our empirical mean and that our target standard deviation is greater than our empirical (sample) standard deviation. In this case, our empirical-based charts would provide a different picture of our process stability. Hence, target-based charts must be interpreted in the overall context of actual operations.

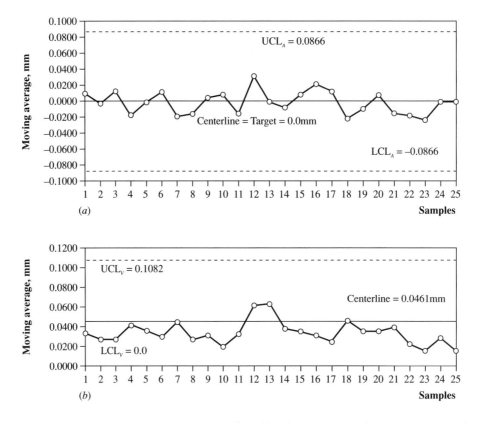

Figure 11.4 EWMA and EWMD charts for solder placement (Example 11.2). (*a*) EWMA chart. (*b*) EWMD chart.

11.6 CuSum Control Charts

The EWMA and EWMD control charts are constructed in such a way that the last few observations of the process are weighted from the present, backward in time, in an exponentially decreasing manner. **The cumulative sum control chart,** frequently called the CuSum chart, introduced in Chapter 10, **was designed to identify slight but sustained shifts in a process.** The statistic to be accumulated and charted is the deviation of an observation from a desired target.

One key to understanding the differences between the X, CuSum, and exponentially weighted control charts rests in knowing how each technique weights the data obtained from the process. Hunter [4] discusses the weighting functions for the X chart, the CuSum chart, and the EWMA chart. A graphical comparison is shown in Figure 11.5. It is clear that **the X chart has no "memory"**—that is, it ignores the immediate history. **The CuSum chart,** however, **gives equal attention to the first data point and the most recent. The EWMA chart gives less and less weight to data as they get older and older.** Along these same lines (in the subgrouped models), the X-bar, R, and S charts focus only on data from the most recent subgroup.

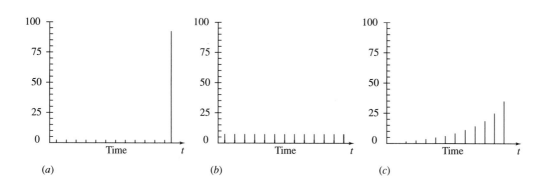

Figure 11.5 Data weighting for the X, CuSum, and EWMA charts. (a) Weighting for X chart. (b) Weighting for CuSum chart. (c) Weighting for EWMA chart. (Adapted with permission from the American Society for Quality Control, from J. S. Hunter, "The Exponentially Weighted Moving Average," *Journal of Quality Technology,* vol. 18, no. 4, p. 205, 1986.)

The CuSum chart for location was introduced in Chapter 10 when we discussed SPC for subgrouped data. We described both the classical V-mask and the tabular CuSum approaches. Here, we concentrate only on the tabular approach. We discuss both two-sided and one-sided tabular CuSum charts for location. To address variation, we can utilize a classical moving range chart, as previously described in this chapter.

TWO-SIDED CUSUM CHARTS

We typically monitor our process metrics on both sides of the target, i.e., in a nominal-is-best format. Here, we are interested in detecting a shift in either direction. **For the two-sided CuSum chart for location, we can use procedures identical to those described in Chapter 10 with the modification of replacing $\bar{x}_j$ with x_j and $\sigma_{\bar{x}_0}$ with σ_0**—we have an equivalent subgroup size of one, $n = 1$.

Hence, for our tabular CuSum chart,

$$L_j^+ = \max[0, x_j - (\mu_0 + k\sigma_0) + L_{j-1}^+] \qquad \textbf{[11.11]}$$

$$L_j^- = \min[0, x_j - (\mu_0 - k\sigma_0) + L_{j-1}^-] \qquad \textbf{[11.12]}$$

where $L_{j=0}^+ = L_{j=0}^- = 0$. Similarly, as in the previous presentation, when

$$L_j^+ > h\sigma_0$$

or

$$L_j^- < -h\sigma_0$$

we declare our process metric to be unstable or out of control, assuming a special cause has created a shift in the process mean. Given that we have encountered a special cause signal and taken appropriate counteraction, we restart the CuSum chart when we restart our process.

As previously, our choice of k and h determines the statistical properties of our CuSum chart. Typically, $k = 0.5$ and $h = 4$ or 5 are recommended. Here,

$$k = \frac{1}{2} \frac{|\mu_0 - \mu_1|}{\sigma}$$ [11.13]

where μ_1 represents a shifted mean of critical interest.

Once again, we can use Table 10.6 to select k and h combinations that yield type I error (false alarm) probabilities comparable to a ± 3-sigma Shewhart chart (without runs rules applied). For example, a ± 3-sigma type I error probability is approximately $2[P(Z < -3)] \cong 0.0027$, which roughly translates into an ARL_0 of $1/0.0027 = 370.4$ when $H_0: \mu = \mu_0$ is true. The determination of ARLs (or type I and type II statistical errors) for CuSum charts is complicated. Vance [5], Brook and Evans [6], Siegmund [7], Hawkins [8], and others provide various methods to address CuSum ARLs. We revisit the CuSum ARL issue using simulation technology in Chapter 13.

As before, we can calculate our shift Δ as

$$\Delta = \frac{L_{j=m}^{+}}{m - l} + k\sigma_0 \qquad \text{for } L_{j=m}^{+} > h\sigma_0$$ [11.14a]

$$\Delta = \frac{L_{j=m}^{-}}{m - l} - k\sigma_0 \qquad \text{for } L_{j=m}^{-} < h\sigma_0$$ [11.14b]

when we signal instability at the mth sample observation, and l represents the sample index number just prior to the trend or run that resulted in the instability signal.

Example 11.3

INDIVIDUALS CUSUM CHART   In order to visually compare the individuals data tabular CuSum with the EWMA, use the Example 11.2 data from the PCB manufacturing situation and develop a tabular CuSum chart.

Solution

Here we use identical target values, $\mu_0 = 0$ mm and $\sigma_0 = 0.05$ mm. We select $k = 0.5$ and $h = 4.77$ from Table 10.4 in order to produce roughly a 3-sigma-like chart in terms of ARL. Our data are displayed in Table 11.4, on the left-hand side. We have used Equations (11.11) and (11.12) in order to calculate our L_j^{+} and L_j^{-} statistics. In this solution a spreadsheet was used to facilitate our calculations. The results are shown on the right-hand side of Table 11.4.

Our resulting tabular CuSum chart is depicted in Figure 11.6. Here, we observe a stable process. In this case, the same arguments, as expressed in Example 11.2, as to target versus empirical bases for our CuSum chart hold. For empirical counterparts, we would substitute $\hat{\mu}$ for μ_0 and $\hat{\sigma}$ for σ_0 in Equations (11.11) through (11.14).

It is of interest to compare our individuals charts with our Chapter 10 subgrouped charts, i.e., Figures 10.11 and 10.13 with Figures 11.4 and 11.6. Here, we see the impact of subgrouping as "moving our control limits in" so that we can more easily distinguish small shifts.

It is also of interest to compare our individuals CuSum chart in Figure 11.6 with the individuals EWMA chart in Figure 11.4. We see very similar patterns across the samples. Here, we see two feasible alternatives side by side. Both are effective in monitoring and responding to small sustained shifts. We take up quantitative control chart comparisons using simulation methods later in Chapter 13.

Table 11.4 Printed circuit board location data (Example 11.3).

Sample Number	Board Location Deviation x_j	CuSum L_j	Tabular CuSum Statistics L_j^+	L_j^-
1	0.01760	0.018	0.000	0.000
2	−0.01222	0.005	0.000	0.000
3	0.02591	0.031	0.001	0.000
4	−0.04518	−0.014	0.000	−0.020
5	0.01495	0.001	0.000	0.000
6	0.02429	0.025	0.000	0.000
7	−0.04695	−0.007	0.000	−0.007
8	−0.00873	−0.016	0.000	0.000
9	0.02318	0.007	0.000	0.000
10	0.01314	0.021	0.000	0.000
11	−0.03755	−0.017	0.000	−0.013
12	0.07726	0.060	0.052	0.000
13	−0.03241	0.013	0.000	−0.022
14	−0.01475	−0.001	0.000	−0.012
15	0.02370	0.022	0.000	0.000
16	0.03615	0.058	0.011	0.000
17	0.00409	0.062	0.000	0.000
18	−0.05661	0.006	0.000	−0.032
19	0.00282	0.009	0.000	−0.004
20	0.02587	0.035	0.001	0.000
21	−0.03634	−0.002	0.000	−0.011
22	−0.01946	−0.021	0.000	−0.006
23	−0.02748	−0.049	0.000	−0.008
24	0.02120	−0.028	0.000	0.000
25	0.00218	−0.025	0.000	0.000

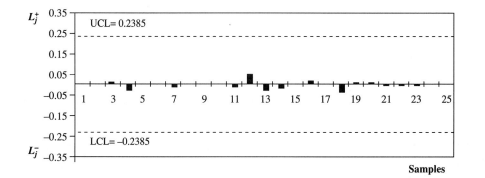

Figure 11.6 Tabular CuSum chart for solder placement (Example 11.3).

ONE-SIDED CuSUM CHARTS

The tabular approach to CuSum location charts can be applied on a one-sided basis when necessary. For example, a minimum purity or net weight requirement would dictate interest primarily on the lower side of the metric. In the one-sided cases, we can apply the appropriate side of the two-sided tabular CuSum chart from Equations (11.11) or (11.12). In the one-sided case, our shift estimate, Δ, would be estimated with the appropriate half of Equation (11.14).

Using the same ± 3-sigma equivalent guidelines for k and h, we would obtain only one-half of our type I error probability—we focus on only one of the two limits. Hence, using the Table 10.6 k and h combinations, we would produce roughly $\alpha = 0.00135$ and $1/0.00135 = 740.7 = \text{ARL}_0$ when $H_0: \mu = \mu_0$ is true.

CuSUM CHARTING VARIATIONS

Several variations of CuSum location monitoring have been developed. One variation allows us to standardize our observations and chart the standardized values. For example, if our process metric is described by random variable X_j, we can define a standardized random variable

$$Y_j = \frac{X_j - \mu}{\sigma}$$

Here, we can chart the y values in a manner similar to that used for our x values. For our two-sided CuSum chart,

$$L_{Sj}^+ = \max[0, y_j - k + L_{Sj-1}^+] \qquad \textbf{[11.15]}$$

$$L_{Sj}^- = \min[0, y_j + k + L_{Sj-1}^-] \qquad \textbf{[11.16]}$$

where $L_{S, j=0}^+ = L_{S, j=0}^- = 0$. Here, if

$$L_{Sj}^+ > h$$

or

$$L_{Sj}^- < -h$$

we declare our process metric to be unstable or out of control, assuming a special cause is present. Our choice of k and h follows the same guidelines as described previously.

The advantage of the standardized charts is that all charts with the same k and h values contain common scales. The disadvantage is that our plotted metric is more abstract than before.

A second variation of the CuSum chart is to use some value other than zero to initiate the chart. For example, $L_{j=1}^+ \neq 0, L_{j=1}^- \neq 0$. **A fast initial response (FIR) CuSum provides a head-start feature.** Lucas and Crosier [9] introduce the FIR concept.

The reasoning for the fast initial response, FIR, CuSum is that upon initiation or restart of the CuSum chart, a special cause may be present; hence, the process mean may not be equal or near the target mean value. **If a special cause remains upon restart, the FIR CuSum will indicate its presence faster than with a conventional restart value of zero.** For example, if we want to protect ourselves for a mean coming in high (or low), we start $L_{j=1}^+ > 0$ (or $L_{j=1}^- < 0$) initially. Then, if the mean indeed comes in high (or low), our chart triggers an instability sooner. If the mean comes in on target, the chart will quickly drift toward the "center" and indicate process stability.

The CuSum chart has not experienced widespread acceptance by practitioners. The tabular form, versus the V mask, simplifies applications but tends to obscure the actual cumulative sum value, L_j, trends. Note that

$$L_j = \sum_{k=1}^{j} (x_k - \mu_0)$$

whereas the tabular charts plot L_j^+ and L_j^- statistics.

11.7 SAMPLING SCHEMES

Developing a sampling scheme for SPC monitoring is challenging. Three major strategies are employed in sample timing:

1. Use of **fixed intervals** between samples.

2. Use of **random intervals** between samples.

3. Use of regular intervals between samples, with **judgmental modifications,** when we think process disruptions are more/less likely to occur.

In general, **available sensor technologies bound our sampling options.** Some sensors may be analog devices embedded in the process and be capable of providing continuous monitoring. Examples of these analog devices include thermocouples, diaphragms, floats, and timing devices. Sensors may be polled on a time-interval basis, varying from fractions of a second to hours, with analog-to-digital conversion used to provide sample data. These data may be used for on-line process control purposes or provided for off-line control models.

In other cases, we may use human-based measurement and/or judgment in sensing the physical process. Within certain ranges, humans can utilize their five senses (vision, touch, hearing, smell, taste) and produce attributes and/or variables data regarding a physical process's status. Calibration in terms of person-to-person accuracy and repeatability over time provide challenges. Nevertheless, in some cases the human sensor remains our most effective option, e.g., food/beverage taste measurement and decor contrast preferences.

Sample timing ultimately comes down to a compromise between (1) **physical process concerns** and (2) **statistical concerns.** In the case of little or no process inertia (independent sample data), Grant and Leavenworth [10], Deming [11], and others point out that the statistical performance of the X-bar, R chart pair is enhanced when we take samples close together in time within the subgroup and allow more time-space between subgroups. When we use this grouped spacing method, we seek to capture only common cause variation within our subgroup, thus maximizing our ability to detect heterogeneity between subgroups, e.g., process changes as manifested through special cause variation. Many piece-part production processes have little if any process inertia, and hence these processes lend themselves to classical Shewhart SPC tools and grouped sampling. But nevertheless we should check for the presence of process inertia using the methods described in Chapter 4.

In some industries, such as chemicals, refining, and plastics, where continuous process flow is encountered and high process inertia exists, tightly grouped samples produce nearly identical measurements for reasons such as local mixing in the flow stream or tank, thermodynamic responses, and so on. Usually the differences found are due more to measurement variation than to product variation. In such situations, we can argue that subgroups of size 1 are justified, provided we can establish that by virtue of the time-space between their collection we obtain "pseudo" independent data. In general, the sampling intervals developed must be compatible with the physical process, yet retain statistical and cost effectiveness. In cases where we cannot obtain a meaningful "pseudo" independent data stream, we must resort to more complicated analyses.

We counter process inertia effects to some degree with our sampling scheme by spreading our sampling interval. However, here we might give up our timeliness component in detecting the presence of special causes. For example, sampling a process whose metric can change drastically in 10 minutes at 8-hour intervals is not an effective strategy. At other times, as described in Chapter 13, we may be able to remove the trend or cyclical component with a time series or covariate model, and assess/monitor the model residuals with traditional SPC tools.

Judgment-based sampling schemes typically rely on operator experience or rules of thumb. For example, we might increase the sampling frequency when points are encountered near the control limits. Or we might use 2-sigma warning limits on our control chart in addition to 3-sigma control limits. **The motivation behind judgmental sampling is to obtain a synergistic effect of statistics, process experience, and engineering judgment.** Hence, specific process knowledge and experience are necessary to effectively use the judgmental sampling approach.

Obviously, **we want to detect process shifts as soon as possible—thus, we concentrate process monitoring near what we expect to be problem periods.** Such periods might result from occurrences such as material lot changes, tooling changes, and so forth. If significant process changes are made, we recalculate our chart parameters to reflect the changes. In the end, **we consider a number of factors when selecting and executing a process control sampling strategy:**

1. The direct **benefits** our control will yield, e.g., early detection of quality problems as well as added process vigilance.

2. The **cost** of sampling and sample evaluation.

3. The **time** it takes to sample and evaluate the sample.

4. **Our customer's goodwill** toward quality-conscious suppliers.

11.8   PRODUCTION SOURCE LEVEL CONTROL

The critical point in process control is to "attach" the sensor to a unique production source that can be controlled by an operator or control equipment in a timely fashion. Whenever product from two sources is mixed and product source identity is lost, our ability to impact quality at the source is lost.

Example 11.4 | **PRODUCTION SOURCE CASE—CORN FLAKES FOR CATTLE**   Beef cattle are typically fattened with rations of corn and other nutrients. A large feed mill produces between 500,000 and 1 million pounds of feed per day, 7 days per week. One method of feed preparation utilizes corn that has been steamed and then flaked. These flakes look similar to breakfast cereal that humans eat, with the exception that they usually have a higher moisture content and are produced under less sanitary conditions.

Figure 11.7 depicts a vertical steam chamber where corn at a moisture level of between 10 and 15% is dropped from the top and steam is fed up from the bottom. In this plant we have three steam chambers, each

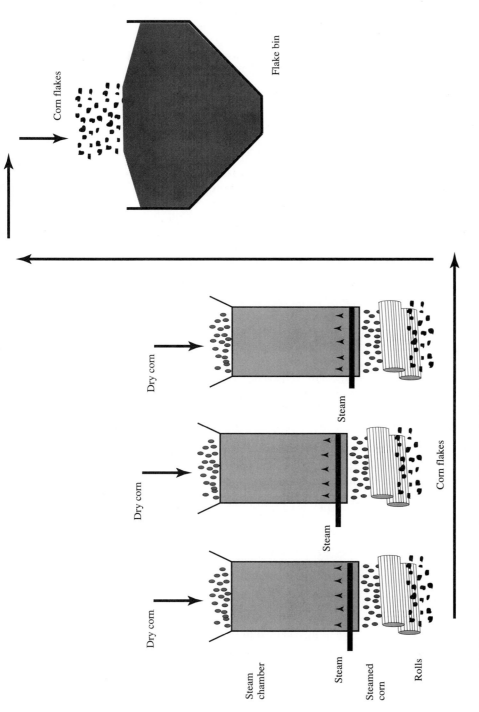

Figure 11.7 Corn flake process flow pictorial (Example 11.4).

with gates and flaking rolls. The corn, when it hits the rolls, should have a moisture level of about 24%. Likewise, the flakes should have a moisture level of about 24%. These three flaking units feed a common conveyor which moves the flakes to a holding bin.

Each steam chamber is large and holds corn for several minutes to nearly an hour in order to bring the dry corn to somewhere near the target moisture level. In general, the dryer the incoming corn, the longer it takes to bring it to a suitable moisture level. Red, the roll master, must regulate both the corn flow and the steam flow. He can adjust both the corn inflow as well as the steam inflow, and does so based on experience. He can also choke the corn flow down at the chamber exit in order to keep it in the chamber longer, but this practice creates problems in keeping the throughput up to schedule. He can open the exit more, but risks choking or overloading the rolls. Hence, we see a rather complex process from a control standpoint, considering the variables and level of instrumentation typically available.

Red and two other people are locked in a discussion as to how best to fulfill the roller process purpose/target regarding producing to a 24% flake moisture target. Hank, the feed mixer, maintains that they should sample the holding bin where all flakes are mixed together. He claims that this sampling point is upstream from feed mixing (with other ingredients) and that these are the flakes the cattle actually see in their feed troughs. Red, on the other hand, wants to sample off each flake roller, claiming this information will help him set the steam/corn flow mix better.

Two types of moisture measurements are possible, fast measurements with an electronic moisture meter (a measurement takes about 2 minutes from sampling to determination) and slow measurements with 24 hour/cycle drying oven weight ratios. In both cases, the sensing equipment is off line, located in an adjacent office cubical. A third party, Jane, the quality control/inspection person, insists that the slow, 24-hour method is more accurate, and should be used. Jane is neutral as to where the sample should be drawn. We want to develop a set of sensible sampling recommendations for these people. Provide valid arguments to support your recommendation.

First, we focus on the process control leverage, so as to hold to target and avoid off-target flakes. From a feedback standpoint, we should sample at the rollers because each roller unit is adjusted independently. Holding each roller unit to target will produce a homogeneous 24% flake all the way down the line. Furthermore, from a feedforward standpoint, we might consider sampling the corn toward the bottom of the steam chamber.

Second, the time it takes to obtain a measurement is critical. Even though we obtain a more accurate determination from the oven, waiting 24 hours for a measurement is too long. We could produce hundreds of thousands of pounds of off-target product before we were aware of the problem. But we should sample and oven dry occasionally in order to estimate/calibrate any bias (lack of accuracy) we experience with the electronic moisture meter.

11.9   **TARGET-BASED CONTROL CHARTS**

For many control charts, we develop control chart parameters regarding upper and lower control limits as well as centerlines from actual process data. **Charts constructed to targets simply substitute target parameters** such as μ_0 and σ_0 **into the equations of Chapters 10 and 11** in place of the empirical process–generated statistics, i.e., $\bar{\bar{x}}, \bar{R},$ and $\bar{s}$. **The logic behind the use of actual process data is simply to reflect what is happening in our process.** On the other hand, **by using target values to set up our SPC charts, we reflect what we want to happen in our process—** not necessarily what is actually happening. Hence, if we use target values, then we should use them carefully.

Target-based charts may help motivate quality improvement teams, provided the teams are empowered to adjust the process to target and/or reduce process variation. On the other hand, they may simply frustrate production teams who are not empowered, i.e., authorized, to change and improve the process. **Stretch targets are meaningful in process improvement, as described in Section 5, but usually demand process changes beyond the scope of process control activities.** The point here is simply that we must use target-based SPC charts with discretion.

11.10 PROCESS CAPABILITY

The goal in process control is to assure a product that consistently meets specifications, minimizing and/or eliminating rework and scrap. Control charts measure process stability in general; they do not measure a process's ability to meet specifications. **It is possible to develop a control chart for a process, show process stability, and at the same time produce an off-specification, defective product.**

When we speak about a capable process, we are inferring that the process outputs meet or exceed our expectations. Usually we express our expectations, as producers, through substitute quality characteristics, e.g., measurable requirements or tolerances. The issue is establishing a reasonable benchmark. The concept of a natural tolerance of ± 3 sigma was considered reasonable. For example, a natural tolerance of ± 3 sigma, using a normal distribution, encompasses roughly 99.73% of the area. Applied to a production process, this implies that on the average about 9973 product units out of 10,000 will meet specifications; or about 2700 parts per million will not meet specifications.

More contemporary expectations are expressed through "zero defects" and "6-sigma" quality. Here, zero defects implies that all product conforms to specifications. Six sigma implies that 2 nonconforming parts per billion is the standard. An alternative, working version of 6-sigma uses 4.3 nonconforming parts per million as the six-sigma standard. Chapter 19 provides more details on zero defect and 6-sigma initiatives.

PROCESS CAPABILITY INDICES

Process capability is a critical performance measure that addresses process results relative to process/product specifications. Process capability measures are widely used in industry to measure a company's own ability or a supplier's ability to meet quality specifications. **Two process capability measures or indices are widely used:** (1) **the C_p index,** an inherent or potential measure of capability, and (2) **the C_{pk} index,** a realized or actual measure of capability; see Kane [12]. Other process capability indices, such as C_{pm} and C_{pmk}, have been developed, discussed, and advocated by Chan et al. [13] and Pearn et al. [14], respectively. An introduction to the C_{pm} and C_{pmk} process capability indices is presented in the appendix to this chapter.

In order to evaluate process capability, it is critical to understand that

1. Process specifications pertain to an individual item's quality characteristics.

2. Capability indices pertain to the population of individual items, with respect to the quality characteristic of interest, when the process of interest is stable.

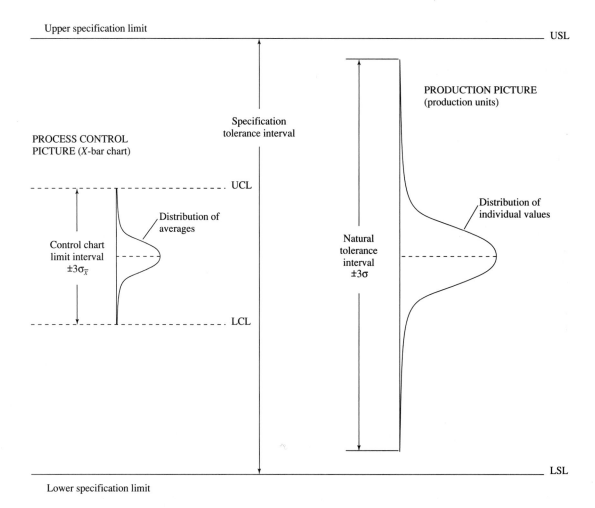

Upper specification limit

USL

PRODUCTION PICTURE
(production units)

Specification
tolerance interval

PROCESS CONTROL
PICTURE (*X*-bar chart)

UCL

Distribution of
individual values

Distribution of
averages

Control chart
limit interval
$\pm 3\sigma_{\bar{x}}$

Natural
tolerance
interval
$\pm 3\sigma$

LCL

LSL

Lower specification limit

Figure 11.8   Global view of the *X*-bar chart, process capability, and specification limits.

3. Subgroup-based control chart limits pertain to the population of subgroups and not to the population of individual items, with respect to the quality characteristic of interest.

Figure 11.8 depicts these critical points, for one possible case, using an *X*-bar chart and a distribution of individual values. A ± 3-sigma natural tolerance interval is marked on the individual value distribution. The distribution of subgroup averages (the distribution of $\bar{X}$, $n \geq 2$) is "tighter" than the individual production unit distribution (the distribution of X). Only in the case of the *X* chart, where $n = 1$, are they equivalent. It should also be noted that **specification limits are set with respect to process/product design (customer) needs, while the process/product dispersion, as measured by $\hat{\sigma}$, is a function of the process, materials, equipment, tooling, operation methods, and so forth.** Since the distribution of individual values may be located anywhere (within physical possibilities), and its dispersion likewise, percentage of product meeting specifications may range from 0 to 100%. **Capability indices link the customer's product design/requirements world with the world of process-related activities/results.**

Potential Capability   The C_p index measures potential or inherent capability of the production process (assuming a stable process) and is defined as

$$C_p = \frac{\text{USL} - \text{LSL}}{6\sigma} \qquad \text{[11.17]}$$

The index is estimated by

$$\hat{C}_p = \frac{\text{USL} - \text{LSL}}{6\hat{\sigma}} \qquad \text{[11.18]}$$

Here, USL and LSL represent the upper and lower specification limits, respectively. If $C_p = 1$, we declare that the process is potentially capable (in a marginal sense). If $C_p < 1$, we declare that the process is potentially incapable, and if $C_p > 1$, we declare that the process is potentially capable.

Actual Capability   The process C_{pk} index measures realized process capability relative to actual production (assuming a stable process) and is defined as

$$C_{pk} = \text{minimum} \left\{ \frac{\mu - \text{LSL}}{3\sigma}, \frac{\text{USL} - \mu}{3\sigma} \right\} \qquad \text{[11.19]}$$

The index is estimated by

$$\hat{C}_{pk} = \text{minimum} \left\{ \frac{\hat{\mu} - \text{LSL}}{3\hat{\sigma}}, \frac{\text{USL} - \hat{\mu}}{3\hat{\sigma}} \right\} \qquad \text{[11.20]}$$

If $C_{pk} = 1$, we declare that the process is marginally capable. If $C_{pk} < 1$, we declare that the process is incapable, and if $C_{pk} > 1$, we declare that the process is capable.

In some cases, product specifications are set on only one side. For example, purity of a product from a chemical process may be required to be at least 98%, i.e., LSL = 98%. As another example, we may state that no more than 3% cracked grain is allowed, i.e., USL = 3%. The process capability concept can be applied to such cases by defining the following two measures:

For processes with only an LSL,

$$C_{pL} = \frac{\mu - \text{LSL}}{3\sigma} \qquad \text{[11.21]}$$

and for processes with only an USL,

$$C_{pU} = \frac{\text{USL} - \mu}{3\sigma} \qquad \text{[11.22]}$$

These two measures are estimated by

$$\hat{C}_{pL} = \frac{\hat{\mu} - \text{LSL}}{3\hat{\sigma}}$$

[11.23]

and

$$\hat{C}_{pU} = \frac{\text{USL} - \hat{\mu}}{3\hat{\sigma}},$$

[11.24]

INTERPRETING CAPABILITY INDICES

Potential capability C_p measures the inherent ability of a process to meet specifications, provided the process can be adjusted to its location-oriented target. It is relevant only to a nominal-is-best (versus smaller-is-better or bigger-is-better) characteristic. **The C_{pk} is developed relative to the process location parameter, it provides a realized measure of actual production.** The C_{pL} and C_{pU} are the single specification counterparts of the C_{pk}. Figure 11.9 depicts relationships between the C_p and C_{pk} measures. We should note that high C_ps do not automatically assure us of a high quality product. In practice, the C_p measure is useful since we can typically adjust to a location target without great effort and expense. However, variance reduction typically is both time-consuming and expensive and usually requires extensive process improvement efforts.

In most cases, we do not know the value of the μ and σ parameters of the production process. We typically use $\hat{\mu}$ and $\hat{\sigma}$ as estimates. These values are usually taken from control chart statistics under stable conditions. Capability analyses based on unstable processes can be expected to produce smaller C_p and C_{pk} indices than those produced in a comparable stable process—instability usually results in greater dispersion estimates.

When μ and σ are estimated using $\hat{\mu}$ and $\hat{\sigma}$, the results are the estimated values $\hat{C}_p$ and $\hat{C}_{pk}$. We expect to see $\hat{C}_p$ and $\hat{C}_{pk}$ greater than 1 in order to support process capability claims. **In some cases, we use a rule of thumb that requires a $\hat{C}_p$ and $\hat{C}_{pk}$ of 1.3** (from a sample size of at least 30 to 50 measurements) **to ensure clear evidence of process capability.**

Capability indices present challenges in statistical inference, as C_p and C_{pk} are themselves random variables with rather complicated probability mass functions. Kushler and Hurley [15] have published a study that focuses on a number of different approximate methods of

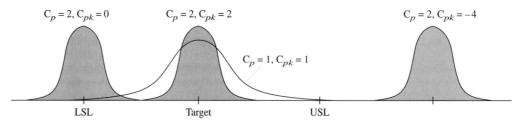

Figure 11.9   Relationships between C_p and C_{pk}.

establishing lower confidence bounds on C_{pk}, assuming that the quality characteristic measure of interest is itself normally distributed. One method discussed results in the approximate $(1 - \alpha)100\%$ lower confidence interval,

$$L_{Cpk,\alpha} \cong \hat{C}_{pk} - Z_\alpha \sqrt{\frac{1}{9n'} + \frac{\hat{C}_{pk}^2}{2n' - 2}} \qquad \text{[11.25]}$$

where $\hat{C}_{pk}$ = estimated C_{pk}, from Equation (11.20).

Z_α = standard normal statistic associated with a right-hand tail area of α.

n' = the sample size used to calculate $\hat{C}_{pk}$.

From Equation (11.25), we can develop what amounts to an approximate lower confidence limit on C_{pk}, and hence, **we can structure an approximate hypothesis test for C_{pk}.** Our test procedure consists of five steps:

Step 1: Set up the hypotheses and select the level of significance for the approximate hypothesis test:

H_0: $C_{pk} \leq C_{pk0}$

H_1: $C_{pk} > C_{pk0}$

Here, we select C_{pk0} on the basis of our needs. We would select $C_{pk0} = 1.0$ if we wanted to represent a marginal level of capability. We choose the desired α level based on our willingness to accept a type I statistical error, where we reject H_0, given that H_0 is true.

Step 2: Determine the sample size, e.g., from a control chart or from a separate process capability study. The sample size is the subgroup size multiplied by the number of subgroups (on the control chart) used to estimate μ and σ.

Step 3: Calculate $\hat{C}_{pk}$ using Equation (11.20).

Step 4: Calculate $L_{Cpk,\alpha}$ using Equation (11.25).

Step 5: If $C_{pk0} < L_{Cpk,\alpha}$, then reject H_0 in favor of the alternate hypothesis at the α level selected, based on the evidence at hand.

METER STICK PROCESS CAPABILITY   Revisiting our meter stick production case and the X-bar and | **Example 11.5**
R control charts set up in Example 10.2, along with product specifications of LSL = 999 mm and USL = 1001 mm,

1. Calculate $\hat{C}_p$ and $\hat{C}_{pk}$. Then, test the following hypotheses at the $\alpha = 0.05$ level:

   H_0: $C_{pk} \leq 1.0$

   H_1: $C_{pk} > 1.0$

   Here, our H_0 implies that we assume the process is incapable until, and unless, we see evidence to the contrary.

2. If our process is not capable at the $\alpha = 0.05$ level, determine if it would be capable if we were sitting exactly on target, e.g., $\bar{\bar{x}} = 1000$ mm—the best possible outcome. Here, we assume that our process is centered, with the same sample size and dispersion as previously described.

Solution 1

From Example 10.2 we know that

$$\bar{\bar{x}} = \hat{\mu} = 999.63 \text{ mm}$$

and

$$\hat{\sigma} = \frac{\bar{R}}{d_2} = \frac{0.500}{2.326} = 0.215 \text{ mm}$$

Hence,

$$\hat{C}_p = \frac{\text{USL} - \text{LSL}}{6\hat{\sigma}} = \frac{1001 - 999}{6(0.215)} = 1.550$$

$$\hat{C}_{pk} = \text{minimum} \left\{ \frac{\hat{\mu} - \text{LSL}}{3\hat{\sigma}}, \frac{\text{USL} - \hat{\mu}}{3\hat{\sigma}} \right\}$$

$$= \text{minimum} \left\{ \frac{999.63 - 999}{3(0.215)}, \frac{1001 - 999.63}{3(0.215)} \right\}$$

$$= \text{minimum} \{0.977, 2.124\} = 0.977$$

For the hypothesis test, from the revised calculations we know that $n = 5$ and $m = 15$. Hence, the sample size for our calculations is $nm = 75$. Since $\hat{C}_{pk} = 0.977$ we have no chance of rejecting our H_0, but we will follow our steps anyway for illustrative purposes:

$$L_{Cpk,0.05} \cong 0.977 - 1.645 \sqrt{\frac{1}{9(75)} + \frac{0.977^2}{2(75) - 2}}$$

$$= 0.977 - 1.645 \sqrt{0.00793} = 0.831$$

Here,

$$C_{pk0} = 1.0 > L_{Cpk,0.05} \cong 0.831$$

Therefore, based on the evidence, we cannot reject the null hypothesis that $C_{pk} \le 1.0$ at the $\alpha = 0.05$ significance level. We conclude that the process is not capable.

Solution 2

Now, since we have established that the current mean of 999.63 mm does not produce a capable process, we examine the best case, assuming that $\bar{\bar{x}} = 1000$ mm.

In this specific "best" case, our

$$\hat{C}_{pk} = \hat{C}_p = 1.550$$

Repeating our hypothesis test procedure with $\hat{C}_{pk} = 1.550$,

$$L_{Cpk,0.05} \cong 1.550 - 1.645 \sqrt{\frac{1}{9(75)} + \frac{1.550^2}{2(75) - 2}} = 1.331$$

Here,

$$C_{pk0} = 1.0 < L_{Cpk,0.05} \cong 1.331$$

Therefore, based on our sample evidence, if we move to, or very near, the target value of 1000 mm, we can expect to see a capable process.

SIX-SIGMA CAPABILITY CASE   Memory Inc. is a semiconductor manufacturing organization that uses processes that require hundreds of sequential steps. In order to produce a high process yield, each step in this sequence must run at a near zero defect level. Memory Inc. has several suppliers that use and understand capability indices. In order to emphasize 6-sigma quality and work with existing suppliers, Memory has undertaken the task of converting conventional capability indices to conform to their 6-sigma expectations. The conversion is presented below.   **Example 11.6**

Our strategy is to use the C_p and C_{pk} indices as previously defined, but to set a new capability threshold. Given that our process is a 6-sigma process, we know that USL − LSL = 6σ + 6σ = 12σ. Hence,

$$C_p = \frac{USL - LSL}{6\sigma} = \frac{12\sigma}{6\sigma} = 2.0$$

$$C_{pk} = \text{minimum}\left\{\frac{\mu - LSL}{3\sigma}, \frac{USL - \mu}{3\sigma}\right\}$$

$$= \text{minimum}\left\{\frac{6\sigma}{3\sigma}, \frac{6\sigma}{3\sigma}\right\} = 2.0$$

Our new threshold is now a capability value of 2 rather than the standard value of 1. In order to communicate this new threshold to our suppliers, we will simply ask them to consider a process capable if C_{pk} is greater than 2. For calculations based on empirical evidence, we will ask them to use our five-step hypothesis test method with $C_{pk0} = 2$.

11.11   GAUGE STUDIES

Previously, in Chapter 4, we discussed the concepts of accuracy and precision; see Figure 4.3. Now, we describe the concept of precision in quantitative terms, as estimated through gauge studies.

Dispersion in measurements has two primary sources: (1) **the failure of our gauge or instrument to exactly repeat itself** and (2) **the failure of an operator or machine to exactly reproduce the measurement technique or method.** We can express the precision of our measurements, assuming independence, as

$$\sigma^2_{\text{measured values}} = \sigma^2_{\text{true dimensions}} + \sigma^2_{\text{gauge repeatability}} + \sigma^2_{\text{operator reproducibility}} \qquad \textbf{[11.26]}$$

Sometimes, the term "measurement error" is defined as

$$\sigma^2_{\text{measurement error}} = \sigma^2_{\text{gauge repeatability}} + \sigma^2_{\text{operator reproducibility}} \qquad \textbf{[11.27]}$$

The precision-to-tolerance (P/T) ratio is sometimes used to assess gauge capability:

$$P/T = \frac{6\hat{\sigma}_{\text{measurement error}}}{\text{USL} - \text{LSL}} \qquad \textbf{[11.28]}$$

Typically, a $P/T \le 0.10$ is considered adequate. However, we can set the P/T threshold at any proportion we believe realistic for our process measurement needs.

Thus we can clearly see that the data we record on our data sheets or in our electronic storage are not the actual dimensions of our process/product. **Our measured values include variation "inflation" over and above the true variation in the population of process/product dimensions.** We usually assume that the three sources of variation in Equation (11.26) are independent. If we wish to estimate this variation inflation, we must conduct a controlled instrument or gauge study. **Through a careful procedure of repeatedly measuring the same process/product dimension, we can extract estimates of gauge repeatability and operator reproducibility.**

If either the gauge repeatability or the operator reproducibility is large in comparison to the true dimensional variance, we usually take action to reduce it. Otherwise, a flawed measurement process limits our ability to assess our product or process. Corrective actions available to us include more precise gauging (better instrumentation), operator training programs, or both.

An on-target, low-dispersion process/product can be obscured by a poor measurement system. For example, a large bias in our measurement system or sensor will falsely indicate that we are off target, while poor measurement precision will falsely indicate that we have excessive variation in our product or process. The following examples serve to demonstrate how we can use our X-bar and R chart technology to estimate $\sigma_{\text{gauge repeatability}}$ and $\sigma_{\text{operator reproducibility}}$. The use of random effects factors in designed experiments offers a more sophisticated method of estimating components of variation; refer to Chapter 5 for details.

Example 11.7 | **ESTIMATION OF VARIATION INFLATION**   Using X-bar and R chart technology, estimate the variation inflation due to gauging for an automatic (no operator) measuring device assigned to our previously described meter stick production process.

1. Estimate $\hat{\sigma}_{\text{gauge repeatability}}$
2. Estimate $\hat{\sigma}_{\text{measured values}}$
3. Estimate $\hat{\sigma}_{\text{true dimensions}}$
4. Calculate the P/T ratio
5. Comment on the ability of the automatic measuring machine to produce precise measurements

Since we have only one automatic measuring machine, we assume that $\sigma_{\text{repeatability}}$ is the only gauge-related variation component of interest. A simple experiment was designed whereby we measured the same meter stick two times, and a total of 20 meter sticks were measured. The data, $\bar{x}$, and range values are summarized in Table 11.5. We use our standard $\hat{\sigma} = \bar{R}/d_2$ calculation on these data.

Table 11.5 Automatic measuring device gauge study data (Example 11.7). (All measurements in millimeters; LSL = 999 mm; USL = 1001 mm.)

Part (Meter Stick) No.	Meas. 1	Meas. 2	$\bar{x}_j$	$\bar{R}_j$
1	1000.158	1000.157	1000.1575	0.0010
2	1000.003	1000.006	1000.0045	0.0030
3	999.722	999.720	999.7210	0.0020
4	1000.303	1000.305	1000.3040	0.0020
5	1000.216	1000.217	1000.2165	0.0010
6	999.636	999.638	999.6370	0.0020
7	1000.153	1000.153	1000.1530	0.0000
8	1000.383	1000.381	1000.3820	0.0020
9	999.754	999.757	999.7555	0.0030
10	1000.237	1000.238	1000.2375	0.0010
11	999.968	999.968	999.9680	0.0000
12	1000.073	1000.075	1000.0740	0.0020
13	1000.186	1000.184	1000.1850	0.0020
14	999.948	999.945	999.9465	0.0030
15	999.868	999.869	999.8685	0.0010
16	999.951	999.952	999.9515	0.0010
17	999.900	999.903	999.9015	0.0030
18	1000.387	1000.387	1000.3870	0.0000
19	999.961	999.962	999.9615	0.0010
20	999.813	999.812	999.8125	0.0010
Averages			1000.0312	0.0015

Solution 1

$$\hat{\sigma}_{\text{gauge repeatability}} = \frac{\bar{R}}{d_2} = \frac{0.0015}{1.128} = 0.0013 \text{ mm}$$

Here, we use $\bar{R}$ from the data in Table 11.5, $n = 2$.

Solution 2

$$\hat{\sigma}_{\text{measured values}} = \frac{\bar{R}}{d_2} = \frac{0.50}{2.326} = 0.215 \text{ mm}$$

Here, we use $\bar{R}$ from our revised production control limits from Example 10.2, $n = 5$.

Solution 3

$$\hat{\sigma}_{\text{true dimensions}} = \sqrt{\hat{\sigma}^2_{\text{measured values}} - \hat{\sigma}^2_{\text{gauge repeatability}}}$$

$$= \sqrt{(0.215)^2 - (0.0013)^2} = 0.2149 \text{ mm}$$

Solution 4

$$P/T = \frac{6\hat{\sigma}_{\text{measurement error}}}{(\text{USL} - \text{LSL})} = \frac{6(0.0013)}{(1001 - 999)} = 0.0039 = 0.39\%$$

Since we have only one automatic measuring device, we assume that $\hat{\sigma}_{\text{measurement error}} = \hat{\sigma}_{\text{gauge repeatability}}$.

Solution 5

The additional variation added by our measuring machine is extremely small relative to our true product dimension variation. We also see a very small P/T ratio, much less than 10%. Hence, we conclude that our automatic measuring device is performing (as far as precision is concerned) very well, relative to the variation inherent in our meter stick production length quality characteristic.

Example 11.8 | **ESTIMATION OF GAUGE REPEATABILITY, OPERATOR REPRODUCIBILITY, AND P/T RATIO**

A gauge study was run on our meter stick process where measurements were obtained manually by three operators using the same gauge. The data are shown in Table 11.6. Use the X-bar and R chart technique to estimate the measures below.

1. $\hat{\sigma}_{\text{gauge repeatability}}$
2. $\hat{\sigma}_{\text{operator reproducibility}}$
3. P/T ratio
4. Comment on the ability of our manual measuring process to supply adequate precision.

Solution 1

Since we are using the same measuring device with all three operators, along with two repeated measurements, $n = 2$,

$$\bar{\bar{R}} = \frac{0.043 + 0.036 + 0.016}{3} = 0.0317 \text{ mm}$$

$$\hat{\sigma}_{\text{gauge repeatability}} = \frac{\bar{\bar{R}}}{d_2} = \frac{0.0317}{1.128} = 0.0281 \text{ mm}$$

Solution 2

Here, we are dealing with three operators. We need a measure of dispersion between operators, $n = 3$,

$$\bar{\bar{X}}_{\text{max}} = \max (1000.040, 999.995, 1000.097) = 1000.097 \text{ mm}$$

$$\bar{\bar{X}}_{\text{min}} = \min (1000.040, 999.995, 1000.097) = 999.995 \text{ mm}$$

$$R_{\bar{\bar{X}}} = 1000.097 - 999.995 = 0.102 \text{ mm}$$

$$\hat{\sigma}_{\text{operator reproducibility}} = \frac{R_{\bar{\bar{X}}}}{d_2} = \frac{0.102}{1.693} = 0.0602 \text{ mm}$$

Solution 3

$$\hat{\sigma}_{\text{measurement error}} = \sqrt{0.0281^2 + 0.0602^2} = 0.0664 \text{ mm}$$

$$P/T = \frac{6(0.0664)}{(1001 - 999)} = 0.1992 = 19.92\%$$

Table 11.6   Manual measuring gauge study data (Example 11.8).
(All measurements In millimeters; LSL = 999 mm; USL = 1001 mm.)

Part (Meter Stick) No.	Operator 1				Operator 2				Operator 3			
	Meas. 1	Meas. 2	$\bar{x}_j$	$\bar{R}_j$	Meas. 1	Meas. 2	$\bar{x}_j$	$\bar{R}_j$	Meas. 1	Meas. 2	$\bar{x}_j$	$\bar{R}_j$
1	1000.16	1000.18	1000.170	0.020	1000.08	999.99	1000.035	0.090	1000.25	1000.28	1000.265	0.030
2	1000.00	999.89	999.945	0.110	999.97	999.93	999.950	0.040	1000.11	1000.09	1000.100	0.020
3	999.74	999.84	999.790	0.100	999.73	999.76	999.745	0.030	999.83	999.87	999.850	0.040
4	1000.29	1000.25	1000.270	0.040	1000.28	1000.29	1000.285	0.010	1000.33	1000.31	1000.320	0.020
5	1000.21	1000.20	1000.205	0.010	1000.12	1000.08	1000.100	0.040	1000.27	1000.29	1000.280	0.020
6	999.74	999.81	999.775	0.070	999.75	999.71	999.730	0.040	999.79	999.80	999.795	0.010
7	1000.16	1000.05	1000.105	0.110	1000.03	1000.08	1000.055	0.050	1000.21	1000.23	1000.220	0.020
8	1000.38	1000.39	1000.385	0.010	1000.33	1000.30	1000.315	0.030	1000.44	1000.42	1000.430	0.020
9	999.85	999.90	999.875	0.050	999.87	999.83	999.850	0.040	999.87	999.87	999.870	0.000
10	1000.24	1000.18	1000.210	0.060	1000.21	1000.20	1000.205	0.010	1000.21	1000.23	1000.220	0.020
11	999.94	999.92	999.930	0.020	999.91	999.92	999.915	0.010	999.99	999.96	999.975	0.030
12	1000.07	1000.08	1000.075	0.010	1000.08	1000.02	1000.050	0.060	1000.12	1000.11	1000.115	0.010
13	1000.19	1000.19	1000.190	0.000	1000.14	1000.11	1000.125	0.030	1000.23	1000.25	1000.240	0.020
14	999.99	1000.02	1000.005	0.030	999.97	999.93	999.950	0.040	1000.13	1000.11	1000.120	0.020
15	899.87	999.82	999.845	0.050	999.85	999.81	999.830	0.040	999.93	999.93	999.930	0.000
16	999.95	999.91	999.930	0.040	999.98	999.94	999.960	0.040	999.98	999.99	999.985	0.010
17	999.90	999.87	999.885	0.030	999.84	999.81	999.825	0.030	999.97	999.96	999.965	0.010
18	1000.40	1000.47	1000.435	0.070	1000.31	1000.34	1000.325	0.030	1000.41	1000.41	1000.410	0.000
19	999.96	999.96	999.960	0.000	999.93	999.95	999.940	0.020	999.95	999.97	999.960	0.020
20	999.79	999.82	999.805	0.030	999.72	999.68	999.700	0.040	999.89	999.90	999.895	0.010
Averages			1000.040	0.043			999.995	0.036			1000.097	0.016

Solution 4

In this case, we have a problem with the measurement error (precision) in that the P/T ratio > 0.10. In other words, the measurement error is relatively large in comparison to our length specification interval. Since our estimated operator reproducibility is over twice the size of our estimated gauge repeatability, we would recommend an examination of the consistency of measurement techniques between operators and, possibly, a training program. The gauge repeatability alone would yield $P/T = 6(0.0281)/(1001 - 999) = 0.0843$. Hence, our gauge instrument appears marginal and may need attention, if the variation in our operator reproducibility cannot be drastically reduced.

REVIEW AND DISCOVERY EXERCISES

REVIEW

11.1. Why is the individuals X chart considered less effective in SPC than the X-bar chart? Explain.

11.2.   Why do we typically prefer small artificial subgroup sizes? Is the preference driven by physical or statistical concerns? Explain.

11.3.   Why is the X chart insensitive to small sustained shifts? Explain.

11.4.   Compare and contrast the X, EWMA, and CuSum location charts in terms of their "memory" characteristics.

11.5.   How do we locate the shift point (after a shift occurs) in time when using a CuSum chart? Explain.

11.6.   What advantages and disadvantages do fixed interval, random, and judgmental sampling schemes offer? Explain.

11.7.   Sensors are fundamental to data collection. What constitutes a sensor? What are desirable properties of sensors? Explain.

11.8.   How does sample timing impact the physical characteristics of process control? Explain.

11.9.   Why is it critical to attach SPC monitoring charts directly to control leverage points? Explain.

11.10. What is the purpose of process capability? Explain.

11.11. How is it possible to have a stable process that is incapable? Explain.

11.12. How are/should process capability indexes be interpreted? Explain.

11.13. What is the purpose of a gauge study? Explain.

11.14. What is the precision-to-tolerance ratio? What does it measure? Explain.

11.15. What is gauge repeatability? Operator repeatability? Explain.

CASE EXERCISES I

The following questions pertain to the case descriptions in Section 7. Case selection list:

Case VII.8:   Door-to-Door—Pizza Delivery (out-of-the-oven temperatures)

Case VII.20:   PCB—Printed Circuit Boards

Case VII.27:   Rainbow—Paint Coating

Any case from Chapter 10 Case Exercises, with the subgrouping ignored

11.16. Select one or more of the above cases. Develop an X, R_M chart set from the data provided in the case description. Use 3-sigma limits. Use an artificial subgroup size of 2. Comment on your findings.

11.17. Select one or more of the above cases. Develop an $\overline{X}_M$, R_M chart set from the data provided in the case description. Use 3-sigma limits. Use an artificial subgroup size of 2. Comment on your findings.

11.18. Select one or more of the above cases. Develop an individuals EWMA and EWMD chart set from the data provided in the case description. Select an appropriate r value. Use appropriate target values to calculate the control limits. Comment on your findings.

11.19. Select one or more of the listed cases. Develop an individuals CuSum chart from the data provided in the case description. Select appropriate k and h values from Table 10.6. Use appropriate target values to establish the chart. Comment on your findings.

11.20. For the cases selected in the previous problems, generate an extended, shifted, data set using the method described in Section 7, Equation (VII.1), and plot the extended data on your existing process control charts. Look for shift points and patterns. Comment on your findings.

CASE EXERCISES II

The following questions pertain to the case descriptions in Section 7. Case selection list:

Case VII.6:   Big Dog—Dog Food Packaging

Case VII.7:   Bushings International—Machining

Case VII.8:   Door-to-Door—Pizza Delivery (out-of-the-oven temperatures)

Case VII.9:   Downtown Bakery—Bread Dough

Case VII.20:   PCB—Printed Circuit Boards

Case VII.21:   Punch-Out—Sheet Metal Fabrication

Case VII.27:   Rainbow—Paint Coating

Case VII.29:   TexRosa—Salsa

11.21. Select a case and perform a capability analysis. Include C_p and C_{pk} analyses. Explain your findings.

11.22. For the case selected and analyzed in problem 11.21, test the hypotheses

$H_0: C_{pk} \leq 1.0$

$H_1: C_{pk} > 1.0$

Interpret your findings, using $\alpha = 0.05$.

CASE EXERCISES III

The following questions pertain to the case descriptions in Section 7. Case selection list:

Case VII.10:   Downtown Bakery—pH Measurement

Case VII.15:   High-Precision—Collar Measurement

Case VII.23:   ReUse—Sensor Precision

11.23. Select a case from the list above and develop a gauge study. Include a numerical breakout of the contributions of each source of variation. Calculate the P/T ratio, if appropriate. Comment on your findings.

11.24. Select a product and identify a simple measurement on the product. Perform a gauge study similar to the ones in Examples 11.7 and 11.8. Analyze your results and comment on the ability of your measurement process to perform in an adequate fashion. How could your measurement process be improved? Explain.

DISCOVERY

11.25. How does the choice of r in the EWMA/EWMD charts impact their ability to respond to a sudden shift in the value of the process metric? Explain.

11.26. If targets, rather than empirical means and standard deviations, are used to design EWMA/EWMD charts, are trial charts necessary? Explain.

11.27. What is a standardized CuSum chart? How does it differ from a regular CuSum chart? Explain.

11.28. How does a fast initial response (FIR) head start feature help speed detection of a special cause in a CuSum chart? Explain.

11.29. How effective is a control chart, e.g., an X, EWMA, CuSum, in detecting the following types of shifts: A small sustained shift? A large sustained shift? A small unsustained shift? A large unsustained shift?

11.30. How are process capability indices constructed? What are the major elements? How are they interpreted? Explain.

11.31. Control charts for X-bar and R are maintained on the breaking (tensile) strength in pounds for deep sea fishing line. The subgroup size is $n = 4$. The values of $\bar{x}_j$ and R_j are computed for each subgroup. After collection of 50 subgroups, $\Sigma\bar{x}_j = 7500$ and $\Sigma R_j = 250$. The control charts use 2.5-sigma limits. One point out of the control limits is the only rule used to determine instability.

    a. Compute the probability of a Type I error for the X-bar chart.

    b. Compute the probability of a Type II error for the X-bar chart, given that the process shifts upward to 156.075 lb. and the process variance does not change.

    c. The specification for the fishing line is set at $150 - 15$ lb. or 135 lb. Is the process a capable process?

11.32. How do we know when a gauge or instrument is capable? Explain.

11.33. Why is the variation from a set of measurements not the true variation of the process/product? How can we obtain a feel for true process/product variation? Explain.

11.34. Criticism has been leveled at both the C_p and C_{pk} capability indices. Alternative indices have been proposed, e.g., the C_{pm} and C_{pmk}. What advantages and disadvantages do these indices offer? Explain after reading the appendix to this chapter.

MORE CHALLENGING CASE EXERCISES

The following questions pertain to the case descriptions in Section 7. *Warning:* These cases contain characteristics that, although realistic, create assumption violations in SPC. Use your results from Chapter 4 to help identify the problem cases and issues. Look for creative solutions to help unlock these cases. Case selection list:

    Case VII.2:   Apple Core—Dehydration

    Case VII.4:   Back-of-the-Moon—Mining

Case VII.8: Door-to-Door—Pizza Delivery (out-of-box temperature)

Case VII.17: LNG—Natural Gas Liquefaction

Case VII.22: ReUse—Recycling

Case VII.30: Tough-Skin—Sheet Metal Welding

11.35. Select one or more of the listed cases. Develop an X, R_M chart set from the data provided in the case description. Use 3-sigma limits. Comment on your findings.

11.36. Select one or more of the listed cases. Develop an individuals EWMA and EWMD chart set from the data provided in the case description. Select an appropriate r value. Use appropriate target values to calculate the control limits. Comment on your findings.

11.37. Select one or more of the listed cases. Develop an individuals CuSum chart from the data provided in the case description. Select appropriate k and h values from Table 10.4. Use appropriate target values to establish the chart. Comment on your findings.

APPENDIX: ADDITIONAL CAPABILITY INDICES

Capability indices are used to express the relationship between technical specifications and production abilities on the factory floor. This relationship is important to both suppliers and purchasers. The C_p and C_{pk} process capability indices, as defined, discussed, and demonstrated earlier, see widespread usage. Two other process capability indices have been developed that also link our technical specifications to our ability to produce (meet) them.

The C_{pm} process capability index was independently proposed by Hsiang and Taguchi [16], as well as by Chan et al. [13]; see Rodriguez [17] for details. The C_{pm} index is expressed as

$$C_{pm} = \frac{\text{USL} - \text{LSL}}{6\sqrt{(\mu - T)^2 + \sigma^2}} \qquad \textbf{[11.29]}$$

where   USL = product upper specification limit

LSL = product lower specification limit

T = product target (e.g., the best value for the quality characteristic of interest)

μ = process mean

σ^2 = process variance

The C_{pm} index measures the degree to which the process output is on target. In the special case where $T = \mu$, by examining Equations (11.17) and (11.29), we can see that $C_{pm} = C_p$. In contrast, the C_{pk}, Equation (11.19), measures the degree to which the process output is within the specification limits. One major advantage of the C_{pm} is that it is applicable to an asymmetrical specification interval, where the target T is not in the middle of the interval. The C_{pm} is compatible with the Taguchi loss function concepts; see Taguchi [18].

Pearn et al. [14] describe the C_{pmk} as a "third-generation" process capability index—C_p is the "first generation" and C_{pk} is the "second generation"—that is structured to include features of both C_{pk} and C_{pm}. The C_{pmk} index is defined as

$$C_{pmk} = \frac{C_{pk}}{\sqrt{1 + \left(\dfrac{\mu - T}{\sigma}\right)^2}}$$

[11.30]

where the components of the index are as described in our discussion of the C_{pm} index.

The C_{pmk} index imposes a penalty when the process mean is not on target. For example, if $\mu = T$, then $C_{pmk} = C_{pk}$. But if $\mu \neq T$, then $C_{pmk} < C_{pk}$.

REFERENCES

1. E. L. Grant and R. S. Leavenworth, *Statistical Quality Control*, 6th ed., New York: McGraw-Hill, 1988.

2. R. E. DeVor, T. Chang, and J. W. Sutherland, *Statistical Quality Design and Control: Contemporary Concepts and Methods*, New York: Macmillan, 1992.

3. A. L. Sweet, "Control Charts Using Coupled Exponentially Weighted Moving Averages," *Transactions of IIE*, vol. 18, no. 1, pp. 26–33, 1986.

4. S. J. Hunter, "The Exponential Weighted Moving Average," *Journal of Quality Technology*, vol. 18, no. 4, pp. 203–210, 1986.

5. L. C. Vance, "Average Run Lengths of Cumulative Sum Control Charts for Controlling Normal Means," *Journal of Quality Technology*, vol. 18, no. 3, pp. 189–193, 1986.

6. D. Brook and D. A. Evans, "An Approach to the Probability Distribution of CuSum Run Length," *Biometrika*, vol. 59, no. 3, pp. 539–549, 1972.

7. D. Siegmund, *Sequential Analysis: Tests and Confidence Intervals*, New York: Springer-Verlag, 1985.

8. D. M. Hawkins, "A Fast Accurate Approximation for Average Run Lengths of CuSum Control Charts," *Journal of Quality Technology*, vol. 24, no. 1, pp. 37–43, 1992.

9. J. M. Lucas, and R. B. Crosier, "Fast Initial Response for CuSum Quality Control Schemes: Give Your CuSum a Head Start," *Technometrics*, vol. 24, pp. 199–205, 1982.

10. E. L. Grant and R. S. Leavenworth, *Statistical Quality Control*, 7th ed., New York: McGraw-Hill, 1996.

11. W. E. Deming, *Out of the Crisis*, Cambridge, MA: MIT Center for Advanced Engineering Studies, 1986.

12. V. E. Kane, "Process Capability Indices," *Journal of Quality Technology*, vol. 18, no. 1, pp. 41–52, January 1986.

13. L. K. Chan, S. W. Cheng, and F. A. Spiring, "A New Measure of Process Capability: C_{pm}," *Journal of Quality Technology*, vol. 20, no. 3, pp. 162–175, July 1988.

14. W. L. Pearn, S. Kotz, and M. L. Johnson, "Distributional and Inferential Properties of Process Capability Indices," *Journal of Quality Technology*, vol. 24, no. 4, pp. 216–231, October 1992.

15. R. H. Kushler and P. Hurley, "Confidence Bounds for Capability Indices," *Journal of Quality Technology*, vol. 24, no. 4, pp. 188–195, October 1992.

16. T. C. Hsiang and G. Taguchi, "A Tutorial on Quality Control and Assurance—The Taguchi Methods," unpublished presentation given at the Annual Meeting of the American Statistical Association, Las Vegas, 1985.

17. R. N. Rodriguez, "Recent Developments in Process Capability Analysis," *Journal of Quality Technology*, vol. 24, no. 4, pp. 176–187, October 1992.

18. G. Taguchi, *Introduction to Quality Engineering: Designing Quality in Products and Processes,* White Plains, NY: Kraus International, UNIPUB (Asian Productivity Organization), 1986.

chapter

12

PROCESS MONITORING—
ATTRIBUTES CONTROL CHARTS
FOR CLASSIFICATION MEASUREMENTS

12.0 INQUIRY

1. How do defects and defectives differ?
2. When do we use an attributes chart? A variables chart?
3. What models are used to develop attributes charts?
4. What is the difference between the P chart and the C, U chart families?
5. How do we link process control to process improvement?

12.1 INTRODUCTION

In our discussion of variables control charts, we pointed out that each chart set can monitor only one process characteristic. We advocate aggressive use of variables charts at process leverage points, especially upstream in our processes. However, we also realize a need for a global control picture of our processes and their products. We know that most processes and products have many quality characteristics. **Sometimes, we prefer to view our process/product across all or at least several process/product characteristics.** Hence, several variables control charts are necessary, unless we can find an alternative way to measure and model our process/product.

   We use attributes-based process control charts to monitor multiple process/product characteristics as well as quality characteristics that are logically defined on a classification scale of measure—the nominal and ordinal scales; refer to Chapter 4. Classification characteristics can include "voids" on a sheet of glass, cracks on an automobile flywheel, surface flaws on

320

sheet metal panels, color inconsistencies on a painted surface, broken potato chips, paperwork errors, and so on.

We can design attributes control charts to cover entire product units involving many process/ product characteristics. For example, a typical die-cast body for a 35-mm camera, made of copper-aluminum alloy, has hundreds of dimensions. Although any one of these dimensions can be monitored with a variables control chart, we usually cannot justify hundreds of charts. Hence, we might use an attributes control chart to cover the entire product unit.

When we use the attributes chart for broad coverage and/or variables charts for more detailed coverage, we want to take advantage of what we have learned in the course of process upsets and the counteractions we have taken to address them. For this purpose, we introduce process logs and subsequent Pareto analyses in this chapter.

12.2 DEFECTS AND DEFECTIVES

Before developing the mechanics of attributes control charts, it is necessary to clearly define a number of important terms. **A defect is an individual nonconformity in a process/product that causes it to fail to meet a specification. A zero defect process/product refers to a process/ product that meets all technical or engineering specifications,** and hence has no defects. **An item or service is defective if it fails to conform to specifications in some respect.** Hence, **an item or service that is defective will contain at least one defect or nonconformity.** We use the terms *defective* and *nonconforming* interchangeably. Also, the words *defect* and *nonconformity* are used interchangeably.

Zero defect products do not necessarily assure customer satisfaction unless the designers have completely captured the customer's true quality characteristics in the product's substitute quality characteristics (through technical specifications). Remember, we typically chart substitute, rather than true, quality characteristics in a production organization. As such, we cannot automatically assume that nonconforming products are without value or pose serious risk to our customers—or that conforming products have value and pose no risk.

It is not uncommon to encounter defective items or services that still satisfy their functions quite well. For example, a scratch on a refrigerator door is not going to reduce its functional value in terms of its refrigeration purpose. But our customer may object to the scratch because he or she doesn't like the way it looks—a form, rather than function, issue. Hence, it is important that we capture true customer needs and expectations in our process/product features and specifications, and that processes/products be evaluated accordingly.

The most difficult aspect of quality characterization by attributes is the precise determination of what constitutes the presence of a particular defect. Since a vast majority of the attribute defects are either visual or cannot be measured precisely because adequate technical measurement tools are not available, some degree of judgment is involved in defect classification. Variation in human judgment tends to be a challenge in attribute characterization. **It is critical that precise operational definitions of a defect be generated and understood.** For example, it is necessary to specify exactly what constitutes a crack, crater, fold, inclusion, pit, seam, splatter, and so on, for manufactured components. In order to declare it a defect, it may be necessary to specify the maximum allowable length or depth of a scratch, size of a surface blemish, and so on.

In general, providing visual standards, such as photographs or samples of defects, helps in reducing the human variation in attribute characterization.

12.3 SPC MODEL RATIONALE FOR ATTRIBUTES DATA

Attributes quality characteristics are classification-oriented. Typically, we deal with only two classes, good and defective product—conforming and nonconforming. **Two general families of charts are used: (1) the binomial-based *P* chart family for monitoring defective-nonconforming processes/products** and **(2) the Poisson-based *C*, *U* chart family for monitoring defects per process/product units.** Single charts are sufficient to monitor process data when they are expressed in an attributes form. The nature of the binomial and Poisson models—specifically their model parameter structure—requires only one control chart.

As stated earlier, we must define what constitutes conformance or nonconformance to specifications or requirements for a *P* chart. For example, we can use the *P* chart as a screening chart and define more than one inspection point in a process/product, e.g., length and width dimensions and appearance characteristics. Hence, the *P* chart can monitor many quality characteristics simultaneously, versus only one measurable characteristic with a variables chart pair. This feature allows broad coverage possibilities, but it does not provide detailed information. Hence, it has advantages and disadvantages.

Likewise, a *C* or *U* chart may track the number of one or more different types of defects encountered. Our physical product and process knowledge helps us to clearly define what constitutes a defect. For example, in cloth quality, weaving defects as well as dying defects may be included in the same *U* chart. We must also determine how we will sense or identify and measure each type of defect once it is defined.

In SPC, it is customary to assume that a process is "in control" or stable until evidence is discovered to the contrary. Since the assumption that the process is stable implies that the attributes quality characteristics of interest are influenced by only common cause variation, we can conceptualize and state a set of statistical hypotheses. For example, **generic hypotheses for the *P* chart are stated below:**

H_0: The process population proportion defective is as stated—*the process is operating under common cause variation, without special cause influence present.*

H_1: The process population proportion defective has shifted from the stated value—*the process is operating under common cause variation, with an additional special cause influence present.*

Basically, **attributes control charts provide a crude, graphical means for assessing the validity of these hypotheses.** This assessment leads us to conclude that the process is stable (when we do not reject H_0) or that the process is unstable (when we reject H_0) on the basis of data collected in our last sample.

Sample sizes in attributes control charts are typically much larger than sample sizes in variables control charts. **We typically see sample sizes an order of magnitude, or more, larger in attributes charts than we see in variables charts.** For example, a *P* chart might have a sample size of 50, whereas an *X*-bar chart might have a subgroup size of 5. Many authors attempt to compare the costs of attributes charts with those of variables charts. Although costs are very important, one chart type is not necessarily a substitute for another. **We choose our model and charts to fit our needs.** The more we know about our process, the wiser we can be in model selection and application.

It is critical to reemphasize that SPC charts (both variables control charts and attributes control charts) are incomplete, in a quality control sense, in that they do not identify special causes, nor do they tell us how to eliminate the effect of a special cause. **Control charts basically serve as statistical warning devices. They do not help us to account for and document physical characteristics, actions taken, and results obtained.** That is the role of process logs and Pareto charts, discussed later in this chapter.

GOLF CLUB SHAFT PRODUCTION PROCESS CASE   Before proceeding with a detailed description of attributes control chart mechanics, we will illustrate the use of an attributes control chart, the P chart. In the following example we use the P chart to monitor the pultrusion process used in manufacturing composite components in the sports equipment industry. | **Example 12.1**

The marketing department of a leading sports equipment manufacturer is concerned about losing its product performance edge in marketing golf clubs, as well as losing its market share. A recent survey indicated that many customers over the last 2 years have been unhappy with the performance of their golf clubs. In addition, a significant number of formerly loyal customers have changed brands. Investigations have determined that most of the dissatisfaction is centered on the strength of the club shaft itself. As a response, the company's strategy is to monitor the production process so as to identify new opportunities for quality improvement.

The basic production process involves pultrusion where fabric is pulled through a thermosetting polymer bath, and then through a long heated steel die. The shaft is cured as it travels through the die. The materials used are a proprietary thermosetting polymer and a proprietary reinforcing fiber. Experience has shown that the major quality considerations for the process include internal voids, broken strands, gaps between successive layers, and microcracks caused by improper curing.

Groups of 10 consecutive shafts were examined every 30 minutes. Each incidence of each type of defect was noted using nondestructive evaluation. The number of shafts with at least one defect present was noted, i.e., the number of defectives in the group. The data for each 8-hour shift (16 groups of 10 shafts each) were collected, $n = 160$, to provide an overall picture of the process performance. A proportion defective control chart, or P chart, was used to monitor the process. The data in Table 12.1 include the number of defective club shafts in each shift, sample of 160. The data were taken over a period of 36 consecutive shifts.

The characteristic of interest here is the process proportion defective. The proportion defective values p_j for each of the shift samples has been calculated, as shown in Table 12.1. For example, 9 defectives were observed among 160 inspected club shafts in shift 1; hence the proportion defective for the shift is $p_1 = 9/160 = 0.05625$.

The p_j values for the 36 successive shifts of sample size 160 are plotted in Figure 12.1. The plot in Figure 12.1a shows a solid line to indicate the average proportion defective for the entire data set. The grand average $\bar{p}$—the average of the 36 proportion defectives—is 0.04844. This is the sum of the defectives observed, $279 = 139 + 140$, divided by the number of items examined, $5760 = 36(160)$.

The individual values for the sample proportion defective p_j vary considerably. But it is too early to determine if the variation about the average proportion defective $\bar{p}$ is due solely to the forces of common cause or if a special cause is present. To help us resolve the problem, we develop control limits relative to the parameter of interest (proportion defective). At this time, we have not introduced P chart mechanics, but we have developed our control limits. Here, we have only a UCL posted on our chart; in this case we have no LCL. The methods for calculating control limits are explained later in this chapter. The limit shown in Figure 12.1b is the trial control limit. The P chart shows no evidence of process instability. Here, all the

Table 12.1   Proportion defective data for 36 shift samples from the golf club pultrusion process, $n = 160$ (Example 12.1).

Shift Number	Number Defective, np_j	Proportion Defective, p_j	Shift Number	Number Defective, np_j	Proportion Defective, p_j
1	9	0.05625	19	6	0.03750
2	6	0.03750	20	12	0.07500
3	8	0.05000	21	8	0.05000
4	14	0.08750	22	5	0.03125
5	7	0.04375	23	9	0.05625
6	5	0.03125	24	15	0.09375
7	7	0.04375	25	6	0.03750
8	9	0.05625	26	8	0.05000
9	5	0.03125	27	4	0.02500
10	9	0.05625	28	7	0.04375
11	1	0.00625	29	2	0.01250
12	7	0.04375	30	6	0.03750
13	9	0.05625	31	9	0.05625
14	14	0.08750	32	11	0.06875
15	7	0.04375	33	8	0.05000
16	8	0.05000	34	9	0.05625
17	4	0.02500	35	7	0.04375
18	10	0.06250	36	8	0.05000
Totals	139		Totals	140	

points plotted are inside the control limit. Our chart indicates that the variation present is due to chance causes alone.

Next, we extend the chart into the future; the dividends from control charts come in the application of the control limits to future production. If the charts indicate a process upset, we seek special causes and bring the process back to a stable condition.

Our control chart here indicates that the process is operating in statistical control—all plotted points are within the control limits. But the estimated process proportion defective value is unsatisfactory, $\bar{p} = 0.04844$. Benchmarking studies suggest that the best-of-the-best manufacturers are operating with less than 1% defective, using similar technology. It is clear that the current production process is not acceptable, since almost 5% of the golf club shafts produced are defective. The following alternatives are available for managerial consideration:

1. Redesign the existing production process so as to reduce the process proportion defective (a process improvement option).

2. Resort to a totally different production process (a process redefinition option).

We should note here that "pushing" operators for a lower proportion defective will do no good in this case, since the process is already operating in statistical control. Better results here call for process improvement or process redefinition, which involves operators but remains primarily the responsibility of management.

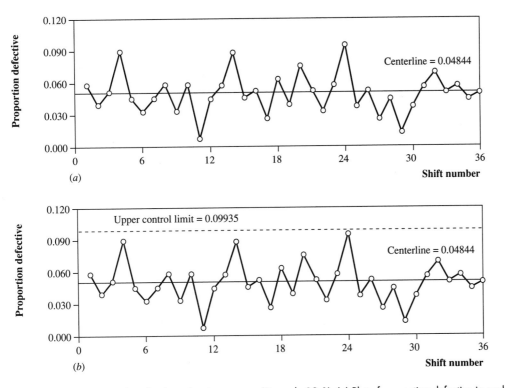

Figure 12.1 *P* chart for the pultrusion process (Example 12.1). (*a*) Plot of proportion defective in each shift sample. (*b*) *P* chart for proportion defective in each shift sample.

12.4 NOTATION FOR ATTRIBUTES SPC MODELS

Two basic models are used for attributes data: (1) **the binomial** and (2) **the Poisson.** A variety of symbols are needed to express our SPC chart mechanics.

GENERAL SYMBOLS

n: The sample size in process/product units

m: The number of samples in the preliminary data set used to set up a control chart

j: The sample index

b: The control limit spread in number of standard deviations of the plotted quantity

P CHART SYMBOLS

p: The population proportion of defective or nonconforming units

p_j: The proportion of defective or nonconforming units in sample j

$\bar{p}$: The mean proportion of defective or nonconforming units in a group of samples

$\text{LCL}_P, \text{UCL}_P$: The lower, upper control limits for the P chart

np: The population mean number of defective or nonconforming units per sample

np_j: Number of observed units in sample j that do not meet specifications (defective or nonconforming units)

C-, U-CHART SYMBOLS

c: The population mean number of defects per inspection unit

c_j: The number of defects observed in sample j; sample size is one inspection unit

$\bar{c}$: The mean number of defects per inspection unit, calculated from a group of samples

$\text{LCL}_C, \text{UCL}_C$: The lower, upper control limits for the C chart

u: The population mean number of defects per inspection unit

u_j: The number of defects observed in sample j; sample size may be more or less than one inspection unit

$\bar{u}_j$: The mean number of defects per inspection unit for a single sample j

$\bar{u}$: The mean number of defects per inspection unit for a group of samples

$\text{LCL}_U, \text{UCL}_U$: The lower, upper control limits for the U chart

BINOMIAL MODEL

The binomial distribution serves as the model for the P chart. The binomial distribution has two parameters, n and p, where p is the probability of failure on each trial and n is the number of independent trials (Bernoulli trials). If we define a random variable X as the number of non-conforming or defective units among n produced units with a probability p of failure for each unit, then the probability mass and cumulative mass functions for the binomial model are

$$f(x;n,p) = C_x^n p^x (1 - p)^{n-x}$$

$$= \frac{n!}{x!(n-x)!} p^x (1 - p)^{n-x} \qquad x = 0, 1, 2, \ldots, n$$

and

$$F(x;n,p) = \sum_{i=0}^{x} C_i^n p^i (1 - p)^{n-i} \qquad x = 0, 1, 2, \ldots, n$$

Mean: $\mu = np$

Variance: $\sigma^2 = np(1 - p)$

In order for the binomial model to be considered a reasonable model for a process/product, the validity of four assumptions must be assessed:

1. The experiment/process run consists of n repeated trials/production units.

2. Each trial/production unit can end in only one of two outcomes, i.e., conformance or non-conformance to specifications.

3. The probability of nonconformance, p, remains constant throughout the experiment/process run.

4. The repeated trials are conducted and respond in an independent manner; i.e., the outcome of one trial/production unit does not influence the outcome of another trial/production unit.

POISSON MODEL

The Poisson distribution (a limiting case of the binomial distribution, where $n \to \infty$, $p \to 0$, and $\mu = np$ remains constant) **serves as the basis for the C and U chart family.** The Poisson distribution has a single parameter, c. If we define the random variable X as the number of defects in a specific "area of opportunity" (1 inspection unit), then the probability density and the cumulative density functions are

$$f(x;c) = \frac{e^{-c}c^x}{x!} \qquad x = 0, 1, 2, \ldots$$

$$F(x;c) = \sum_{i=0}^{x} \frac{e^{-c}c^i}{i!} \qquad x = 0, 1, 2, \ldots$$

Mean: $\mu = c$

Variance: $\sigma^2 = c$

The Poisson model in SPC is typically used to model the number of defects occurring in a given area of opportunity—an inspection unit. We must be able to defend the Poisson as a reasonable model when we use it in SPC work. Several critical properties must be considered:

1. The number of defects occurring in one specified region of opportunity is independent of the number that occurs in any other disjoint region of opportunity.

2. The probability that a single defect will occur in a small region is proportional to the size of the region and does not depend on the number of outcomes occurring outside this region.

3. The probability that more than one defect will occur in such a small region is negligible.

12.5 P CONTROL CHART CONCEPTS AND MECHANICS

The control chart concepts and mechanics described here for attributes data are applicable when the process in question is considered stable. **If instabilities are present—as judged by points that fall beyond the 3-sigma control limits or in distinct trends or patterns—we should pursue engineering work to stabilize the process.** Setting up charts using data from unstable processes yields misleading chart dimensions and, hence, misleading process control indications.

We identify special causes, eliminate their effects, and develop revised control chart limits that represent a stable process. In this section, we describe SPC models useful for dealing with attributes data, expressed in proportion defective—the fraction of product or process results that fails to meet specifications.

P-CHART MECHANICS

When we select samples of size n from a process, each sample indicates how the process is behaving at one point in time. If the process is subject only to chance/common causes, the sample proportions defective p_j's should be randomly distributed within certain probabilistic limits. Occasionally, the process may experience some real change or upset due to a special cause. **The P chart indicates a process upset/special cause by exceeding the control chart limits or by showing a nonrandom pattern.**

As a matter of chance alone, variations in the number of defective items encountered are inevitable from sample to sample. The nonconforming proportion in the sample may vary considerably. As long as the nonconforming proportion in the process as a whole remains unchanged, the relative frequencies of nonconforming proportions in the samples may be expected to follow the binomial model. The population or process proportion defective p can be estimated by $\bar{p}$ as long as we have a sufficiently large number of samples, e.g., $m \geq 20$. Assuming the binomial model is valid, the sample proportions defective p_j's follow a binomial model with mean p and standard deviation

$$\sigma_p = \sqrt{\frac{p(1-p)}{n}}$$

The general form to construct the P-chart control limits, assuming all samples are of the same size n and the process proportion defective p is known, is

$$\left.\begin{array}{c} \text{UCL}_P \\ \text{LCL}_P \end{array}\right\} = p \pm b\sqrt{\frac{p(1-p)}{n}} \qquad \textbf{[12.1]}$$

$$\text{Centerline} = p$$

If the lower control limit calculated is less than 0, we say that no lower limit exists or set it equal to 0—a proportion less than 0 is not realistic. Negative numbers are sometimes encountered in the LCL_P calculation since we are using the traditional $\pm b$-sigma limits (from the centerline) with typically a nonsymmetric distribution—the binomial model is symmetric only when $p = 0.5$.

It is common practice in the United States to use 3-sigma control chart limits, where we set b equal to 3. We usually estimate p from $\bar{p}$. As Grant and Leavenworth [1] point out, it is very important to note that the X-bar chart is the only instance in the application of the Shewhart control chart models for which the distribution of the random variable can be shown to tend toward the normal distribution. In all other cases, such as the R chart, S chart, and P chart, it is appropriate to say, for a stable process, that the occurrence of a point falling outside 3-sigma limits at random would be very unlikely, rather than stating an explicit probability as was done for the X-bar chart in Chapter 10. This is not to say that it is impossible to associate probability values with points outside control limits for charts other than the X-bar chart. But such calculations are tedious and time-consuming and contribute little to the application of the basic decision rule; see Grant and Leavenworth [1].

The 3-sigma control limits for a P chart, where p is estimated from data obtained from a stable process with constant sample size n, are

$$\left.\begin{array}{c} \text{UCL}_P \\ \text{LCL}_P \end{array}\right\} = \bar{p} \pm 3\sqrt{\frac{\bar{p}(1-\bar{p})}{n}} \qquad\qquad \textbf{[12.2]}$$

$$\text{Centerline} = \bar{p}$$

Again, if the LCL calculated is less than 0, no lower limit exists. Here,

$$\bar{p} = \frac{\sum\limits_{j=1}^{m} np_j}{mn}$$

where j is used as a sample index.

Once the P chart is set up for the computed control limits and centerline, and stability is indicated, we calculate p_j after each sample is taken and plot the values. We connect the points together with a solid line and use the chart to monitor the process.

We calculate p_j as

$$p_j = \frac{np_j}{n} = \frac{\text{number of observed defective items in sample } j}{\text{sample size}}$$

GOLF CLUB SHAFT PRODUCTION PROCESS CASE REVISITED   We now return to our golf club | **Example 12.2**
shaft pultrusion process as described in Example 12.1. At one point in the process the completed golf club shaft was examined for the presence of molding defects. Recall that the data for 36 shift samples of size $n = 160$ are shown in Table 12.1 and the proportion defective values are plotted in Figure 12.1. Given the data in Table 12.1 and the P-chart mechanics, the centerline and the 3-sigma control limits for the P chart can be calculated as follows:

$$\bar{p} = \frac{\sum\limits_{j=1}^{m} np_j}{mn} = \frac{279}{5760} = 0.04844$$

For $\bar{p} = 0.04844$ and $n = 160$,

$$\hat{\sigma} = \sqrt{\frac{\bar{p}(1-\bar{p})}{n}} = \sqrt{\frac{(0.04844)(1-0.04844)}{160}} = 0.01697$$

From Equation (12.2), the 3-sigma control limits for the P chart are

$$\text{UCL}_P = \bar{p} + 3\sqrt{\frac{\bar{p}(1-\bar{p})}{n}} = 0.04844 + 3\,(0.01697) = 0.09935$$

$$\mathrm{LCL}_P = \bar{p} - 3\sqrt{\frac{\bar{p}(1 - \bar{p})}{n}} = 0.04844 - 3\,(0.01697) < 0$$

Here, LCL_p, as calculated is negative; hence, no lower limit exists.

$$\mathrm{Centerline} = \bar{p} = 0.04844$$

The control limit for the P chart is shown in Figure 12.1b. We have noted all along that to construct meaningful control chart limits, the process under consideration must be in statistical control. Figure 12.1 shows that there are no signals to indicate the process is unstable. As was mentioned earlier, while the process seems to be stable over the period studied, the proportion defective rate observed was considered too high and process improvement action was taken.

The process improvement team used a Pareto diagram tool and found that microcracks due to improper curing accounted for about 70% of the defects. The team decided to redesign the steel die such that curing takes place more uniformly and at a faster rate as the product moves through the die. In order to further enhance the process quality, it was decided to implement the following changes:

1. Insulate the die to maintain uniform temperatures.

2. Increase the percentage of catalyst in the polymer so as to increase the stiffness of the club.

3. Modify the puller so as to create a more uniform tension on the product.

After incorporating these changes, the data collection process was resumed.

Table 12.2 provides the proportion defective data for 24 shifts, following the process improvement changes. These data are plotted in Figure 12.2, as a continuation of the original chart (Figure 12.1), using the same control limits. The general appearance of the chart indicates that a significant change has occurred in the production process, and that the control limits must be revised in order to reflect the new process.

The revised control limits for the P chart are given below:

Revised UCL_p = 0.05841

Revised LCL_p < 0   hence, no lower limit exists

Revised centerline = 0.02292

Detailed calculations are not presented, our readers may want to verify the accuracy of these revised limits.

Figure 12.3 presents the revised control limits along with past data and past control limits. It is evident from the plot that the process seems to be in statistical control and does not present any nonrandom patterns.

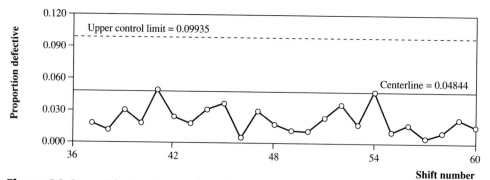

Figure 12.2   *P*-chart continuation after pultrusion process improvement (Example 12.2).

Table 12.2 Proportion defective data for 24 shifts after process improvement, $n = 160$ (Example 12.2)

Shift Number	Number Defective, np_j	Proportion Defective, p_j
37	3	0.01875
38	2	0.01250
39	5	0.03125
40	3	0.01875
41	8	0.05000
42	4	0.02500
43	3	0.01875
44	5	0.03125
45	6	0.03750
46	1	0.00625
47	5	0.03125
48	3	0.01875
49	2	0.01250
50	2	0.01250
51	4	0.02500
52	6	0.03750
53	3	0.01875
54	8	0.05000
55	2	0.01250
56	3	0.01875
57	1	0.00625
58	2	0.01250
59	4	0.02500
60	3	0.01875
Totals	88	

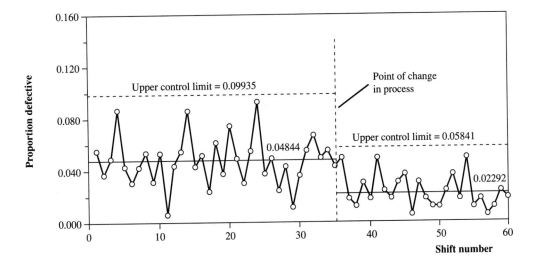

Figure 12.3 *P* chart for the entire pultrusion process (Example 12.2).

The proportion defective rate was reduced dramatically, to less than half of its original level. We have to remember that, while the control chart showed process stability even during the first 36 shifts of operation, it indicated an unacceptable operating level. The problem faced was not one of special causes but of system performance itself. We can also see that it was the improvement team (the users of the control chart) that identified opportunities for process improvement—not the control chart.

In some cases, we may want to examine all of our product created during a shift or workday. Or, perhaps we want to examine a certain percent of our daily production. **In some cases, the sample size may vary from one sample to the next.** In other words, we are dealing with an n_j, rather than a constant sample size of n. It was stated earlier that the sample proportion defective follows a binomial model with mean p and standard deviation $\sqrt{p(1-p)/n}$ as long as we have a constant sample size. This general form can be modified to incorporate variable sample sizes. The sample proportion defective still follows a binomial model with mean p, but now the standard deviation is $\sqrt{p(1-p)/n_j}$. Here, we see that the control limits will change with each different sample size. Hence, we will see "stair-stepped" control limits on the chart and must judge each sample's calculated p_j on the basis of its specific control limits—based on its specific n_j.

The 3-sigma control limits for a P chart, where p is estimated from data obtained from a stable process with variable sample size n_j, are

$$\left.\begin{array}{c} \text{UCL}_{Pj} \\ \text{LCL}_{Pj} \end{array}\right\} = \bar{p} \pm 3\sqrt{\frac{\bar{p}(1-\bar{p})}{n_j}} \qquad \textbf{[12.3]}$$

$$\text{Centerline} = \bar{p}$$

where

$$\bar{p} = \frac{\sum\limits_{j=1}^{m} np_j}{\sum\limits_{j=1}^{m} n_j}$$

Here j is used as the sample index. Note the centerline is constant, but the control limits change when the sample size changes. Again, if the lower control limit calculated is less than 0, a lower control limit does not exist.

Once the P chart is set up using the computed centerline, we calculate p_j and its control limits after each sample is taken, and then we plot the values. We connect the points with a solid line for p_j's and a dashed line for the control limits. Provided stability is indicated, we use the chart to monitor the process. We calculate p_j as

$$p_j = \frac{np_j}{n} = \frac{\text{number of observed defective items in sample } j}{\text{sample size } j}$$

Example 12.3 | **35-MM CAMERA BODY PRODUCTION PROCESS CASE**   The production process of a photographic equipment manufacturer is facing problems with its new die-cast body for a 35-mm camera, made of copper-aluminum alloy. For the last 3 months, the proportion defective was consistently high at 3.5%. As

Table 12.3 Number of defects and proportion defective for 20 shifts of camera body production (Example 12.3).

Shift Number	Sample Size, n_j	Number of defects			Number Defective, np_j	Proportion Defective, p_j
		Voids	Marks	Flashes		
1	140	3	2	2	5	0.0357
2	132	7	3	3	9	0.0682
3	98	2	0	1	1	0.0102
4	102	4	4	8	4	0.0392
5	40	0	0	3	3	0.0750
6	126	2	3	2	5	0.0397
7	132	9	5	2	6	0.0455
8	132	2	1	9	4	0.0303
9	156	5	4	3	9	0.0577
10	190	9	3	5	7	0.0368
11	210	12	6	6	8	0.0381
12	167	7	4	3	1	0.0060
13	134	7	5	4	6	0.0448
14	120	1	2	0	3	0.0250
15	110	0	0	5	3	0.0273
16	134	1	1	3	7	0.0522
17	167	8	3	7	6	0.0359
18	187	3	4	3	4	0.0214
19	123	0	0	1	1	0.0081
20	135	4	3	6	5	0.0370
Totals	2735	86	53	76	97	

a response, the process control strategy was to monitor the production process so as to identify new opportunities for quality improvement and to better control the production process.

The basic production process involves cold-chamber die casting. The molten metal is poured into an injection cylinder or shot chamber, and then the metal is forced into the die cavity at high pressure and is held under pressure until it solidifies in the die. Major quality considerations for the process include internal voids, ejector marks, and small amounts of flash (thin material squeezed out between the dies).

The process production rate changes from shift to shift. The production team decided to sample (inspect) 4% of the shift production. Each incidence of each type of defect was noted, and the values are shown for 20 consecutive shifts in Table 12.3.

Given the data in Table 12.3 and the results above, based on the binomial model, the centerline can be calculated as follows:

$$\bar{p} = \frac{\sum_{j=1}^{20} np_j}{\sum_{j=1}^{20} n_j} = \frac{97}{2735} = 0.0355$$

As was discussed earlier, the control limits will be a function of the sample size. The 3-sigma control limits for shift sample $j = 1$ can be calculated from Equation (12.3) as follows:

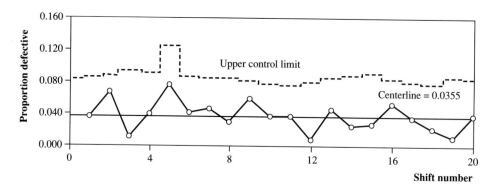

Figure 12.4   *P* chart for the die-casting process (Example 12.3).

$$\text{UCL}_{PI} = \bar{p} + 3\sqrt{\frac{\bar{p}(1 - \bar{p})}{n_1}} = 0.0355 + 3\sqrt{\frac{0.0355(1 - 0.0355)}{140}} = 0.0824$$

$$\text{LCL}_{PI} = \bar{p} - 3\sqrt{\frac{\bar{p}(1 - \bar{p})}{n_1}} = 0.0355 - 3\sqrt{\frac{0.0355(1 - 0.0355)}{140}} < 0$$

Hence, no lower limit exists for sample $j = 1$. In a similar fashion, we can calculate the control limits for the other 19 shift samples. These control limits for the *P* chart are shown in Figure 12.4.

The control chart shows that there are no signals present that would indicate process instability, suggesting that the calculated control limits are valid. Even though the process seems to be stable over the period studied, the proportion defective was considered too high, and process improvement actions were deemed necessary.

12.6 *C* AND *U* CONTROL CHART CONCEPTS AND MECHANICS

The *P* control charts deal with the notion of a defective process/product. We recognize, however, that **any one of a number of possible defects/nonconformities will qualify a product as defective.** For example, a part with 10 defects, any one of which makes it a defective, is on an equal footing with a part with only 1 defect, in terms of being defective. In this section, we describe SPC models for attributes data expressed in terms of number of defects encountered in our samples.

In general, **the opportunities for nonconformities or defects may be numerous, even though the chances of a nonconformity occurring in any one location in a process/product are relatively small.** The Poisson model is appropriate in this case and serves as a basis for the *C* chart. For example, in the case of an automobile, we may find that the opportunity for the occurrence of a defect is quite large in an entire automobile, while the probability of occurrence of a defect in any one location on the automobile is very small.

C-CHART MECHANICS

It is important to note that **the area of opportunity for defects to occur must be constant from sample to sample when we apply the C chart.** For a generic C chart where we assume that c is known,

$$\left. \begin{array}{c} \text{UCL}_C \\ \text{LCL}_C \end{array} \right\} = c \pm b \sqrt{c} \qquad\qquad \textbf{[12.4]}$$

$$\text{Centerline} = c$$

An inspection unit is defined by the chart designer. For example, an inspection unit could be 1 automobile, 2 automobiles, 1 computer, 3 computers, 100 m^2 of cloth, and so on; however, once defined, an inspection unit must remain fixed. In other words, the area of opportunity for defects to occur is constant from sample to sample. The U chart relaxes this assumption. As in the case of any of the other control charts, if the calculated LCL is less than 0, we assume that the lower limit does not exist—we have no LCL.

For a 3-sigma C chart where c is estimated from our data,

$$\left. \begin{array}{c} \text{UCL}_C \\ \text{LCL}_C \end{array} \right\} = \bar{c} \pm b \sqrt{\bar{c}} \qquad\qquad \textbf{[12.5]}$$

$$\text{Centerline} = \bar{c}$$

where

$$\bar{c} = \frac{\sum\limits_{j=1}^{m} c_j}{m}$$

Here, j is used as a sample index, and m should be at least equal to 20.

Once the C chart is set up using the computed control limits and centerline, we plot the c_j **values. We connect the points with a solid line and use the chart to monitor the process (once stability is established).** Here, c_j is the observed number of defects in the (one) inspection unit.

TRACTOR PRODUCTION PROCESS CASE   A tractor manufacturer desires to use a C chart to monitor the tractor production process. Two tractors are selected each day for assessment; 1 inspection unit is made up of 2 tractors. The data obtained from 30 inspection units (60 tractors) are shown in Table 12.4. Given the data in Table 12.4, the centerline and the 3-sigma control limits for the C chart are calculated below:

Example 12.4

$$\text{Centerline} = \bar{c} = \frac{\sum\limits_{i=1}^{30} c_j}{30} = \frac{135}{30} = 4.500 \text{ defects per inspection unit}$$

Table 12.4 Number of defects for 30 inspection units of tractor production (Example 12.4).

Date	Sample Number	Number of Defects, c_j
1 May	1	2
2 May	2	14
3 May	3	1
4 May	4	2
5 May	5	2
8 May	6	2
9 May	7	11
10 May	8	5
11 May	9	4
12 May	10	1
15 May	11	5
16 May	12	6
17 May	13	2
18 May	14	5
19 May	15	4
22 May	16	7
23 May	17	13
24 May	18	0
25 May	19	4
26 May	20	6
29 May	21	8
30 May	22	12
31 May	23	1
1 Jun	24	4
2 Jun	25	0
5 Jun	26	2
6 Jun	27	4
7 Jun	28	5
8 Jun	29	0
9 Jun	30	3
Totals		135

and from Equation (12.5),

$$\text{UCL}_C = \bar{c} + 3 \sqrt{\bar{c}} = 4.500 + 3 \sqrt{4.500} = 10.864$$

$$\text{LCL}_C = \bar{c} - 3 \sqrt{\bar{c}} = 4.500 - 3 \sqrt{4.500} < 0$$

Hence, no lower limit exists.

These control limits for the C chart, along with the 30 sample points, are shown in Figure 12.5. Figure 12.5 makes it clear that the tractor production process is not in statistical control. In fact, there are four sample points outside the trial control limits (samples 2, 7, 17, and 22). Hence, the calculated trial limits are not meaningful for future production.

Since the process seems to be unstable, the current variation in the data collected is likely due to both the variation created by chance or common causes and that created by special causes. To arrive at control

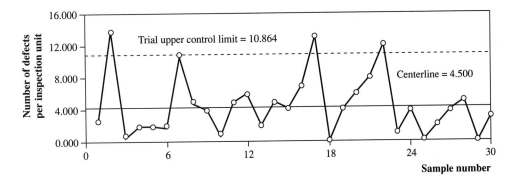

Figure 12.5 Trial C control chart for the tractor production process (Example 12.4).

chart limits that represent a stable process, we must find and eliminate the effect of the special causes. At this point in our example, we will assume these causes were identified and their effect eliminated. Hence, we will eliminate subgroups 2, 7, 17, and 22, and we will recompute our trial control limits. Here, we are now left with 26 samples ($m = 26$).

The revised control limits for the tractor production process C chart are calculated as follows. The revised centerline is

$$\bar{c} = \frac{\sum_{j=1}^{26} c_j}{m} = \frac{85}{26} = 3.269 \qquad \text{defects per inspection unit}$$

and, from Equation (12.7),

$$\text{Revised UCL}_C = \bar{c} + 3\sqrt{\bar{c}} = 3.269 + 3\sqrt{3.269} = 8.693$$

$$\text{Revised LCL}_C = \bar{c} - 3\sqrt{\bar{c}} = 3.269 - 3\sqrt{3.269} < 0$$

Hence, no lower limit exists.

These revised control limits are shown in Figure 12.6. All 26 sample points are now within the control limits. Therefore, it is safe to extend the control limits into the future to monitor tractor production.

U-CHART MECHANICS

When the opportunity space for the occurrence of defects per sample changes from sample to sample, we consider it a violation of the constant "area of opportunity" assumption upon which the C chart is based. In this case, we can apply a U chart. **For the U chart, we must define a standard measure for the area of opportunity and chart the average number of defects for this measure.** The symbol $\bar{u}_j$ is used to denote the sample statistic of the average nonconformities per inspection unit u_j/n_j, where u_j is the count of nonconformities found in n_j inspection units in sample j. In a U chart we plot the statistic $\bar{u}_j$—the average number of nonconformities per inspection unit. The control limits for a generic U control chart with known u and variable sample size n_j are

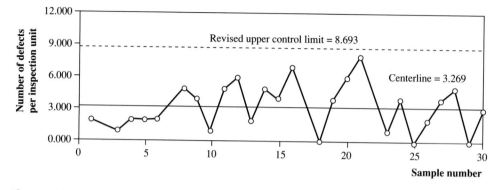

Figure 12.6   Revised C control chart for the tractor production process (Example 12.4).

$$\left.\begin{array}{c} \text{UCL}_{U_j} \\ \text{LCL}_{U_j} \end{array}\right\} = u \pm b \sqrt{\frac{u}{n_j}} \qquad \textbf{[12.6]}$$

$$\text{Centerline} = u$$

This chart has a constant centerline, but the control limits vary as a function of the sample size. Hence, the control chart limits appear stair-stepped.

For a 3-sigma U chart where u is estimated from the data, the control limits are

$$\left.\begin{array}{c} \text{UCL}_{U_j} \\ \text{LCL}_{U_j} \end{array}\right\} = \bar{u} \pm 3 \sqrt{\frac{\bar{u}}{n_j}} \qquad \textbf{[12.7]}$$

$$\text{Centerline} = \bar{u}$$

Again, if the calculated lower limit is less than 0, we assume that we have no LCL.

Here,

$$\bar{\mu} = \frac{\sum\limits_{j=1}^{m} u_j}{\sum\limits_{j=1}^{m} n_j}$$

Here, j is used as a sample index, and m should be at least equal to 20.

Once the U chart is set up for the computed centerline, we calculate $\bar{u}_j$ and its control limits after each sample is taken, and then we plot the control limits and their respective $\bar{u}_j$ values. Next, **we connect the points and use the chart to monitor the process (assuming stability has been established).** We calculate $\bar{u}_j$ as

$$\bar{u}_j = \frac{u_j}{n_j} = \frac{\text{number of observed defects in sample } j}{\text{size, in inspection units, of sample } j}$$

As an added note, Grant and Leavenworth [1] point out that the statistic u does not actually follow the Poisson distribution; however, the statistic nu does.

35-mm CAMERA BODY PRODUCTION PROCESS REVISITED   We will now return to our camera body die-casting process of Example 12.3. We have been examining a cold-chamber process for making camera bodies using copper-aluminum alloy. At one point in the process, the completed 35-mm camera bodies are examined for the presence of casting defects. Recall that the data for 20 shift samples with sizes equal to 4% of the shift production are given in Table 12.3, and the proportion defective values are plotted in Figure 12.4. It was concluded in Example 12.3 that there exists no evidence that the production process is unstable.

| **Example 12.5** |

Even though the process appears to be stable over the period studied, the proportion defective was considered too high, and the team concluded that process improvement actions are necessary to compete with the best-of-the-best in the industry. To gain a further understanding of the production process, and to identify new opportunities for quality improvement, the team decided to maintain a control chart to monitor the total number of defects. The U control chart is appropriate, since the sample sizes are changing from shift to shift. Table 12.3 presents information regarding the number of defects in each of the three categories for the 20 shift samples collected. It was decided to treat 40 camera bodies as 1 inspection unit for the U chart. Average number of defects per inspection unit are calculated for the same 20 shift samples and are shown in Table 12.5.

Table 12.5   Average number of defects per inspection unit of 40 camera bodies for 20 shifts of camera body production (Example 12.5).

Shift Number	Camera Bodies	Number of Inspection Units n_j	Number of Defects Voids	Marks	Flashes	Total Defects u_j	Ave. Defects per Unit $\bar{u}_j$	Control Limits UCL$_U$	LCL$_U$
1	140	140/40 = 3.500	3	2	2	7	7/3.5 = 2.0000	5.987	0.301
2	132	132/40 = 3.300	7	3	3	13	13/3.3 = 3.9394	6.072	0.216
3	98	2.450	2	0	1	3	1.2245	6.542	0.000
4	102	2.550	4	4	8	16	6.2745	6.475	0.000
5	40	1.000	0	0	3	3	3.000	8.463	0.000
6	126	3.150	2	3	2	7	2.2222	6.141	0.147
7	132	3.300	9	5	2	16	4.8485	6.072	0.216
8	132	3.300	2	1	9	12	3.6364	6.072	0.216
9	156	3.900	5	4	3	12	3.0769	5.838	0.450
10	190	4.750	9	3	5	17	3.5789	5.585	0.703
11	210	5.250	12	6	6	24	4.5714	5.466	0.822
12	167	4.175	7	4	3	14	3.3533	5.747	0.541
13	134	3.350	7	5	4	16	4.7761	6.050	0.238
14	120	3.000	1	2	0	3	1.0000	6.215	0.073
15	110	2.750	0	0	5	5	1.8182	6.352	0.000
16	134	3.350	1	1	3	5	1.4925	6.050	0.238
17	167	4.175	8	3	7	18	4.3114	5.747	0.541
18	187	4.675	3	4	3	10	2.1390	5.604	0.684
19	123	3.075	0	0	1	1	0.3252	6.177	0.111
20	135	3.375	4	3	6	13	3.8519	6.040	0.248
Totals	2735	68.375	86	53	76	215	61.4403		

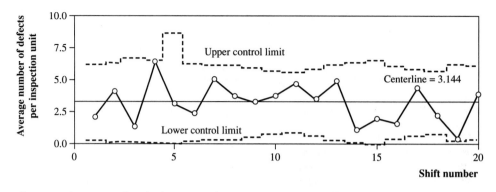

Figure 12.7   U chart for the camera die-casting process (Example 12.5).

From the mechanics presented earlier, the 3-sigma control limits for the U chart can be calculated as follows:

$$\text{Centerline} = \bar{u} = \frac{\sum_{j=1}^{20} u_j}{\sum_{j=1}^{20} n_j} = \frac{7 + 13 + 3 + \ldots + 13}{3.500 + 3.300 + \ldots + 3.375} = \frac{215}{68.375} = 3.144$$

From Equation 12.7, we have for shift sample $j = 1$,

$$\text{UCL}_{U_1} = \bar{u} + 3\sqrt{\frac{\bar{u}}{n_1}} = 3.144 + 3\sqrt{\frac{3.144}{3.500}} = 5.987$$

$$\text{LCL}_{U_1} = \bar{u} - 3\sqrt{\frac{\bar{u}}{n_1}} = 3.144 - 3\sqrt{\frac{3.144}{3.500}} = 0.301$$

where $n_1 = 140/40 = 3.500$

In a similar fashion we can calculate the control limits for each of the remaining 19 shift samples. The calculated control limits are shown in Table 12.5. These control limits, along with sample points from the 20 shifts, are plotted in Figure 12.7. All 20 samples are within the control limits, and the control chart shows no apparent special cause signals. Our chart suggests that the process is in statistical control. But the average number of defects per production unit places the manufacturer in a high-cost situation. The most viable alternative for improving the process quality is to make fundamental improvements in the production process.

A team of engineers enhanced the quality of the lubricant (parting agent) applied on the die surfaces and redesigned the die so as to include more taper (draft) to allow easy removal of the casting, and thus avoided ejector-pin marks. The process engineers agreed to increase the pressure at which the molten metal is forced into the die cavity in order to reduce voids. The tooling group accepted the responsibility to improve the die design so as to reduce wear and tear and thereby lower the chance of flashes. Through these actions the production process and final product quality were significantly improved.

12.7   PROCESS LOGS AND PARETO CHARTS

A targeted and stable process indicates the "mastery" of process operations. At the operations level, once process stability is reached, our process performance is as good as it can be. **Further improvement, such as process variance reduction, requires process changes rather than corrective action.** These improvements may involve changes in process inputs, in the process itself, or in both.

Achieving and sustaining process stability is a major accomplishment that deserves recognition. However, **process stability is seldom permanent.** Process upsets of all imaginable types—and some that are impossible to imagine—are eventually experienced. **Process upsets are usually unpleasant and costly experiences.** However, **they offer us a chance to learn more about our processes and products.** Documentation is essential in order to maximize our learning experience.

A comprehensive, but brief, process log that addresses the what, when, where, why, how, and who of critical operations is necessary. The purpose of the process log is to educate and increase awareness. **The process log supplies relevant facts regarding cause, effect, counteraction, and results. It should describe both successes and failures.** Successes are easily described; failures are more difficult to describe. Nevertheless, failures offer valuable lessons, even though we do not choose to repeat them. In other cases, the difference between success and failure is not great—sometimes minor modifications to actions that failed may lead to huge successes.

The Pareto chart is an excellent tool for classifying process upset causes. The rank ordering in the Pareto chart automatically isolates and focuses our attention on the most frequent cause (if ranked by frequency), the most wasteful cause (if ranked by off-specification product volume), the most expensive process upset category (if ranked by cost or downtime), and so on.

A wide variety of effective log styles can be developed. A representative process log for a torque-controlled machining center is shown in Figure 12.8. The point is to match the level of log detail to the level of process sophistication and loss potential present. For example, complicated processes that are not well understood justify more detail than simple processes that are well understood. Processes with marginal capability must be logged very carefully.

The Pareto analysis data sheet shown in Figure 12.9a basically serves as a stratified process log summary and provides an overview of process problems. The Pareto chart in Figure 12.9b is a graphical representation of special cause occurrences. **The Pareto chart is a useful tool for prioritizing process improvement efforts.** For example, we direct intensive process improvement efforts toward the problem classes resulting in the highest losses. In this case, the torque sensor, workpiece consistency, setup, and tooling appear to warrant further analysis.

Machining Center — Process Log

Date	Time	Counteraction Indicator	Special Cause	Counteraction Taken	Results Obtained	Lessons Learned
10/8	13:15	High torque requirement at spindle	Dull cutting tool — position 8	Replace cutting tool — position 8	Torque back within limits	None
10/10	10:45	High torque requirement/ tool chatter	Chipped cutting edge on carbide insert drill — position 12	Replace drill — position 12 (carbide insert drill)	Torque back within limits/quiet operation	None
10/10	10:49	High torque requirement/ tool chatter	Chipped cutting edge on carbide insert drill — position 12	Replace drill — position 12 (carbide insert drill); Increase spindle rpm from 2500 to 3000	Torque back within limits/quiet operation	Feed problem with the carbide drill, decrease feed/revolution, increasing rpm — result longer tool life with no loss in productivity

Figure 12.8   Process log layout (simplified).

Cause / Result	Tally	Total	Percent
Wrong tool	ⅣⅥ ⅣⅥ Ⅱ	12	12
Chucking	ⅣⅥ	5	5
Setup	ⅣⅥ ⅣⅥ Ⅲ	13	13
Feed / speed	ⅣⅥ Ⅲ Ⅲ	9	9
Torque sensor	ⅣⅥ ⅣⅥ ⅣⅥ ⅣⅥ ⅣⅥ ⅣⅥ Ⅰ	31	31
Workpiece defect	ⅣⅥ ⅣⅥ ⅣⅥ ⅢⅢ	19	19
Wrong workpiece	Ⅱ	2	2
Fixture problem	ⅣⅥ Ⅰ	6	6
Other	Ⅲ	3	3
	Total	100	

(*a*)

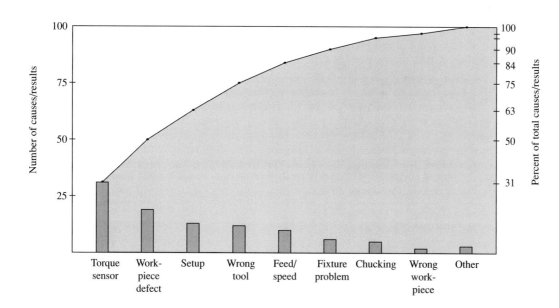

(*b*)

Figure 12.9 Process improvement Pareto analysis for a machining center subprocess. (a) Pareto analysis data sheet with special causes listed. (b) Pareto chart for special causes.

REVIEW AND DISCOVERY EXERCISES

REVIEW

12.1. Why do some process/product characteristics require the use of the nominal or ordinal scales of measure? Give one example of each.

12.2. Explain the difference between a defect and a defective. Provide an example.

12.3. Does a zero defect product assure customer satisfaction? A safe product? Explain.

12.4. Why is it necessary to provide precise guidance in defect classification when nominal or ordinal scales are used? Explain.

12.5. How much information do attributes-based SPC charts provide as to the physical nature of a special cause influence? Explain.

12.6. How can a binomial process model, i.e., one resulting in a P chart, be justified? Explain.

12.7. How can a Poisson process model, i.e., one resulting in a C chart, be justified? Explain.

12.8. Why is a point falling below a lower control limit (indicating an unusually low proportion defective) on a P chart considered an instability? Explain.

12.9. Why are negative numbers sometimes encountered when calculating the lower control limit on a P chart? Explain.

12.10. Why is it that the control limits change when the sample size n changes on a P chart? Explain.

12.11. Why is it that the number of defectives does not always equal the number of defects detected? See Table 12.3 for an example. Explain.

CASE EXERCISES

The following questions pertain to the case descriptions in Section 7. Most cases contain both defective and defect data; develop appropriate SPC models. Case selection list:

Case VII.11:   Fix-Up—Automobile Repair

Case VII.13:   Health Assist—Service

Case VII.24:   Silver Bird—Baggage

Case VII.25:   Snappy—Plastic Injection Molding

12.12. Select an appropriate case from above. Develop a P chart from the defective-related data provided in the case description. Use 3-sigma limits. Comment on your findings.

12.13. Select an appropriate case from above. Develop a C chart from the defect-related data provided in the case description. Use 3-sigma limits. Comment on your findings.

12.14. Select an appropriate case from above. Develop a U chart from the defect-related data provided in the case description. Use 3-sigma limits. Comment on your findings.

DISCOVERY

12.15. Could we convert a variables data set to an attributes data set? Could we convert an attributes data set to a variables data set? Explain.

12.16. Why can more than one process/product characteristic be monitored by an attributes metric? Explain.

12.17. Is it possible to develop a C or U chart for a true zero defect process? A six-sigma process? Explain.

REFERENCE

1. E. L. Grant and R. S. Leavenworth, *Statistical Quality Control*, 7th ed., New York: McGraw-Hill, 1996.

13

PROCESS MONITORING— NONTRADITIONAL SPC CONCEPTS AND MODELS

13.0 INQUIRY

1. What constitutes proper/improper application of SPC?
2. How can SPC performance be assessed and interpreted?
3. What constitutes good SPC performance? Bad SPC performance?
4. How can complex processes with inertia be monitored with traditional SPC?
5. How can multiple characteristics be monitored, together, with SPC?

13.1 INTRODUCTION

Up to this point in our discussions of variables-based SPC, we have assumed that our process data stream was essentially normally distributed, with independent measurements. For brevity, we refer to these data as *nid*. We pointed out in Chapter 10 that, according to the Central Limit Theorem, we could relax the assumption of normality regarding our location chart when we use subgrouped data. But, even with this relaxation we need to see independent data. Here, we refer to identically distributed independent data as *iid*. **Under the *nid* assumption, with limited departures from normality, SPC is a highly effective and efficient monitoring tool.** All examples studied in Chapters 10 and 11 have more or less conformed to the *nid* assumptions.

    Each physical process/product characteristic presents a unique data stream that may or may not be *nid* or even *iid* in its natural state. For example, physical process inertia generally produces an autocorrelated data stream, since physical principles such as heat transfer and mass transfer require time to manifest themselves in the data. Since departures from ideal SPC assumptions are so common, it is necessary to illustrate both our options and means to assess statistical performance of our SPC tools. The first half of this chapter is devoted to this illustration.

Another major issue in SPC applications is that of multiple data streams, e.g., multivariate data versus univariate data. **Univariate SPC consists of a single response model,** just as we have discussed in Chapters 10 and 11. **Multivariate SPC consists of multiple responses, collected together, and modeled together in one model.** We introduce what are termed *multivariate SPC models* in the second half of this chapter.

13.2 SPC MODEL PERFORMANCE EVALUATION

The sampling nature of SPC modeling produces type I and type II statistical errors. Here, a type I error is termed a *false alarm* and a type II error is termed a *failure to detect* a process shift or instability. In Chapter 10, in order to calibrate SPC performance, we described and illustrated both OC curves and average run lengths (ARLs). Our illustrations were based on highly simplified models, under *nid* assumptions. In this chapter, we relax these assumptions and use an SPC simulation approach in performance analysis.

The simulation approach taken in these analyses is essentially the only approach that is feasible, considering the complexity of the cases we develop. **A simulation approach allows us to introduce a shift at a known point in time or production order, and then allows us to assess SPC performance in terms of run lengths until our SPC model produces an out-of-control signal.** For example, if we do not introduce a shift, and our SPC model indicates an instability, e.g., one point outside the control limits, 2 out of 3 in the same outer zone, 4 out of 5 in the same middle zone, then we have encountered a false alarm on the first sample to meet the instability criteria. Here, obviously, since no shift was introduced, a long run length is better than a short run length. **We refer to the run length and average run length in the case of no shift** (i.e., when the null hypothesis H_0 is true—no process shift) **as RL_0 and ARL_0, respectively.**

On the other hand, once we introduce a shift in our simulation model, we simulate our SPC performance until we encounter a shift signal, e.g., one point outside the control limits, 2 out of 3 in the same outer zone, or 4 out of 5 beyond the same middle zone. The run length to shift detection here represents the time or production until we detected the shift. In this case, a run of one sample is the best we can do. Here, **we refer to the run length and average run length in the case of a shifted process** (i.e., when some alternative hypothesis H_1 is true—a process shift of some type and magnitude) **as RL_1 and ARL_1, respectively.**

In either case, we simulate our situation over and over, e.g., 100, 500, 1000 times, depending on the amount of computer time and speed of simulation. Thus, we develop an empirical run length data set, perhaps containing 300 RLs as in our Chapter 13 cases. This data set is then summarized with an average run length (ARL), a standard deviation of the run lengths collected (SRL), a standard error of the ARL (SE_{ARL}), and finally a histogram of the data set of run lengths. With these outputs, we can readily assess our simulated SPC performance by comparing it to theoretical benchmarks and/or assessing it subjectively, relative to our physical and economic situation, e.g., the consequences of a false alarm and/or a failure to detect. **SPC performance analyses help us to tailor and/or predict the SPC performance we are likely to see in actual practice.** These simulations are also useful to study SPC effects in situations that depart from the *nid* assumptions.

The simulation tool used in this demonstration is SPClab; see Karim [1]. This simulation tool is a specialized tool capable of simulating SPC performance for the X-bar, X, EWMA, and tabular CuSum location charts. In addition, it is capable of assessing the performance of the R, S, S^2, R_M, and EWMD charts. SPClab is capable of simulating both subgrouped as well as individuals data from a univariate data stream. The basic nature of each of these SPC models has been described in Chapters 10 and 11.

Three cases are presented in order to demonstrate appropriate and inappropriate applications of SPC technology. **The appropriateness of SPC application has two general dimensions: (1) the physical context of the process** and (2) **the statistical level of performance.** The physical context has previously been addressed in our discussion of physical analyses—cause and effect, process leverage, and process levers; refer to Chapter 4. **We stress applying SPC at critical process points so that process counteraction can be taken, when necessary, in a timely manner.**

We focus on the statistical performance as judged by our exposure to false alarms and failures to detect process changes under different process scenarios. The SPClab tool allows us to model and assess detailed SPC strategies and tactics, with few limiting assumptions. Such detailed analyses are useful to demonstrate SPC performance in realistic cases. We can quantitatively assess SPC performance in false alarms and failures to detect, and roughly tailor/choose strategies and tactics that offer the level of performance we demand.

13.3   PERFORMANCE ASSESSMENT WITH *NID/IID* DATA STREAMS

When we encounter *nid* or near *nid* data streams, we focus on choosing an SPC model that will limit the relative frequency of false alarms as well as failures to detect shifts, e.g., the presence of special cause. We can choose from several SPC models associated with both subgrouped and individuals data. These models are described in Chapters 10 and 11, respectively. Within each model, we choose and tailor our model parameters such as subgroup size, sigma limits, runs rules, and weighting factors.

In general, the literature has established rules of thumb regarding model performance; see Montgomery [2], Grant and Leavenworth [3], DeVor et al. [4], and Duncan [5]. However, with a tool such as SPClab, we can move to new levels of detail in performance assessment. The Case 1 sequence provides an illustration of the level of detail we can extract from simulated SPC performance analysis.

Example 13.1 | **METER STICK PRODUCTION PROCESS CASE REVISITED AND SIMULATED—APPROPRIATE SPC APPLICATION**   The meter stick case originally described and introduced in Chapter 4 and followed up in Chapter 10 is revisited here in a simulation format. Several subcases are provided and assessed. **All Example 13.1 subcases use *nid* data streams** with mean 1000 mm and standard deviation 0.2 mm. This data stream was previously introduced and characterized in Chapter 4; see Figure 4.5.

Subcase 1 involves a proper SPC application of the X-bar, R, EWMA, EWMD, and tabular CuSum charts, with no mean or standard deviation shifts. Subgrouped data (a subgroup size $n = 4$) were used for this subcase. Subcase 2 involves the same data stream characteristics with individuals data, $n = 1$, and the X, R_M, EWMA, EWMD, and tabular CuSum charts. Subcases 3 and 4 involve a small sustained location shift of 0.1 mm, or 0.5 standard deviations, and utilize the X-bar, R, X, R_M, EWMA, EWMD, and tabular CuSum charts. Subcases 5 and 6 involve a larger sustained location shift of 0.2 mm, or 1.0 standard deviation. They simulate the performance of the same SPC models as in the previous subcases. Performance results are summarized in Table 13.1. In Subcases 1 through 6, we produce X-bar and X charts both without and with runs rules applied. Here, we use only the 2 out of 3 and 4 out of 5 runs rules. Runs rules are discussed in Chapter 10.

Table 13.1 Simulated meter stick SPC performance summary (Example 13.1).

SPC Model (All ARLs/SRLs Based on 300 Observations)	No Shift	Location Shift (shift = 0.5σ = 0.1)	Location Shift (shift = 1σ = 0.2)	Dispersion Shift (shift = 1.12σ = 0.224)
Subgrouped Data ($n = 4$)	Subcase 1*	Subcase 3*	Subcase 5*	Subcase 7
X-bar chart (3-sigma limits) (without runs rules)	$ARL_0 = 360.5$ $SRL_0 = 410.5$ $SE_{ARL} = 23.7$	$ARL_1 = 44.4$ $SRL_1 = 47.0$ $SE_{ARL} = 2.7$	$ARL_1 = 6.5$ $SRL_1 = 6.5$ $SE_{ARL} = 0.4$	$ARL_0 = 20.3$ $SRL_0 = 18.8$ $SE_{ARL} = 1.1$
R chart (3-sigma limits)	$ARL_0 = 207.0$ $SRL_0 = 215.3$ $SE_{ARL} = 12.4$	$ARL_0 = 224.5$ $SRL_0 = 260.3$ $SE_{ARL} = 15.0$	$ARL_0 = 224.5$ $SRL_0 = 260.3$ $SE_{ARL} = 15.0$	$ARL_1 = 8.4$ $SRL_1 = 8.9$ $SE_{ARL} = 0.5$
X-bar chart (3-sigma limits) (with 2/3 and 4/5 runs rules)	$ARL_0 = 239.6$ $SRL_0 = 252.8$ $SE_{ARL} = 14.6$	$ARL_1 = 21.0$ $SRL_1 = 21.0$ $SE_{ARL} = 1.2$	$ARL_1 = 3.3$ $SRL_1 = 2.9$ $SE_{ARL} = 0.2$	
R chart (3-sigma limits)	$ARL_0 = 224.5$ $SRL_0 = 260.3$ $SE_{ARL} = 15.0$	$ARL_0 = 224.5$ $SRL_0 = 260.3$ $SE_{ARL} = 15.0$	$ARL_0 = 224.5$ $SRL_0 = 260.3$ $SE_{ARL} = 15.0$	
EWMA chart ($r = 0.4$)	$ARL_0 = 460.0$ $SRL_0 = 491.7$ $SE_{ARL} = 28.4$	$ARL_1 = 15.0$ $SRL_1 = 13.9$ $SE_{ARL} = 0.8$	$ARL_1 = 3.4$ $SRL_1 = 1.8$ $SE_{ARL} = 0.1$	
EWMD chart	$ARL_0 = 193.9$ $SRL_0 = 214.3$ $SE_{ARL} = 12.4$	$ARL_0 = 186.1$ $SRL_0 = 208.6$ $SE_{ARL} = 12.0$	$ARL_0 = 150.6$ $SRL_0 = 203.0$ $SE_{ARL} = 11.7$	
CuSum chart ($h = 4.77, k = 0.5$)	$ARL_0 = 350.2$ $SRL_0 = 324.4$ $SE_{ARL} = 18.7$	$ARL_1 = 9.4$ $SRL_1 = 5.5$ $SE_{ARL} = 0.3$	$ARL_1 = 3.6$ $SRL_1 = 1.3$ $SE_{ARL} = 0.1$	
Individuals Data	Subcase 2*	Subcase 4*	Subcase 6*	Subcase 7
X chart (without runs rules)	$ARL_0 = 356.3$ $SRL_0 = 344.9$ $SE_{ARL} = 19.9$	$ARL_1 = 161.3$ $SRL_1 = 168.1$ $SE_{ARL} = 9.7$	$ARL_1 = 44.8$ $SRL_1 = 43.8$ $SE_{ARL} = 2.5$	$ARL_0 = 23.0$ $SRL_0 = 27.0$ $SE_{ARL} = 1.6$
R_M chart ($n_a = 2$)	$ARL_0 = 123.8$ $SRL_0 = 127.8$ $SE_{ARL} = 7.4$	$ARL_0 = 122.8$ $SRL_0 = 132.7$ $SE_{ARL} = 7.7$	$ARL_0 = 119.9$ $SRL_0 = 133.5$ $SE_{ARL} = 7.7$	$ARL_1 = 13.3$ $SRL_1 = 12.8$ $SE_{ARL} = 0.7$
X chart (with 2/3, 4/5 runs rules)	$ARL_0 = 150.4$ $SRL_0 = 161.0$ $SE_{ARL} = 9.3$	$ARL_1 = 55.1$ $SRL_1 = 51.1$ $SE_{ARL} = 3.0$	$ARL_1 = 15.2$ $SRL_1 = 13.7$ $SE_{ARL} = 0.8$	
R_M chart ($n_a = 2$)	$ARL_0 = 122.6$ $SRL_0 = 132.8$ $SE_{ARL} = 7.7$	$ARL_0 = 122.8$ $SRL_0 = 132.7$ $SE_{ARL} = 7.7$	$ARL_0 = 119.9$ $SRL_0 = 133.5$ $SE_{ARL} = 7.7$	

(continued)

Table 13.1 *(concluded)*   Simulated meter stick SPC performance summary (Example 13.1).

SPC Model (All ARLs/SRLs Based on 300 Observations)	No Shift	Location Shift (shift = 0.5σ = 0.1)	Location Shift (shift = 1σ = 0.2)	Dispersion Shift (shift = 1.12σ = 0.224)
Individuals data	Subcase 2*	Subcase 4*	Subcase 6*	Subcase 7
EWMA chart ($r = 0.4$)	$ARL_0 = 432.0$ $SRL_0 = 427.7$ $SE_{ARL} = 24.7$	$ARL_1 = 63.3$ $SRL_1 = 60.5$ $SE_{ARL} = 3.5$	$ARL_1 = 13.6$ $SRL_1 = 10.3$ $SE_{ARL} = 0.6$	
EWMD chart	$ARL_0 = 183.9$ $SRL_0 = 201.0$ $SE_{ARL} = 11.6$	$ARL_0 = 184.5$ $SRL_0 = 190.7$ $SE_{ARL} = 11.0$	$ARL_0 = 177.3$ $SRL_0 = 190.6$ $SE_{ARL} = 11.0$	
CuSum chart ($h = 4.77, k = 0.5$)	$ARL_0 = 377.0$ $SRL_0 = 421.8$ $SE_{ARL} = 24.4$	$ARL_1 = 34.3$ $SRL_1 = 27.6$ $SE_{ARL} = 1.6$	$ARL_1 = 9.4$ $SRL_1 = 5.6$ $SE_{ARL} = 0.3$	

ARL: Average of all 300 run lengths.
SRL: Standard deviation of the sample of 300 run lengths.
SE_{ARL}: Standard error of the ARL (mean).
*Dispersion results are identical or nearly identical due to controlled random number streams, from subcase to subcase and within subcases.

Selected graphics are presented relative to Subcases 1 through 6. Figure 13.1 presents the *X*-bar and *X* chart pictures under the simulated small shift, along with their respective run length distributions, both without and with runs rules applied. Figure 13.2 depicts the counterpart EWMA and tabular CuSum charts and run length histograms.

Subcase 7, see Table 13.1, uses the *X*-bar and *R* and the *X* and R_M chart models to illustrate a sustained shift in the dispersion of 0.224 mm, i.e., a 112% increase in the standard deviation. Figure 13.3 depicts the corresponding *X*-bar and *X* charts, as well as the run length histograms for both. Here, again, a subgroup size $n = 4$ was used.

Example 13.1 Summary

From our previous theoretical development of OC curves and ARL curves in the Chapter 10 Appendix, we established that a 3-sigma *X*-bar or *X* chart should possess an ARL_0 of 370 for a true H_0 of no process shift, assuming an *nid* case and no runs rules. We can compare our Subcases 1 and 2, Table 13.1 and Figure 13.1, with this theoretical benchmark. Considering our location charts, with the exception of those using runs rules, we see an ARL_0 range from about 350 to 460, with 360.5 and 356.3 for the *X*-bar and *X* charts, respectively. Likewise, our sample size of 300 simulated RLs provides us with standard errors of the ARLs ranging from about 19 to 28. Hence, we are seeing a good bit of variation, but we see our simulated results in the ballpark, with respect to the theoretical value of 370. When we add the 2/3 and 4/5 runs rules, we see a large drop-off in the ARL_0's in subcases 1 and 2. We see 240 for the *X*-bar and 150 for the *X* chart models. Here, we expect to see a higher false alarm, type I, error probability, and our simulation supports our expectation.

Our dispersion charts in Subcases 1 and 2, although modeled after 3-sigma charts, tend to produce much lower ARLs. Here, we see a range from 123 to 225, with standard errors ranging from 7 to 15. Hence, we see more frequent false alarms than we might ordinarily expect, when compared to our theoretical 3-sigma ARL_0 of 370, for normal-based data.

Subcases 3 and 4 work with a small location shift of 0.5 standard deviations in the simulated data. Here, we are focusing primarily on the location charts, as the shift was induced only in location. An $ARL_1 = 1$ in our location charts is ideal. We see large drop-offs in location chart ARL_1's. We see a low of 9.4 in the

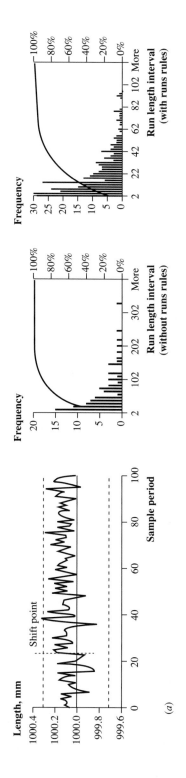

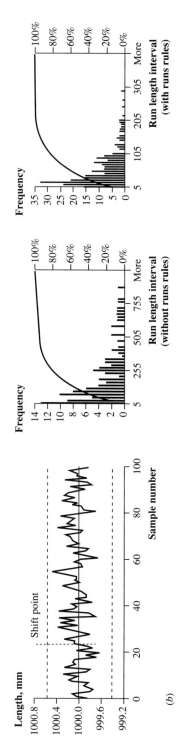

Figure 13.1　Selected Xbar and X chart model performance using simulated *nid* data streams, shifted location (Example 13.1). (a) Xbar chart model performance without and with 2 out of 3 and 4 out of 5 runs rules applied small location shift, *n* = 4. (b) X chart model performance without and with 2 out of 3 and 4 out of 5 runs rules applied, small location shift, *n* = 1.

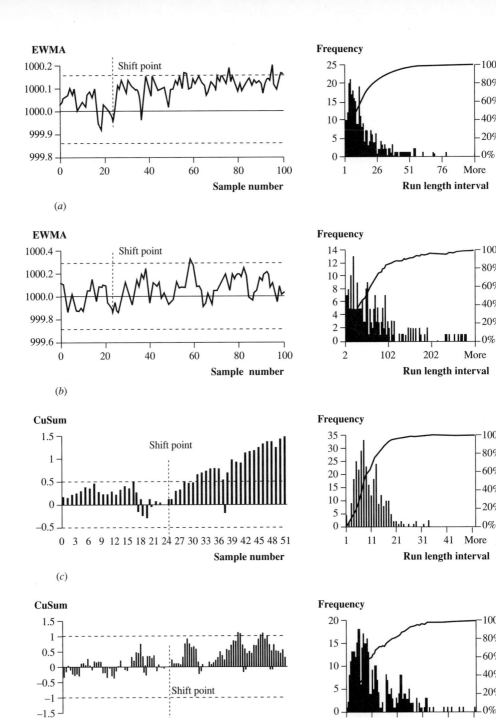

Figure 13.2 Selected EWMA and CuSum model performances using simulated *nid* data streams, shifted location (Example 13.1). (*a*) EWMA performance, small location shift, subgrouped data, $n = 4$. (*b*) EWMA performance, small location shift, individuals data, $n = 1$. (*c*) CuSum performance, small location shift, subgrouped data, $n = 4$. (*d*) CuSum performance, small location shift, individuals data, $n = 1$.

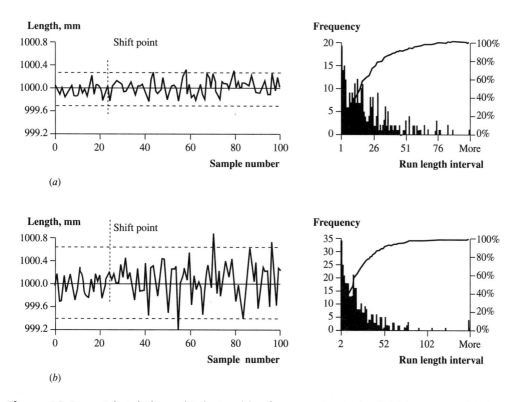

Figure 13.3   Selected *X*-bar and *X* chart model performance using simulated *nid* data streams, shifted dispersion (Example 13.1). (*a*) *X*-bar chart model performance, dispersion shift, subgrouped data, *n* = 4. (*b*) *X* chart model performance, dispersion shift, individuals data, *n* = 1.

subgrouped CuSum model and a high of 161.3 in the *X* chart model without runs rules applied. For a small sustained shift, we can see our CuSum, EWMA, and *X*-bar and *X* models with runs rules performing reasonably well.

Our plots in Figure 13.1 reinforce our quantitative findings. Here, the *X*-bar and *X* chart plots represent only a small fraction of the simulated data. These charts help us to visualize the general nature of our data. Each chart has been produced with a shift point, so that we can visualize how such a shift typically looks. The run length histograms provide interesting perspectives as to the distributions that our RLs take. Here, we see a good bit of dispersion, but a tendency to cluster at the lower run lengths, with several observations falling off to the right at the higher run lengths.

Selected EWMA and CuSum counterpart models under the condition of a small sustained shift are graphically illustrated in Figure 13.2. Here, we show both subgrouped and individuals performance. Our representative EWMA and CuSum charts contain marked shift points, where we can readily pick up the shifts. Our run length histograms provide us with information as to the nature of the run length distributions we have experienced.

As expected, the larger sustained shift in location was more readily picked up by our charts. In these subcases, i.e., Subcases 5 and 6, we see ARL_1's in the subgrouped models in the single digits. In the individuals data models, we see correspondingly lower ARL_1's, as compared to Subcases 3 and 4, but higher than in our corresponding subgrouped cases. All models, except our *X*-bar and *X* models, without runs rules, performed reasonably well. In Subcases 1 through 6 we see about the same ARL levels in the dispersion charts.

This is entirely expected, as we induced our shift only in the location parameter and controlled our random number streams to facilitate comparisons in the location models.

Subcase 7 provides a different look at our X-bar and X model performance. Here, we held our location parameter constant but shifted our dispersion parameter upward by 112%, a large dispersion shift. This shift manifested itself in both the dispersion and the location charts. When compared to the corresponding models in Subcases 1 and 2, we see much lower ARLs, a trend that we expect to see. We see from Figure 13.3 large dispersion shifts in our location charts, as we would expect. For example, the shift pushes our data toward the control limits and hence, produces the lower ARLs in the location charts, as well as in the dispersion charts (not shown in Figure 13.3).

In general, **Example 13.1 provides us with a number of empirical findings that fall in line with our theoretical expectations.** It clearly demonstrates the value of SPC simulation as a tool to help us proactively assess the performance of our SPC models. We should keep in mind that **every subcase studied, as summarized in Table 13.1, is a proper SPC application from the perspective that we used an *nid* random number stream to feed our simulations.** In other words, process inertia was not a factor. In Examples 13.2 and 13.3, we use high-inertia data streams to illustrate several pitfalls that we face in SPC practice.

13.4   PERFORMANCE ASSESSMENT WITH NON-*NID* DATA STREAMS

When we are faced with a non-*nid* data set, characterized by a high level of autocorrelation, our objective is to transform this non-*nid* data set into a "near-*nid*" data set. We then use this transformed data set to track process performance, looking for signs of special causes or shifts, in a consistent time/production sequence. Here, **we have two basic options: (1) model with respect to a time series** or (2) **model with respect to physical variables** that have some influence on the process data stream.

The necessity for finding and using a suitable transformation of our original data set is a result of our inability to distinguish between common and special cause influences. In the case of *nid* data, special causes are manifested by shifts and outliers. These shifts and outliers are readily detectable with standard SPC models. In cases of high autocorrelation, we cannot readily separate common and special cause variations. Here, natural variability manifests itself in ways that are not in general detectable with traditional SPC models. Hence, what is common cause in the process itself may be portrayed by the SPC model as special cause—our SPC model is generally not capable of accommodating natural variations, other than random noise.

In many cases, process inertia has a mean or pattern structure such that when SPC is applied, the natural or common cause variation appears as special cause variation. Or, in other cases, special cause variation appears as common cause variation. With process inertia, in general, our SPC model's ability to measure common cause variation is sensitive to our sampling scheme, resulting in very different looking and performing SPC charts, depending on the duration of our sampling interval. **We provide two simulated cases to demonstrate these oddities.** Such oddities, when encountered, tend to frustrate people using standard SPC models. Such a frustration was described in the Natural Gas Production Case (Example 4.3) described in Chapter 4.

SHREDDER CASE—SIMULATED APPLICATIONS AND MISAPPLICATIONS OF SPC   This | **Example 13.2**
case involves a shredder, for shredding recycled paper, where we are monitoring the energy used to drive the
shredder as a substitute measure of the blade cutting edge condition. Here, we start with sharp blades and
continue to shred until our blades become dull or damaged. The blade edge wear is measured indirectly by
the energy requirements to drive the shredder unit. As the blades' edges dull or chip, the energy requirements
increase. At some point, we regrind or replace the blades and thus begin a new wear cycle. Figure 13.4a
provides an energy–blade condition profile versus operating time for this process characteristic. We see a

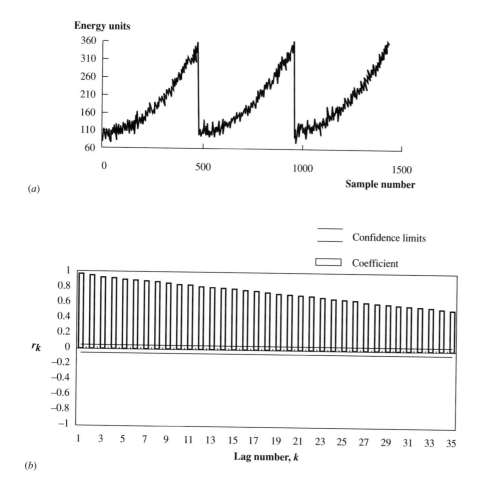

(a)

(b)

Figure 13.4   Data characterization for shredder energy demand, simulated non-*nid* data
(Example 13.2). (*a*) Runs chart for three cycles of simulated shredder data.
(*b*) Autocorrelation correlogram for simulated shredder data.

sawtoothed data pattern, with the dip in the sawtooth as a result of blade maintenance, i.e., grinding or replacement. Here, we assume that our physical process feed rate (mass flow) remains constant across time.

First, we will assess the nature of this data stream, as discussed in Chapter 4. Figure 13.4*b* provides a correlogram. Both physical reasoning and the correlogram point to the fact that we see significant process inertia—we have a highly non-*nid* data stream characterized by a high degree of autocorrelation. A correlogram such as that shown in Figure 13.4*b* is an indication that a traditional SPC model, applied directly to the data stream, is not appropriate in the sense that we encounter excessive false alarms or observe excessive failures to detect special causes, depending on our sampling scheme.

Subcase 1 illustrates a standard misapplication of our SPC tools to a sawtooth or wear-based non-*nid* data stream. Here, we sample frequently, with a sampling interval of one time unit, with no special cause, i.e., no shift. And then, we sample infrequently, every 240 time units, with no special cause present. From a physical standpoint, the every-time-unit sampling scheme is acceptable, while the every 240-time-unit sampling scheme is clearly too infrequent to be of help in our physical process.

Figure 13.5*a* depicts results from this subcase regarding the X chart model (without runs rules). On the left, we see an X chart that is constantly signaling an out-of-control condition, when there is no such influence present in our process. On the right, with the infrequent sampling scheme, we see an X chart that is always in control, as it should be in this case, but the limits are moved out to the point that we would be hard-pressed to detect process shifts. In Figure 13.5*b*, we see counterpart EWMA chart models. The results are about the same as in the X chart subcase. Finally, in Figure 13.5*c*, we see counterpart CuSum chart models. The every-unit sampling scheme on the left produces similar results to its counterparts above. The infrequent sampling scheme, sampled every 240 units, produces an unusual CuSum chart that eventually drifts out of control. From these two particular extreme sampling schemes, and our shredder data stream, we see a frustrating misapplication of SPC technology. The failure is not due to the SPC technology; rather the failure is due to a misapplication of the technology.

Subcase 2 takes the strategy of fitting an energy-wear model to the data stream to help explain the wear trend. Then, we utilize SPC tools with the fitted model residuals. Here, the residuals are more or less nid. Hence, it is reasonable to use SPC on the residuals, since any special causes will reflect themselves in the residuals as shifts or patterns.

We begin with a regression modeling strategy (refer to Chapter 6), whereby we want to "explain" wear and subsequent increasing energy demands. By regressing energy usage with time in service, within each cycle, we obtain the fitted regression model below.

$$\hat{\mu} = \hat{Y} = b_0 + b_1 X + b_2 X^2 = 99.85 + 0.0224(X) + 0.00995\ (X^2)$$

Here, X represents time in service, our substitute measure for blade edge wear. Our residuals, e_j, and terms from our regression analysis are extracted. At this point we assess them with respect to our *nid* assumption. This step is omitted for brevity. Then, we apply corresponding 3-sigma X (without runs rules and with 2/3 and 4/5 runs rules), EWMA, and tabular CuSum SPC models. Model performances were assessed for both the case of no shift and the case of a sustained small shift, an increase of 25 energy units. Results are summarized in Table 13.2.

Examination of the statistics in Table 13.2 indicates that we have approached our expected 3-sigma chart performance when no shift was present. Here, we would expect to see an $\text{ARL}_0 = 370$, i.e., a false alarm about every 370 samples, given no shifts in the process. In this simulated case, a shift of 25 energy units was picked up very fast. Here, we see ARL_1's between 1 and 4. Hence, our SPC appears to work effectively when we use our residuals. From a practical standpoint, a shift of 25 energy units is reasonably small, considering our data within a cycle (see Figure 13.4*a*) tend to range from about 100 up to 360 energy units.

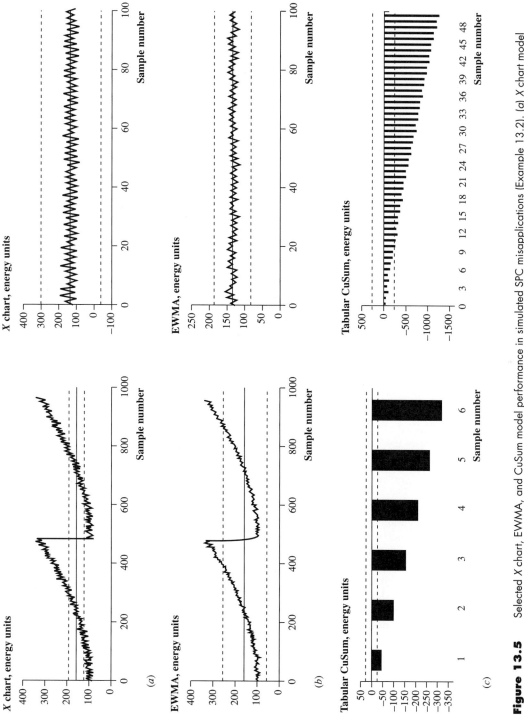

X chart, energy units

EWMA, energy units

Tabular CuSum, energy units

(a)

X chart, energy units

EWMA, energy units

Tabular CuSum, energy units

(b)

(c)

Sample number

Figure 13.5   Selected X chart, EWMA, and CuSum model performance in simulated SPC misapplications (Example 13.2). (a) X chart model performance from sampling intervals of 1 time unit and 240 time units, no shift, $n = 1$. (b) EWMA model performance from sampling intervals of 1 time unit and 240 time units, no shift, $n = 1$. (c) Tabular CuSum model performance from sampling intervals of 1 time unit and 240 time units, no shift, $n = 1$.

Table 13.2   Simulated shredder residuals SPC performance summary (Example 13.2).

	Individuals Data	
SPC Model (All ARLs/SRLs Based on 300 Observations)	**No Shift***	**Location Shift* (25 units)**
X chart (without runs rules)	$ARL_0 = 368.2$ $SRL_0 = 380.5$ $SE_{ARL} = 22.0$	$ARL_1 = 3.4$ $SRL_1 = 2.8$ $SE_{ARL} = 0.2$
R_M chart ($n_a = 2$)	$ARL_0 = 115.3$ $SRL_0 = 118.9$ $SE_{ARL} = 6.9$	$ARL_0 = 97.6$ $SRL_0 = 120.0$ $SE_{ARL} = 6.9$
X chart (with 2/3, 4/5 runs rules)	$ARL_0 = 159.4$ $SRL_0 = 161.1$ $SE_{ARL} = 9.3$	$ARL_1 = 2.3$ $SRL_1 = 1.3$ $SE_{ARL} = 0.1$
R_M chart ($n_a = 2$)	$ARL_0 = 115.3$ $SRL_0 = 118.9$ $SE_{ARL} = 6.9$	$ARL_0 = 97.6$ $SRL_0 = 120.0$ $SE_{ARL} = 6.9$
EWMA chart ($r = 0.4$)	$ARL_0 = 395.2$ $SRL_0 = 393.4$ $SE_{ARL} = 22.7$	$ARL_1 = 2.5$ $SRL_1 = 1.1$ $SE_{ARL} = 0.1$
EWMD chart	$ARL_0 = 183.9$ $SRL_0 = 197.4$ $SE_{ARL} = 11.4$	$ARL_0 = 124.1$ $SRL_0 = 176.2$ $SE_{ARL} = 10.2$
CuSum chart ($h = 1.99, k = 1.25$)	$ARL_0 = 363.0$ $SRL_0 = 352.1$ $SE_{ARL} = 20.3$	$ARL_1 = 2.2$ $SRL_1 = 1.1$ $SE_{ARL} = 0.1$

ARL: Average of all 300 run lengths.
SRL: Standard deviation of the sample of 300 run lengths.
SE_{ARL}: Standard error of the ARL (mean).
*Dispersion results are identical or nearly identical due to controlled random number streams within subcases.

Selected shredder performance SPC charts developed from the residuals and corresponding run length histograms are provided in Figure 13.6. The shift points are marked on each model's SPC chart in order to provide a graphical assessment of the magnitude of the shift in terms of the residual values. Figure 13.6a depicts the X chart performance without and with the two runs rules applied. Figure 13.6b and c depicts the EWMA and tabular CuSum performances, respectively.

In summary, our shredder case has illustrated an interesting use of model fitting and residual usage. Of course, if we cannot find a suitable model that will allow us to extract *nid* or near *nid* residuals, then our residual SPC models will not be effective. We illustrated a standard regression model, but other specialized models such as time series have been shown to be effective; see Montgomery [2] for a summarized discussion of time series applications. **In general, we desire to use the simplest model possible that will adequately yield near-*nid* residuals.**

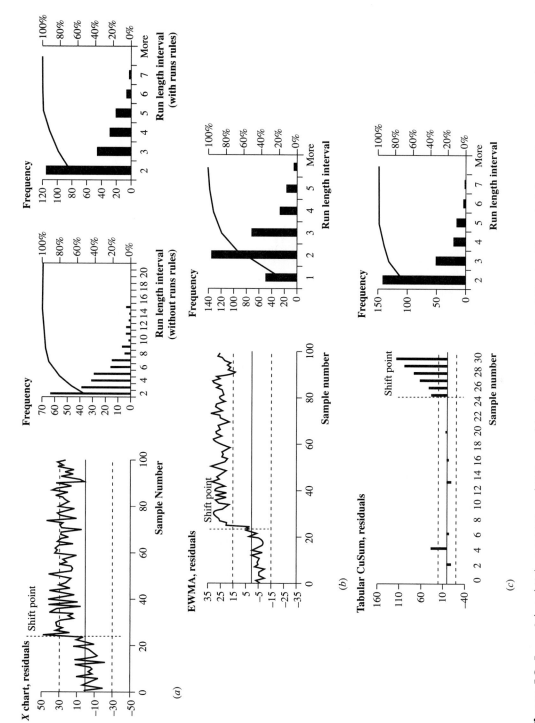

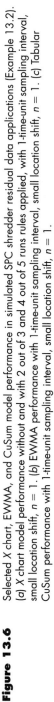

Figure 13.6 Selected X chart, EWMA, and CuSum model performance in simulated SPC shredder residual data applications (Example 13.2). (a) X chart model performance without and with 2 out of 3 and 4 out of 5 runs rules applied, with 1-time-unit sampling interval, small location shift, $n = 1$. (b) EWMA performance with 1-time-unit sampling interval, small location shift, $n = 1$. (c) Tabular CuSum performance with 1-time-unit sampling interval, small location shift, $n = 1$.

Example 13.3   **NATURAL GAS CHILLER CASE—SIMULATED APPLICATIONS AND MISAPPLICATIONS OF SPC**   The chiller case is an extension of the Natural Gas Production Case involving data characterization described in Chapter 4 (Example 4.3). Here, our chiller temperature data stream experiences a sinusoidlike pattern due primarily to day/night temperature patterns. For example, the cool of the night allows the chiller to operate at a lower temperature, while the heat of the day serves to increase the chiller temperature. Figure 13.7 depicts the nature of the simulated chiller data, along with a correlogram that provides a picture of the autocorrelation in the chiller data stream. Here, we see a high degree of process inertia displayed in the

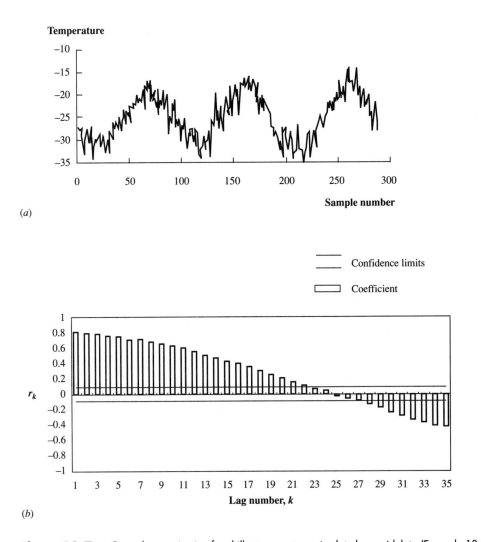

(a)

(b)

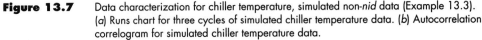

Figure 13.7   Data characterization for chiller temperature, simulated non-*nid* data (Example 13.3). (*a*) Runs chart for three cycles of simulated chiller temperature data. (*b*) Autocorrelation correlogram for simulated chiller temperature data.

autocorrelation structure. Subcases are provided to illustrate how we can address SPC modeling with this process data stream.

Subcase 1 provides inappropriate SPC applications. These applications apply traditional SPC tools directly to the chiller data stream, not unlike what might be attempted in practice. In Subcase 1 a small sampling interval of 1 time unit is used, as well as a large sampling interval of 24 time units. Here, we can associate each time unit to 1 hour of operations. The poor SPC performance in Subcase 1 is obvious when we examine the general nature of the SPC charts obtained. These charts are depicted in Figure 13.8.

In every case, the X, EWMA, and CuSum, we see erratic behavior in the presence of process stability, i.e., no process shifts. The small sampling intervals produce high levels of false alarms, while the large sampling intervals produce essentially no false alarms, but will be highly insensitive to sustained shifts. The small sampling intervals are physically useful, but the long intervals, one per day, lack timeliness value, considering the physical process. It is not uncommon in practice to encounter one or both of these misapplications.

Subcase 2 takes the strategy of attempting to explain the natural variation in the chiller temperature in terms of the ambient temperature surrounding the exposed/outdoor chiller unit. Previously, in Chapter 4, we determined that the ambient temperature was a prospective covariate that could be used to explain at least part of the chiller temperature fluctuation. This clue was obtained by physical cause effect reasoning and was supported through a correlation analysis.

At this point, we continue with the analysis, using our regression strategy on the simulated data in an attempt to extract near-*nid* residuals. We fit the regression model

$$\hat{\mu} = \hat{Y} = b_0 + b_1 X = -66.842 + 0.5691X$$

where X represents ambient temperature.

Our residuals are extracted and appear to be near-nid. The corresponding 3-sigma X (without runs rules and with 2/3 and 4/5 runs rules), EWMA, and tabular CuSum SPC model performances are assessed for both the case of no shift and the case of a small shift, i.e., an increase of 2 temperature units. Results are summarized in Table 13.3.

These SPC statistics appear to be adequate when we compare the ARL_0 results obtained with no process shift to our 3-sigma benchmark, $ARL_0 = 370$. Then, when we compare our no-shift statistics to our 2-temperature-unit-shifted data stream results, we see a large drop in our ARL, as we would expect in a proper SPC application. Here, with the exception of the X chart model without runs rules, we see ARL_1's in the 9 to 17 sample ranges. These results appear to support our regression modeling as a successful endeavor.

Selected chiller performance SPC charts and run length histograms, for the 2-temperature-unit-shifted data, are provided in Figure 13.9. Figure 13.9a depicts the X chart model performance without and with the two runs rules. Figure 13.9b and 13.9c depicts corresponding EWMA and tabular CuSum performances, respectively. All parts of the figure include a control chart plot, with a shift point, as well as a run length histogram.

The 2-unit shift is evident in the residual-based SPC plots. However, in the original data plots it would be all but obscured by the movement in the chiller temperature. Hence, our regression/residual modeling approach to convert the simulated chiller data stream from non-*nid* to near-*nid* was a success. We should note that other models, such as time series models, may prove to be as, or more, successful. In other words, **several models may ultimately provide successful residual data sets. Or, in some cases, no model may provide adequate residuals.**

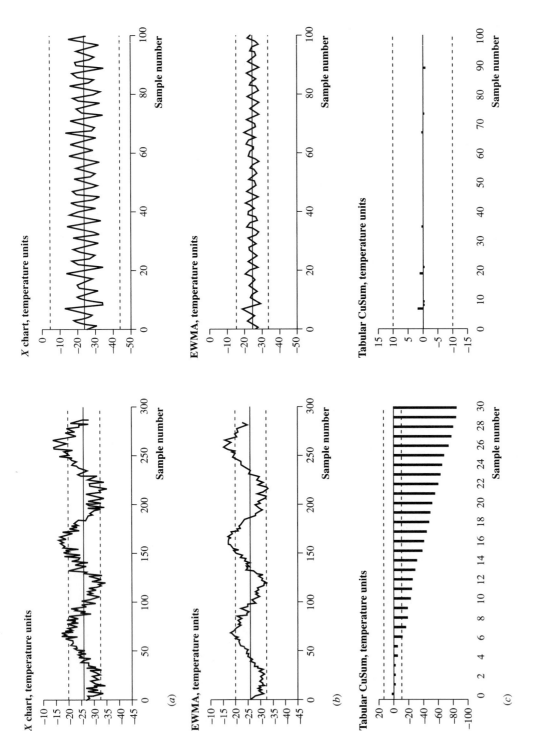

Figure 13.8 Selected X chart, EWMA, and CuSum model performance in simulated SPC misapplications (Example 13.3). (a) X chart model performance from sampling intervals of 1 time unit and 24 time units, no shift, $n = 1$. (b) EWMA model performance from sampling intervals of 1 time unit and 24 time units, no shift, $n = 1$. (c) Tabular CuSum model performance from sampling intervals of 1 time unit and 24 time units, no shift, $n = 1$.

Table 13.3 Simulated chiller temperature residuals SPC performance summary (Example 13.3).

	Individuals Data	
SPC Model (All ARLs/SRLs Based on 300 Observations)	**No Shift***	**Location Shift* (2 units)**
X chart (without runs rules)	$ARL_0 = 375.6$ $SRL_0 = 400.6$ $SE_{ARL} = 23.1$	$ARL_1 = 46.7$ $SRL_1 = 40.6$ $SE_{ARL} = 2.3$
R_M chart ($n_a = 2$)	$ARL_0 = 109.7$ $SRL_0 = 131.5$ $SE_{ARL} = 7.6$	$ARL_0 = 107.0$ $SRL_0 = 131.7$ $SE_{ARL} = 7.6$
X chart (with 2/3, 4/5 runs rules)	$ARL_0 = 158.4$ $SRL_0 = 155.9$ $SE_{ARL} = 9.0$	$ARL_1 = 16.5$ $SRL_1 = 14.9$ $SE_{ARL} = 0.9$
R_M chart ($n_a = 2$)	$ARL_0 = 109.7$ $SRL_0 = 131.5$ $SE_{ARL} = 7.6$	$ARL_0 = 107.0$ $SRL_0 = 131.7$ $SE_{ARL} = 7.6$
EWMA chart ($r = 0.4$)	$ARL_0 = 422.4$ $SRL_0 = 445.7$ $SE_{ARL} = 25.7$	$ARL_1 = 15.4$ $SRL_1 = 13.3$ $SE_{ARL} = 0.8$
EWMD chart	$ARL_0 = 175.0$ $SRL_0 = 189.2$ $SE_{ARL} = 10.9$	$ARL_0 = 169.3$ $SRL_0 = 190.9$ $SE_{ARL} = 11.0$
CuSum chart ($h = 4.77, k = 0.5$)	$ARL_0 = 277.1$ $SRL_0 = 263.6$ $SE_{ARL} = 15.2$	$ARL_1 = 9.2$ $SRL_1 = 5.4$ $SE_{ARL} = 0.3$

ARL: Average of all 300 run lengths.
SRL: Standard deviation of the sample of 300 run lengths.
SE_{ARL}: Standard error of the ARL (mean).
*Dispersion results are identical or nearly identical due to controlled random number streams within subcases.

In general, our success in transformed data stream SPC will be directly proportional to our ability to model and extract near-*nid* residuals. As mentioned previously, we have an almost infinite number of possible modeling formats from which to choose. We prefer to start with a physical-based regression format, rather than a time series–based format. The physical basis is preferred since it is likely that some sort of physical phenomenon is driving our process metric, and physical-based regression modeling is typically simpler than time-based modeling. If the physical basis is not satisfactory, then the more complex time series models should be attempted.

Another advantage, in addition to direct cause-effect relationships, to physical-based models over time-based models lies in sampling flexibility. When we use a physical covariate, rather than a time covariate, we can obtain sampling flexibility. For example, we may not want to sample in a constant time order sequence as typically required in time series models. Or our physical process may not operate in a continuous manner over time.

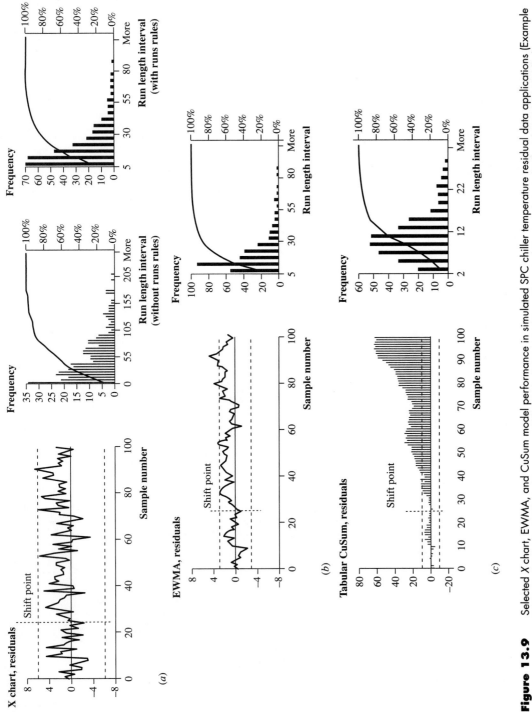

Figure 13.9 Selected X chart, EWMA, and CuSum model performance in simulated SPC chiller temperature residual data applications (Example 13.3). (a) X chart model performance without and with 2 out of 3 and 4 out of 5 runs rules applied, with 1-time-unit sampling interval, small location shift, $n = 1$. (b) EWMA model performance with 1-time-unit sampling interval, small location shift, $n = 1$. (c) Tabular CuSum model performance of 1-time-unit sampling interval, small location shift, $n = 1$.

13.5 INTRODUCTION TO MULTIVARIATE SPC MODELS

Up to this point in our discussion of SPC, we have focused exclusively on univariate SPC. In univariate SPC, we consider only one process response characteristic at a time. Obviously, most processes and products have a number of quality characteristics associated with them. Hence, a major concern emerges regarding how best to model multiple process characteristics.

When multiple process/product characteristics are independent of each other, multiple univariate control charts are typically used, but are not entirely equivalent to a multivariate model. If we encounter two physically related process characteristics, say X_1 and X_2, then we become concerned that perhaps two univariate SPC charts may not adequately reflect the activity of both variables, together. Hence, we consider a multivariate SPC model.

The purpose of this section is to introduce several basic multivariate models useful for modeling multiple process characteristics. We focus on both subgrouped and individuals data in our discussion and examples. We provide several references for more in-depth study of multivariate SPC.

MULTIVARIATE SPC CHARACTERIZATION

In order to choose an SPC strategy for multiple process characteristics, we resort to our physical and statistical discussions in Chapter 4. The basics of Chapter 4 regarding process characteristics hold in general for the multivariate case. For example, physical cause-effect and autocorrelation are meaningful. However, several extensions are necessary to deal with our multiple process control variables as a set. Here, we add scatter plots and correlation analyses for variable pairs.

In general, **we assume that our multiple variables are multivariate normal in distribution and each independent across time** (i.e., not significantly autocorrelated). We term this condition *mnid* in our subsequent discussion. As a set of p multivariate process characteristics, we describe our process variables set in terms of location and dispersion. For example, when we consider location we think in terms of means,

$$\boldsymbol{\mu} = \begin{bmatrix} \mu_1 \\ \mu_2 \\ \cdot \\ \cdot \\ \cdot \\ \mu_p \end{bmatrix} \qquad\qquad \textbf{[13.1]}$$

and

$$\hat{\boldsymbol{\mu}} = \overline{\mathbf{X}} = \begin{bmatrix} \overline{x}_1 \\ \overline{x}_2 \\ \cdot \\ \cdot \\ \cdot \\ \overline{x}_p \end{bmatrix} \qquad\qquad \textbf{[13.2]}$$

In terms of variation, our covariance matrix covers the general case. We introduce the concept of the covariance matrix (variance-covariance characteristics) where

$$\Sigma = \begin{bmatrix} \sigma_1^2 & \sigma_{12} \cdots \sigma_{1p} \\ & \sigma_2^2 \cdots \sigma_{2p} \\ & & \ddots \ \sigma_p^2 \end{bmatrix} \qquad \textbf{[13.3]}$$

and

$$\hat{\Sigma} = \mathbf{S} = \begin{bmatrix} s_1^2 & s_{12} \cdots s_{1p} \\ & s_2^2 \cdots s_{2p} \\ & & \ddots \\ & & \ s_p^2 \end{bmatrix} \qquad \textbf{[13.4]}$$

Here, the diagonal elements of the covariance matrix, Equation (13.3), represent the variances of each variable, while the off-diagonal elements represent the covariance between pairs of variables. Equation (13.4) represents an estimate of Equation (13.3). If the off-diagonal elements (covariances) are zero in Equation (13.3), we treat our variables as independent variables; otherwise, we assume dependence. **In the independent case, from a practical standpoint, we typically resort to univariate SPC charts.** However, **especially in the presence of dependence, we should consider multivariate SPC in order to capture the nature of our dependence in our model.**

The concepts of individual and joint process control regions for a location chart are depicted in Figure 13.10. Here, we present a bivariate (two variables X_1 and X_2) case conceptually. The two individual control limit sets in Figure 13.10a are projected to the individual joint region, independently. Here, we see a rectangular region, whereby a point falling outside the rectangle represents an out-of-control point. This case represents two univariate control charts used together, independently, to form the rectangular control region.

A joint process control region is depicted in Figure 13.10b. Here, we see an ellipse structure that takes its shape from the nature of our bivariate dependence. For example, in our bivariate case, if we assume a subgroup size n and that parameters μ_1, μ_2, σ_1^2, σ_2^2, and σ_{12} (i.e., means, variances, and covariance) are known, then the control region is on and inside the ellipse given by the equation

$$\frac{n}{\sigma_1^2 \sigma_2^2 - \sigma_{12}^2} [\sigma_2^2 (\bar{x}_1 - \mu_1)^2 + \sigma_1^2 (\bar{x}_2 - \mu_2)^2 - 2\sigma_{12} (\bar{x}_1 - \mu_1)(\bar{x}_2 - \mu_2)] \leq \chi_{\alpha,2}^2 \qquad \textbf{[13.5]}$$

where $\chi_{\alpha,2}^2$ is the chi-squared statistic, Table VIII.3, Section 8, with a right-hand tail area of α and 2 degrees of freedom.

In any case, even when σ_{12} is zero, the control region is not a rectangle, i.e., different from the Figure 13.10a case.

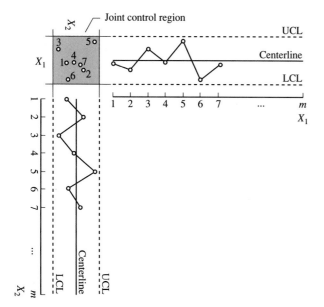

(a)

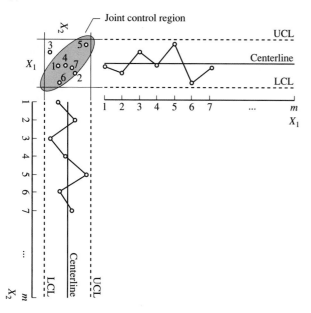

(b)

Figure 13.10 Multivariate control region concepts. (a) Joint control region for two individual univariate control charts. (b) Joint control region for a bivariate control chart, positive correlation.

χ^2 MULTIVARIATE LOCATION CHART—SUBGROUPED DATA

For more than two process characteristics, $X_1, X_2, \ldots, X_p$, we can develop a multivariate χ^2 control chart for location. To construct the χ^2 control chart we set our chart limits at

$$\text{LCL}_{\chi 2} = 0$$

$$\text{UCL}_{\chi 2} = \chi^2_{\alpha,2} \qquad \qquad \textbf{[13.6]}$$

We plot our χ^2_j statistic

$$\chi^2_j = n\,(\overline{\mathbf{x}} - \boldsymbol{\mu})^{\mathrm{T}}\,\boldsymbol{\Sigma}^{-1}\,(\overline{\mathbf{x}} - \boldsymbol{\mu}) \qquad \qquad \textbf{[13.7]}$$

horizontally, as in the univariate chart case.

From Equation (13.7) and the matrix inversion term, it is obvious that our calculations will require the aid of a computer for all but the simplest formats, i.e., a bivariate case. In the bivariate case we can invert our 2×2 matrix readily, since

$$\mathbf{M}^{-1} = \begin{bmatrix} a & b \\ c & d \end{bmatrix}^{-1} = \begin{bmatrix} d/D & -b/D \\ -c/D & a/D \end{bmatrix}$$

where $D = ad - bc$ is the determinate of the 2×2 matrix **M.**

Now, in our case

$$\boldsymbol{\Sigma}^{-1} = \begin{bmatrix} \sigma_1^2 & \sigma_{12} \\ \sigma_{12} & \sigma_2^2 \end{bmatrix}^{-1} = \begin{bmatrix} \dfrac{\sigma_2^2}{D} & \dfrac{-\sigma_{12}}{D} \\ \dfrac{-\sigma_{12}}{D} & \dfrac{\sigma_1^2}{D} \end{bmatrix}$$

where $D = \sigma_1^2 \sigma_2^2 - \sigma_{12}^2$

HOTELLING T^2 MULTIVARIATE LOCATION CHART—SUBGROUPED DATA

In the event that we do not have $\boldsymbol{\mu}$ and $\boldsymbol{\Sigma}$ targets available (i.e., μ_1, μ_2, σ_1^2, σ_2^2, and σ_{12} from the bivariate normal), we can develop empirical counterparts from our data and obtain the Hotelling T^2 chart. Here, we work with

$$\hat{\boldsymbol{\mu}} = \overline{\overline{\mathbf{x}}} \text{ and } \hat{\boldsymbol{\Sigma}} = \mathbf{S}$$

We plot

$$T^2_j = n\,(\overline{\mathbf{x}}_j - \overline{\overline{\mathbf{x}}})^{\mathrm{T}}\,\mathbf{S}^{-1}\,(\overline{\mathbf{x}}_j - \overline{\overline{\mathbf{x}}}) \qquad \qquad \textbf{[13.8]}$$

where T^2 is known as the Hotelling T^2, the multivariate counterpart to Student's t.

The control limits for the T^2 chart are sensitive to how the chart is used. Two phases of chart usage are identified: Phase I and Phase II; see Alt [6], Montgomery [2], and Lowry and Montgomery [7]. Phase I refers to the phase we go through to establish process control, i.e., a historic picture of our process. Phase II corresponds to usage of the chart for monitoring future production.

Phase I T^2 control chart limits are calculated as

$$\text{LCL}_{T^2} = 0$$

$$\text{UCL}_{T^2} = \frac{p(m-1)(n-1)}{mn-m-p+1} F_{\alpha,\, p,\, mn-m-p+1} \qquad \textbf{[13.9]}$$

where

   p = number of variables we are monitoring

   m = number of subgroups used to set the limits

   n = subgroup size

   F_α = the F statistic corresponding to a right-hand tail probability of α

Phase II T^2 control chart limits are calculated as

$$\text{LCL}_{T^2} = 0$$

$$\text{UCL}_{T^2} = \frac{p(m+1)(n-1)}{mn-m-p+1} F_{\alpha,\, p,\, mn-m-p+1} \qquad \textbf{[13.10]}$$

Phase I and II also both apply in univariate SPC charting. Phase I refers to trial limit calculations, while Phase II refers to chart usage in the future. In the univariate case, $m \geq 20$ is typical and usually considered adequate for both Phase I and II control chart limits. Multivariate charts do not have such a straightforward rule of thumb. In general, using

$$\text{UCL} = \chi^2_{\alpha,\, p} \qquad \textbf{[13.11]}$$

is acceptable for very large values of m; see Lowry and Montgomery [7]. They provide tables in terms of $p, n,$ and m and identify "safe" m values. These values range from 20 to several hundred. In general, it is more advisable to use Equations (13.9) and (13.10) rather than the simpler Equation (13.11) form.

In order to develop the T^2 chart, we estimate μ and Σ. When we use subgrouped data our estimates are straightforward. Essentially, we estimate each element of the mean and covariance matrix structure. We begin by working within each subgrouped sample.

First, we calculate

$$\bar{\mathbf{x}}_j = [\bar{x}_{1j}, \bar{x}_{2j}, \ldots, \bar{x}_{pj}] \qquad j = 1, 2, \ldots, m$$

where

$$\overline{x}_{1j} = \frac{\sum\limits_{i=1}^{n} x_{1ij}}{n}$$

$$\vdots$$

[13.12]

$$\overline{x}_{pj} = \frac{\sum\limits_{i=1}^{n} x_{pij}}{n}$$

Next, we calculate corresponding sample variances for each element:

$$s_{1j}^2 = \frac{\sum\limits_{i=1}^{n} (x_{1ij} - \overline{x}_{1j})^2}{n-1}$$

$$\vdots$$

$$s_{pj}^2 = \frac{\sum\limits_{i=1}^{n} (x_{pij} - \overline{x}_{pj})^2}{n-1}$$

[13.13]

followed by the covariance estimates for each pair (combination) of variables:

$$s_{12j} = \frac{\sum\limits_{i=1}^{n} (x_{1ij} - \overline{x}_{1j})(x_{2ij} - \overline{x}_{2j})}{n-1}$$

$$s_{13j} = \frac{\sum\limits_{i=1}^{n} (x_{1ij} - \overline{x}_{1j})(x_{3ij} - \overline{x}_{3j})}{n-1}$$

$$\vdots$$

$$s_{(p-1)pj} = \frac{\sum\limits_{i=1}^{n} (x_{(p-1)ij} - \overline{x}_{(p-1)j})(x_{pij} - \overline{x}_{pj})}{n-1}$$

[13.14]

Then, we calculate our corresponding averages for the $\overline{\overline{\mathbf{x}}}$ vector

$$\overline{\overline{x}}_1 = \frac{\sum\limits_{j=1}^{m} \overline{x}_{1j}}{m}$$

$$\vdots$$

$$\overline{\overline{x}}_p = \frac{\sum\limits_{j=1}^{m} \overline{x}_{pj}}{m} \qquad \textbf{[13.15]}$$

Next, we focus on the variances and covariances:

$$\overline{s}_1^2 = \frac{\sum\limits_{j=1}^{m} s_{1j}^2}{m}$$

$$\vdots$$

$$\overline{s}_p^2 = \frac{\sum\limits_{j=1}^{m} s_{pj}^2}{m} \qquad \textbf{[13.16]}$$

and

$$\overline{s}_{12} = \frac{\sum\limits_{j=1}^{m} s_{12j}}{m}$$

$$\overline{s}_{13} = \frac{\sum\limits_{j=1}^{m} s_{13j}}{m}$$

$$\vdots$$

$$\overline{s}_{(p-1)p} = \frac{\sum\limits_{j=1}^{m} s_{(p-1)pj}}{m} \qquad \textbf{[13.17]}$$

Finally,

$$\hat{\Sigma} = \mathbf{S} = \begin{bmatrix} \bar{s}_1^2 & \bar{s}_{12} & \bar{s}_{13} \cdots \bar{s}_{1p} \\ & \bar{s}_2^2 & \bar{s}_{23} \cdots \bar{s}_{2p} \\ & & \bar{s}_3^2 \cdots \bar{s}_{3p} \\ & & & \cdot & \cdot \\ & & & & \cdot & \cdot \\ & & & & & \cdot & \cdot \\ & & & & & & \bar{s}_p^2 \end{bmatrix}$$ **[13.18]**

At this point, we have everything we need to develop our Phase I chart. **Once we have Phase I in control, e.g., all points fall between the LCL and UCL, we shift to Phase II and work out our T^2 control chart limits and apply them to future production.**

Example 13.4

PANEL ASSEMBLY DEVIATIONS   Assembly operations in which we fit parts and panels together present interesting challenges that are multivariate in nature. For example, in automobile assembly, body panels are assembled to floor, firewall, and roof panels on moving assembly lines. These panels are relatively large and must be aligned properly in all respects, so that after the welding operations we obtain body dimensions that allow for proper fit. If the panels are out of alignment, doors and other parts do not fit properly. In many cases, we see automated or semiautomated equipment used to perform these assembly subprocesses.

Once the assembly and welds are made, we use measuring machines to assess how our assembly subprocesses performed, compared to target dimensions. This type of assembly presents a large multivariate statistical process control challenge. What we want to assess is when to seek out special causes. Multivariate SPC can be used to help meet this challenge.

Our example presents a simplified assembly case where we introduce only two of many possible variables. Here, we introduce variables X_1 and X_2 representing two deviations from target. For example, X_1 could be thought of as Z directional (in-out) fit and position at the top of the panel and X_2 as the corresponding fit deviation at the bottom of the panel. Here, we can use physical-based reasoning to justify a multivariate SPC model rather than two independent SPC models. We reason that the two dimensions are possibly related to each other but are basically independent from one panel to the next. Hence, we have a good candidate for multivariate SPC.

Solution

Using a subgroup size of $n = 5$ panels, we present our data for $m = 25$ subgroups. The data are displayed in Table 13.4. We have produced a scatter plot for our 125 data points. The scatter plot appears in Figure 13.11. Here, we can see a tendency for the data to produce an upward trend or positive correlation. We notice that the sample correlation coefficient between X_1 and X_2 is 0.434.

We have used the T^2 equations and a computer spreadsheet to make our calculations. Our Phase I calculations involve Equations (13.8) and (13.9), as well as the supporting Equations (13.12) through (13.18).

Table 13.4 Body panel deviation data for the T^2 model (Example 13.4).

Sub-group j	Panel x_{1ij} 1	2	3	4	5	$\bar{x}_{1j}$	s^2_{1j}	Panel x_{2ij} 1	2	3	4	5	$\bar{x}_{2j}$	s^2_{2j}	s_{12j}	T^2	$|S|$
1	-0.934	1.381	2.633	-0.629	3.178	1.1258	3.4676	-1.331	4.074	0.194	-0.537	0.419	0.5638	4.3203	1.4690	1.0230	12.8233
2	-1.421	-0.633	-7.406	-4.426	2.671	-2.2430	14.7093	-3.246	-0.735	-1.308	-2.540	0.967	-1.3724	2.6892	3.2996	5.5920	28.6694
3	4.901	-2.580	2.198	-0.581	1.774	1.1424	8.1167	2.880	-0.022	2.258	-2.900	2.446	0.9324	5.8599	4.9582	1.5699	22.9785
4	-2.131	-0.811	-0.584	-4.532	1.075	-1.3966	9.8255	-2.415	-2.087	-1.712	-1.652	-2.993	-2.1718	0.3056	-0.7644	4.5623	0.7512
5	-5.440	-0.803	3.242	0.402	-0.333	-0.5864	4.6594	-3.435	-0.462	2.294	3.385	-2.831	-0.2098	9.1320	7.0457	0.6559	40.0841
6	4.531	-1.245	0.854	1.951	2.796	1.7774	4.6594	0.811	-0.933	1.817	3.915	2.234	1.5688	3.2105	1.5826	4.0000	12.4543
7	-0.406	-5.200	-2.973	2.468	-0.510	-1.3242	8.4068	-0.201	-2.704	2.294	0.920	-6.723	-1.2828	12.6079	1.1321	2.4547	104.7106
8	1.793	-1.120	3.094	-1.240	0.260	0.5574	3.5235	0.645	-0.363	-0.040	0.759	1.602	0.5206	0.5842	-0.1341	0.5363	2.0404
9	2.836	-0.542	6.317	1.999	-1.500	1.8220	9.4722	-1.189	-0.502	0.134	-1.957	-2.814	-1.2656	1.3559	2.3963	5.5186	7.1007
10	-0.433	4.631	-1.914	-2.522	0.176	-0.0124	7.9219	2.279	1.943	-2.982	-3.120	0.572	-0.2616	6.8952	5.4180	0.0407	25.2690
11	0.689	-1.263	4.402	-2.305	2.144	0.7334	7.1670	-4.256	3.495	0.974	2.451	-0.307	0.4714	9.0669	-2.7739	0.5421	57.2883
12	2.150	0.436	4.164	-0.529	-0.290	1.1862	3.8700	-2.615	3.237	2.661	-1.226	0.211	0.4537	6.2295	1.1919	1.0167	22.6875
13	0.706	-0.841	2.886	5.396	4.166	2.4626	6.4213	0.722	1.156	0.946	0.533	4.044	1.4802	2.1089	0.9413	5.6357	12.6557
14	-1.610	-1.943	1.391	0.959	-3.250	-0.8906	3.9546	-1.867	-1.710	0.253	1.223	-0.797	-0.5796	1.7313	1.9656	1.0911	2.9829
15	2.246	-2.408	1.385	0.994	-0.636	0.3162	3.4135	0.921	0.111	-1.199	3.035	-2.978	-0.0220	5.0939	1.2716	0.0377	15.7708
16	-3.756	-4.205	-2.393	-1.234	-4.409	-3.1994	1.8243	-1.156	-2.226	-3.035	-3.752	-1.255	-2.2848	1.2634	-1.3554	11.4983	0.4677
17	2.801	-0.431	-1.093	3.889	0.880	1.2092	4.4535	0.719	-0.560	-3.294	4.563	-0.085	0.2686	8.0371	5.4756	0.9648	5.8110
18	3.259	1.296	-0.611	0.747	1.214	1.1810	1.9330	-2.117	0.902	-5.598	1.130	-0.054	-1.1474	7.8325	1.3110	2.8043	13.4216
19	-2.348	-0.437	1.159	0.657	-1.877	-0.5692	2.3455	-2.903	0.614	-2.232	-0.191	2.108	-0.5208	4.2276	-0.4008	0.5423	9.7553
20	3.863	-1.620	-0.738	1.185	1.808	0.8996	4.6796	2.520	-3.751	-1.841	0.569	-1.771	-0.8548	5.9041	4.6218	1.3789	6.2680
21	0.205	-2.831	-0.141	-1.294	-1.817	-1.1756	1.5356	3.444	-3.861	-2.847	-1.918	-0.839	-1.2042	8.0015	2.2415	2.0276	7.2624
22	-0.973	-1.374	1.060	0.109	1.386	0.0416	1.4706	4.030	-2.340	-1.650	-0.054	2.280	0.4532	7.1533	0.1512	0.6151	10.4969
23	2.562	0.029	2.883	2.566	1.607	1.9294	1.3576	1.160	-2.742	0.160	-1.971	0.025	-0.6736	2.6266	1.2086	4.1246	2.1051
24	2.437	0.826	-4.513	-0.827	3.779	0.3404	10.3435	-1.539	3.180	-2.170	2.402	2.856	0.9458	6.6609	3.9665	1.4191	53.1633
25	-1.132	-4.436	1.208	1.378	0.814	-0.4336	6.0101	-0.523	-0.279	1.385	0.078	3.964	0.9250	3.4277	2.2106	2.4908	15.7141
26	6.242	6.812	2.560	1.424	3.794	4.1664	5.3868	4.638	-2.738	0.245	-0.486	-4.263	-0.1208	8.5880	1.0412	16.6055	45.1780
27	9.528	10.314	7.829	7.548	4.296	7.9030	5.3989	-0.103	1.238	-2.825	1.289	-4.291	-0.9384	6.2988	4.5116	68.8483	13.6515
28	3.254	3.768	2.981	0.904	7.627	3.7068	5.9898	-0.334	-1.479	-3.823	4.233	-0.786	-0.4378	8.6230	-3.0275	13.9199	42.4843
29	4.187	6.259	7.281	7.036	9.218	6.7962	3.3135	-4.759	-0.460	-1.107	-2.584	0.120	-1.7580	3.8309	2.9496	57.8047	3.9936
30	3.323	5.838	3.710	2.642	6.454	4.3934	2.7533	-2.705	1.462	-1.485	-0.749	0.729	-0.5496	2.8105	2.2091	20.1760	2.8580

Subgroups 26 through 30 include a 5-unit upward shift in the X_1 variable only.

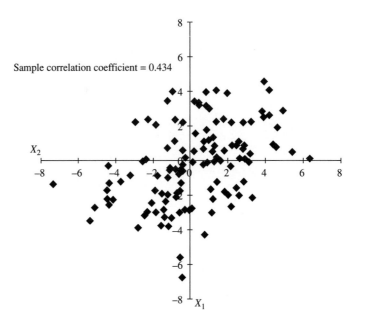

Figure 13.11   Panel deviation scatter plot (Example 13.4).

$$\overline{\overline{x}}_1 = \frac{\sum\limits_{j=1}^{m} \overline{x}_{1j}}{m} = 0.1957$$

$$\overline{\overline{x}}_2 = \frac{\sum\limits_{j=1}^{m} \overline{x}_{2j}}{m} = -0.2107$$

$$\overline{s}_1^2 = \frac{\sum\limits_{j=1}^{m} s_{1j}^2}{m} = 5.4101$$

$$\overline{s}_2^2 = \frac{\sum\limits_{j=1}^{m} s_{2j}^2}{m} = 5.0530$$

$$\overline{s}_{12} = \frac{\sum\limits_{j=1}^{m} s_{12j}}{m} = 1.9372$$

$$\hat{\Sigma} = \mathbf{S} = \begin{bmatrix} 5.4101 & 1.9372 \\ 1.9372 & 5.0530 \end{bmatrix}$$

We calculate our Phase I control limits for a type I error of $\alpha = 0.001$ as

$$\text{UCL}_{T^2} = \frac{p(m-1)(n-1)}{mn - m - p + 1} F_{\alpha, p, mn - m - p + 1} = \frac{2(25-1)(5-1)}{125 - 25 - 2 + 1} 7.413 = 14.3767$$

$$\text{LCL}_{T^2} = 0$$

where $F_{\alpha, p, mn - m - p + 1} = F_{0.001, 2, 99} = 7.413$.

Our T^2 calculations are produced here in the expanded form of Equation (13.8) for $p = 2$:

$$\frac{n}{\bar{s}_1^2 \bar{s}_2^2 - \bar{s}_{12}^2} [\bar{s}_2^2 (\bar{x}_{1j} - \bar{\bar{x}}_1)^2 - 2\bar{s}_{12} (\bar{x}_{1j} - \bar{\bar{x}}_1)(\bar{x}_{2j} - \bar{\bar{x}}_2) + \bar{s}_1^2 (\bar{x}_{2j} - \bar{\bar{x}}_2)^2] = T_j^2$$

For $j = 1$

$$\frac{5}{5.4101 (5.0530) - 1.9372^2} \{5.0530 (1.1258 - 0.1957)^2 - 2(1.9372)(1.1258 - 0.1957)$$

$$[0.5638 - (-0.2107)] + 5.4101 [0.5638 - (-0.2107)]^2\} = 1.0230$$

Our Phase I results are plotted in Figure 13.12a. All 25 samples appear to be well within our control limits.

Since our Phase I results appear to be in control, we calculate our Phase II T^2 chart limits using $\alpha = 0.001$ as

$$\text{UCL}_{T^2} = \frac{p(m+1)(n-1)}{mn - m - p + 1} F_{\alpha, p, mn - m - p + 1} = \frac{2(25+1)(5-1)}{125 - 25 - 2 + 1} (7.413) = 15.5748$$

$$\text{LCL}_{T^2} = 0$$

In Figure 13.12b, we have reproduced our original 25 subgroups with our Phase II control limits. Additionally, we have added 5 more subgroups (see the bottom portion of Table 13.4, subgroups 26 to 30), where we have shifted (using simulation) our X_1 values up by 5 units and left our X_2 location values intact. Here, with this large shift in the X_1 values, we can readily pick up our shift on the T^2 chart in Figure 13.12b.

SAMPLE GENERALIZED VARIANCE MULTIVARIATE DISPERSION |S| CHART—SUBGROUPED DATA

The sample generalized variance chart is but one of several options available to develop a multivariate dispersion chart; see Alt [6]. This model is based on the determinate of the sample covariance matrix, |S|, whereby we use the expected value of |S|, E[|S|], and variance of |S|, V[|S|]. Since most of the probability distribution of |S| falls within $E[|S|] \pm 3\sqrt{V[|S|]}$, relationships exist such that

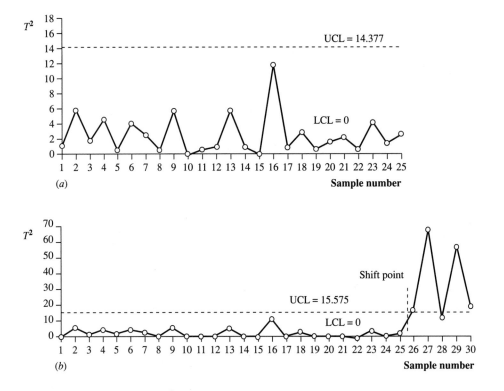

Figure 13.12   Hotelling T^2 control charts for assembly deviations (Example 13.4). (a) Phase I chart. (b) Phase II chart with shifted data added (location shift in X_1 only).

$$E[|S|] = b_1 \, |\Sigma_0|$$ **[13.19]**

and

$$V[|S|] = b_2 \, |\Sigma_0|^2$$ **[13.20]**

where

$$b_1 = \frac{\prod_{i=1}^{p}(n - i)}{(n - 1)^p}$$

and

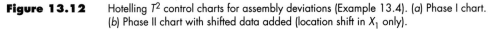

$$b_2 = \frac{\prod_{i=1}^{p}(n - i)\left[\prod_{j=1}^{p}(n - j + 2) - \prod_{j=1}^{p}(n - j)\right]}{(n - 1)^{2p}}$$

Hence, **the parameters for the |S| control chart are**

$$\text{LCL}_{|S|} = b_1|\boldsymbol{\Sigma}_0| - 3|\boldsymbol{\Sigma}_0|(b_2)^{1/2}$$

$$\text{UCL}_{|S|} = b_1|\boldsymbol{\Sigma}_0| + 3|\boldsymbol{\Sigma}_0|(b_2)^{1/2} \qquad \textbf{[13.21]}$$

$$\text{Centerline} = b_1|\boldsymbol{\Sigma}_0|$$

If $\text{LCL}_{|S|}$ from above is less than zero, set $\text{LCL}_{|S|} = 0$. We plot corresponding $|S_j|$ values on our $|S|$ control chart, where

$$|S_j| = s_{1j}^2 s_{2j}^2 - s_{12j}s_{12j}$$

Here, we use $|\boldsymbol{\Sigma}_0|$ directly, if a target for $\boldsymbol{\Sigma}$ is used or available. Otherwise, we use $|\hat{\boldsymbol{\Sigma}}| = |S|/b_1$ if estimates for $\boldsymbol{\Sigma}$ are used, and simply substitute $|S|/b_1$ for $|\boldsymbol{\Sigma}_0|$ in Equation (13.21).

PANEL ASSEMBLY DEVIATION (CONTINUED)   Develop and plot generalized sample variance | **Example 13.5**
charts for the panel assembly example data featured in Table 13.4.

Solution

Our $|S|$ calculations are shown on the right-hand side in Table 13.4. Here, we have $p = 2$, a 2×2 matrix. We use an empirical estimate for $\boldsymbol{\Sigma}$, and substitute $|S|/b_1$, for $|\boldsymbol{\Sigma}_0|$ in both Phases I and II.

$$S = \begin{bmatrix} \bar{s}_1^2 & \bar{s}_{12} \\ \bar{s}_{12} & \bar{s}_2^2 \end{bmatrix}$$

$$|S| = \bar{s}_1^2\bar{s}_2^2 - \bar{s}_{12}\bar{s}_{12} = 5.4101\,(5.0530) - (1.9372)^2 = 23.585$$

For our control limits,

$$b_1 = \frac{\displaystyle\prod_{i=1}^{p}(n - i)}{(n - i)^p} = \frac{4(3)}{4^2} = 0.75$$

and

$$b_2 = \frac{\displaystyle\prod_{i=1}^{p}(n - i)\left[\prod_{j=1}^{p}(n - j + 2) - \prod_{j=1}^{p}(n - j)\right]}{(n - 1)^{2p}} = \frac{4(3)[6(5)-4(3)]}{4^{2(2)}} = 0.8438$$

$$\text{LCL}_{|S|} = b_1\left(\frac{|S|}{b_1}\right) - 3\left(\frac{|S|}{b_1}\right)(b_2)^{1/2} = -63.074; \text{ therefore, set } \text{LCL}_{|S|} = 0$$

$$\text{UCL}_{|S|} = b_1\left(\frac{|S|}{b_1}\right) + 3\left(\frac{|S|}{b_1}\right)(b_2)^{1/2} = 110.243$$

$$\text{Centerline} = b_1\left(\frac{|S|}{b_1}\right) = 23.585$$

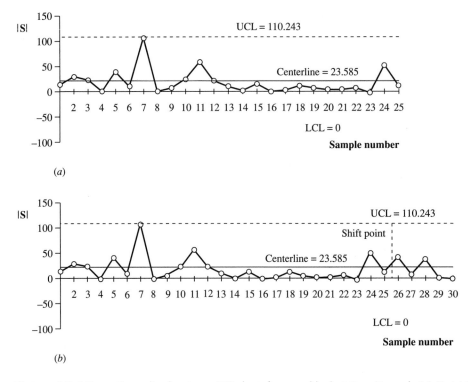

Figure 13.13   Generalized variance |**S**| charts for assembly deviations (Example 13.5). (*a*) Phase I chart. (*b*) Phase II chart with shifted data added (with location shift in X_1 only).

We plot corresponding |S_j| values on our |S| control chart, where, for $j = 1$,

$$|S_1| = \bar{s}_{11}^2 \bar{s}_{21}^2 - \bar{s}_{121}\bar{s}_{121} = 3.4676(4.3203) - (1.4690)^2 = 12.823$$

Our resulting generalized sample variance |S| chart is depicted in Figure 13.13. In Figure 13.13*a*, we observe that the multivariate process variation appears to be in control. Figure 13.13*b* repeats the original data and adds the five shifted subgroups as shown in the bottom of Table 13.4. Here, we shifted the X_1 variable up by 5 units, and notice no disturbance in our |S| chart.

Our generalized sample variance chart tends to be useful, but because it considers several variables simultaneously, it should be used with caution. For example, any number of S matrices can yield the same value of |S|. Hence, **we usually plot corresponding univariate R or S charts along with the multivariate chart to help understand what is happening to and in our processes.** The rationale here is simply that we take counteraction regarding process variables one at a time in the physical world, even though we monitor them as a group (multivariate) statistically.

HOTELLING T^2 MULTIVARIATE LOCATION CHART— INDIVIDUALS DATA

Given individuals data ($n = 1$), we can rewrite Equation (13.8) and develop a T^2 multivariate control chart where

$$T_j^2 = (\mathbf{x}_j - \bar{\mathbf{x}})^T \mathbf{S}^{-1} (\mathbf{x}_j - \bar{\mathbf{x}}) \qquad \textbf{[13.22]}$$

Our challenge here deals with estimates of $\mathbf{S}$. Previously, we used our within-subgroup data to estimate the elements of $\mathbf{S}$. Now, we have samples of $n = 1$. The classical argument of common cause and special cause variation, and possible process variable shifts due to special causes between samples, is still relevant. These concepts were discussed in our univariate control chart chapters, Chapters 10 and 11. The only difference here is the multivariate nature of the challenge.

Previously we introduced the moving range R_M model in Chapter 11 to deal with the challenge of estimating our dispersion component when using individuals data, $n = 1$. Here, we introduce its multivariate counterpart as described by Tracy, Young, and Mason [8]; also see Sullivan and Woodall [9]. They describe a moving range counterpart developed from our individuals data,

$$\mathbf{S} = \frac{\mathbf{V}^T \mathbf{V}}{2(m - 1)} \qquad \textbf{[13.23]}$$

where the elements of the $\mathbf{V}$ matrix are constructed by using

$$\mathbf{V} = \begin{bmatrix} \mathbf{v}_1 \\ \mathbf{v}_2 \\ . \\ . \\ . \\ \mathbf{v}_{m-1} \end{bmatrix}$$

and

$$\mathbf{v}_j = [(x_{1j+1} - x_{1j}), (x_{2j+1} - x_{2j}), \dots, (x_{pj+1} - x_{pj})] \qquad j = 1, 2, \dots, m - 1$$

The Phase I SPC chart, for the start-up phase, presents several problems in establishing T^2 control limits for individuals. Tracy, Young, and Mason [8] suggest the following T^2 (for individuals) chart parameters.

$$\mathrm{LCL}_{T^2} = \frac{(m - 1)^2}{m} [B_{1-(a/2), p/2, (m-p-1)/2}]$$

$$\mathrm{UCL}_{T^2} = \frac{(m - 1)^2}{m} [B_{a/2, p/2, (m-p-1)/2}] \qquad \textbf{[13.24]}$$

where $B(.)$ refers to a beta distribution. When beta distribution statistics are not available, they use a comparable form based on the F distribution:

$$\text{LCL}_{T^2} = \frac{(m-1)^2}{m} \left\{ \frac{[p/(m-p-1)] \, F_{1-(\alpha/2), \, p, \, m-p-1}}{1 + [p/(m-p-1)] \, F_{1-(\alpha/2), \, p, \, m-p-1}} \right\}$$

$$\text{UCL}_{T^2} = \frac{(m-1)^2}{m} \left\{ \frac{[p/(m-p-1)] \, F_{\alpha/2, \, p, \, m-p-1}}{1 + [p/(m-p-1)] \, F_{\alpha/2, \, p, \, m-p-1}} \right\} \qquad \textbf{[13.25]}$$

For Phase II, we use the Tracy, Young, and Mason [8] formulation

$$\text{LCL}_{T^2} = \frac{p(m+1)(m-1)}{m^2 - mp} \, F_{1-(\alpha/2), \, p, \, m-p}$$

$$\text{UCL}_{T^2} = \frac{p(m+1)(m-1)}{m^2 - mp} \, F_{\alpha/2, \, p, \, m-p} \qquad \textbf{[13.26]}$$

As usual, we plot our T_j^2 values on the T^2 chart for individuals, along with the appropriate chart control limits.

INTERPRETATION OF MULTIVARIATE SPC CHARTS

The univariate SPC chart approach poses several concerns. First, taken as a group of independent control charts with individual α, type I error probabilities, we face a combined type I error probability for p independent individual univariate charts of

$$\alpha_{\text{overall}} = 1 - (1 - \alpha_{\text{individual chart}})^p$$

Additionally, multiple individual univariate charts produce control regions, together, that are not equivalent to the multivariate control region; see Figure 13.10.

Multivarite SPC addresses these concerns, but presents its own concerns. We track p process variables simultaneously with our multivariate SPC models. Hence, when an out-of-control signal is encountered, we do not know which of the p variables produced the signal. One approach to this dilemma is to plot the corresponding set of p univariate control charts and examine each one for unusual patterns or points within or out of the corresponding control limits.

Several other multivariate-based approaches to locating the variables acting with special causes present have been suggested. Approaches include Bonferroni-type control limits, exact simultaneous confidence intervals, principal components, and T^2 decomposition; see Alt [6], Hayter and Tsui [10], Prins and Mader [11], and Mason et al. [12]. An overview of many of these approaches is provided by Mason et al. [13].

The T^2 decomposition method is interesting and of potential use. Here, we calculate the contribution to the overall T^2 statistic made by each of our individual p variables or combinations thereof. In other words, we calculate partial T^2 statistics, leaving out each of the p variables or combinations thereof. Then, we difference these partial T^2 values with the total T^2 to obtain differences, i.e., the contribution of each variable or combination to the total T^2.

The larger differences are associated with the most likely variable candidates for examination in our physical process counteraction investigation. In this case, our differences are calculated as

$$D_1 = T^2 - T^2_{2,3,\ldots,p}$$
$$D_2 = T^2 - T^2_{1,3,\ldots,p}$$

$$\cdot$$
$$\cdot$$
$$\cdot$$

$$D_p = T^2 - T^2_{1,2,\ldots,p-1}$$

$$\cdot$$
$$\cdot$$
$$\cdot$$

Any other combination of interest

Conceptually, this method is simple, but mathematically it is calculation-intensive and requires computer aids to be practical for large p's.

OTHER MULTIVARIATE MODELS

Multivariate process control is an active topic in both practice and theory. It has been in existence for decades; e.g., Hotelling's T^2 model dates back to the 1940s; Hotelling [14]. The advent of fast, and relatively cheap, computing power and more complicated physical processes both stimulate interest in multivariate process control models.

A wealth of publications has been produced dealing with multivariate quality control. Lowry and Montgomery [7] provide a review of multivariate control charts. Crosier [15] and Pignatiello and Runger [16] provide discussion regarding multivariate CuSum models. Multivariate EWMA models are described by Lowry, Woodall, Champ, and Rigdon [17]. Computer aids in multivariate quality control are emerging; however, Montgomery [2] suggests that we approach these aids cautiously and verify their integrity before using them.

REVIEW AND DISCOVERY EXERCISES

REVIEW

13.1. Why and how is the nid property of data streams critical to SPC modeling? Explain.

13.2. What is the difference between an ARL_0 and an ARL_1? Explain.

13.3. In a simulated SPC model, what constitutes a false alarm? A failure to detect a shift? Explain.

13.4. In Subcases 1, 3, and 5, Table 13.1, why did the dispersion charts' performances remain about the same across the shifts? Explain.

13.5. Across Subcases 1, 3, and 5 in Table 13.1, how did the shifts impact the average run lengths? How would this translate into type I and II error probabilities? Explain.

13.6. In Subcases 2, 4, and 6, Table 13.1, why did the dispersion charts' performances remain about the same across the shifts? Explain.

13.7. Across Subcases 2, 4, and 6 in Table 13.1, how did the shifts impact the average run lengths? How would this translate into type I and II error probabilities? Explain.

13.8. Why does Subcase 7 in Table 13.1, under both X-bar, R, and X, R_M charts, show low ARL_1's and ARL_0's (when only the variance shifted)? Explain.

13.9. How are the run length histograms featured in Figures 13.1, 13.2, and 13.3 helpful in assessing run length characteristics? What do they add that the ARL cannot address? Explain.

13.10. After reviewing the physical nature of the shredder described in Example 13.2, what general process characteristics are apparent? What constitutes stability and common causes in a process of this nature? What constitutes instability and special causes? Explain.

13.11. Characterize the misapplication results depicted in Figure 13.5. Could we have selected a sampling scheme that would have produced acceptable results? Explain.

13.12. Compare and contrast Figures 13.5 and 13.6. How effective was the regression/residuals approach in modeling the shredder data? Explain.

13.13. Summarize the no shift–shift SPC chart performances described in Table 13.2 regarding shredder process control. Is residuals modeling an acceptable form of SPC charting? Explain.

13.14. After reviewing the physical nature of the chiller described in Example 13.3, what general process characteristics are apparent? What constitutes stability and common causes in a process of this nature? What constitutes instability and special causes? Explain.

13.15. Characterize the misapplication results depicted in Figure 13.8. Could we have selected a sampling scheme that would have produced acceptable results? Explain.

13.16. Compare and contrast Figures 13.8 and 13.9. How effective was the regression/residuals approach in modeling the chiller data? Explain.

13.17. Summarize the no shift–shift SPC chart performances described in Table 13.3 regarding chiller process control. Is residuals modeling an acceptable form of SPC charting? Explain.

13.18. How is multivariate SPC different from and the same as univariate SPC? What is to be gained by a multivariate SPC model, versus say several univariate models? Explain.

13.19. Why are multivariate SPC charts that indicate a special cause difficult to interpret? Explain.

CASE EXERCISES I

The following questions pertain to the case descriptions in Section 7. Case selection list:

Case VII. 9: Downtown Bakery—Bread Dough

Case VII. 21: Punch-Out—Sheet Metal Fabrication

Case VII. 29: TexRosa—Salsa

13.20. Select a case and a corresponding set of variables. Develop a subgrouped multivariate control chart. Comment on your findings.

13.21. For a case selected in problem 13.20, generate an extended, shifted, data set using the method described in Section 7, Equation (VII.1), and plot the extended data on your existing process control charts. Look for shift points and patterns. Comment on your findings.

DISCOVERY

13.22. What is creating the difficulty in SPC modeling that we see in the Example 13.2 misapplication results in Figure 13.5? Explain in terms of common and special cause manifestation.

13.23. What is creating the difficulty in SPC modeling that we see in the Example 13.3 misapplication results in Figure 13.8? Explain in terms of common and special cause manifestation.

13.24. How does an approach such as regression modeling help to sort special from common cause variation? Explain.

CASE EXERCISES II

The following questions pertain to the case descriptions in Section 7. *Warning:* Some of these cases contain characteristics that, although highly realistic, create assumption violations in SPC. Use your results from Chapter 4 to help identify the problem cases and issues. Look for creative solutions in terms of regression/residuals. Case selection list:

Case VII.2: Apple Core—Dehydration

Case VII.17: LNG—Natural Gas Liquefaction

Case VII.22: ReUse—Recycling

Case VII.30: Tough-Skin—Sheet Metal Welding

13.25. Select one or more of the above cases. Develop an X, R_M chart set from the data provided in the case description. Use 3-sigma limits. Try a regression/residual approach to transform the chart data to near-*nid* if possible. Comment on your findings.

13.26. Select one or more of the above cases. Develop an individuals EWMA and EWMD chart set from the data provided in the case description. Select an appropriate r value. Try a regression/residual approach to transform the chart data to near-*nid* if possible. Use appropriate target or empirical values to calculate the control limits. Comment on your findings.

13.27. Select one or more of the above cases. Develop an individuals CuSum chart from the data provided in the case description. Select appropriate k and h values from Table 10.4. Try a regression/residual approach to transform the chart data to near-*nid* if possible. Use appropriate target or empirical values to calculate the control limits. Comment on your findings.

REFERENCES

1. Karim, M., "Statistical Process Control Performance Measurement," M.S. thesis, Texas Tech University, 1997 (W. Kolarik, research adviser).

2. Montgomery, D. C., *Introduction to Statistical Quality Control*, 3rd ed., New York: Wiley, 1996.

3. Grant, E. L., and R. S. Leavenworth, *Statistical Quality Control*, 7th ed., New York: McGraw-Hill, 1996.

4. DeVor, R. E., T. Chang, and J. W. Sutherland, *Statistical Quality Design and Control: Contemporary Concepts and Methods*, New York: Macmillan, 1992.

5. Duncan, A. J., *Quality Control and Industrial Statistics*, 5th ed., Homewood, IL: Irwin, 1986.

6. Alt, F. B., "Multivariate Quality Control," *Encyclopedia of Statistical Sciences*, vol. 6, edited by N. L. Johnson and S. Kotz, New York: Wiley, pp. 110–122, 1985.

7. Lowry, C., and D. Montgomery, "A Review of Multivariate Control Charts," *IIE Transactions*, vol. 27, pp. 800–810, 1995.

8. Tracy, N., J. Young, and R. Mason, "Multivariate Control Charts for Individual Observations," *Journal of Quality Technology*, vol. 24, no. 2, pp. 88–95, 1992.

9. Sullivan, J., and W. Woodall, "A Comparison of Multivariate Control Charts for Individual Observations," *Journal of Quality Technology*, vol. 28, no. 4, pp. 398–408, 1996.

10. Hayter, A., and K. Tsui, "Identification and Quantification in Multivariate Quality Control Problems," *Journal of Quality Technology*, vol. 26, no. 3, pp. 197–208, 1994.

11. Prins, J., and D. Mader, "Multivariate Control Charts for Grouped and Individual Observations," *Quality Engineering*, vol. 10, no. 1, pp. 49–57, 1997-1998.

12. Mason, R., N. Tracy, and J. Young, "Decomposition of T^2 for Multivariate Control Chart Interpretation," *Journal of Quality Technology*, vol. 27, pp. 99–108, 1995.

13. Mason, R., C. Champ, N. Tracy, S. Wierda, and J. Young, "Assessment of Multivariate Process Control Techniques," *Journal of Quality Technology*, vol. 29, no. 2, 1997.

14. Hotelling, H., "Multivariate Quality Control," *Techniques of Statistical Analysis*, edited by Eisenhart et al., New York: Wiley, 1947.

15. Crosier, R., "Multivariate Generalizations of Cumulative Sum Quality Control Schemes," *Technometrics*, vol. 30, pp. 291–303, 1988.

16. Pignatiello, J., and G. Runger, "Comparisons of Multivariate CUSUM Charts," *Journal of Quality Technology*, vol. 22, pp. 173–186, 1990.

17. Lowry, C., W. Woodall, C. Champ, and S. Rigdon, "A Multivariate Exponentially Weighted Moving Average Control Chart," *Technometrics,* vol. 34, pp. 46–53, 1992.

chapter
14

PROCESS ADJUSTMENT—
INTRODUCTION TO AUTOMATIC
PROCESS CONTROL,
CONVENTIONAL MODELS

14.0 INQUIRY

1. What conditions must be met for successful automatic process control?
2. What options exist in automatic process control?
3. How does discrete automatic process control work?
4. How does continuous automatic process control work?
5. How are manual-control processes moved toward automatic process control?

14.1 INTRODUCTION

Process adjustment focuses on detecting when our process output has deviated, is deviating, or will likely deviate from target, and taking appropriate counteraction with an input or transformation variable to counter the deviation—bringing or holding the response to target. Adjustment means span from human, judgmental intervention to fully specified equipment-based control. This chapter and the next introduce process adjustment through automatic means. **Automatic means of process adjustment are based on quantitative models that reflect logical approaches.** If we strip the model equations out of automatic process control (APC) and replace them with the human's ability to reason and respond, we see amazing correspondences in operational strategies. For example, **on/off control, proportional, integral, and derivative**

control, artificial neural networks, and expert systems, to one degree or another, mimic human reasoning and responses. We introduce APC concepts intuitively, as well as mathematically, so that our readers can more easily see and appreciate the similarities. This approach not only explains the essence of automatic process adjustment—it explains in part human-based process adjustment.

The concepts of open- and closed-loop processes were introduced in Chapter 3. Chapter 9 expanded these concepts. Transformation from open-loop to closed-loop processes was described in terms of growth in the level of process calibration and measurement as well as growth in process understanding. Figure 9.3 introduced three basic strategic branches in process control systems: (1) subjective-based process control, (2) objective-based, SPC-aided process control, and (3) equipment-based, automatic process control. Chapters 10 through 13 focused on objective-based process control regarding process monitoring. In this chapter, we focus on process adjustment.

Our specific purpose here is to provide sufficient fundamentals so that both human-based and equipment-based process control adjustment options are clear. **In process definition, redefinition, and improvement efforts we want to identify feasible options for process control, using the most appropriate option in each application—considering physical, economic, timeliness, and customer service performance.** For example, if we consider only human-based alternatives for all applications, we may lose both effectiveness and efficiency. Or, if we consider equipment-based alternatives without high levels of process inertia and/or definite counteractions, we may waste resources and compromise process effectiveness.

Equipment-based process control possesses the same basic goal as human-based process control, that of keeping a process on target and working as intended. Automatic applications, however, demand that sensing, measurement, comparison, and correction all be imbedded in equipment. Early applications such as steam engine governors (or fly balls) used mechanical means to sense, measure, compare, and modify engine operations. Today, we see electronics, software, electromechanical, and other technologies integrated with sophisticated mathematical models, exercising control over a wide range of processes. Technical advantages include both faster response times and control fidelity.

Some processes lend themselves to automatic control and others do not. Typically, the presence of strong process inertia and definitive counteraction are prerequisites for successful automatic process control, whereas weak process inertia and indefinite counteraction situations require human intervention in the form of technical reasoning—in detection of need for control action as well as the determination of appropriate action.

The focus of this chapter is on the conventional branch of APC, depicted in Figure 9.3. One subbranch of conventional APC consists of discontinuous/discrete control technology, such as on/off switching. A second subbranch of conventional APC includes continuous control, including proportional, integral, and/or derivative control. Unconventional APC, including artificial neural networks and expert systems, is described in Chapter 15.

14.2 CLASSICAL CONTROL CONCEPTS

Modern APC technology has evolved from general control theory. The body of knowledge in this area is extensive. **The analytical nature of APC is quite different from that of SPC.** Here, **we see a related, but different, vocabulary as well as a more extensive focus on process modeling**

in a quantitative sense. Whereas we see primarily output data analysis in SPC, focused on both location and dispersion, we see input-output process modeling in APC, focused mainly on location. APC applications range from simple on/off applications such as thermostats to telemetric guidance control systems for space exploration.

Before we study different APC techniques to control a process, it is essential that we understand the process in terms of its characteristics and its behavior. **Responses are of two types: transient and steady-state.** In control systems, **transient response refers to the part of the response that approaches zero as time becomes very large.** Thus, if $y_t(t)$ is the system's transient response at time t,

$$\lim_{t \to \infty} y_t(t) = 0 \qquad\qquad [14.1]$$

The steady-state response of the system is the part of the total response that remains after the transient has died out. Thus, if the steady-state response of a system is denoted by $y_{ss}(t)$, the system output $y(t)$ can be written as

$$y(t) = y_t(t) + y_{ss}(t) \qquad\qquad [14.2]$$

Transients are usually observed in all physical systems. For example, physical inertia prevents the response associated with a typical control system from responding to changes in the input instantaneously. Transience is a significant part of the dynamic behavior of the system. **Steady-state response of a control system indicates where the system output will be as time increases.** The final accuracy of an automatic control system is assessed by comparing the steady-state response with the desired or target response. The system is said to have a steady-state error (also known as offset) if the final system output does not agree with the target exactly.

Systems are termed *stable* **or** *unstable*. **A stable system is one in which the transient response gradually moves toward a steady-state value and remains there.** In contrast, **an unstable system is one in which the response transient continues to increase with time.** If the system is unstable, either the output will reach some limiting value or the system "self-destructs." Figure 14.1 illustrates the concept of system stability and instability. Here, Figure 14.1a depicts a stable system, as its response approaches the target value, as time increases. The system in Figure 14.1b is unstable in that its response becomes more erratic and does not approach the target over time.

Obviously, system stability is a desirable property. In addition, the time to stabilize is also critical. For example, when counteraction is taken, the response is expected to stabilize at the target in a timely manner. Figure 14.2 illustrates this point graphically. Here, we introduce an input change Δ_{input} at time t_0. The physical nature of our control system requires that the process output "stabilize" within a tolerance interval $\pm\delta$ about the target, within a time period of t_s. In our illustration, our response easily meets our tolerance interval before time $t_0 + t_s$. The actual time to meet the response tolerance interval is depicted as $\Delta t_{stability}$. In general, stability timeliness is a critical characteristic in APC applications.

A system is referred to as a *dynamic* **system when one or more of its aspects change with time.** A dynamic system is typically described by two types of models: (1) input-output models and (2) state-space models. If $u(t)$ is the system's input at time instant t and $y(t)$ is the system's output at time t, then an input-output model assumes that the system output depends not only on

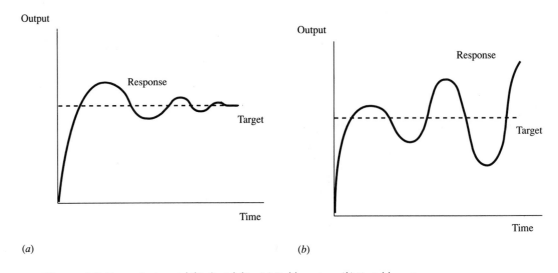

Figure 14.1   System stability/instability. (a) Stable system. (b) Unstable system.

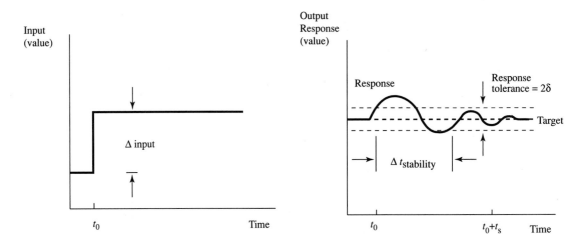

Figure 14.2   Timeliness in process stability.

$u(t)$ but also on the past history of the system's inputs and outputs. The input-output model of a single-input, single-output system can be represented as

$$y(t) = h(y(t - 1), y(t - 2), \ldots, y(t - n), u(t - 1), u(t - 2), \ldots, u(t - m)) \quad \textbf{[14.3]}$$

where positive integers n and m are the number of past outputs (also called the order of the system) and the number of the past inputs, respectively. The function h maps the past inputs and outputs to a new output. The input-state-output representation of a dynamic system can be given as

$$\frac{d\mathbf{x}(t)}{dt} = g(\mathbf{x}(t), \mathbf{u}(t)) \qquad \textbf{[14.4]}$$

$$\mathbf{y}(t) = h(\mathbf{x}(t))$$

where

$\mathbf{x}(t) = [x_1(t), x_2(t), \ldots, x_n(t)]^T \in \Re^n$ is the n-component state vector of the system
$\mathbf{u}(t) = [u_1(t), u_2(t), \ldots, u_p(t)]^T \in \Re^p$ is the system input vector
$\mathbf{y}(t) = [y_1(t), y_2(t), \ldots, y_m(t)]^T \in \Re^m$ is the system output vector
g and h = mappings defined as $g: \Re^n \times \Re^p \to \Re^n$ and $h: \Re^n \to \Re^m$

The vector $\mathbf{u}$ is the input to the dynamic system and contains both controllable and uncontrollable elements. The former are referred to as control inputs. The vector $\mathbf{x}(t)$ represents the state of the system at time t. The elements of $\mathbf{x}(t)$, $x_i(t)$ ($i = 1, 2, \ldots, n$) are called state variables. The state $\mathbf{x}(t)$ at time t is determined by the time t and the input u defined over the interval $[t_0, t)$. The output $\mathbf{y}(t)$ is determined by the time t as well as the state $\mathbf{x}(t)$ at time t. The discrete-time systems represented by difference equations corresponding to the differential equations in (14.4) take the form

$$\mathbf{x}(k + 1) = g[\mathbf{x}(k), \mathbf{u}(k)] \qquad \textbf{[14.5]}$$

$$\mathbf{y}(k) = h[\mathbf{x}(k)]$$

where $\mathbf{u}(\cdot)$, $\mathbf{x}(\cdot)$, and $\mathbf{y}(\cdot)$ are discrete time sequences.

When the functions g and h are not known, the system must be identified. Identification of the unknown system involves constructing a suitable identification model that, when subjected to the same input $\mathbf{u}(k)$ as the system, produces an output $\hat{\mathbf{y}}(k)$ that approximates $\mathbf{y}(k)$. If the functions g and h are known, control of the system involves designing a controller that generates the desired control input $\mathbf{u}(k)$ based on all the information available at instant k; see Narendra and Parthasarathy [1].

Systems are termed *linear* or *nonlinear*. If the principle of superposition applies, a system is said to be linear. The principle of superposition states that the response produced by the simultaneous application of two different inputs is the sum of the two individual responses. **A system is said to be nonlinear if the principle of superposition does not apply.** Thus, for a linear system, the response of a system receiving multiple inputs can be calculated by treating each input individually and adding the results. A similar addition of individual responses to get the system response cannot be performed for a nonlinear system. An example of a linear differential equation is

$$y = x + \frac{dx}{dt} + 3\frac{d^2x}{dt^2} \qquad \textbf{[14.6]}$$

An example of a nonlinear differential equation is

$$y = x + \left(\frac{dx}{dt}\right)^2 + \frac{d^2x}{dt^2} \qquad \textbf{[14.7]}$$

Most physical systems are nonlinear, and even so-called "linear" systems are really linear only in limited operating ranges; see Ogata [2]. For example, consider a tank containing a liquid that is being heated by steam flowing through a jacket around the tank. As the amount of steam

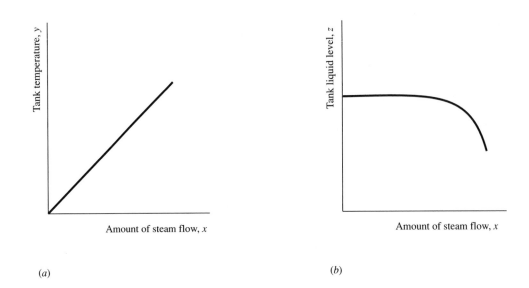

Figure 14.3   Example of linear and nonlinear curves (from a liquid temperature and level example).
(a) Linear system. (b) Nonlinear system.

(control variable x) passing through the jacket is increased, the temperature (controlled variable y) of the liquid in the tank rises proportionately. This linear relationship may be described as depicted in Figure 14.3a. Now, consider that the liquid continues to be heated by the steam. The level of the liquid in the tank first remains stable and may rise a little due to expansion, but as soon as the liquid reaches its boiling point, it begins to evaporate. From that point on the liquid level (controlled variable z) will begin to fall rapidly. This nonlinear relationship depicted in Figure 14.3b.

Example 14.1   **SALSA PRODUCTION PROCESS CASE—BASICS**   TexRosa specializes in the production of salsa. TexRosa commenced operations using simple, manually controlled processes. The company was marginally profitable for small orders. Rapid growth in the salsa market required modernization of their production process. Management proposed redefining their production system using a process-based format in order to address both throughput growth and competitive edge in product and process—business success. Figure 14.4a depicts a global view of TexRosa's production system, broken out by processes. Our focus in this case is the salsa holding/filling tank depicted in Figure 14.4b.

Preheated salsa enters the tank at the top and is released into jars at the bottom. Control valve V_1 is used to control and maintain the salsa level in the tank. The salsa temperature in the tank is maintained by steam that flows through the jacket of the tank and is then released into the drain at the bottom of the tank. Control valve V_2 allows for control of the amount of steam entering the jacket. A paddle continuously stirs the salsa in the tank. The paddle performs three functions: (1) it mixes solid and liquid contents, (2) it minimizes clogging in the salsa outlet, and (3) it distributes the heat from the steam jacket.

Two primary targets exist for the holding/filling tank contents: (1) a salsa height target of 3 meters and (2) a salsa temperature target of 84°C. The rate of salsa flow into the tank as well as steam flow through the jacket are manipulated constantly to address these targets. Many factors—different sizes of jars/jugs, packing jars/jugs not arriving at fixed intervals in time, and so on—cause the rate at which the salsa leaves the tank to vary. The salsa temperature, in the holding/filling tank, is impacted by two factors. First, as the

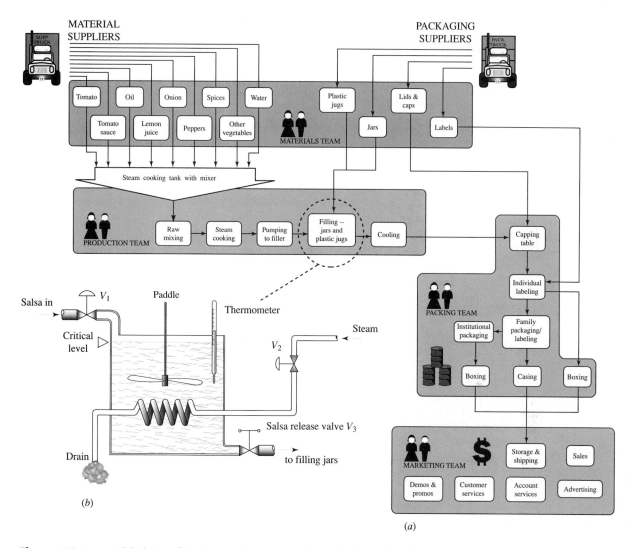

Figure 14.4   Global view of TexRosa production system (Example 14.1). (*a*) Production system overview. (*b*) Schematic representation of holding/filling tank.

throughput of salsa increases (decreases) the temperature decreases (increases) unless more (less) steam is introduced. Second, when the salsa level falls below (above) the target height level, the temperature increases (decreases) unless the steam is adjusted accordingly. It is critical to maintain the salsa temperature within 80°C to 88°C.

In the original manual-control method, shown in Figure 14.4*b*, an operator continually checks the temperature of the salsa in the tank before and during each filling operation. If the temperature falls below the target value (84°C), the operator turns on the steam flow through valve V_2 to reheat the tank. If the temperature rises over the target value (84°C), the operator turns off the steam flow through valve V_2. The operator also keeps an eye on the level of salsa in the tank. The operator opens the inflow valve V_1 when the salsa level in the tank falls below 3m. The operator closes the inflow valve V_1 when the salsa level in the tank exceeds 3m. After several weeks of experience, an operator becomes reasonably proficient with this manual control process [MS].

14.3   DISCONTINUOUS/DISCRETE CONTROL ACTION

All control actions can be classified as one of two types: discontinuous/discrete or continuous. Discontinuous/discrete control action is implemented through mechanical and electronic triggers. It represents a relatively crude but economical means for controlling simple processes, i.e., relatively few inputs and outputs, with well-understood control actions. Continuous controls afford more control fidelity than discontinuous controls, but require more sophisticated process models and technology. Both options see widespread applications.

In discontinuous control, produced by a discrete controller, the control action is one of two or more discrete values. Two-position control and multiposition control constitute discontinuous control. In contrast, continuous control, e.g., that provided by an analog controller, gives, theoretically, an infinite number of control action choices within a given range.

Two-position control, also known as on/off control, is the most widely used type of control for both industrial and domestic service. A familiar application is found in the operation of a domestic water heater. The controller turns the heater off if the water temperature rises above a maximum setpoint and on if the temperature falls below a minimum setpoint. The system cycles back and forth between the on and off positions. The obvious advantage of on/off control is its simplicity and low cost.

At time instant t, let the controller output be $v(t)$ and the error be $e(t)$. **In an on/off control scheme, the signal $v(t)$ remains at a maximum or minimum value, depending on whether the error signal is positive or negative.** In this case, the error $e(t)$ is defined as the difference between the actual response and its target value:

$$e(t) = \text{response } (t) - \text{setpoint } (t) \qquad \textbf{[14.8]}$$

Then,

$$v(t) = m_1 \qquad \text{for } e(t) > 0$$

$$v(t) = m_2 \qquad \text{for } e(t) < 0$$

where m_1 and m_2 are constants. For two-position control, we can define m_1 and m_2 above as 0 and 1, respectively, in our process control model.

When we pursue two-position control, we develop two-position logic to link our input variables to our output variables. In model building, we define appropriate output variable responses for all possible cases of input variable values. In simple cases, which do not involve timers and/or counters, we use truth tables and Boolean logic expressions to model our control logic. Typically, we produce ladder logic diagrams—compatible with programmable logic controllers; see Asfahl [3] for a more detailed description of two-position APC.

SALSA PRODUCTION PROCESS CASE—DISCONTINUOUS CONTROL   Revisiting our salsa **Example 14.2**
container filling subprocess, we have several options: (1) develop setpoints for our operator, (2) use discontinuous control, (3) use continuous control. For example, on our temperature sensor display we can mark off 84°C and instruct our operator to turn the steam valve V_2 to "on" when the temperature drops below the 84°C mark and then to off when it rises above 84°C; refer to Figure 14.4b. Likewise, we can set a 3-meter salsa tank head/height as a target. Then, we can instruct our operator to regulate V_1 accordingly.

Assume that the salsa level is maintained at a constant level and that the steam inflow rate q is either zero (m_1) or a positive constant (m_2). This kind of control action maintains the salsa temperature around our target, 84°C, by allowing the steam to flow through the valve when the temperature is below the setpoint and closing off the flow when the temperature exceeds the setpoint.

A plot of salsa temperature with the corresponding valve position is shown in Figure 14.5. When the salsa temperature exceeds the setpoint, the valve is closed, causing a gradual drop in temperature. Process inertia produces dead time and lags and causes the temperature to continue to rise for a short time, even after the valve has been turned off. Alternatively, when the temperature is below the setpoint, the valve is opened and the temperature gradually rises. Again, physical inertia causes a delay before the temperature shows an upward trend. **On/off control does not allow the operator/controller to maintain the response at the**

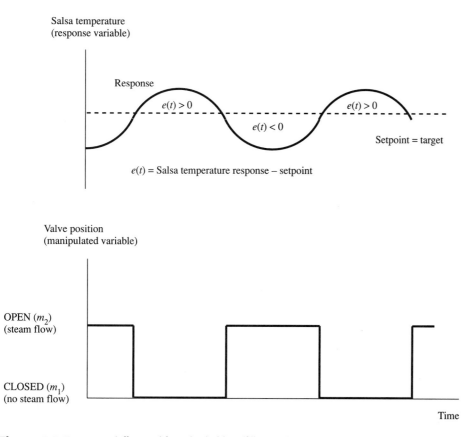

Figure 14.5   On/off control for salsa holding/filling tank temperature (Example 14.2).

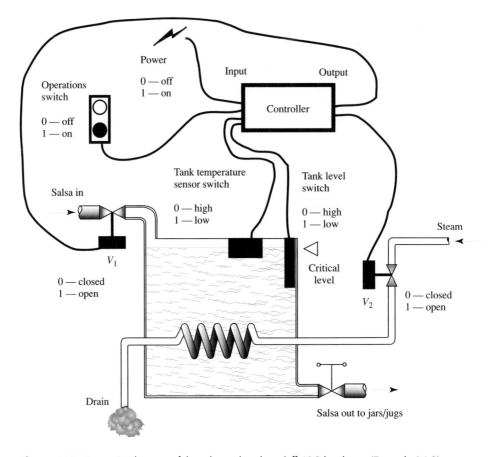

Figure 14.6   A schematic of the salsa tank with on/off APC hardware (Example 14.2).

setpoint exactly. The cycle of rising and falling temperature continues throughout the production process. But this fluctuation may be acceptable, provided we can hold our process response within an acceptable tolerance, e.g., 80°C to 88°C.

At this point, we describe a simplified APC subprocess for our salsa tank temperature and head control requirements of 84°C and 3 meters. Figure 14.6 provides a basic physical depiction of our salsa tank in terms of on/off APC. We define four input variables, electrical power, an on/off operations switch, a salsa level limit switch adjusted in the tank to trigger at the 3-meter level, and a temperature sensor-switch adjusted to trigger at the 84°C temperature level. The 0-1 input characteristics are as defined in Figure 14.6.

On the output variable side, we have two valves, V_1 controlling the salsa inlet flow and V_2 controlling the steam flow. In this case, the valves have only two positions, completely closed, denoted by 0, and wide open, denoted by 1. The controller is responsible for sensing inputs and providing counteractions as outputs to the valves.

Our next step is to define appropriate output responses for each input possibility. In this case, we use a truth table to relate our outputs to our inputs. The 2^4 or 16 combinations shown in Table 14.1 cover all possible input variable states. Now, we relate our outputs to our inputs. Here, we match our valve open/closed

Table 14.1   Input/output variables and relationships for the salsa tank controller (Example 14.2).

Input variables				Output variables	
Power (P)	Operations switch (S)	Tank temperature switch (T)	Tank level switch (L)	Salsa inlet valve (V_1)	Steam inlet valve (V_2)
0	0	0	0	0	0
0	0	0	1	0	0
0	0	1	0	0	0
0	0	1	1	0	0
0	1	0	0	0	0
0	1	0	1	0	0
0	1	1	0	0	0
0	1	1	1	0	0
1	0	0	0	0	0
1	0	0	1	0	0
1	0	1	0	0	0
1	0	1	1	0	0
1	1	0	0	0	0
1	1	0	1	1	0
1	1	1	0	0	1
1	1	1	1	1	1

possibilities to our input combinations on the basis of what action/counteraction our controller should relay to our valve actuators.

We distribute our 0's and 1's to the output variables according to how we want them to respond to the input states. Essentially we want our valves open only in a few cases, when the level is below 3 meters and/or the temperature is below 84°C. We do not want our valves open when the power is off or when our operations switch is off.

We next convert our truth table to Boolean logic equations. We develop one equation for each output. We express our logic in terms of valve open activity. Table 14.2 provides basic Boolean logic symbols and operations relevant to APC applications. Our equations are shown below along with logical manipulations that simplify our expressions.

$$V_1 = P \cdot S \cdot \overline{T} \cdot L + P \cdot S \cdot T \cdot L = P \cdot S \cdot L(\overline{T} + T) = P \cdot S \cdot L$$

$$V_2 = P \cdot S \cdot T \cdot \overline{L} + P \cdot S \cdot T \cdot L = P \cdot S \cdot T(\overline{L} + L) = P \cdot S \cdot T$$

Now, we convert our general logic models to ladder logic diagrams. These diagrams are compatible with two-position controllers, i.e., a programmable logic controller (PLC). Our ladder diagrams serve to "tell" our controller what to do on the output side, given a set of inputs. The ladder logic simply expresses our Boolean logic in a more useful form—useful for our controller. Our ladder logic diagram is depicted in Figure 14.7. We read our logic from left to right, one rung at a time.

Each rung on our ladder is equivalent to its respective Boolean equation. For example, the top rung expresses the fact that V_1 is to be open when P and S are in the on position and L is in the low position. Consequently, we can infer from the top rung that V_1 will close, go to the zero position, when either P, S, or L take "off" positions—i.e., are read at the 0 level. The lower rung is interpreted in similar fashion. Our controller essentially scans the inputs at some chosen frequency, reading input positions as 0's or 1's, and taking/maintaining output action accordingly, as 0's or 1's.

In order to choose between manual and automated technologies, we assess the benefits and burdens of each technology, i.e., physical, economic, timeliness, and customer service performance. In this

Table 14.2   Selected Boolean theorems and laws for discrete controls

Operators	Symbols/Examples	Comments
AND	$A \cdot B = AB$	Implies that both A and B must be in the on position before the expression is true
OR	$A + B$	Implies that either A or B or both must be on before the expression is true
NOT	$\bar{A}$	Implies not A, in terms that if A is true, the expression is false

Characteristic theorems:
$$A \cdot 0 = 0$$
$$A \cdot 1 = A$$
$$A + 0 = A$$
$$A + 1 = 1$$

Commutative law:
$$A + B = B + A$$
$$A \cdot B = B \cdot A$$

Associative law:
$$A + B + C = A + (B + C)$$
$$= (A + B) + C$$
$$A \cdot B \cdot C = A \cdot (B \cdot C)$$
$$= (A \cdot B) \cdot C$$

Distributive law:
$$(A + B) \cdot (C + D) = AC + AD + BC + BD$$
$$A \cdot B + A \cdot C = A \cdot (B + C)$$

Idempotent theorems:
$$A + A = A$$
$$A \cdot A = A$$

Negation theorem:
$$\bar{\bar{A}} = A$$

Inclusion theorems:
$$A \cdot \bar{A} = 0$$
$$A + \bar{A} = 1$$

Absorptive laws:
$$A + AB = A$$
$$A \cdot (A + B) = A$$

Reflective theorems:
$$A + \bar{A}B = A + B$$
$$A \cdot (\bar{A} + B) = AB$$
$$AB + \bar{A}BC = AB + BC$$

Consistency theorems:
$$AB + A\bar{B} = A$$
$$(A + B) \cdot (A + \bar{B}) = A$$

DeMorgan's laws:
$$\overline{AB} = \bar{A} + \bar{B}$$
$$\overline{A + B} = \bar{A}\,\bar{B}$$

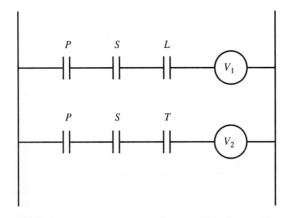

Figure 14.7   Ladder logic diagram (Example 14.2).

case, choosing between manual and on/off automatic alternatives, it is likely that we would opt for the automated on/off control system. Here, we would see similar physical, timeliness, and customer service performance. The economics would be the deciding factor—replacing the operator with relatively inexpensive and reliable electromechanical devices.

14.4 CONTINUOUS CONTROL ACTION

Computers are commonly used in process automation, for monitoring and control action generation. Although the processes are themselves continuous, the computer-based controller typically obtains knowledge of the process outputs at discrete intervals in time. For this reason, it is preferable to develop a mathematical model of the process in discrete form. Based on the control algorithm used, **continuous control actions can be classified into three basic categories:** (1) **proportional,** (2) **integral, and** (3) **derivative.**

PROPORTIONAL CONTROL

The simplest form of continuous control action is proportional control in which the controllable process input $v(t)$ at time instant t is proportional to the error $e(t)$. This can be mathematically stated as

$$v(t) = m + K_p e(t) \qquad \textbf{[14.9]}$$

where m is the initial value, the manual reset constant, and K_p is the gain or the proportionality constant of the controller. Writing the discrete form of the control algorithm in Equation (14.9), we have

$$v_n = m + K_p e_n \qquad \textbf{[14.10]}$$

where n refers to the time period.

Shifting this equation by one sample period, we have

$$v_{n-1} = m + K_p e_{n-1} \qquad \textbf{[14.11]}$$

The incremental change made to the controllable variable at the nth time period is obtained by subtracting Equation (14.11) from Equation (14.10):

$$\Delta v_n = K_p(e_n - e_{n-1}) \qquad \textbf{[14.12]}$$

SALSA PRODUCTION PROCESS CASE—PROPORTIONAL CONTROL   Proportional control **actions generally provide rapid adjustment responses and result in relatively stable systems.** Proportional control for the salsa temperature will result in a smoother response than that offered by on/off control. The valve V_2 is changed to a proportioning valve, which can be positioned to any degree of opening, from fully open to fully closed. The operator or on/off switching device is replaced with an automatic proportional controller that consists of a temperature sensor and a valve actuator mechanism, usually either pneumatic or electric. For each rate of flow of salsa in and out of the tank, there exists an ideal rate of flow of steam,

Example 14.3

corresponding to valve position *m*, in the jacket that maintains the salsa temperature at the setpoint. That is, *m*, also known as the manual reset constant, is the position of the valve when the error *e* is zero.

Proportional control involves setting the valve V_2 position to *m* and then changing the valve setting *v* by an amount proportional to the error value (i.e., the deviation of the salsa temperature response from its set-point). **The amount of adjustment/change action taken in the manipulated variable for a given error depends on the value chosen for K_p, the proportional gain of the controller.** Figure 14.8 illustrates the amount of valve change for three different values of the proportional gain ($K_p = 1$, $K_p < 1$, and $K_p > 1$).

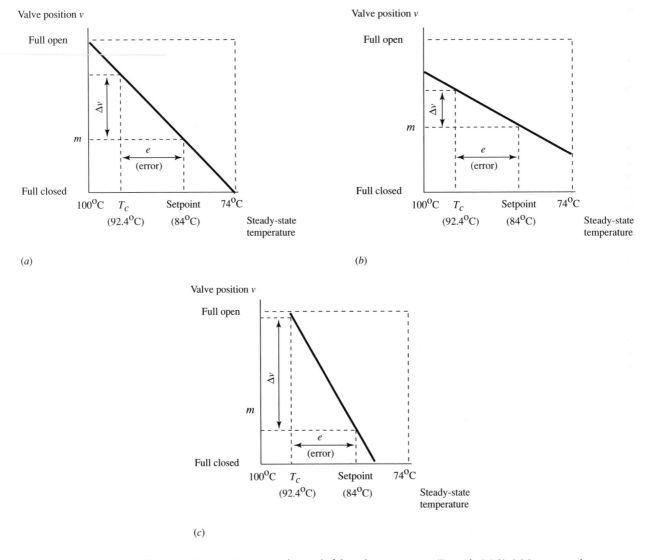

Figure 14.8   Proportional control of the salsa temperature (Example 14.3). (*a*) Proportional gain $K_p = 1$. (*b*) Proportional gain $K_p < 1$. (*c*) Proportional gain $K_p > 1$.

For a constant salsa level, when the valve is fully open, the temperature reaches a steady state of 100°C, and when it is fully closed, the steady state temperature is 74°C. Assuming a setpoint of 84°C, if the current temperature T_C of the salsa is, say, 92.4°C, then the valve is moved toward the closed position by an amount Δv. For $K_p = 1$ in Figure 14.8a, Δv equals e. Since the current temperature is 10% higher than the setpoint, Δv will equal 10% of its full travel toward the closed position. When a low gain, $K_p < 1$, is selected in Figure 14.8b, for the same error of 10%, the valve is moved by a smaller amount than before. In this case, a large error value requires a small change in the valve position. When a high gain $K_p > 1$ is selected, a small error value causes a large change in the valve position.

The percentage change in error required to move the valve/adjustment full scale is known as the *proportional band* **(PB):**

$$PB = \frac{1}{K_p} \times 100 \qquad\qquad \textbf{[14.13]}$$

The proportional band provides an intuitive feel for how small an error causes full corrective action. High percentages of PB, termed wide bands, correspond to less sensitive responses; and low percentages of PB, termed narrow bands, correspond to more sensitive responses.

A major disadvantage of proportional control action is that at steady state, it exhibits offset. That is, at steady state, there is a difference between the setpoint/target and the actual value of the controlled/response variable. To understand how offset occurs, we consider a sudden increase in the rate of flow of salsa into the tank from 0.2 m³/s to 0.25 m³/s. This causes the temperature to drop below the target 84°C. The proportional controller changes the valve position according to Equations (14.11) and (14.12), and more steam is allowed into the jacket. But, the manual reset constant was set for a lower flow rate of 0.2 m³/s. So, unless the value of m is adjusted manually, the steady-state temperature (for the new flow rate of 0.25 m³/s) will not reach 84°C.

The manual reset constant can be readjusted to remove the offset every time the flow rate changes or the setpoint changes, but the task requires considerable effort. Proportional control, alone, is not very practical when disturbances occur frequently or there are frequent changes in the setpoint. Figure 14.9 illustrates a plot of the temperature versus time of the salsa in the holding/filling tank. Here, the flow rate shift on the top strip drives the temperature response variable down, below the setpoint. Then, at discrete points in time, we see the controller open the valve in proportion to the change in the error value. In other words, as the salsa temperature begins to fall, and since observations of the process response are made at discrete time steps, the error is not noticed until time t_1, when the next observation is made. At time t_1, error e_1 is observed and a change in valve position v, proportional to the change in error, $e_1 - e_0$, is made. At time t_2, the valve is moved by an amount proportional to the change in error, $e_2 - e_1$. The control operation continues until the system stabilizes, i.e., the change in error is negligible.

In summary, even though the valve opens, it does not open enough, in the proportional scheme, to move the response temperature back to the setpoint/target. Rather, it flattens out below the target but above the steady-state response temperature that would have resulted had no control been exercised.

An intuitive approach to handle the offset problem is to change the value of the manipulated variable, the valve position, at a rate proportional to the error. This approach is known as integral control action.

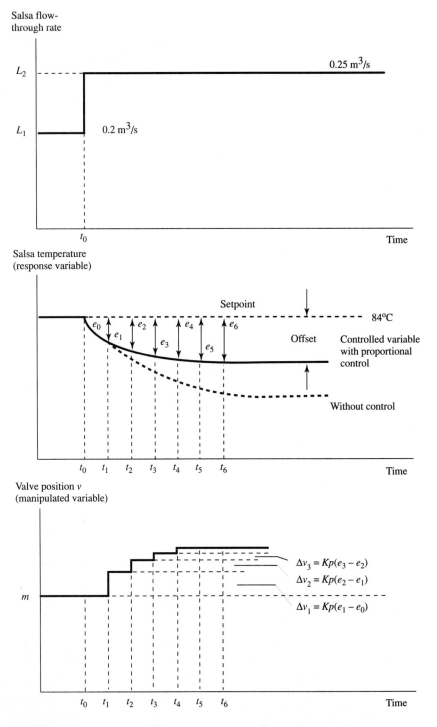

Figure 14.9   Proportional control in discrete time periods for the salsa filling tank (Example 14.3).

INTEGRAL CONTROL

In an integral control algorithm, the controllable process input $v(t)$ is the time integral of the error $e(t)$:

$$v(t) = m + K_i \int_0^t e(\tau) \, d\tau \qquad \text{[14.14]}$$

where K_i is the gain for the integral controller. The discrete form of this equation is

$$v_n = m + K_i \sum_{j=1}^{n} \Delta t e_j \qquad \text{[14.15]}$$

where Δt is the size of the sample interval. The controllable variable at one sample period earlier is

$$v_{n-1} = m + K_i \sum_{j=1}^{n-1} \Delta t e_j \qquad \text{[14.16]}$$

We can rewrite Equation (14.15) as

$$m_n = m + K_i \sum_{j=1}^{n-1} \Delta t e_j + K_i \Delta t e_n \qquad \text{[14.17]}$$

The increment in the controllable variable at time period n can be obtained by subtracting Equation (14.16) from Equation (14.17):

$$\Delta v_n = K_i \Delta t e_n \qquad \text{[14.18]}$$

Since the controllable variable is changed at a rate proportional to the time-integrated error, if this value is twice the previous value, the control element is moved twice as far. If the time-integrated error is zero, the control element is not moved. This means that **in the presence of integral control action, there is no offset at steady state; that is, the steady state error is zero.**

Proportional and integral control algorithms are usually combined to form what is known as proportional-plus-integral or PI control. The composite control manipulation is done according to the equation

$$\Delta v_n = (\Delta v_n)_p + (\Delta v_n)_i \qquad \text{[14.19]}$$

where the subscripts p and i refer to proportional and integral, respectively.

Figure 14.10 shows a plot of the valve position versus time for a step function change in error. At zero error, the valve position remains fixed at m. But as soon as there is a change in error, the valve position changes by an amount $K_p e$ due to the proportional control, and simultaneously, the integral portion of the controller moves the valve at a rate proportional to the error. **The final position of the manipulated variable/valve is therefore the sum of the position changes due to the two individual modes of control—the proportional control and the integral control. By combining integral control with proportional control, the offset is eliminated.** Because now two tuning adjustments are to be made, tuning a PI controller is more difficult than a proportional controller, which requires only one tuning adjustment, i.e., to set values for the gains, K_p and K_i.

Valve position v
(manipulated variable)

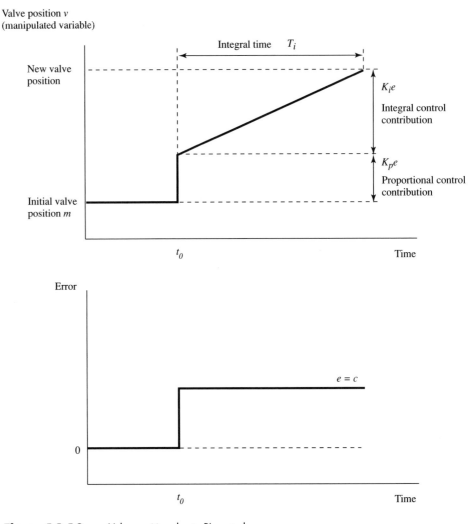

Figure 14.10   Valve position due to PI control.

Example 14.4 | **SALSA PRODUCTION PROCESS CASE—PI CONTROL**   Given an upward shift in our salsa inflow, indicated on the top strip in Figure 14.11, our response temperature on the middle strip tends to drop. Integral control action, on a discrete basis, such as executed by a digital computer, involves a time-sequenced concept (see the middle strip of Figure 14.11). We integrate (calculate the area between our setpoint/target and measured response curve) over time, and adjust our valve position on the bottom strip in proportion to our areas $(a_1, a_2, \dots)$. In our illustration, we see integral control first opening the steam valve, then closing the steam valve—and using this opening and closing concept, finally converging to the target setpoint.

In other words, from Figure 14.11, **under integral control, the manipulated variable/valve is moved by an amount proportional to the area under the error curve associated with the last time period.** No change is made at time t_0. At time t_1, the valve is opened by an amount proportional to a_1. This control action causes the response curve to deviate from the "no control" curve but physical inertia continues to increase

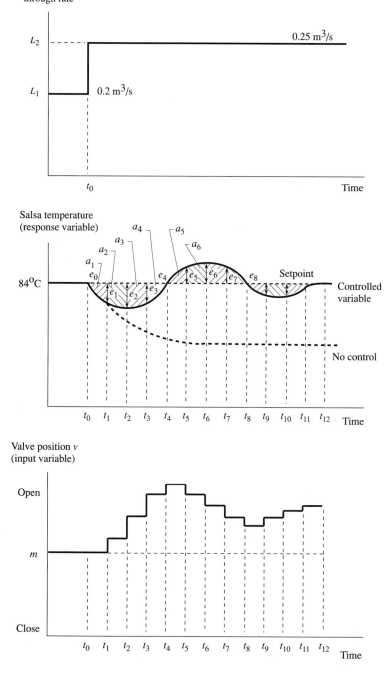

Figure 14.11 Integral control in discrete time periods for the salsa filling tank (Example 14.4).

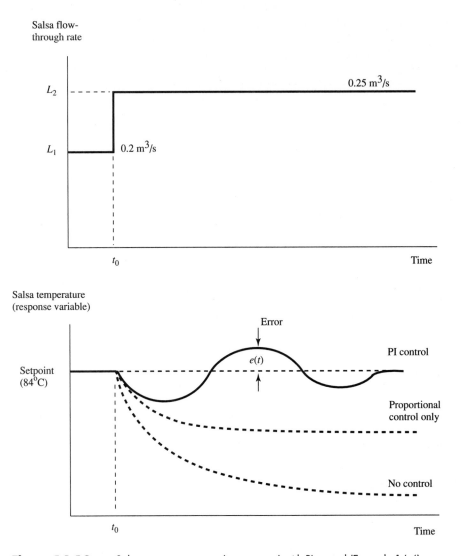

Figure 14.12   Salsa process response (temperature) with PI control (Example 14.4).

the error. At time t_2, the valve is opened by an amount proportional to a_2, causing the temperature to rise. As the temperature continues to rise, the area under the error curve decreases and the valve is opened by smaller amounts. By time t_5, inertia has caused the response to overshoot the setpoint, and the valve is closed by an amount proportional to a_5. As the temperature oscillates above and below the setpoint, the area under the error curve decreases over time, until finally, the temperature stabilizes at the setpoint.

Assuming we use PI control in our container-filling subprocess, we can expect to see enhanced performance to target. Figure 14.12 describes the process response using PI control. The plot also shows the process responses when only proportional control is implemented and when no control is implemented. **There is an obvious improvement in control due to the addition of the integral control mode. The PI control removes the offset by returning the error to zero.**

DERIVATIVE CONTROL

In general, process adjustment is based on error compensation. This compensation is achieved by taking into account the nature of the error change. Adjustment can be accomplished by proportional and/or integral controller actions, as previously described. Derivative control action serves to complement proportional and integral control action. Derivative action focuses on the timeliness nature of the counteraction/adjustment.

An effective and efficient control action involves two aspects: the size of the control action/adjustment, and the speed of the control action/adjustment. Proportional and integral control deal with specific approaches in the determination of counteraction on the input variable, as a function of the magnitude of change exhibited by the output variable, relative to the target/setpoint. Derivative control adjusts the counteraction taken in the input variable to the rate of change in the response, relative to the target/setpoint. In other words, **proportional and integral control action are based on the magnitude of the error, while derivative control action is based on the rate of change of the error**. Therefore, we can say that proportional and integral control on one side and derivative control on the other side pertain to two different control concepts. However, as we have stated previously, these two concepts are complementary in achieving an effective and efficient control action.

The derivative control algorithm uses the derivative or the time rate of change of the error to update the control action. The control action is given by

$$v(t) = m + K_d \frac{de(t)}{dt} \qquad [14.20]$$

where K_d is the controller's gain. In discrete form, assuming that Δt is a very small increment of time,

$$v_n = m + K_d \sum_{j=1}^{n} \frac{\Delta e_j}{\Delta t_j} \qquad [14.21]$$

and

$$v_{n-1} = m + K_d \sum_{j=1}^{n-1} \frac{\Delta e_j}{\Delta t_j} \qquad [14.22]$$

The increment in the controllable variable at time period n is therefore

$$\Delta v_n = K_d \frac{\Delta e_n}{\Delta t_n} \qquad [14.23]$$

Although it is conceivable, in theory, to have a control action that is solely based on derivative control, it is not practical. If the error is unchanging, no change will be made to the control action, even though the error may be high. Derivative control is usually combined with proportional control. The combination is called proportional-plus-derivative control, or PD control. The incremental control action in PD control is given by

$$\Delta v_n = K_p(e_n - e_{n-1}) + K_d \frac{de_n}{dt} \qquad [14.24]$$

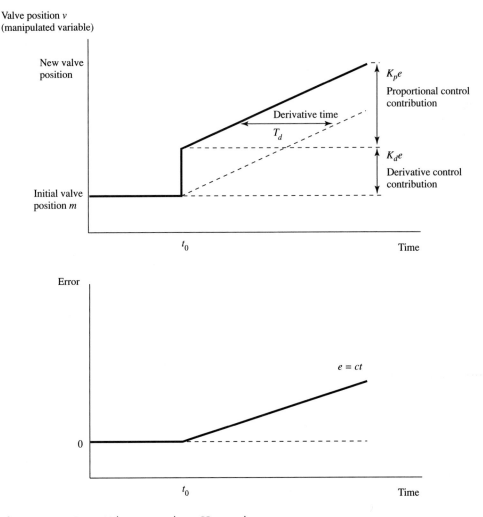

Figure 14.13   Valve position due to PD control.

Figure 14.13 depicts a plot of the valve position versus time when PD control is implemented. The error is initially zero and then begins to increase at a rate

$$\frac{\Delta e}{\Delta t} = ct \qquad \textbf{[14.25]}$$

where c is a constant and t is the time. This type of error function is known as a ramp function. The derivative portion of the control contributes an immediate, constant valve change that is proportional to the rate of change of the error ct. The proportional portion of the control contributes a valve change that is proportional to the error. Here, the slope of the valve position change is the same as the slope c of the error change.

PD control results in improved control in various applications, especially when the rate of change in error is very important, such as temperature control. Hughes [4] points out that in temperature loops, large time delays generally occur between the application of corrective action and the process response; therefore, derivative action is used to control steep temperature changes.

SALSA PRODUCTION PROCESS CASE—PD CONTROL   Using the approximate form for the   **Example 14.5**
derivative control scheme in Equation (14.23), we can develop an approximate valve change diagram for the salsa tank, similar to those developed for previous salsa cases. The derivative control case is depicted in Figure 14.14. Here, we see our usual shift in salsa flow at t_0, followed by the temperature response and valve response at t_1. Our derivative response is based on a point-to-point slope, i.e., a crude derivative form, immediately before time point t_1. Likewise, additional valve changes are shown for t_2, t_3, ...

As our temperature response flattens, our error slope decreases, and hence, our derivative control action approaches zero—even though the error is clearly not zero. Hence, we see a crude graphical demonstration of the limitations of pure derivative control. Nevertheless, we see that derivative control action is effective. Its impact is especially pronounced for a sharp change in error. For rapid error changes, we obtain a large slope/derivative, and hence our valve responds rapidly. For PD control, Equation (14.24), we combine proportional and derivative action at each point in time.

PID Control

The combination of all three modes of control, proportional-plus-integral-plus-derivative control, results in PID control—a widely used control algorithm. The composite control manipulation obtained by combining the effects of the three basic control actions is given by

$$\Delta v_n = (\Delta v_n)_p + (\Delta v_n)_i + (\Delta v_n)_d \qquad \textbf{[14.26]}$$

where the subscripts p, i, and d refer to proportional, integral, and derivative, respectively.

The PID controller, or its simplified version the PI controller, is the most common algorithm in use today. It exhibits no offset and provides rapid responses. Sometimes modified versions of the standard algorithms described above are used.

14.5 Controller Tuning

Up to this point, our discussion has been conceptual. For example, we described basic control alternatives and how they work in a generic sense. **Once we have decided what algorithm is to be used to generate the control action, the next phase in the implementation of the controller is the selection of the numerical values of the constants in the algorithm, otherwise known as tuning.** In our salsa example, tuning involves calculating and setting values for the parameters K_p, K_i, and K_d so that the controller will behave in the required manner, i.e., perform PID control action.

One of the first methods proposed for tuning feedback controllers was the Ziegler-Nichols method. Ziegler and Nichols proposed rules for determining values of the proportional, integral,

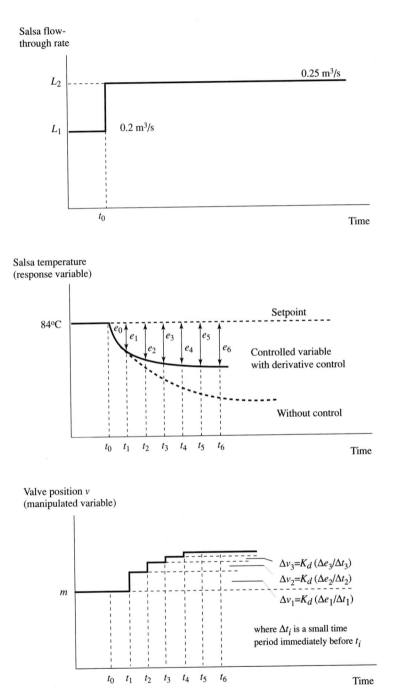

Figure 14.14   Derivative control in discrete time periods for the salsa filling tank (Example 14.5).

Table 14.3   Ziegler-Nichols tuning rules—based on ultimate sensitivity S_u and ultimate period P_u.

Type of controller	K_p	K_i	K_d
P	$0.5S_u$	0	0
PI	$0.45S_u$	$0.54\,S_u/P_u$	0
PID	$0.6S_u$	$1.2\,S_u/P_u$	$0.075\,S_uP_u$

and derivative constants based on the transient response characteristics of a given process. Engineers use Ziegler-Nichols and other tuning rules on line by performing experiments on the process. These rules require no knowledge of the process. The constants K_p, K_i, and K_d are set to zero; K_p is increased until the system is slightly unstable.

For a closed-loop system, when the controller gain is increased, the loop will tend to oscillate. As the gain is continuously increased, the oscillations grow larger and larger until the system becomes unstable. The ultimate sensitivity is the maximum acceptable value of gain (for a controller with only a proportional mode in operation) before which the closed-loop system becomes unstable. The period of the sustained oscillations is called the ultimate period. Table 14.3 presents the Ziegler-Nichols tuning rules based on ultimate sensitivity S_u and ultimate period P_u. Ziegler-Nichols and other rules are presented and explained in Ogata [2] and Ellis [5].

If we imagine our process had a "knob" for each parameter to be adjusted, the proportional controller would have one knob, K_p, the PI controller would have two knobs, K_p and K_i, and the PID controller would have three, K_p, K_i, and K_d. In more practical terms, **tuning refers to calculating how much each knob needs to be turned and in which direction.**

Obviously, **the difficulty in tuning the controller increases as the number of parameters increases.** PID controllers have three knobs that have to be tuned simultaneously. Despite their difficulty in tuning, PID controllers are popular because when good tuning is implemented, they respond rapidly and provide good control.

14.6   TRANSFER FUNCTIONS AND BLOCK DIAGRAM REPRESENTATION

The transfer function of a system or a component of a system is defined as the ratio of its output to its input, with all initial conditions set to zero. The transfer function of a system is a mathematical model which expresses the differential equation that relates the system's output variables to the input variables. In its simplest form, a transfer function can be a multiplying operator.

Consider the process of buying candy bars as an example. We pay, say, 80 cents (process input) and we receive one candy bar (process output). If the input unit is in cents and the output unit is in number of candy bars, the transfer function of this process is simply 1/80. Unfortunately most systems are not as simple as this illustration, and they therefore involve more complex transfer functions. Laplace transforms for continuous processes and z transforms for discrete time systems are utilized to simplify the transfer functions. Detailed explanations of Laplace and z transforms can be found in introductory control systems books including Isermann [6] and Warwick [7].

The discrete transfer function is obtained from difference equations by using the backward shift operator B. Using the operator B, a variable y at time n is represented by y_n; at time $n - 1$ is represented by $B^1 y_n$ or simply By_n; and at time $n - j$ is represented by $B^j y_n$. To demonstrate the application of the backward shift operator, we rewrite Equation (14.12) as

$$v_n - v_{n-1} = K_p(e_n - e_{n-1})$$ **[14.27]**

Introducing the B operator yields

$$(1 - B)v_n = K_p(1 - B)e_n$$ **[14.28]**

Example 14.6 | **SALSA PRODUCTION PROCESS CASE—TRANSFER FUNCTION**   We revisit the salsa production subprocess where the preheated salsa is poured into the top of the filling tank and then released from the bottom of the tank; see Figure 14.4b. Figure 14.15 shows a simplified version of the tank, the salsa inflow, and the salsa outflow. The area A of the tank is 4.9 m² and the hydraulic resistance R offered by the pipe is 1.2 s/m². The rate of flow out of the tank, x_2, is 0.0556 m³/s, and the rate of flow into the tank, x_1, can be varied between 0 and 0.3 m³/s.

We desire to maintain the level h of salsa in the tank at a fixed height (3 m). To control the level h by manipulating the rate of flow of salsa x_1 (through valve V_1), we treat x_1 as the system input and the head h as the system output. The equation governing this flow process is given by

$$x_1 - x_2 = A\,\frac{dh}{dt}$$ **[14.29]**

and

$$x_2 = \frac{h}{R}$$ **[14.30]**

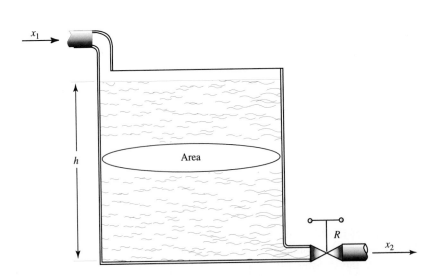

Figure 14.15   A schematic of salsa flow into and out of the filling tank (Example 14.6).

Substituting Equation (14.30) in Equation (14.29), we have

$$x_1 = A \frac{dh}{dt} + \frac{h}{R} \qquad \textbf{[14.31]}$$

The discrete form of Equation (14.31) is

$$x_1 = \frac{A[h(t) - h(t-1)]}{T} + \frac{h(t)}{R} \qquad \textbf{[14.32]}$$

where T is the time period and the other notations are the same as before. The B operator can be introduced to yield

$$x_1 = \frac{AR(1-B)h + Th}{TR} \qquad \textbf{[14.33]}$$

The transfer function of the process, represented by $G(B)$, is obtained by taking the ratio of the process output to the process input:

$$G(B) = \frac{h}{x_1} = \frac{TR}{AR(1-B) + T} \qquad \textbf{[14.34]}$$

Assuming a time period of 1 second and substituting the values of the variables into Equation (14.34), we obtain the transfer function

$$G(B) = \frac{1(1.2)}{4.9[1.2(1-B)] + 1} \qquad \textbf{[14.35]}$$

[SP]

APC block diagrams are simple input-output schematic diagrams. They typically consist of rectangular blocks, arrows, and circles. The blocks represent a component or a combination of components of a system, the arrows represent information flow, and the circle represents an operator. The box symbol with e input and y output, as depicted in Figure 14.16, represents the equation

$$y = Ge$$

In other words, the output y is equal to the operator G, operating on the input e. G is the transfer function for the component shown. The nodal symbol with inputs r and y and output e represents the equation

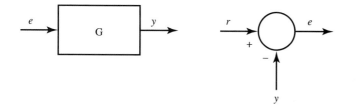

Figure 14.16   APC blocks and nodes.

$$e = r - y \qquad\qquad\qquad \textbf{[14.36]}$$

The variables y and r enter the circle where their difference (calculated using the algebraic symbols) is assigned to the variable e.

Example 14.7   **BLOCK DIAGRAMS**   Consider the following equation in which $x_1, x_2, \ldots, x_n$ are variables and $a_1, a_2, \ldots, a_n$ are mathematical operators or general coefficients:

$$x_3 = a_1 x_1 + a_2 x_2 - 0.8$$

To draw the block diagram for this equation, we identify all blocks, inputs, and outputs. Here, x_3 is the output and the terms on the right-hand side of the equation are combined at a summing point, as shown in the upper portion of Figure 14.17.

In the $a_1 x_1$ term, a_1 may represent any mathematical operation. Therefore, $a_1 x_1$ can be represented by a single block, with x_1 as the input, $a_1 x_1$ as the output, and the coefficient a_1 inside the block. Representing the term $a_2 x_2$ in the same manner, the block diagram for the entire equation can be represented as shown in the lower portion of Figure 14.17.

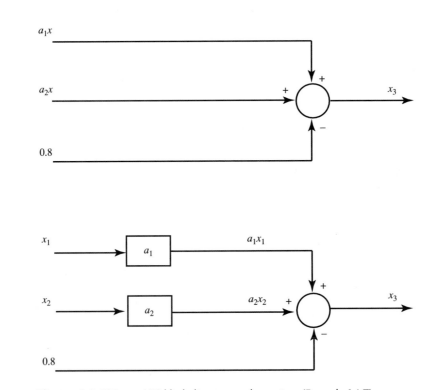

Figure 14.17   APC block diagrams and equations (Example 14.7).

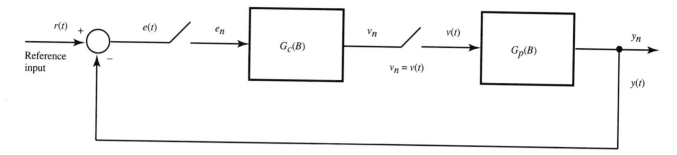

Figure 14.18 Example block diagram for a closed-loop process control system (Example 14.8).

Since transfer functions are input-output mathematical concepts and block diagrams are input-output pictorial representations, control engineers often use them in conjunction with each other to portray the interrelationships between the components of a system. The block diagram for a closed-loop process control is shown in Figure 14.18.

To obtain the closed-loop transfer function $G(B)$ for the block diagram in Figure 14.18, given that

$$G_c(B) = k$$

and

$$G_p(B) = \frac{TB}{(1 - B)}$$

we take the following steps:

The system transfer function $G(B)$ is given by the ratio y_n/r_n. From the block diagram, we obtain the transfer function for the process component as

$$G_p(B) = \frac{y_n}{v_n} = \frac{TB}{(1 - B)} \qquad \textbf{[14.37]}$$

and the transfer function for the controller component as

$$G_c(B) = \frac{v_n}{e_n} = k \qquad \textbf{[14.38]}$$

These two equations yield

$$\frac{y_n}{e_n} = G_p(B)G_c(B) \qquad \textbf{[14.39]}$$

The block diagram also gives us the equation

$$e_n = r_n - y_n \qquad \textbf{[14.40]}$$

We begin with the equation

$$y_n = \frac{y_n}{e_n}\, e_n \tag{14.41}$$

Using Equation (14.40) to replace e_n in the numerator, we get

$$y_n = \frac{y_n(r_n - y_n)}{e_n} \tag{14.42}$$

Substituting Equation (14.39) in Equation (14.42), we get

$$y_n = G_c(B)G_p(B)(r_n - y_n) \tag{14.43}$$

This yields

$$\frac{r_n}{y_n} = \frac{1}{G_c(B)G_p(B)} + 1 \tag{14.44}$$

from which we obtain the general form of the system transfer function as

$$G(B) = \frac{y_n}{r_n} = \frac{G_c(B)G_p(B)}{1 + G_c(B)G_p(B)} \tag{14.45}$$

Substituting Equation (14.37) and Equation (14.38) in Equation (14.45), we obtain the closed-loop system transfer function as

$$G(B) = \frac{kTB}{(1 - B)kTB} \tag{14.46}$$

[SP]

Example 14.8 | **TRANSFER FUNCTION AND BLOCK DIAGRAM**   Consider the block diagram shown in Figure 14.18. The general form of the system transfer function from Equation (14.45) is

$$y(t) = \frac{G_c(B)G_p(B)}{1 + G_c(B)G_p(B)}\, r(t) \tag{14.47}$$

Assuming the controller to be a proportional controller, we have

$$v(t) = K_p e(t) \tag{14.48}$$

Since $v(t)$ is the result of the transformation $G_c(B)$ applied to the error $e(t)$, we now have

$$v(t) = G_c(B)e(t) \tag{14.49}$$

which is identical to

$$v_n = G_c(B)e_n \tag{14.50}$$

Consider as an example the discrete-time process component transfer function

$$G_p(B) = \frac{0.5(B - 0.6)}{(B - 0.2)(B + 0.3)} \tag{14.51}$$

First we need to consider steady-state conditions that can be obtained by setting B to unity in $G_p(B)$. For a reference or target input of say, 3 units, the steady-state output can be obtained from Equation (14.47) as follows:

$$\lim_{t \to \infty} y(t) = \frac{G_c(1)G_p(1)}{1 + G_c(1)G_p(1)} (3) \qquad \textbf{[14.52]}$$

Since $G_c(B) = K_p$, we have $G_c(1) = K_p$. Now

$$G_p(1) = \frac{0.5(1 - 0.6)}{(1 - 0.2)(1 + 0.3)} = \frac{1}{5.2} \qquad \textbf{[14.53]}$$

Hence the steady-state value of $y(t)$ is

$$\frac{G_c(1)G_p(1)}{1 + G_c(1)G_p(1)} (3) = \frac{K_p(3)}{5.2 + K_p} \qquad \textbf{[14.54]}$$

And, for $K_p \gg 5.2$, Equation (14.54) approaches a constant value of 3.

Different values of K_p will result in different values of steady-state output. In order for the steady-state output to equal the reference value of 3 units, K_p must be very large. Else, the steady-state offset will be large. For example, if $K_p = 5$, then the steady-state output can be obtained from Equation (14.54) as $y(t) = 1.47$, resulting in an offset of $3 - 1.47 = 1.53$ units. [SP]

Review and Discovery Exercises

Review

14.1. How are the goals of automatic and human-based process control the same? Different? Explain.

14.2. How does the process characteristic of inertia facilitate the application of adjustment and automatic process control? What other process characteristics facilitate adjustment/automatic process control? Explain.

14.3. Explain the difference between a stable and an unstable system in relation to adjustment/APC and in relation to monitoring/SPC. How are they the same? Different?

14.4. What is the difference between discontinuous and continuous control actions? Explain.

14.5. In order to automate the manual flow and temperature control activities at the salsa tank, described in Example 14.1, the Salsa Production Process Case—Basics, what, in general, must be accomplished? Explain.

14.6. Why is it virtually impossible to hold exactly to a nominal target (setpoint) in on/off control? Explain.

14.7. How does proportional control work? Explain the concept.

14.8. How does integral control work? Explain the concept.

14.9. What advantage does a combination of proportional and integral control offer? Explain.

14.10. How does derivative control work? Explain the concept.

14.11. What advantages does a combination of proportional and derivative control offer? Explain.

14.12. What advantages does PID control offer? Explain.

14.13. What is controller tuning? Why/how is it done? Explain.

14.14. What is a transfer function? What purpose does it serve? Explain.

14.15. What purpose does an APC block diagram serve? How are they constructed? Explain.

DISCOVERY

14.16. Why is feedback control so widely used? Are there any situations where feedback control would not be useful? Explain.

14.17. Given a proportional gain of $K_p = 1$, how much will a valve open or close, in terms of its full travel, if a response is reduced from its target at 100 to a value of 75?

14.18. How does PI control eliminate offset? Explain.

14.19. Why is derivative control typically not attempted in a pure sense? Explain.

14.20. A liquid-level control system for a tank, as shown in Figure 14.15, is equipped with a proportional controller to control the head/liquid level in the tank. The controller senses the head change using a float-transducer device. If a change occurs it takes appropriate action, closing or opening the input valve. We know that at the operating setpoint level $H_0 = 5$ m, the valve is initially 50% open, $m = 50\%$, and when the valve is fully open, the inflow rate is 0.30 m^3/s. We also know that the controller sampling rate is set at 20-second intervals, t, and its gain, $K_p = 800$. The tank dimensions are radius $r = 3.0$ m and height $h = 7$ m. We assume that the percentile change in valve position will determine an equal percentile change in the inflow (a linear displacement valve). The initial outflow is $Q_0 = 0.15$ m^3/s, and for this value, the tank level is maintained at the H_0 value. At the beginning of a sampling period, a sudden outflow increase of $\Delta q_0 = 0.1$ m^3/s occurs lasting 60 sec.; after that the outflow decreases and stabilizes at 0.20 m^3/s.

   a. Considering our assumptions, determine the approximate inflow valve action. Using strip plots as a function of time, show the head and valve inflow values at the end of each sampling period for five consecutive periods. Pattern your quantitative response related to tank head according to our qualitative depiction related to temperature in Figure 14.9. Draw outflow, head response and error, and inflow valve position diagrams as a function of time. Label values on the diagrams.

   b. Determine the offset at the end of the fifth sampling period.

14.21. A liquid-level control system for a tank, as shown in Figure 14.15, is equipped with an integral controller to control the head/liquid level in the tank. The controller senses the head change using a float-transducer device. If a change occurs it takes appropriate action, closing or opening the input valve. We know that at the operating setpoint level $H_0 = 4.5$ m, the valve is 33.33% open, $m = 33\%$; and when the valve is fully open, the inflow rate is 0.15 m^3/s. We also know that the controller sampling rate is set at 30-second intervals and its gain $K_i = 30$. The tank dimensions are radius $r = 2.0$ m and height $h = 6$ m. The following relation describes the valve position as a percentage of full capacity:

$$v(t)\% = 33.33\% + K_i e(t)$$

We assume that the percentile change in valve position will determine an equal percentile change in the inflow. When the valve is opening, the inflow can be increased up to the maximum value of 0.15 m³/s; and when the valve is closing the inflow can change from the actual value to zero.

  a. If the initial outflow is $Q_0 = 0.05$ m³/s and a sudden, sustained outflow increase of $\Delta q_0 = 0.03$ m³/s occurs at the beginning of the first sampling period, determine the head and inflow values at the end of each sampling period for seven consecutive periods. Pattern your quantitative response related to tank head according to our qualitative depiction related to temperature in Figure 14.11. Use simple geometric approximations to determine the time-integrated error; i.e., use sequences of triangles and rectangles. Express your responses in graphical form, providing strip plots for the outflow, head response and error, and inflow valve position, as a function of time.

  b. Is there any offset from the initial value of the head level? Compare and contrast your results here with those in Problem 14.20?

14.22. For the process of guiding/steering an automobile, describe the nature of the process and the nature of the control used by the driver, i.e., inherent discrete, continuous methods. Explain the concept of control system stability in the context of driving an automobile.

14.23. Select a production system of your choice. Briefly describe/characterize a process within this system. Identify the critical control points. Which points lend themselves to human-based control? Why? To automated control? Why? Explain.

14.24. Select a production system of your choice. Briefly describe/characterize a process within this system and identify a critical control point that currently uses human-based control. Describe what it would take to convert this control point to APC.

REFERENCES

1. Narendra, K. S., and K. Parthasarathy, "Identification and Control of Dynamical Systems Using Neural Networks," *IEEE Transactions on Neural Networks*, vol. 1, no. 1, pp. 4–27, 1990.

2. Ogata, K., *Modern Control Engineering*, 3rd ed., Upper Saddle River, NJ: Prentice Hall, 1997.

3. Asfahl, C. R., *Robots and Manufacturing Automation*, 2nd ed., New York: Wiley, 1992.

4. Hughes, T. A., *Measurement and Control Basics*, 2nd ed., Research Triangle Park, NC: Instrument Society of America, 1995.

5. Ellis, G., *Control System Design Guide—Using Your Computer to Develop and Diagnose Feedback Controllers*, New York: Academic Press, 1991.

6. Isermann, R., *Digital Control Systems*, vol. 1, New York: Springer-Verlag, 1989.

7. Warwick, K., *Control Systems—An Introduction*, Englewood Cliffs, NJ: Prentice Hall, 1989.

chapter

15

PROCESS ADJUSTMENT— INTRODUCTION TO AUTOMATIC PROCESS CONTROL, UNCONVENTIONAL MODELS

15.0 INQUIRY

1. What limitations exist for traditional APC?
2. How is APC technology expanding to meet contemporary process needs?
3. What constitutes a nonparametric APC model?
4. How are APC and SPC technologies used together?

15.1 INTRODUCTION

Traditional APC models involving discrete controls and continuous proportional, integral, and derivative control models have found many uses in modern production processes. **Typically, traditional APC technology requires that explicit process models be developed,** as described in Chapter 14. In our previous discussions, we focused exclusively on feedback control systems with highly specified mathematical models.

Most traditional APC applications view the process world as static and deterministic, rather than dynamic and probabilistic. However, we know that the process world is probabilistic and dynamic in general. Hence, **in order to provide process controls that have the ability to track and adjust complex processes, higher levels of model flexibility and model intelligence are required.** The focus of this chapter is to overview several extensions of traditional APC, as well as to describe several nontraditional APC technologies that have emerged and show promise in process control work. At the same time, we need to be aware of the process adjustment nature of advanced APC technologies and their relationship to human-based process control strategies.

15.2 ADVANCED CONCEPTS IN CONVENTIONAL APC

The basic APC methods described so far imply single, simple control loops. Many processes are far from simple. **Advanced APC concepts, which utilize APC basics in different forms, exist to allow for more effective process control in complex physical systems.** Three concepts are briefly described in this section: cascade control, ratio control, and feedforward control.

CASCADE CONTROL

When a very slow process is being controlled, errors due to disturbances entering the system may not be noticed for a relatively long time. This timeliness issue causes delays in the application of corrective action. Also, a relatively long period of time may be required before the result of corrective action is noticeable. **Cascade control is a technique used to deal with process control where large response time lags are expected.**

In cascade control, we locate our primary controlled variable and determine our primary/main process control loop. Then, we locate secondary controlled variables within the primary process control loop. We then control these secondary control variables with our APC technology.

In cascade control, we nest feedback loops within other larger feedback loops. An intermediate variable within the outer process loop is identified and used as the controlled variable for the inner loop. In essence, we divide or split the time lag into parts. This provides the opportunity to detect disturbances and take corrective actions more promptly. For more information on cascaded process control see Hughes [1].

RATIO CONTROL

Ratio control is a control structure that maintains a desired ratio between two or more process variables. When an operation involves mixing two or more streams together, continuously, to maintain a steady composition in the resulting mixture, e.g., in chemical industries, a conventional flow controller can be used for one stream (called the primary or wild flow) and a ratio controller can be used to maintain the second stream flow at some fraction of the primary flow rate.

The control action selected for the ratio controller is normally of the PI type. The signal from the wild flow is used as the ratio input to the ratio controller. This signal is multiplied by a ratio setting to determine the setpoint of the flow controller. **Unlike a cascade control system, a ratio control system's loops are designed individually.**

FEEDFORWARD CONTROL

The feedforward control concept was initially introduced in Chapter 3. Feedforward process control is simple in concept but difficult to execute in practice. Here, we apply a sensor to a variable outside the immediate process, i.e., an upstream uncontrollable variable. We feed this information into our process model, along with the usual process desired result/setpoint. **In feedforward control, the controller takes action on the basis of the "anticipated" error—in order to manipulate a controllable variable at the process input,** e.g., a valve.

Ideally, feedforward APC allows us to hold to target "perfectly" over time. It represents the ultimate means of process control. Practically, difficulty in feedforward APC comes from the fact that we must predict what effect the outside variable will have on our process result and then take appropriate counteraction. Prediction includes physical as well as time lag information content regarding process impact. This element of anticipation, both physically and timewise, provides a significant challenge in process modeling. Nevertheless, humans utilize an intuitive form of feedforward control in performing activities throughout the day, e.g., in walking, driving, and so on.

15.3   PROCESS IDENTIFICATION AND NONPARAMETRIC MODELS

Finding a solution to a control problem typically requires insights into the process that is to be controlled. **Process identification involves two basic steps: (1) postulating a process model** and (2) **determining the model parameters.** A model supplies a functional relationship between an input and an output of a process. Knowledge of this relationship is essential in developing a controller. We first identify the process by approximating/associating it with a basic model form. Then, we "fit" the model parameters and develop a unique process model.

In practical terms, **the process model allows us to predict process outputs (responses), given different levels of process inputs (values of controllable variables). Process models can be expressed as fully specified mathematical models, termed parametric models or less than fully specified mathematical models, termed nonparametric models.**

MATHEMATICAL MODELS

A mathematical model is an expression or formula that enables prediction of the values of the dependent variables (i.e., outputs) when values of the independent variables (i.e., inputs) are plugged into it. **Mathematical models of physical processes can be derived from fundamental principles and assumptions about the process, known as analytical models, or they can be formulated from experimental data collected from observing the process, known as empirical models.**

Derivation of analytical models usually begins with stating a fundamental law, i.e., writing down the conservation laws involving momentum, materials, or energy balances. An example is Newton's second law, which states that the force acting on a body is the product of the mass of the body and its acceleration. This can be stated mathematically as

$$F = m\,\frac{d^2y}{dt^2}$$

[15.1]

where F is the force, m the mass, and d^2y/dt^2 the acceleration.

In developing an empirical model, less concern is given to the process's underlying principles. Data collected from the process are used to develop a model that best fits the process. Linear and nonlinear regression analysis methods provide tools to fit models. The Taylor tool-life equation

$$T = \left(\frac{C}{V}\right)^{1/n}$$
[15.2]

is an example of a model developed through empirical methods. In the equation, V is the cutting speed, T is the tool life, and n and C are constants. For a particular tool, a given work material, and given cutting conditions, data are collected. Then, values of n and C are fit or determined so that the model will "best" fit the data. Best in these cases usually refers to the simplest model that will adequately explain the data, as measured by the sum of squared error values. The error is usually defined as the difference between the empirical observation and its predicted (from the model) value.

Sometimes analytical and empirical methods of modeling are combined to develop a mathematical model of a process. In such cases, a general form of the model is postulated from theoretical considerations. Then, the model is validated and its constants evaluated from observations of the system's performance. For example, if we had a process with two inputs u_1 and u_2 and one output y, we might postulate a general second-order response surface model of the form

$$y = b_0 + b_1u_1 + b_2u_2 + b_3u_1u_2 + b_4u_1^2 + b_5u_2^2$$
[15.3]

The coefficients $b_0, b_1, \ldots, b_5$ can be estimated by using regression analysis. Regression analysis is introduced in Chapter 6.

DYNAMIC PROCESS MODELING

Inertia or lag is inherent in many processes. The presence of process inertia causes a delay in the process' response, given a sudden change in the input. Since the output doesn't follow immediately, there will be a time lag before the full impact of the change in input is observed. **To control a dynamic process effectively when input changes occur, we need to know**

- The **magnitude of the change** in the output.
- The **direction of the change** in output.
- The **time lag,** or how long the output takes to change.
- The **pattern of the output variation** with change.

To take the dynamics of the process into account, the mathematical expressions can be extended with values from the past. Equation (15.3), for example, can be extended as

$$y(t) = b_0 + b_1u_1(t) + b_2u_1(t-1) + \cdots + b_{m-1}u_2^2(t) + b_mu_2^2(t-1)$$
[15.4]

It is obvious that the number of parameters increases very rapidly. Only one lag term was considered in Equation (15.4), but it still has a large number of parameters. We apply great care when using this method. It may be possible to reduce the number of terms in the model if the dynamic behavior of the process is known to a certain extent.

NONPARAMETRIC MODELS

The high speed of computation provided by computers has motivated the development of a large number of nonparametric techniques for process identification. Some common examples are neural networks, expert systems, and evolutionary computation. Researchers working with "intelligent" techniques typically attempt to develop methods that are motivated by form, representation, and reasoning in humans and natural/biological systems. They often heuristically construct controllers that are nonlinear and perhaps adaptive.

Intelligent control techniques are used in situations of high process complexity and demanding performance specifications. A key attribute of an intelligent system is the ability to learn. The term "learning control system" is defined by Farrel and Baker, as cited in Antsaklis and Passino [2]:

> A learning control system is one that has the ability to improve its performance in the future, based on experiential information it has gained in the past, through closed-loop interactions with the plant and its environment.

A learning control system has autonomous capability. It has the ability to improve its own performance; it is dynamic. Since **it can change and update its model or structure,** it incorporates memory. It is critical to retain past information in order to improve future performance—the learning control system receives performance feedback information based on an objective function—an expression that it seeks to optimize.

Higher levels of control, regarding multivariable systems involving a high degree of nonlinearity, generally operate with less complete information, thus requiring learning and intelligent techniques. Adaptive control is based on a fixed model that responds to some degree to environmental changes. Intelligent, autonomous controllers are enhanced adaptive controllers, in the sense that they can adapt to more significant global changes in the process and its environment than conventional adaptive controllers, while meeting more stringent performance requirements.

Learning control correlates past experiences with past situations and exploits this empirical information to improve future performance. Learning results in the automatic synthesis of iteratively improved mappings used within a control system architecture. Two such mappings are (1) control mapping that relates command and measured process outputs to a desirable control action and (2) model mapping that relates the process operating condition to an accurate set of model or controller parameters. In a typical learning control application, the desired mapping is expressed in terms of an objective function involving the outputs of both the process and the learning system.

The objective function is used to provide performance feedback to the learning system, which then associates this feedback with specific adjustable elements of the mapping. The underlying idea is that performance feedback can be used to improve the mapping furnished by the learning system. The application of learning control techniques to complex manufacturing processes provides the capability to adapt to changes in materials and process characteristics over time.

ARTIFICIAL NEURAL NETWORKS

Neural networks are nonparametric computing systems characterized by the ability to learn from examples. Hence, the behavior of complex systems can be modeled and predicted

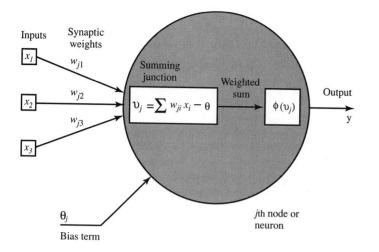

Figure 15.1 A typical node in an artificial neural network.

without a priori information about the systems' structures or parameters. **Neural networks consist of multiple nodes that communicate with each other through weighted connections called synapses.** The network produces an output pattern in response to an input pattern. Knowledge of the process model is stored in the synapses by updating the weights so as to minimize the difference between the network's output and the process output. A neural network generally consists of an input layer, one or more hidden layers, and an output layer. Figure 15.1 depicts the structure of a single node or neuron from an arbitrary network.

The neuron in the figure, designated the jth neuron, accepts inputs from and sends output to other neurons. A typical neuron gets its inputs from interconnections with the outputs of other neurons. These interconnections are known as synapses, a term borrowed from biology. Each synaptic connection's strength is expressed by a numeric value called a weight. When the ith neuron sends a signal to the jth neuron, that signal is multiplied by the weighting, w_{ji} on the ith synapse. If the output of the ith neuron is designated as x_i, then the input to the jth neuron from the ith neuron is $w_{ji}x_i$. Summing the weighted inputs to the jth neuron:

$$v_j = \left(\sum_i w_{ji}x_i \right) - \theta_j \qquad \textbf{[15.5]}$$

where θ_j is a bias term. The bias term is the product of the bias input (which is given a value of either $+1$ or -1) and the weight on the connection between the bias node and the jth neuron. Each θ_j provides a threshold that determines whether the jth neuron is excited or remains inhibited. The summing of the weighted inputs is carried out by a processor within the neuron. The sum, v_j, is called the activation of the neuron.

The activation is a purely internal state of the neuron. The synaptic weights and the inputs can be either positive or negative and hence the activation can be positive, zero, or negative. Any weighted input that makes a positive contribution to the activation represents a stimulus (tending to turn the jth neuron on), whereas making a negative contribution represents an inhibition (tending to turn the jth neuron off).

After the activation is determined, the neuron applies a signal transfer function to determine an output. This transfer function can be a simple step function, a hard limiting threshold, or a linear threshold function. A very common formula for determining a neuron's output signal makes use of the logistic function

$$\phi(v_j) = \frac{1}{1 + e^{-\lambda v_j}} \qquad \textbf{[15.6]}$$

where λ is the slope parameter and v_j is the activation. This function belongs to the class of sigmoidal functions, and it has characteristics that are advantageous within many paradigms. These characteristics include the fact that it is continuous, that it has a derivative at all points, and that it is monotonically increasing, and asymptotic to 0 and $+1$, as its arguments go to $-\infty$ and $+\infty$, respectively.

A neural network derives its power through its massively parallel distributed structure and its ability to learn and generalize, i.e., produce reasonable outputs for inputs not encountered before. Learning corresponds to parameter weight changes. As a result of exposure to the environment, the network assimilates information that can later be recalled by the user.

NEURAL NETWORK ARCHITECTURES

The manner in which the neurons in a neural network are structured is known as the architecture of the neural network. Two commonly used architectures are the feedforward network and the recurrent network. Once an architecture is chosen for a particular application, there is no hard and fast rule for selecting the topology (the number of layers and the number of nodes in each layer). The designer relies on past experience to make such decisions. The literature provides conflicting information regarding the choice of the number of hidden layers. While some claim one hidden layer is adequate to model any given continuous nonlinearity, others claim that two are needed. This decision is best made heuristically: if one hidden layer is insufficient then use two; Murray [3]. As for the "optimal" number of neurons per hidden layer, trial and error typically produces acceptable results.

FEEDFORWARD NETWORKS

In a feedforward network, the layers project strictly in the forward direction. Figure 15.2 illustrates a multilayer, partially connected network. When every node is connected to every node in the adjacent forward layer, it is known as a fully connected network. A partially connected network is one in which some of the connections are missing. One or more hidden layers enable the network to extract higher-order statistics; see Haykin [4].

RECURRENT NETWORKS

When a network has at least one feedback loop, it is known as a feedback or recurrent network. A recurrent network may or may not have self-feedback. Self-feedback occurs when the output of a neuron is fed back to its own input. Feedback loops involve the use of branches composed of delay elements that result in nonlinear dynamic behavior.

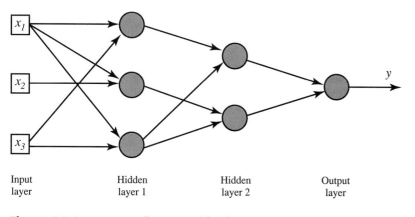

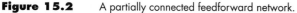

Figure 15.2 A partially connected feedforward network.

LEARNING IN NEURAL NETWORKS

The knowledge represented by the network typically increases with each iteration. The ability of the network to learn from its environment, and to improve its performance, is of primary significance. Haykin [4] defines learning in neural networks as

> A process by which the free parameters of a neural network are adapted through a continuing process of stimulation by the environment through an iterative process of adjustments applied to its synaptic weights and thresholds.

The methodology used to train neural networks can be categorized into two different paradigms: supervised and unsupervised. Supervised learning requires a set of training patterns of known classification and an external teaching procedure. During the teaching procedure, the network parameters are adapted according to the network's response to the training patterns. Supervised learning is also called reinforced learning, learning with a teacher, and associative learning; see Heileman et al. [5]. Typically, during supervised learning accurate responses are rewarded and inaccurate responses are punished.

When the desired response is not known, e.g., when incomplete information is available regarding which inputs produced a particular output, the unsupervised learning paradigm is applied. Since the desired response is not known, explicit error information cannot be used to improve network behavior. In this mode, learning is accomplished on the basis of observations of responses to inputs that are not fully understood.

APPLICATIONS OF NEURAL NETWORKS IN CONTROL

In the last decade the use of neural networks in a control system framework has been an active topic in research. When implemented for control applications, **neural networks may be used either as part of an integrated control system,** i.e., the process identifier, **or for the entire control system.** Neural networks are universal approximators. That is, a feedforward network with at least one hidden layer is capable of approximating uniformly any continuous multivariate function to any degree of accuracy; see Hassoun [6].

The universal function approximation property of neural networks makes them good candidates to perform process identification. Neural networks have been successfully tested for process model identification; see Anderson et al. [7] and Haesloop and Holt [8]. The neural network model of the process to be controlled may be utilized by an "optimizer" to generate a control action for the process. Attempts have also been made to use neural networks for direct control; see Chinnam and Kolarik [9] and Narendra and Mukhopadhyay [10].

Several shortcomings of neural networks limit their control applications. The backpropagation learning algorithm requires gradient information about the search space. For some problem domains, such as reinforcement learning, gradient information is unavailable or costly to obtain; see Whitley et al. [11].

Most networks in supervised learning use feedforward networks with sigmoidal transfer functions or radial basis functions. These choices make gradient information relatively easy to obtain, but if more complex transfer neurodes, such as product neurodes, are used, or if fully recurrent networks are trained, then computing gradient information becomes far more costly.

The backpropagation learning algorithm cannot be used for direct control of a nonlinear process with an unknown model since the process dynamics are unknown and cannot be used to generate the desired partial derivative; see Narendra and Parthasarathy [12]. Even if the process model has been identified, a neural network controller will risk convergence to local optimal solutions because, during the weight optimization, the search begins with a single point in the search space and therefore has a high risk of getting stuck at a local minima or maxima. This "suboptimization" is not a drawback if the solution space is unimodal, but advanced manufacturing systems are typically complex, resulting in the solution space being multimodal. Hence, using a neural network to control such processes will not always result in efficient control. Neural networks are therefore often used in combination with other techniques. Examples of such techniques may be found in Gupta and Sinha [13], Lu [14], and Patro [15].

EXPERT SYSTEMS

An expert system is designed to emulate a human's skill in a specific problem domain. It consists of an explicit representation of the application's domain knowledge. The knowledge is usually in the form of if-then rules. When an expert system is designed to emulate the expertise of a human in performing control activities, it is known as an expert controller. **The expert system's knowledge base is constructed by extracting rules from available human experts.** Detailed process knowledge is stored in the database and used to make control decisions. An expert controller interprets process outputs and reference inputs, reasons about alternative control strategies, and generates inputs to the process to improve the closed-loop performance of the system. Figure 15.3 shows an expert control system consisting of the process being controlled and the expert controller.

The expert controller consists of a knowledge base (knowledge obtained from the human control expert) **and an inference engine that emulates the human expert's decision-making process.** This decision process involves collecting process outputs and target values, comparing them, and reasoning regarding what control inputs to suggest for the process. With a rule base and a capability to reason, an expert system's operation is similar to human problem solving. Typically, the focus is not on whether a good model of the human expert was obtained, but

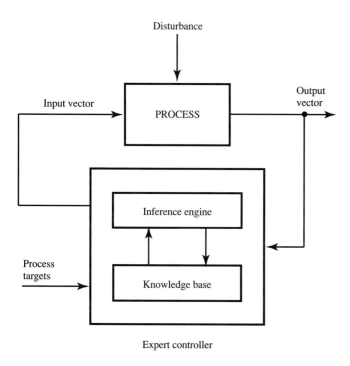

Figure 15.3   An expert control system structure.

whether the heuristically designed expert controller performs adequately when used to control the process.

The key factors in the design of an expert control system, as in any expert system, **are knowledge acquisition and knowledge representation.** In an environment with incomplete and/or imprecise information, knowledge acquisition and learning become major bottlenecks in developing expert systems. Knowledge can be acquired through the following channels; see Lu [14]:

- Knowledge about process characteristics and behavior—from process engineers
- Knowledge about controller parameter tuning, stability, nonlinearity, controllability, etc.— from control experts
- Knowledge about process response time, gain, etc.—from operators
- Other system knowledge—from data and related information sources such as statistical and dynamic production data

The extracted knowledge can be represented in various forms such as mathematical algorithms, production rules, fuzzy rules, and even neural networks.

Expert systems are usually problem-specific—it is difficult to have a generic methodology for the design of an expert control system. Design of an expert control system depends on the desired functions and the information available. Expert systems and other learning techniques for control applications are discussed in Gupta and Sinha [13].

E VOLUTIONARY C OMPUTATION

We recognize that intelligence displayed by living creatures is impacted by evolution, and therefore it would be reasonable to model the evolutionary process in order to create entities capable of intelligent behavior. The term used for such simulation of evolution on a computer is evolutionary computation. A population of potential solutions to the problem at hand is created. In process modeling, the population consists of potential models to represent the process. These individuals compete with each other and strive for survival. The individual models are evaluated in terms of how well they fit the past data. A selection scheme, biased toward fitter individuals, selects individuals to be the parents for the next generation. The parents then undergo transformations to produce offspring for the next generation. The cycle of evaluation, selection, and regeneration continues until a satisfactory solution is obtained.

Three decades of research in this area have shown that modeling a search process associated with natural evolution can yield very robust, direct computer algorithms. **Evolutionary algorithms are based on the collective learning processes within a population of "individuals," each of which represents a point in the space of potential solutions to a given problem.** Emulating the optimization process inherent in natural selection results in iteratively improving solution quality. The rate of improvement depends on the shape of the response surface, but many studies, such as Fogel [16], Goldberg [17], and Pham and Yang [18], have shown that the procedures generally converge to near-optimal solutions despite difficult-to-optimize response surfaces. These stochastic optimization techniques often out-perform classical methods of optimization when applied to real-world problems; see Fogel [19, 20].

Conventional search techniques, such as hill-climbing, are often incapable of finding the global optimal solution of nonlinear or multimodal functions. In such cases, a random search technique is generally required. However, undirected search techniques are extremely inefficient for large search spaces. **Theoretical analyses suggest that evolutionary techniques can rapidly locate high-performance regions in extremely large and complex search spaces.** Because of the distributed and repeated sampling in the population, evolutionary methods have a propensity to be insensitive to noisy feedback.

Evolutionary techniques can contribute to process control applications in numerous ways. They have been used **to define controller structures** and/or **to tune controller parameters.** Their ability to locate global optimal solutions may be used to find the best control action out of a finite, or infinite, number of potential control actions within a bounded region. For such an application, a reliable process model, either parametric or nonparametric, is required. The process model is used to evaluate individual solutions in the populations.

Evolutionary techniques may be used **for process identification.** In such an application, the search consists of finding the optimal process model. The search begins with a random population of models—belonging to different model classes and consisting of different numbers of parameters. The individuals are evaluated by fitting past data.

Finally, evolutionary techniques may be used **in process control as aides to other techniques.** For example, they may be used as the learning algorithm in neural networks. The "best" weights for the network may be found by performing an evolutionary search on the weight space.

15.4 SELF-TUNING CONTROL

A linear control loop is tuned once, and it is tuned forever. However, continuous adjustments are required to tune a nonlinear control loop. Self-tuning or adaptive tuning is based on the concept that controllers should be capable of tuning themselves, i.e., automatic tuning. **Self-tuning control involves two basic tasks being performed on line, in a feedback loop: (1) process identification** and (2) **process control.** First, information about the process behavior is gathered by continuously determining the actual condition of the process on the basis of measured process input, process output, and state signals. This information is used to identify the process on line. The second task is based on control performance criterion optimization. Depending on the performance objective, decisions are made as to new controller parameters. The new controller is then used to generate the control action for the process.

On-line process models are adjusted to keep up with the gradual changes in the process. The model is used to predict the output variables on the basis of the measured input variables. The deviations between the predicted output and the actual output are decreased by an iterative algorithm. The deviation, δ, can be calculated as

$$\delta = \sum_{i=1}^{k} (Y_{i,\,\text{model}} - Y_{i,\,\text{process}})^2 \qquad \textbf{[15.7]}$$

or

$$\delta = \sum_{i=1}^{k} |Y_{i,\,\text{model}} - Y_{i,\,\text{process}}| \qquad \textbf{[15.8]}$$

where k represents the total number of output variables.

PHOTOVOLTAIC SYSTEM CASE—ADAPTIVE CONTROLLER For most processes, the design | **Example 15.1**
engineer selects a specific "best" value for the controlled output—the target or setpoint. If this value holds constant over time, a conventional controller, e.g., proportional or integral, can be used successfully to maintain the parameter value close to the target.

A special situation occurs if the target value sits at one of the parameter's physical limits, maximum or minimum, and also varies with time. In this case, a conventional controller, as previously described in Chapter 14, cannot fulfill the process goals effectively or efficiently. Here, the design engineer determines a domain where the target should be located during the process, rather than a clearly defined target for "all" time.

For processes characterized by target location uncertainty, a self-tuning controller is useful. This type of controller is expected to track and determine the best target position. One self-tuning controller, well known in the field of photovoltaics applications, is the maximum power point tracker (MPPT). Photovoltaics technology converts the sun's radiation energy to electricity. The conversion element is the solar cell, which is actually a large-area silicon diode. A simplified conversion system connects a photovoltaic generator and a load, as depicted in Figure 15.4a.

Experimentally, the photovoltaic generator current-voltage characteristic (I-V) curve, depicted in Figure 15.4b, is obtained as the value of the external load resistor is varied from zero up to large values. The power output of such a generator is dependent on the illumination level Φ and on the operating temperature T, symbolically expressed as

$$P_{PV} = f(\Phi, T)$$

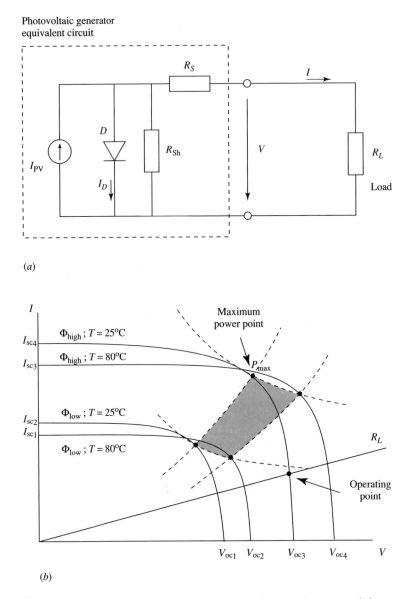

(a)

(b)

Figure 15.4   Photovoltaic conversion system scheme and operational characteristics (Example 15.1). (a) Simplified representation of the photovoltaic system. (b) Parametric I-V curve of the photovoltaic generator and the linear characteristic of the resistive load.

For a certain generator, each Φ, T pair defines an *I-V* curve that has a maximum power point at some current and voltage:

$$P_{\max} = I_m V_m$$

During outdoor system operations, the luminous flux Φ and the generator temperature T are variables, depending on the weather conditions, clear or cloudy sky, and on the season and other conditions, summer, winter, windy, dry, etc.—the generator output power varies accordingly. The maximum power point of the possible *I-V* curve families lies in a location domain—the shaded area in Figure 15.4*b*—rather than at several discrete positions.

The power available in the external circuit is a function of the load resistance. Its maximum is achieved when the load resistance equals the generator internal resistance, as stated by the maximum power transfer theorem. The type and the nature of the load constitute another source of power variation, depending on the user's needs.

The operating point of the conversion system is defined by the intersection between the operating characteristics, *I-V* curves, of both generator and load. In most situations, this operating point differs from the maximum power point. In the photovoltaic energy case, this is a major issue, as the cost of installed power per watt ranges between $4 and $6.

For overcoming this power loss, a self-tuning controller—the MPPT—was designed. Basically, the controller is a "chopper" device placed between the photovoltaic system and the load. It generates a pulse-modulated voltage, which seeks to match the maximum power point–related voltage. The control principle is based on power measurement at two successive locations of the *I-V* curve. The error is computed as their difference, and the error sign determines the position of the maximum power point, relative to the actual position of the operating point.

This type of self-tuning controller has a variable target—the maximum power point—and contains a variable reference, represented by the last measured power value. The control action is fast and accurate, resulting in tight tracking of the power point during the operating cycle. The relative complexity and the possibility of targeting local maximum points, if they exist, are the main disadvantages of this self-tuning controller. [IG]

15.5 APC/SPC MODEL COMBINATIONS

SPC and APC technologies evolved from different sources and address related, but different, needs as explained in Chapter 9. SPC's heritage goes back to piece part, or discrete part, process output, where discrete part measurements can often be considered independent of each other in a physical sense. This physical characteristic of independence manifests itself in variables data as time independence—no autocorrelation. Chapters 10, 11, and 12 focus exclusively on time- or order-independent variables data and conventional SPC models, addressing special cause/common cause issues. Chapter 13 focuses on both time- and order-independent and autocorrelated data with respect to unconventional SPC models, likewise addressing the special cause/common cause issues.

APC's heritage goes back to general control theory, specifically applied to continuous flow processes. Here, the usual case is to observe some level of sustained physical influence or mixing occurring in the output, thus creating process inertia. Measurements from these processes tend to reflect process inertia, which manifests itself statistically as autocorrelation in the data. Chapter 14 introduces APC models that thrive on autocorrelated data, as they address process adjustment issues.

It is insightful to recognize that **both SPC and APC technologies address the same general goals and objectives: (1) get/keep the process response/output on target** and (2) **maintain process response/output stability.** These same goals and objectives pertain to all process control forms—subjective-, objective-, and equipment-based, SPC and APC-related. Hence, common ground is clear. At issue is the determination of effective and efficient applications of the control technologies, together.

In the subjective case, we make both special cause-related and adjustment-related decisions on the basis of qualitative experience as it relates to process behavior. As we observed in the Natural Gas Production Case (Example 4.3) in Chapter 4, subjective means were effective in removing the wild disruptions in chiller performance. Here, process understanding through experience was exploited. However, through process understanding and APC-like modeling, we may be able to capture a higher level of process output/throughput performance in more effective adjustments. Hence, **SPC and APC combinations offer attractive options.**

The inherent assumption in SPC monitoring is that the process is stable, and essentially on target, until significant evidence—judged in a probabilistic sense—is offered to suggest otherwise. Once we have the evidence, we are justified in pursuing a special cause influence and neutralizing its effect on our process response. Here, we examine both controllable and uncontrollable variables, find one or more appropriate variables, and take counteraction according to either a model or our judgment.

Using our APC technology, we seek to move/adjust the response, output, to target. However, at the same time, we introduce variation in our related input variable by changing its level—making it more unstable. For example, the room temperature output variable is related to a fuel flow input variable in the space heating cases described in Chapters 3, Section 3.4, and 9, Example 9.1. Furthermore, we typically seek to develop a Δ output to Δ input relationship—a transfer function. In general, every time we make an observation on our output variable, comparing it to target, using our error term, we automatically pull our input lever and execute some degree of adjustment. Here, we see our adjustments as counteractions, through input variables, to a source typically unknown to the system.

Traditionally, in APC, the location and dispersion of our input leverage variable is not as critical as the location and dispersion of our output variable. However, we must live within basic physical limits. For example, in the room temperature case, our fuel, fuel delivery, and burner offer capacity restrictions. The rate of fuel flow and ultimately amount of fuel delivered have an impact on both physical performance as well as economics. Hence, in general we cannot say that the behavior of the input lever variable manipulated is unimportant.

If we consider the nature of SPC and APC, together, we see enough common ground to believe that strategic combinations may ultimately lead to enhanced results in terms of physical, economic, timeliness, and customer service performance. We know that, over time, processes are constantly changing because of forces of nature and are constantly changed by people seeking improved performance.

Interest in designing process control tools that provide both special cause detection, monitoring, and adjustment features, together, in a holistic manner, is apparent. Currently, a number of perspectives exist; see Montgomery and Woodall [21]. Historically, these perspectives range from economic-based SPC models proposed as early as the 1950s (see Duncan [22]) to forecast-based EWMA models. Box and Luceno [23] describe more recent cost-based process control feedback schemes that consider costs of being off-target, costs of process changes, and costs of sampling/testing. They also describe combined SPC/APC models using Cuscore statistics to provide efficient detection of signals of interest regarding process disruptions within proportional and integral control schemes.

Although current efforts are impressive in their level of mathematical sophistication, they fall short of a holistic process control scheme. However, with growth in computer power and reductions in computer costs, real-time hybrid control models with the ability to effectively and efficiently deal with measurement, comparison, and correction in terms of both process disruptions and adjustments in an optimal manner will undoubtedly emerge.

REVIEW AND DISCOVERY EXERCISES

REVIEW

15.1. What advantages and difficulties does feedforward control present? Explain.

15.2. What is process identification? How is it accomplished? Explain.

15.3. Explain the difference between an analytical and an empirical model in terms of APC. What is the purpose of each? Explain.

15.4. Why does dynamic process modeling present so many challenges? Explain.

15.5. What is the difference between parametric and nonparametric process models? Explain.

15.6. What is a neural network? How does it work? How does it learn? Explain.

15.7. Explain the difference between supervised and unsupervised learning in neural networks.

15.8. Why are neural network control models attractive in process control? Explain.

15.9 What is an expert system? How does it work? Explain.

15.10. What advantages and challenges does evolutionary computation offer in APC? Explain.

15.11. Why are self-tuning process controllers desirable? What functions must be addressed in self-tuning controllers? Explain.

15.13. What is cascade control? When is it useful? Provide an example process. Explain.

15.14. What is ratio control? When is it useful? Provide an example process. Explain.

DISCOVERY

15.15. Search the quality control research literature for combinations of SPC/APC. In the literature, APC is sometimes referred to as *engineered process control,* EPC. Sources such as the *Journal of Quality Technology* and *Quality Engineering* are good resources. Describe a model or application. Consider both conceptual as well as quantitative aspects.

REFERENCES

1. Hughes, T. A., *Measurement and Control Basics*, 2nd ed., Research Triangle Park, NC: Instrument Society of America, 1995.

2. Antsaklis, P. J. and K. M. Passino, *An Introduction to Intelligent and Autonomous Control*, Boston: Kluwer Academic Publishers, 1993.

3. Murray, A. F., *Applications of Neural Networks*, Boston: Kluwer Academic Publishers, 1995.

4. Haykin, S., *Neural Networks: A Comprehensive Foundation*, New York: Macmillan, 1994.

5. Heileman, G. L., M. Georgiopoulos, H. R. Myler, and G. M. Papadourakis, "Improved Backpropagation Learning Algorithms for Neural Networks," *Advances in Artificial Intelligence Research*, vol. 2, pp. 177–211, 1992.

6. Hassoun, M. H., *Fundamentals of Artificial Neural Networks*, Cambridge, MA: MIT Press, 1995.

7. Anderson, C. W., J. A. Franklin, and R. S. Sutton, "Learning a Nonlinear Model of a Manufacturing Process Using Multilayer Connectionist Networks," *Proceedings of the IEEE International Symposium on Intelligent Control*, pp. 404–408, 1990.

8. Haesloop, D. and B. R. Holt, "Neural Networks for Process Identification," *International Joint Conference on Neural Networks*, vol. 3, pp. III-429–III-434, 1990.

9. Chinnam, R. B., and W. J. Kolarik, "Quality Controller for Automated Systems," *Proceedings of the 48th Annual Quality Congress*, ASQC, pp. 430–438, 1994.

10. Narendra, K. S., and S. Mukhopadhyay, "Adaptive Control of Nonlinear Multivariable Systems Using Neural Networks," *Neural Networks*, vol. 7, no. 5, pp. 737–752, 1994.

11. Whitley, D., S. Dominic, and R. Das, "Genetic Reinforcement Learning for Neurocontrol Problems," *Machine Learning*, vol. 13, pp. 259–284, 1993.

12. Narendra, K. S. and K. Parthasarathy, "Identification and Control of Dynamical Systems Using Neural Networks," *IEEE Transactions on Neural Networks*, vol. 1, no. 1, pp. 4–27, 1990.

13. Gupta, M. M., and N. K. Sinha, *Intelligent Control Systems: Theory and Applications*, Piscataway, NJ: IEEE Press, 1996.

14. Lu, Y., *Industrial Intelligent Control: Fundamentals and Applications*, New York: Wiley, 1996.

15. Patro, S., *Neural Networks and Evolutionary Computation for Real-Time Quality Control*, Ph.D. dissertation, Texas Tech University (W. J. Kolarik, research adviser), 1997.

16. Fogel, D. B., *Evolutionary Computation—Toward a New Philosophy of Machine Intelligence*, New York: IEEE Press, 1995.

17. Goldberg, D. E., *Genetic Algorithms in Search, Optimization and Machine Learning*, Reading, MA: Addison-Wesley, 1989.

18. Pham, D. T., and Y. Yang, "Optimization of Multi-Modal Discrete Functions Using Genetic Algorithms," *Proceedings of the Institute of Mechanical Engineers-D, Journal of Automobile Engineering*, vol. 207, pp. 53–59, 1993.

19. Fogel, B., "An Evolutionary Approach to the Traveling Salesman Problem," *Biological Cybernaetics*, vol. 60, pp. 139–144, 1988.

20. Fogel, D. B., "An Introduction to Simulated Evolutionary Optimization," *IEEE Transactions on Neural Networks*, vol. 5, no. 1, pp. 3–14, 1994.

21. Montgomery, D. C., and W. H. Woodall, "A Discussion on Statistically Based Process Monitoring and Control," *Journal of Quality Technology*, vol. 29, no. 2, pp. 121–162, 1997.

22. A. J. Duncan, "The Economic Design of $\bar{X}$-Charts To Maintain Current Control of a Process," *Journal of American Statistical Association*, vol. 51, pp. 228–242, 1956.

23. Box, G. E. P., and A. Luceno, *Statistical Control by Monitoring Feedback Adjustment*, New York: Wiley, 1997.

PROCESS ANALYSIS AND IMPROVEMENT

The purpose of Section 5 is to identify and describe the elements of process improvement in the context of a production system.

Chapter 16: Process Improvement—Questioning Perspectives

The purpose of Chapter 16 is to overview the elements of process improvement and to address the nature of process improvement with regard to opportunity.

Chapter 17: Process Improvement—Analysis and Implementation Perspectives

The purpose of Chapter 17 is to address the nature of process improvement with regard to describing and implementing process change.

chapter

16

PROCESS IMPROVEMENT—
QUESTIONING PERSPECTIVES

16.0 INQUIRY

1. How do process definition/redefinition and improvement interface?
2. What constitutes process improvement?
3. How are processes systematically improved?
4. How do we decide between improvement and redefinition?

16.1 INTRODUCTION

We make gains in process performance in three fundamental ways. First, we define or redefine our process in a strategic sense. Second, once defined or redefined, we commence process operations and use process control methods to target and stabilize our process. Third, we use process improvement methods, as described in this section, along with process control to fully exploit our process technology. In short, our objective is to climb to the top of the process definition bar, i.e., the dotted-line portion in Figure 3.2, as quickly as possible.

Process improvement is focused primarily in our subprocesses and sub-subprocesses. **Process leverage is the key to process improvement initiatives.** Our process characterization efforts provide clues as to which variables are leverage variables, and support us in questioning: What do we do? Who does it? How do we do it? Where do we do it? When do we do it? Why do we do it?

Process improvement involves questioning, analysis, and implementation, as depicted in Figure 16.1. Over the years, **people have developed and advocated several general initiatives and a host of tools for addressing improvement in production systems.** Selected initiatives and

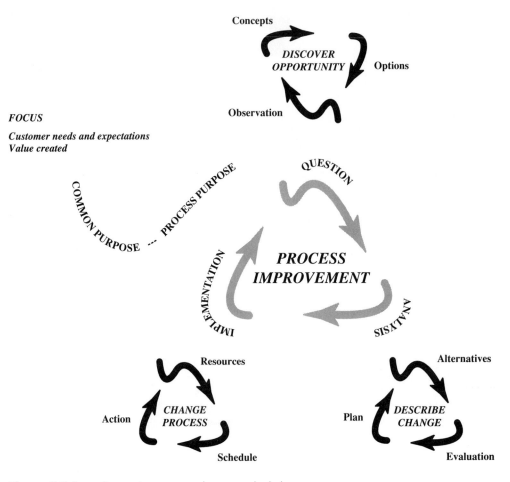

Figure 16.1　　Process improvement elements and subelements.

tools are listed in Table 7.1 along with comments relative to each initiative's and tool's general usefulness in process improvement. Each initiative and tool listed is briefly described in Section 6.

The complexity within even small organizations, in people, products, and processes, creates significant challenges in effectively and efficiently using these initiatives and tools. In this section we use some of the features of these initiatives and tools to illustrate their merit in process improvement. **We place process purposes in the foreground and initiatives and tools in the background as facilitators to help accomplish process purpose.** Initiatives and tools are not the ends we are seeking; results/outcomes in physical, economic, timeliness, and customer service performance matter.

Process boundaries are set by our process purpose and our process definition. **Process improvement is initiated within our existing process boundaries.** For example, in a fast-food restaurant, if we define our cooking process around a frying technology, then we provide process improvements within our frying technology. On the other hand, if we are considering changing to a broiling technology, then we are likely faced with extensive change, impacting our external customers, and a process redefinition may be required.

The major distinction between redefinition and improvement is the scope of the effort. Redefinition provides broad impact that reaches beyond current process definitions, e.g., to external customers and significant production system impact. Redefinitions involve high-level leadership and management in decision. **Improvement provides focused impact.** Improvement efforts vary from localized sub-subprocess impact only, to broad-impact projects involving several processes—process improvement is tactical in nature. Redefinition is strategic in nature.

16.2 CRITICAL ELEMENTS

Just as in process definition/redefinition, we provide for systematic and thorough efforts in process improvement. Process improvement as depicted in Figure 16.1 breaks the question, analysis, and implementation elements into their constituent parts. In the questioning element we think in terms of opportunity within our current process, subprocess, and sub-subprocess purposes. **We identify process improvement opportunities through process understanding and observation.** We establish benchmarks relative to what others are doing, and goals and objectives relative to results we want to accomplish.

Once our questioning has reoriented our process thinking, **we develop an explicit improvement purpose statement expressing our intentions,** i.e., desired outcomes and commitment relative to our defined process goals and objectives. Our specific improvement purpose is typically tied to one or more process/subprocess/sub-subprocess metrics, e.g., assembly time, scrap rate, . . . , associated with our process/subprocess/sub-subprocess. In other words, **our improvement purpose complements our existing process/subprocess/sub-subprocess purposes** as described in Section 3.

Given an explicit improvement purpose, we focus on concepts. We examine both the concepts involved in the current process/subprocess/sub-subprocess, as well as concepts that hold promise for improvement. Our concept assessment leads us to options. Two basic option categories exist— an improvement option or a redefinition option. Our choice depends on the potential gains and risks involved in both the short term and long term. In the redefinition case, we break out of the current process definition and use Section 3 methods. In the improvement case we stay within the current process definition and use our improvement methods described within this section.

Within the analysis element, we focus on facts, ideas, and details, and describe process change. First, we collect relevant facts and figures, and then generate creative improvement ideas. We transform these ideas into feasible alternatives, evaluate our alternatives, and choose a course of action. In exercising this choice, we assess the technical and business impacts associated with our alternatives. Once we choose a course of action, we build our improvement plan. Here, we review and modify our current process team structures, knowledge and skill bases, metrics, targets, specifications, and control points, and reassess or realign our process value chain model.

Finally, we implement our plan—change our process. Here, we identify needed resources, schedule our plan, and then act. We undertake a limited-scale implementation. We test, tune, and mistakeproof our improvement. Then, when we are satisfied that we have a workable and productive improvement, we pursue full-scale implementation and follow-up using process control methods, as described in Section 4, Chapters 9 to 15, so as to provide additional incremental gains.

Example 16.1 | **HIGH LIFT CASE—BACKGROUND**　　High Lift is a $400 million annual sales company that offers its customers solutions to automatic warehousing challenges. Solutions include entire materials, storage, and retrieval systems. Each system is custom-designed, but many components are common to each custom application. One division of High Lift manufactures storage and retrieval lifts that can be configured to run on paths within the rack systems under semiautomatic control; i.e., to store and retrieve palletized materials without the aid of a human operator. This same product can also be configured to serve as a standard forklift, provided a manual guidance and control system is installed.

High Lift is a successful material handling systems provider, with a recognized competitive edge in system integration within customer facilities. A summarized layout of High Lift processes is depicted in Figure 16.2. Here, we see fundamental High Lift processes. High Lift maintains extensive product definition, design, and development processes to service each customer in their unique storage and retrieval needs. Production processes are extensive, but not unique, in that common structural elements are used within each customer's unique capacity and layout requirements. Production serves to support the product definition and design processes in producing to their specifications. Field installations are made by local or regional third-party contractors, aided by High Lift field engineers, working out of the design and development processes.

High Lift maintains manufacturing facilities that produce both structural and material handling equipment for their integrated systems. One of these facilities is responsible for producing High Lifters. The High Lifter can be built to automated specifications as well as to manual operating specifications. This dual use characteristic is the result of innovative product design, whereby common frames, motors, and transaxles are utilized. Basically, the wheel, control, and mast configurations distinguish each High Lifter.

As an entire production system, High Lift has a record of profitable operations. Historically, High Lift has generated most of its profits and reputation from the definition, design, and development processes by adding to customer profitability in reducing material handling costs. The actual manufacturing and production process maintained by High Lift has lagged in its contribution to profits. To address this performance gap, High Lift redefined their design and production processes as a whole. As one concept, in redefinition, they considered divesting their manufacturing/production capabilities in order to focus on their core engineering services through their definition, design, and development processes.

The results from the divestment study produced two basic findings. First, benchmarking High Lift manufacturing plants with other comparable manufacturing plants showed that High Lift production costs were about 15% higher. This production cost gap clearly favored outsourcing. Second, High Lift customers have very little patience with installation delays; installation of customer systems must be completed as rapidly as possible to provide minimal production and warehousing upsets. Because of this time-critical demand, High Lift would have to order ahead and hold excessive inventory to assure no component or project delays. This stocking need was found to apply both to new installations and to service, parts, and maintenance demands by High Lift customers.

The timeliness issue was ultimately considered critical to market share and product reputation. Hence, a commitment to in-house manufacturing was made, on a conditional basis. The condition is that manufacturing and production process costs must be reduced to levels comparable to manufacturers of similar equipment within a time frame of 2 years.

Management at the High Lifter facility set a goal of reducing production costs. Their target was set at a 20% reduction, with no compromises to product integrity or delivery performance. When mapped on their process value chain (PVC) model, this reduction goal fit into the right-hand side, within the business outcomes. In order to support this goal with action in their processes and subprocesses, they sponsored several process redefinition initiatives. One such initiative involved the interfacing of the procurement, product design, and production processes. Several significant findings were produced by this redefinition team. First, although the product, as designed, met and exceeded customer demands, it was difficult and time-consuming to fabricate and assemble. Studies of the assembly subprocess indicated that a high proportion of production costs was being incurred in the assembly area.

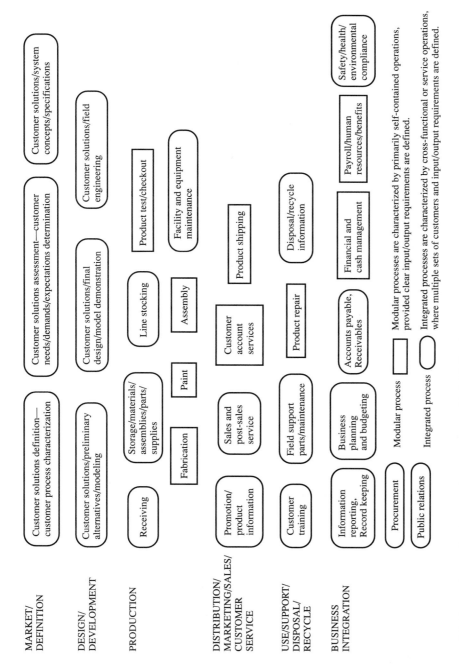

PROCESS TYPE

PROCESSES/SUBPROCESSES

MARKET/ DEFINITION

- Customer solutions definition— customer process characterization
- Customer solutions assessment—customer needs/demands/expectations determination
- Customer solutions/system concepts/specifications

DESIGN/ DEVELOPMENT

- Customer solutions/preliminary alternatives/modeling
- Customer solutions/final design/model demonstration
- Customer solutions/field engineering

PRODUCTION

- Receiving
- Fabrication
- Paint
- Storage/materials/ assemblies/parts/ supplies
- Line stocking
- Assembly
- Product test/checkout
- Facility and equipment maintenance

DISTRIBUTION/ MARKETING/SALES/ CUSTOMER SERVICE

- Promotion/ product information
- Sales and post-sales service
- Customer account services
- Product shipping

USE/SUPPORT/ DISPOSAL/ RECYCLE

- Customer training
- Field support parts/maintenance
- Product repair
- Disposal/recycle information

BUSINESS INTEGRATION

- Information reporting, Record keeping
- Business planning and budgeting
- Accounts payable, Receivables
- Financial and cash management
- Payroll/human resources/benefits
- Safety/health/ environmental compliance

- Procurement
- Public relations

[ ] Modular process

( ) Integrated process

Modular processes are characterized by primarily self-contained operations, provided clear input/output requirements are defined.

Integrated processes are characterized by cross-functional or service operations, where multiple sets of customers and input/output requirements are defined.

Figure 16.2   Systems-level process layout overview (High Lift Case).

441

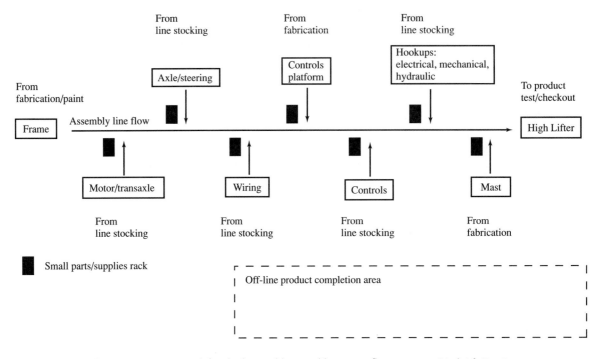

Figure 16.3   Redefined subassembly-assembly process flow overview (High Lift Case).

Direct observation indicated a cluttered production area with partially finished production units pulled off line, waiting to be completed. All the units waiting lacked a few critical parts. Further study indicated that many parts, especially purchased parts, had to be airfreighted into the plant, creating high delivery costs. Line shortages, even on small parts such as fasteners, were creating stoppages and requiring a high level of off-line assembly on an overtime basis. These limitations were creating production bottlenecks, and thereby increasing production costs. The net result was long cycle times, demanding longer lead or order times, as well as excessive shipping and expediting costs, relative to comparable production facilities.

In order to address these issues, a modular design was developed for the product. The goal was to reduce production assembly complexity. This modular design was integrated with a straight-line flow final assembly concept. Here, major assemblies are built up off line and flow across to the main assembly line. A portion of this modular assembly concept as installed is depicted in Figure 16.3.

We see a fabricated/welded forklift frame entering the final assembly area. Then, on opposite sides we see major assemblies moving to the main assembly line. These assemblies are married to the frame, with an assembled forklift emerging at the end of the assembly line. Along the line, racks and bins are installed for line parts, e.g., fasteners, couplings, and hoses. A factory computer network was designed for order and control of outside materials, assemblies, parts, and supplies. The result was centralized, computer-scheduled ordering for all module lines and the final assembly line, with shop e-mail communications to expedite parts when shortages occur. The computer scheduling system was redesigned to be a part of the procurement subprocess, located within the business integration process; see Figure 16.2.

It took about 18 months to implement this new design and corresponding production process. Several months were required for product and process redefinition and redesign, and several months were required for installation of new equipment. During this period, production operations were maintained on the original lines, using the original design and assembly methods. Key people from the original assembly teams were reassigned to the redefined and redesigned process for 3 months of limited-scale production, including on-the-job training and testing, tuning, and mistakeproofing. At the present time, full-scale production is underway using the redefined processes. Currently about 2 months of full-scale production experience exists.

16.3 PROCESS IMPROVEMENT OPPORTUNITY

Process improvement is initiated from questions. There are always many aspects to any process that can be questioned. We question what we do, why we do it, how we do it, who does it, when we do it, and where we do it—all in the context of creating short-term and long-term gains for our customers, employees, and stakeholders.

In order to intelligently question the nature of a process, we need contact with it as an observer or participant—a supplier, customer, or stakeholder. **We question the effectiveness and efficiency of existing processes from the perspective of our experiences and measured process performance,** i.e., physical, economic, timeliness, customer service.

Again, just as in the case of process definition (Section 3), people are responsible for process improvement. **Process improvement team selection guidelines are straightforward—involvement is the key.** We include people affected by the improvement, people knowledgeable about the improvement technology, and people who will be involved with the implementation and operations.

OBSERVATION

Process improvement addresses opportunities. Observation, perception, and measurement are critical in recognizing opportunities. **Opportunities may emerge reactively or proactively. Reactively, opportunity springs from hindsight** (problems), **while proactive opportunity springs from foresight.** Details useful in process observation are presented in Table 16.1.

Process improvement encompasses observation and review of input, transformation, and output parameters as well as actual process transformation mechanics. Our actual results are calibrated through our critical metrics. Here, physical, economic, timeliness, and customer service performance parameters, as well as encompassing business metrics and ratios, are useful to mark our current positions.

In a reactive sense, our competitors play a role in establishing our own expectations. Initiatives such as benchmarking are used to gather information to help us set process-relevant expectations. Benchmarking other parts of our own organization, competing organizations, and also noncompeting organizations with similar process purposes all adds to our knowledge base, which influences our expectations in process outputs, transformations, and inputs.

We bring our actual and expected performance together in order to assess process gap effects reactively, e.g., shortfalls where our expectations exceed our actual performance. We focus on expression of opportunity around these gaps. These gaps offer a starting point for reactive improvement thinking. At this point, we are primarily interested in how much gap closure offers in terms of technical and business results. If this offering is minimal, then we do not pursue

Table 16.1   Details related to observation.

Action	Description

Review purpose/expectations:
    Proactively
    Reactively
	Revisit production system vision, mission, and core values. Review process and subprocess competition and benchmarks. Review process and subprocess purpose, definition, goals, objectives, targets, and specifications.

Review actual operations and results:
    Our processes/subprocesses
    Other processes/subprocesses
	Understand what's happening—observe transformation mechanisms, measure output, transformation, and input parameters.

Compare actual with expected	Look for possible performance gains. Assess process mechanisms. Locate leverage points for performance enhancement.
	Look for possible performance gaps. Compare actual process measurements with expected process measurements.
Express opportunity	Express process gains and process gaps in the context of process mechanisms and metrics, respectively. Explain in general terms what the gain/gap opportunity is in the context of physical, economic, timeliness, and customer service performance enhancement.

further activities. If the offering is judged to be significant, then we continue our process improvement pursuit. For now, we are not highly concerned about exactly how we will close the gap—**our focus is on what closing the gap will mean to our production system as a whole.**

In a proactive sense, we focus on potential process gains. These gains are tied to our vision of improvement, rather than an explicit shortfall or gap. Hence, gain thinking is handled somewhat differently than gap thinking. **Using gain thinking, we typically start with some sort of reasonably explicit process characteristic regarding process improvement,** i.e., in our process transformation itself. Then, we project this improvement concept to outputs and results and to inputs and resource requirements. If the contribution appears to be significant, then we continue our improvement thinking and activities; otherwise we abandon this particular effort and focus our attention on other opportunities.

In both the reactive and proactive approaches, **we want to clearly express and communicate opportunities that hold promise in order to facilitate further development.** And we want to abandon efforts that do not hold promise to close gaps or generate gains that impact technical and business results in the short term and long term. **We tie our expressions to our process layouts and PVC model, locating general impact areas in each.** Isolating opportunities is highly exploratory and speculative, generating challenges for all personnel involved. Depending on the specific process purpose and line of questioning or observation, several initiatives may be useful. A general overview of usefulness is presented in Table 7.1. In general, benchmarking and theory of constraints-related initiatives are useful in observation. Tools such as the flow chart, PVC model analysis, relations diagram, cause-effect diagram, and root cause analysis are universally useful.

HIGH LIFT CASE—RECEIVING/STOCKING SUBPROCESS OBSERVATION Revisiting the | **Example 16.2**
High Lift Case, after several months of full-scale operations, production costs have been reduced, but not as much as expected. Direct observation of the redefined assembly subprocess, as depicted in Figure 16.3, indicates smoother operations than before redefinition. However, unanticipated line shortages and delays remain. In order to investigate the nature of these problems we move upstream to the receiving, storage, and stocking subprocesses.

The receiving subprocess is highly dependent on the procurement subprocess, which utilizes computer-generated orders, scheduled from a master schedule of product delivery dates. The line orders for specific materials, assemblies, parts, and supplies are driven by the production scheduling sub-subprocess, a part of the procurement subprocess. Figure 16.4 depicts the receiving, storage, and stocking subprocesses currently in operation.

Several customer-side and supplier-side performance metrics for the last week are provided in Figure 16.4. The metrics and observations point toward symptoms of late deliveries, both inbound and outbound. Recent performance indicates a subprocess inbound on-time percentage of 53% and a corresponding outbound (to the assembly line) statistic of 69%. In both cases, targets of 95% seemed reasonable during the redefinition phase several months ago.

Slow outbound deliveries lead to line shortages and sometimes production delays, creating excessive costs in both order expediting and production delays. To some degree, slow inbound orders create slow outbound orders, but not always. The higher outbound on-time statistic indicates that expediting line stock orders can help to get them back on or near schedule, but additional costs are incurred.

Observation indicates that there is no particular pattern to the late deliveries and expedited orders. However, several vendors have 100% on-time delivery records. Further investigation into the operations of these on-time vendors indicates that they also supply several other manufacturing organizations, one of which demands on-time delivery and has trained its vendors to deliver on time to their production schedule. Apparently, the discipline of this training is showing up in High Lift orders as well. One other surprising observation is that a relatively high proportion of the expedited line orders involves small, generic piece parts such as fasteners.

CONCEPTS

Once we have discovered process improvement opportunities and their respective challenges, we describe the opportunities in enough detail to address root causes in reactive efforts and address process levers in proactive efforts. Here, we explore concepts related to process transformation that impact our results and outputs and/or resources and inputs. We look for bases to establish performance enhancement in physical, economic, timeliness, and/or customer service dimensions. Process characterization, as described in Chapter 4, is helpful as a starting point. Details of process improvement conceptualization are described in Table 16.2.

The proactive approach is significantly different from the reactive approach in initiation. Here, we typically do not begin at a metric shortfall because we may already be meeting our target and showing well against our benchmark. **In the proactive approach, we begin somewhere in the physical process,** e.g., at a subprocess, where we question our process transformation or input and respective resources in a "what if" manner. For example, we might observe a natural gas liquefying subprocess, directly observing its criticality in the process flow, and begin our line of questioning in the physical process itself, rather than initiating our questioning at a

Receiving subprocess purpose: To accept, unload, and verify contents of arriving items in a timely manner.

Storage subprocess purpose: To accurately and rapidly store, protect, and retrieve all items upon request.

Stocking subprocess purpose: To deliver the right item, at the right place, at the right time, all the time.

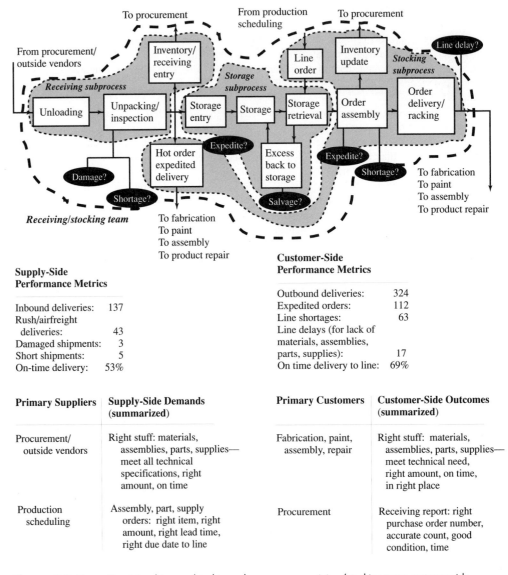

Supply-Side Performance Metrics

Inbound deliveries:	137
Rush/airfreight deliveries:	43
Damaged shipments:	3
Short shipments:	5
On-time delivery:	53%

Customer-Side Performance Metrics

Outbound deliveries:	324
Expedited orders:	112
Line shortages:	63
Line delays (for lack of materials, assemblies, parts, supplies):	17
On time delivery to line:	69%

Primary Suppliers	Supply-Side Demands (summarized)	Primary Customers	Customer-Side Outcomes (summarized)
Procurement/ outside vendors	Right stuff: materials, assemblies, parts, supplies— meet all technical specifications, right amount, on time	Fabrication, paint, assembly, repair	Right stuff: materials, assemblies, parts, supplies— meet technical need, right amount, on time, in right place
Production scheduling	Assembly, part, supply orders: right item, right amount, right lead time, right due date to line	Procurement	Receiving report: right purchase order number, accurate count, good condition, time

Figure 16.4   Receiving/storage/stocking subprocesses, receiving/stocking team customer-side outcomes, and supplier-side demands and performance, summarized (High Lift Case).

Table 16.2   Details related to concepts

Action	Description
Proactive Conceptualization	
Isolate process leverage points	Starting with process/subprocess observation, layouts, and the PVC model, isolate the technical essence of a gain effect. Determine potential leverage points.
Isolate process levers	Using a cause-effect argument, isolate the major drivers (variables) associated with the process leverage points—those that can produce the process gain. In other words, explain why we can produce the gain.
Identify gain concept	Identify the principles that allow us to make the gain.
Assess secondary impacts	Identify other processes/subprocesses that connect to the process levers.
Reactive Conceptualization	
Isolate process limitations	Starting with process/subprocess observation, layouts, and the PVC model, isolate the technical essence of a gap effect. Determine potential limitations.
Isolate root causes	Using a cause-effect argument, isolate the major drivers (variables) associated with the process limitation that together can close the process gap. In other words, explain why we can close the gap.
Identify gap-closing concept	Identify the principles that allow us to close the gap.
Assess secondary impacts	Identify other processes/subprocesses that connect to the root cause.

historic metric. Here, we might consider thermodynamic principles that eventually lead us to a more efficient process, with a higher yield and lower cost. Why do we do it?

Starting from a conceptual physical process perspective, e.g., somewhere in transformation, we work in both directions. First, we might look toward the result end to assess our output leverage. For example, if we think we can improve the natural gas liquefying subprocess, we estimate the gain in results likely from our process enhancement. If the potential result gain looks promising, then we look in the other direction, upstream, to isolate a specific cause-effect or concept/principle that we can exploit for the gain.

Or, alternatively, we might look upstream to a subprocess lever and then speculate on possible gains available if we can "pull" the lever to a different position. Here, subprocess levers will typically consist of controllable variables. We seek to identify the most important variables and go to work on them, so as to affect our subprocess and yield our anticipated process gains. For example, we might isolate the lever variables associated with our natural gas stream chiller temperature. And we might postulate that, by moving the lever, we can lower the chiller temperature, increase our yield, and decrease our cycle time, but with an additional energy requirement. Hence, at this stage, we have an improvement focal point. Refer to the Energy Cost/Operations Cases (Examples 5.6, 6.1, and 6.2) in Chapters 5 and 6 for brief examples.

In the reactive approach, we initially focus on our product-process metrics. When we are off target or off benchmark, we naturally question our process. Here, we tend to look for process result gaps and the physical bottlenecks that create them. We then describe them in terms of limitations. Initial clues are usually picked up in the process metrics from a historical perspective. We might pick up additional clues to identify our limitation anywhere along the PVC structure, e.g., upstream in our subprocess blocks.

For example, a cycle time that is above target or benchmark might be identified. Further investigation might lead back to excessive rework and scrap in a subprocess, i.e., the bottleneck.

Next, we typically continue our questioning, headed toward identifying the limitation or root cause. For example, in a manufacturing assembly subprocess we might locate equipment that does not produce a highly repeatable result. At this point, we have started to uncover our limitation.

Our improvement efforts are applied so as to eliminate the effect of this limitation, and open the bottleneck. Here, in order to eliminate the effect of our limitation, we examine the equipment, materials, operating methods, and so forth, in order to isolate root causes. Root causes will likely be associated with the variables acting on the subprocess. For example, in the Process Variation Case (Example 5.1) of Chapter 5, our physical characterization and C-E diagram led us to the roll pressure variable. We designed and executed a single-factor experiment and observed that the roll pressure was responsible for about 50% of the variation in blank length. Now, we can address this issue through process improvement, with the confidence we are addressing a root cause.

Both proactive and reactive approaches are productive and active in our workplaces. The former is a physically driven approach, encouraged in our workplaces by engineering leadership-based initiatives. **The latter is a data-driven approach,** encouraged in our workplaces by management-based initiatives.

A clear description of the improvement opportunity in conceptual terms as soon as possible, but not sooner than possible, i.e., prematurely, avoids several pitfalls. First, we avoid "hip shooting" or jumping to a "solution" without describing the problem, or trying "something" new without considering potential impacts. Second, we avoid needless "head-butting" conflicts—arguing over or championing "solutions" to ill-defined issues—that waste time and create unnecessary discord in the workplace.

We tie opportunities to process variables, through cause-effect reasoning and concepts, as quickly as possible. Once we have focused our energies collectively and formed some level of consensus as to the underpinnings of our opportunity, **we collectively address resolving our process gap or capturing our process gain.**

Example 16.3	**HIGH LIFT CASE—PROCUREMENT, RECEIVING, STORAGE, AND STOCKING SUBPROCESSES IMPROVEMENT CONCEPTS** Currently the High Lift production system is operating better than ever before. Our previous redefinition in our production process has enhanced our performance in general. Our product is now more modular than ever, allowing for faster service and maintenance in the field. Our product assembly is faster than ever. We see a smooth flow of materials, assemblies, parts, and supplies coming together to assemble our High Lifters; see Figure 16.3. Our product's performance is better than ever, our customer service is as good as ever, our costs are reduced, and our product is delivered in a more timely fashion.

Nevertheless, observations in our High Lifter production process clearly indicate that our economic and timeliness performance, although enhanced, is falling short of our targets. These concerns have led to the formation of an improvement team. This team is an interdisciplinary team made up of people from the fabrication, paint, assembly, product repair, field support, procurement, and receiving, storage, and stocking subprocesses. The constitution of this team was deliberate, so that all directly affected subprocesses are represented with respect to the general issues brought out in previous observations.

The challenges for this group are the timeliness issues reflected in the on-time delivery statistics and the cost issues reflected in the expediting and line shortages issues. The team focus is on limitations and root causes creating these bottlenecks, holding High Lift back from outstanding production system performance. Although this focus is clearly reactive, i.e., to missed timeliness and cost targets, several proactive opportunities are expected to emerge.

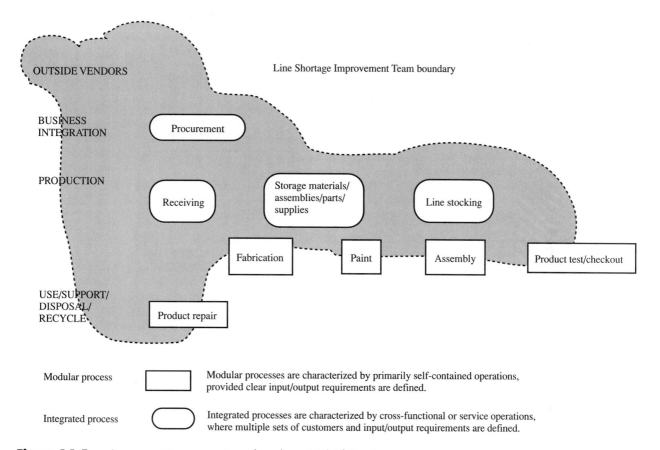

Figure 16.5 Line support improvement team boundaries (High Lift Case).

Our High Lifter improvement team has agreed conceptually to take a wide-angle view of the issues. On the basis of the flowchart and statistics shown in Figure 16.4, our team has decided to establish boundaries around their problem that encompass outside vendors, as well as internal procurement, receiving, storage, stocking, product repair, and production line subprocesses. Figure 16.5 depicts these boundaries.

Justification of these broad boundaries lies in the fact that the problem at hand, although manifesting itself primarily at the assembly points, appears to be back upstream. It is not certain at this point if we have a single root cause or multiple root causes. Several observations point to multiple root causes. Hence, broad coverage seems wise at the outset of our improvement initiative.

In order to break into the opportunity, we enter through the bottleneck of late deliveries, line shortages, and excessive costs in the areas of expediting materials, assemblies, parts, and supplies to the assembly areas. At this point, we set the issue of line shortages up as our bottleneck and work accordingly. Figure 16.6 provides a summarized relationship diagram, aligned with a theory of constraints–like initiative; see Chapter 19. We note here that our basic observations, facts, and figures are clustered, i.e., an affinity concept, into several groups. Then, we relate the groups to our line shortage focal point; refer to the discussion of the relationship diagram tool in Chapter 20 (20.21, "Relations Diagram").

The result from this conceptual investigation of line shortages does not indicate how we intend to resolve our problem; rather it brings to the forefront issues that, when pursued, should lead to creative

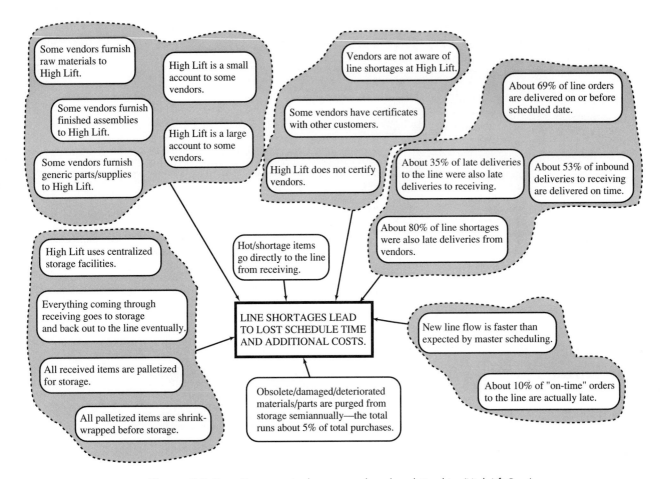

Figure 16.6   Line support subprocess and vendor relationships (High Lift Case).

resolution. Our strategy here is to obtain a wide-angle look at our bottleneck and the upstream limitations that are at its core. Our issues summary consists of three primary "why" issues:

1. Our vendors are not aware of, or responsive to, our line shortage problems—why?

2. Line flow is faster than the master schedule algorithm reflects in order issuance—why?

3. Everything that enters receiving goes through the storage area, with the exception of "hot" items that are needed to resolve a line shortage—why?

These issues are the product of team investigation in fact finding, data analysis, and interviews with vendors, as well as direct observations of the respective High Lift activities. They constitute a consensus starting point for root-cause analysis. Further discussion, based on experience and judgment, matched possible root causes with these issues. Table 16.3 summarizes the issues, possible root causes, and general impact. It is clear that our observations and conceptual assessment have produced both challenges and potential rewards for meeting these challenges. But we are not at the solution and resolution stage yet.

Table 16.3   Issues, possible root causes, and general impact summary (High Lift Case).

Issue	Possible Root Causes	General Impact
1. Our vendors are not aware of, or responsive to, our line shortage problems—why?	We do not communicate as well as we should with our vendors. Our vendors do not see High Lift as a large account. Our vendors are not capable of providing better service to us under current conditions.	More prompt deliveries from our vendors could decrease our airfreight costs and reduce our line shortages, speeding up/smoothing out our assembly subprocess. Estimated savings potential: $1.1 million per year
2. Line flow is faster than the master schedule algorithm reflects in order issuance—why?	Our manufacturing time estimates that drive several parts of our master scheduling system were made using time estimates from our former/pre-redefinition product/production processes. Our redefined product/production processes flow better than we anticipated/estimated—provided materials, assemblies, parts, and supplies are readily available.	A lack of current reality of our present redefined processes within our scheduling algorithm is holding our production process back from realizing its full potential. Present mismatches are putting brakes on potential assembly improvement on the lines. Estimated savings potential: $1 to $10 million per year
3. Everything that enters receiving goes through the storage area, with the exception of "hot" items that are needed to resolve a line shortage—why?	High Lift supplies centralized storage/inventory system solutions to its customers. This concept is a part of High Lift culture and reflected in current operations. Centralized storage for all items is questionable.	Centralized storage capital as well as operational costs are running about $1.2 million per year. Material, assembly, part, and supply obsolesce costs are running at about 5% of purchased part costs or about $2 million dollars per year. Potential customers are brought in to observe the technical operations of the High Lift storage system. This demonstration is viewed as a decisive element in customers choosing High Lift. Such observation is involved with about 45% of system sales.

OPTIONS

Our basic options take one of three general forms: (1) **process improvement,** (2) **process redefinition,** or (3) **other opportunities,** e.g., abandoning this particular opportunity for lack of potential contributions. Table 16.4 provides a detailed description of option assessment considerations.

First, we review our original process/subprocess purpose and definition. Then, we position the opportunity at hand as a process improvement within the current process definition. Here, our position is one of evolutionary/tactical change aimed at closing an identified process results gap or realizing an identified process results gain.

Table 16.4   Details related to options.

Action	Description

Express three options:
    Improvement
    Redefinition
    Other opportunities
	Visualize the process gain/gap in three contexts: (1) a tactical improvement context within the current process/subprocess purpose and definition and (2) a strategic process/subprocess redefinition context, and (3) abandonment/pursue other opportunities..

Assess option performance:
    Physical
    Economic
    Timeliness
    Customer service
    Flexibility
	Using an appropriate assessment format, compare and contrast the options, considering the short term and long term.

Choose approach	From the perspective of the whole production system, choose the strategic, the tactical, or pursue other opportunities option.
Set and communicate purpose	For the option selected, express a razor-sharp purpose and proceed.

Next, we think beyond the evolutionary improvement option to a revolutionary/strategic option involving process redefinition. Here, we typically work within our process purpose—as long as it is still valid. Validity centers on creating value for our customers and stakeholders. Redefinition may encompass new or different technologies for process transformation, and hence, perhaps different input resources and/or different product and by-product characteristics.

Our redefinition thinking at this point is primarily conceptual and does not contain the level of detail described in Section 3, Chapters 7 and 8. However, the elements of process definition/ redefinition in Section 3 are relevant and help us lay out our redefinition option. Here, our re-definition option serves as a background for weighing the merit of our improvement option. **We do not want to pursue an extensive improvement option within an outdated or misconceived process definition.**

We assess our options, improvement, redefinition, or pursue other opportunities, with respect to each other, so that we can channel our resources in the best interest of our production system, in the short term and long term. Five general performance categories are of interest: physical, economic, timeliness, customer service, and flexibility.

Physical performance encompasses the function, form, and fit associated with both the process as well as the process output—the product and by-products. Economics relate to costs and revenues with respect to volume or throughput associated with the process. Timeliness issues include cycle times associated with the process. Customer service includes the attention that customers receive relative to satisfying their needs and expectations. Flexibility refers to how readily the process can be adapted to changes in product specifications and volume of production.

Our analysis will be rather crude. Ballpark economic estimates and general qualitative descriptions of benefits and burdens based on experiences and technical judgment are typical. Benchmarking information describing similar experiences of other organizations is usually available in addressing process gap effects. Assessing proactive process gain effects typically requires a high degree of technical and business judgment because of sparse data.

In all cases, we use established analysis procedures that we can defend as being thorough and objective. In most cases, the improvement option will prevail. However, we must give due consideration to redefinition so that we do not lock in on short-term solutions and overlook long-term opportunity. In breakthrough and creative thinking terms, we consider the purposes principle, the systems principle, and the solution after next principle; see Table 18.8.

Once we choose our general course of action—improvement, redefinition, or pursue other opportunities—**we commit.** We express our commitment in the form of a purpose statement regarding our choice. **Our purpose should be a consensus expression that directs our attention and thinking toward the betterment of our production system as a whole.** If it is an improvement purpose, we pursue the remainder of the process improvement cycle elements in Figure 16.1. If it is a redefinition purpose, we revert to the redefinition elements. If it is a pursue other opportunities choice, we stop and focus our time and energy on some other endeavor.

HIGH LIFT CASE—PROCUREMENT, RECEIVING, STORAGE, AND STOCKING SUBPROCESSES ENHANCEMENT OPTIONS | **Example 16.4**

From our previous observations and conceptual cause analyses, it is clear that attention to our line shortage issue, as summarized in Table 16.3, holds great potential for closing our production gap, compared to other manufacturers. It is also clear that "how" we address these issues will be rather broad in terms of the bottlenecks and limitations we have identified.

We have two basic paths from which to choose: (1) we can continue on the improvement path or (2) we can channel our effort along a redefinition path. Our choice in this case is not clear. Arguments can be developed to support either path. The decisive elements in path choice lie in the impact on customers, i.e., our storage system customers on the outside. Should we choose the redefinition path, we would expand our team expertise. We would see a larger team and typically a longer cycle for team action. Should we decide to continue the improvement effort that has produced our findings to date, we would reevaluate our team expertise and continue our work, as described in Chapter 17.

In our assessment of path choice, our key is to examine the process/subprocess purposes and technologies. From this perspective, we see that our work clearly lies within the current boundaries regarding a modular product design and a straight-line assembly flow. However, we also see that a centralized storage technology and centralized computer scheduling technology are subjects of critical study, as they appear within our issues and potential root causes as laid out in Table 16.3. The concept of selling our customers on automated storage by viewing our showcase system also enters our decision picture. In some respects, we are faced with the choice of a small redefinition effort or a large improvement effort.

Although not as clear-cut as we would like it, our choice is to pursue an improvement project. Our primary rationale is that our boundaries for the improvement initiative, as depicted in Figure 16.5, are clear and within the scope of an improvement initiative. Furthermore, we are comfortable that we have the right people on the improvement team. We also believe that resolution of our line shortage problems can be addressed through benchmarking and creative swiping from other manufacturers.

The one doubt remaining centers around our centralized storage technology, i.e., our in-plant demonstrations, and its ability to draw customers. Here, if we should find that significant decentralization is necessary for our own production technologies, we may put our premier product, a centralized storage system, at some risk. After reflecting on this issue, leadership has decided to support the improvement initiative, without any constraints as to how our High Lift storage subprocess should look.

Working on the basis of this decision, a crisp process improvement purpose is developed:

Line support improvement purpose: Working with our line customers and vendors, within the procurement, receiving, storage, and stocking subprocesses, we intend to eliminate line shortages and reduce related production costs to their lowest possible levels.

The idea here is to resolve our internal production bottlenecks and at the same time to stimulate the creativity of our organization from within—thus possibly challenging our product definition teams to become even more involved in their customer's storage solutions, and perhaps offering external customers more distributed solutions and more savings in the end. For example, long-term results for High Lift might generate a broader range of products to market to external customers.

REVIEW AND DISCOVERY EXERCISES

REVIEW

16.1. Why is process improvement necessary? Explain.

16.2. What is the major distinction between process improvement and process redefinition? Explain.

16.3. What relationship exists between a process purpose and a process improvement purpose? Explain.

16.4. Why, and on what basis, do we bound process improvement efforts? Explain.

16.5. Compare and contrast the concepts of modular and integrated processes. What distinguishing characteristics does each one possess? Explain.

16.6. Who should be involved in process improvement efforts? Why?

16.7. Compare and contrast proactive and reactive strategies in process improvement, regarding actuals and expecteds, relative to process improvement opportunity identification.

16.8. How are process levers isolated in proactive process improvement efforts? Explain.

16.9. How are root causes isolated in reactive process improvement efforts? Explain.

16.10. What parameters of performance are relevant for assessing whether to improve or redefine? Explain.

DISCOVERY

16.11. How do process control points help in locating process improvement opportunities?

16.12. Why are initiatives and tools secondary to process purpose/improvement? Explain.

16.13. Compare and contrast the High Lift Case system-level process layout overview, Figure 16.2, with the overview from the Downtown Bakery Case shown in Figure 7.2. Why are the fundamental processes broken out differently? Explain.

16.14. Why are process control points and their metrics critical in isolating reactive process improvement opportunities? Explain.

16.15. Why are process leverage points critical in conceptualizing proactive process improvements? Explain.

16.16. Why are process-related principles and technologies critical in process improvement/redefinition conceptualizations? Explain.

16.17. Why is it critical to assess secondary impacts with regard to process levers in proactive process improvement efforts? With regard to root causes in reactive efforts? Explain.

16.18. Why do we focus our questioning process element toward the three basic options of improvement, redefinition, or pursue other opportunities? Explain.

The following project challenges are best worked as team projects. In academic exercises with college students, we have found small teams of from four to seven students work best. For practicing professionals, teams sized according to the project scale are best. For example, an extensive project that requires a variety of expertise typically requires a larger team. As an outcome, we expect to see a well-functioning team produce a logical and meaningful process improvement. Each team is responsible for both effective and efficient oral presentations as well as written presentations consisting of both graphical and prose communication elements. Teamwork fundamentals are discussed in Chapter 18.

SECTION 5, PROJECT 1

Identify an operational organization of your choice, e.g., service, hardgood, or basic material producer. Identify a fundamental process or subprocess and develop a process improvement. Here, we start with an existing process description and improve the process, using our process improvement elements, as expanded and discussed in Chapters 16 and 17. Our expanded materials and the High Lift Case Examples 16.1 to 17.4 provide a guide. Feel free to expand and enhance the level of detail in your work beyond that specifically illustrated in the High Lift Case examples.

SECTION 5, PROJECT 2

Identify a fundamental process or subprocess of your choice in your personal life, e.g., associated with studying, cleaning your domicile, or a sports activity. Briefly document your current process using the methods of Sections 2 and 3 as guidelines. Then, develop a process/subprocess improvement purpose. Use our process improvement elements, as expanded and discussed in Chapters 16 and 17, to improve your identified process/subprocess. Our expanded materials and the High Lift Case Examples 16.1 to 17.4 provide a guide. Feel free to expand and enhance the level of detail in your work beyond that specifically illustrated in the High Lift Case examples.

chapter
17

PROCESS IMPROVEMENT—ANALYSIS AND IMPLEMENTATION PERSPECTIVES

17.0 INQUIRY

1. How are alternatives for change described?
2. How are process gaps and their root courses isolated and addressed?
3. How are process gains, leverage points, and process levers discovered and addressed?
4. How does implementation of change differ from planning for change?
5. How does process improvement interface with process control and redefinition?

17.1 INTRODUCTION

After we commit to an improvement option, we pursue a detailed analysis, focused on generating specific alternatives, alternative evaluation and choice, and change plan formulation. **The analysis element of our process improvement triad,** Figure 16.1, **focuses on creativity.** Facts, ideas, and details are all critical for success. Implementation of process improvements constitutes the final leg of our process improvement triad. **Implementation brings plans for change to life.**

17.2 PROCESS CHANGE DESCRIPTION

The formulation and description of process change provides the means through which we identify and screen the multitude of possibilities available for process improvement. We

proceed in a systematic fashion to avoid oversights that may handicap our effectiveness and efficiency in eventual implementation and operation. **We involve people who will be affected by the process improvement, people who have knowledge and experience to offer, and people who will be instrumental in implementing and operating the improvement.** We use a participative format so that we can unlock both leadership and creative potentials within each person and ultimately produce the best possible result for our production system in terms of both plans and commitment for implementation.

ALTERNATIVES

At the outset of alternative formulation we reaffirm our purpose. Any misunderstanding of purpose will surely lead to unnecessary confusion and conflict later on. Hence, our details in Table 17.1 begin with reviewing our purpose hierarchy—common purpose, process purpose, and our specific process improvement purpose.

The seven breakthrough principles and the creative thinking concepts described in Section 6 (see Chapter 18) provide guidelines for dealing with alternative formulation. **Our alternatives are forged out of facts and ideas related to both technical and business concepts.**

The questioning activity described in Chapter 16 is driven by general facts and figures. A wealth of facts and figures—details—is usually already available at this point. Here, cause-effect relationships associated with our process/subprocesses, as well as interrelationships with the balance of the production system, i.e., other processes/subprocesses, are reused if applicable.

Ideas begin with initial thoughts. Some thoughts are more original than others. For example, thoughts offered toward our process improvement purpose may include a spectrum of originality— copying others, creative swiping, and truly original. At this point, we collect all thoughts (embryonic ideas) for further refinement. We refine our thoughts both individually and as a group

Table 17.1   Details related to alternatives.

Action	Description
Restate improvement purpose— build consensus	Our direction and specific purpose must be clear and acceptable to all involved in the process improvement effort. Purpose should be razor sharp—focused—and crystal clear before we proceed.
Collect facts and figures	Historic data regarding critical metrics should be available. Special studies and possibly experiments may be required. Further benchmarking may be required.
Generate and express ideas	All ideas are welcome. At this point, do not critically assess ideas; include copying and creative swiping in idea collection.
Critically assess ideas and formulate a list of feasible improvement alternatives	From the ideas (above), form feasible alternatives. Critical analysis, led by champions and critics of ideas, is necessary. Modify and combine ideas as appropriate.
Formulate basic process/subprocess layouts for each improvement alternative	Detail each improvement alternative. Position each feasible process improvement alternative in the overall context of the production system—integrate alternatives in process layouts.

by adding details as we weigh them against our creditable facts, opinions, and feelings. **We take special care to avoid judging or evaluating our thoughts and ideas prematurely.**

Once we have an idea portfolio—ranging from copying, to creative swiping, to novel ideas—we are ready for critical evaluation. We can evaluate ideas in several ways. One way is to play the roles of both critic and champion of the ideas, and scrutinize both the advantages and disadvantages in both technical and business dimensions.

Our objective is to work our ideas into legitimate process improvement alternatives. Here, we freely add to, take away from, and combine ideas in order to fit the ideas to our needs. We position each alternative in the context of our process/subprocess definitions and production system. Here, we look for interrelationships with potential impacts on other processes.

This ideation element is the primary creative element in process improvement. Up to this point, we have talked about closing gaps and capturing gains. So far, no gaps have been closed and no gains captured—just a lot of talk and notes. **We rely on our alternatives and their implementation to actually close our gaps or capture our gains.** If our ideas and alternatives lack this potential, further implementation/operation efforts will be handicapped. In the end, we are typically left with several feasible alternatives.

It is possible to successfully close a gap or capture a gain in one process/subprocess, all at the expense of another process/subprocess, and therefore create a marginal gain or a net loss for our production system as a whole. We want to avoid this possibility of suboptimization. Hence, **we take each alternative and position it within the context of the production system.** Our process and subprocess layouts provide a structure for this placement. Here, we examine supplier-customer and input-output relationships that will be affected.

We use available facts and figures as a starting point. We collect other facts and figures, and verify the validity of the ones on hand as well. Collection mechanisms may include cause-effect analysis, experiments, benchmarking other similar processes, and so forth. We take care to separate facts, opinions, and feelings. The source and creditability of each should be established and questioned, respectively.

Example 17.1	**HIGH LIFT CASE—LINE SUPPORT IMPROVEMENT ALTERNATIVES** Previously, High Lift leadership chose the improvement path to address line shortages and excessive costs in the production process. An interdisciplinary team was chartered. The improvement team boundaries at the outset were as shown in Figure 16.5. Here, the specific focus was set to include outside vendors, procurement, receiving, storage, and stocking, with involvement from fabrication, paint, assembly, and repair. Relevant purposes are restated.

Line support improvement purpose: Working with our line customers and vendors, within the procurement, receiving, storage, and stocking subprocesses, we intend to eliminate line shortages and reduce related production costs to their lowest possible levels.

Procurement subprocess purpose: To locate and coordinate capable vendors, produce accurate delivery and production schedules, and maintain an adequate inventory.

Receiving subprocess purpose: To accept, unload, and verify contents of arriving items in a timely manner.

Storage subprocess purpose: To accurately and rapidly store, protect, and retrieve all items upon request.

Stocking subprocess purpose: To deliver the right item, at the right place, at the right time, all the time.

After reviewing the purposes, the team consensus was that the purposes accurately reflect the situation and are adequate to direct future improvement activities.

Several critical facts and figures were uncovered during the conceptual phase of the improvement effort. These facts and figures appear in Figure 16.4, Table 16.3, and Figure 16.6. Additional facts and figures were gathered through benchmarking/observations of several other production systems. One particular system observed was an automotive assembly plant, where a visual approach to stocking and inventory was under development and test. Here, the production system was evolving toward keeping inventory immediately off line, staged to move into position when needed. This pseudo just-in-time system required vendors to make smaller and more frequent deliveries to the line. In this transformation, the plant visited had several vendor training initiatives active; all addressed better relationships with the vendors so that the materials would indeed arrive in time to avoid line shortages.

Regarding the High Lift team, two basic alternatives were developed. The first alternative involved fine-tuning the existing procurement/scheduling system, applying an incentive system to encourage timely vendor deliveries, and adding a line customer service team that would meet to assess performance of the improvements and help to communicate High Lift changes, demands, and concerns to vendors. Keys to success for this alternative are summarized below:

1. An update to more accurate production time estimates and more frequent updates of the master delivery schedule should lead to more accurate and timely issuance of orders.

2. A financial incentive to vendors, e.g., rewards for on-time deliveries and penalties for late deliveries, should capture the attention of vendors and improve their performance.

3. The line customer service team concept should enhance coordination and communication, so that High Lift can impress upon its vendors the importance of timely deliveries, and so that line customers can communicate with their internal procurement, receiving, storage, and stocking suppliers likewise.

The second alternative, the visual production approach, involved a far different approach to addressing the line shortage and cost issues. Here, a creative swiping approach was developed, based in large part on the benchmarked automobile assembly plant. Noting several distinct differences in volume and product, the High Lift improvement team produced a conceptual design for incorporating a visual inventory, reduced inventory system. This alternative fits within the current purposes, as stated above, but requires several grass-roots changes in the way each subprocess is executed. Figure 17.1 depicts the second alternative as it would be manifested in the assembly area.

Two major changes are apparent in this depiction. First, materials, assemblies, parts, and supplies are stored immediately off line. Assemblers can see clearly how much inventory they have and get to their needed items quickly. Second, all materials, assemblies, parts, and supplies are racked or binned so that they are protected from damage and they are convenient to pull into the line. This arrangement was devised so that all current off-line assembly, e.g., units waiting for missing parts, will cease to exist, freeing up the floor space necessary for these staged items.

In order to allow the visual assembly area alternative to become a reality, an interface between the vendors and the assembly area was defined. This definition differed significantly from the current interface—after several modifications and discussions a sensible interface emerged. This interface is depicted graphically in Figure 17.2.

Here, we see a traditional pallet and shrink-wrap method of transport and storage as obsolete. Instead, we see custom-designed racks and bins to fit our specific materials, assemblies, parts, and supplies. These racks and bins are used as transport vehicles for all items from vendors to High Lift. For example, since each vendor typically delivers to High Lift, loaded racks and bins are received, while empty racks and bins are loaded out and sent back to the vendor's facility. The cycle essentially moves the items to the line and recycles the racks and bins as they are emptied. As an additional feature, each rack and bin is designed to reduce assembly time, as well as protect its cargo.

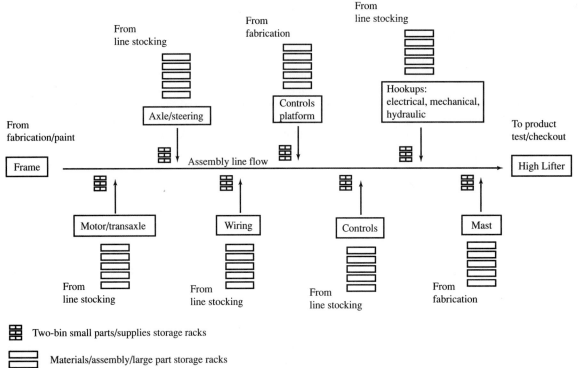

 Two-bin small parts/supplies storage racks

Materials/assembly/large part storage racks

Figure 17.1   Visual alternative—improved subassembly-assembly process flow overview (High Lift Case).

Under this visual production alternative, each rack and bin is considered a unit. As such, reordering is done on a unit basis, with bar-coded cards moving and passing along to keep the stock flowing. Each item, with the exception of the small inexpensive items, is inventoried on the basis of units, varying from size one to more than one, depending on needs. The small items are binned on a two-bin system, whereby when one bin is exhausted, the second bin in the rack is tapped; a card is passed back, along with the bin; and a new reloaded bin replaces the exhausted bin.

Although the second alternative is definitely a departure from the traditional system, it is viewed as workable, with several distinct advantages:

1. Its simplicity and visual cues should allow everyone involved, from vendors to receivers to stockers to assemblers, to be aware of holding a modest- (i.e., reduced-) cost inventory, and to avoid both holding and obsolescence costs related to larger inventories.

2. The integrity of inbound materials, assemblies, parts, and supplies should be protected, thus avoiding losses from damaged materials.

3. The handling costs from receiving through assembly should be reduced since the racks and bins are specifically designed to avoid rehandling and repacking, as well as to expedite moving from rack to line.

4. The racks and bins should serve as an effective means of back-shipping defective materials. If an assembly is judged as nonconforming, it can be shipped back right on the rack, providing the vendor a visual cue immediately and avoiding the current lengthy assessment and adjustment for such items.

RECEIVING

Loaded/inbound

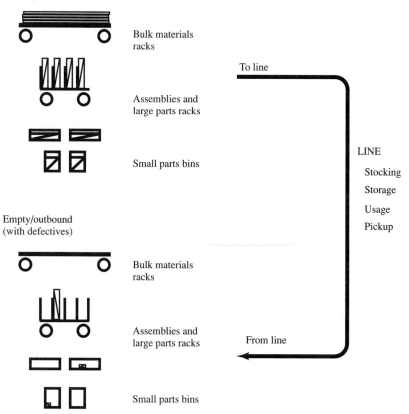

Bulk materials racks

Assemblies and large parts racks

Small parts bins

To line

LINE

Stocking

Storage

Usage

Pickup

Empty/outbound (with defectives)

Bulk materials racks

Assemblies and large parts racks

From line

Small parts bins

Figure 17.2   Visual alternative—improved receiving/storage/stocking flow overview (High Lift Case).

5. The integrated vendor-to-line order system will issue and coordinate vendor orders with the High Lift production schedule, as opposed to the present system, which has two distinct modules, one for the vendors and one for internal production orders. This integrated/singular communication-coordination mechanism should capture the attention of the vendors, thus involving them in High Lift activities, adding to High Lift effectiveness and efficiency.

At this point, we have esssentially three alternatives: (1) do nothing and continue as we are, (2) provide enhancements to the present system, and (3) pursue the visual production alternative. These alternatives seem to be feasible and within the expertise of High Lift personnel. In the last two cases, changes are called for that require more study in both impact and decision support, as well as in design detail, before implementation. All alternatives fit within current production process definitions, and no alternative requires product modification as a condition for implementation.

Table 17.2   Details related to evaluation.

Action	Description
Form baseline for comparison.	Describe present process/subprocess results. Describe the impact of the present process/subprocess on output and input, customers and suppliers, to form a baseline for assessing improvement alternatives.
Describe each feasible improvement alternative in the baseline context.	Estimate the impact of each process alternative on output and input, customers and suppliers.
Assess significant output and input differences for each alternative versus present process/subprocess methods.	Consider outputs (physical, economic, timeliness, customer service, and flexibility) and inputs (people, machines, tools, facilities, etc.). Express differences between each alternative and the present situation.
Choose a process improvement alternative.	Set up an acceptable analysis format. Choose the most appropriate alternative.

EVALUATION

At this point, we have a set of process improvement alternatives, each described in general and positioned within our production system. Now, **we evaluate our alternatives and choose one alternative to fully develop and perhaps implement.** The details in Table 17.2 provide a general format useful for moving us through this evaluation.

Evaluation in process improvement is similar to, but usually simpler than, process definition/redefinition evaluation—since most improvement alternatives have narrower scopes than most definition/redefinition alternatives. Improvement alternatives in most cases, especially within modular processes/subprocesses, tend to be self-contained and hence relatively simple to evaluate. Improvement alternatives in integrated processes/subprocesses are typically more far-reaching and require a more complex evaluation, involving impact on other processes/subprocesses. For example, improvement in a modular fabrication subprocess is typically self-contained, whereas in an integrated maintenance process it may impact several other processes.

Independent of the alternatives themselves, we establish a baseline for comparison. This baseline is built around the present processes/subprocesses, i.e., the ones we are calling to question. Here, we start with the process/subprocess purposes and definition information available. Essentially, we seek to fulfill our improvement purpose—within the defined process technology—more effectively with higher quality and more efficiently with higher productivity.

Our evaluation is developed by assessing differences between our alternatives and our baseline. Since several different dimensions are relevant, a univariate comparison, e.g., dollars, is usually not totally sufficient to support our decision. For example, we usually consider factors such as flexibility in product and volume changes, changes to customer service, and so on, in a semiquantitative fashion in their natural units, other than economic units.

We tend to see many process improvement initiatives versus only a few definition/redefinition initiatives. For reasons of both simplicity and high volume, we streamline improvement evaluations. **We typically develop pro-forma evaluation templates and apply them to our alternatives—noting any unique evaluation characteristics that are not included in the standard format.**

In any specific case, usually one or a few categories will influence our decision more than the others. However, we must be aware of all significant categories and their impact to assure a thorough and systematic analysis, to minimize unpleasant surprises later.

HIGH LIFT CASE—LINE SUPPORT IMPROVEMENT EVALUATION Revisiting our High Lift line **Example 17.2** support improvement case, we have three alternatives to consider. First, we can maintain present operations, i.e., the "status quo alternative." Second, we can pursue an improvement of our current centralized receiving, storage, and stocking subprocesses, i.e., the "centralized alternative." And, third, we can pursue our visual decentralized receiving, storage, and stocking improvement alternative, i.e., the "decentralized alternative." The physical nature of each of these alternatives was summarized previously. Now, we will summarize the decision-relevant characteristics of each of our alternatives.

We establish our status quo alternative as a baseline alternative from which to develop relative comparisons of the other two alternatives. The use of our status quo alternative is reasonable as a baseline because we are currently using it in our operations—and have some history regarding its effectiveness and efficiency. We use this historic basis as a starting point to set up each of the other two alternatives. Table 17.3 provides an organized summary of significant economic differences in our three alternatives.

Working with the information in Table 17.3, we set up cash flow models for evaluating each alternative with respect to the baseline, status quo alternative. These cash flow models allow us to assess the economic feasibility of each alternative relative to the baseline. We have omitted tax considerations for brevity. We use a minimum attractive rate of return set at 15%. This value represents the earnings potential of High Lift money. In general, any investment that earns less than 15 percent is deemed unattractive. Hence, we use 15% as our discount rate in our net present value evaluation. We use a 5-year time horizon for our evaluation, as this period of time seems reasonable considering the physical nature of our alternatives. Tables 17.4 and 17.5 provide cash flow models for the centralized alternative versus the baseline and for the decentralized alternative versus the baseline, respectively.

The net present value is calculated as

$$PV_n = \text{total cash flow entry } [(1 + i)^{-n}]$$

$$NPV = \sum_{\text{all } n} PV_n$$

[17.1]

where PV_n = present value associated with the cash flow from time period n

NPV = net present value at the present time, sometimes called year zero

i = minimum attractive rate of return (MARR)

n = time period index

Additional details regarding discounted cash flow modeling are available in Chapter 20.

Since both of these alternatives involve significant savings—i.e., they both contain cost and savings cash flows—we develop a supplemental classical payback period analysis. In the payback analysis, we do not take the time value of money into consideration in our model. The payback is defined as the time period necessary for a stream of cash flows to turn from negative to positive.

In our economic model results, we see two excellent improvement projects, compared to the status quo alternative. Our cash flow models indicate that both challenger alternatives are better than the status quo or baseline, i.e., the do nothing alternative. In both cases the NPV is positive, over $2 million in the centralized case and over $5 million in the decentralized case, with both projected over the same 5-year time horizon. In the NPV method, an NPV of zero indicates a marginal project, just barely earning the MARR. Because of its higher NPV, the visual, decentralized alternative, although more radical in terms of change, clearly has the edge over the improved centralized alternative. Both payback periods are extremely short, less than 1 year.

Factors other than economics are summarized with each alternative. Economic and timeliness factors are driving this improvement project. Both are reasonably well reflected and summarized in our economic evaluation. We are concerned that the product integrity and customer service remain at high levels. This concern seems resolved in that we are not affecting the product in our improvement project. If anything, the product quality should be enhanced by reducing the current level of off-line assembly.

Table 17.3   Summary table—decision-relevant characteristics (High Lift Case).

Status Quo System Baseline	Improved Centralized System versus the Baseline	Visual Decentralized System versus the Baseline
Existing		
Current system investment (1 year ago): $7 million	No change	Salvage system: $850,000
Current system maintenance: $300,000 per year	No change	Reduce by 100%: $300,000
Current system obsolete/damaged parts losses: $2 million per year	Reduce by 25% or $500,000 per year	Reduce by 50% or $1 million per year
Current system expedited order cost (in house only): $150,000 per year	Reduce by 50% or $75,000 per year	Reduce by 75% or $112,500 per year
Current system expedited order cost (inbound airfreight, etc.): $225,000 per year	Reduce by 50% or $112,500 per year	Reduce by 90% or $202,500
Inventory carrying costs: $300,000 per year	Reduce by 25% or $75,000 per year by faster turnover	Reduce by 50% or $150,000 per year by faster turnover
Line stoppage/overtime off-line fabrication/assembly cost: $430,000 per year	Reduce by 50% or $215,000 per year	Reduce by 85% or $365,500 per year
Additional		
	Vendor relations costs: $50,000 per year	Facility modifications: $425,000
	One percent vendor incentive for on-time delivery (based on 65% participation): $260,000 per year	New system/equipment (racks/bins, etc.): $900,000
	Database update/modifications: $65,000 year 1, $25,000 per year maintenance starting in year 2	Operation/maintenance on new system/equipment: $250,000 per year
	Scheduling code update/ modifications: $85,000 year 1, $35,000 per year maintenance starting in year 2	Vendor certification: $85,000 year 1, $15,000 per year maintenance starting in year 2
	Labor savings (one person): $40,000 per year	Personnel training: $45,000 year 1, $12,000 per year maintenance starting in year 2
		Inventory to racks/bins switch-out: $265,000 year zero

Major concerns involve risk regarding technical feasibility and estimated savings. In the case of the centralized alternative, our estimates appear to be soft but realistic. Regarding the decentralized alternative, we have observed similar savings in our benchmarked automobile assembly plant. We have developed this improvement project around a creative swiping theme, and we believe that we can make our decentralized concept work in our plant. But the technical feasibility question persists.

Table 17.4 Cash flow model—do nothing versus improved centralized system alternative (High Lift Case).

Parameters of Difference	0	Time Frame—End of Year				
		1	2	3	4	5
Vendor relations	0	$ (50,000)	$ (50,000)	$ (50,000)	$ (50,000)	$ (50,000)
Vendor on-time incentives	0	(260,000)	(260,000)	(260,000)	(260,000)	(260,000)
Database update/maintenance	0	(65,000)	(25,000)	(25,000)	(25,000)	(25,000)
Scheduling software update/maintenance	0	(85,000)	(35,000)	(35,000)	(35,000)	(35,000)
Inventory carrying cost savings	0	75,000	75,000	75,000	75,000	75,000
Labor savings	0	40,000	40,000	40,000	40,000	40,000
Obsolete/damaged parts savings	0	500,000	500,000	500,000	500,000	500,000
Expediting savings (in house)	0	75,000	75,000	75,000	75,000	75,000
Expediting savings (inbound)	0	112,500	112,500	112,500	112,500	112,500
Line stoppage/off-line assembly savings	0	215,000	215,000	215,000	215,000	215,000
Total	$ —	$ 557,500	$ 647,500	$ 647,500	$ 647,500	$ 647,500
Present value (15%)	$ —	$ 484,783	$ 489,603	$ 425,742	$ 370,210	$ 321,922
Net present value (year 0*)	$2,092,260					
MARR discount rate	0.15					
Payback analysis:						
Cumulative cash flow	$ —	$ 557,500	$1,205,000	$1,852,500	$2,500,000	$3,147,500
Payback period	Less than 1 year					

*Year 0 is defined as the present time.
Other critical parameters:
    Inventory turnover: Inventory turnover is expected to be about 25% faster.
    Risk: It is not clear how well the 1% cash incentive to vendors will work; we project 65% participation.
    Risk: Line stoppage savings, while real, are difficult to estimate and are contingent on vendor performance.
NOTES: 1. This alternative is attractive because of the limited capital and minimal change it demands.
        2. Two major contributors to the effectiveness of this plan, on-time incentives and line stoppage savings, are the best estimates available, but to some degree speculative. Obsolete/damaged savings estimates are solid.

    Regarding technical feasibility, an experiment was developed to assess proof of concept in the High Lift plant. This experiment is described in detail in Chapter 5; see the Assembly Improvement Case, Examples 5.2 to 5.5. Empirical evidence indicates that the rack/bin concept is feasible and will provide significant reductions in assembly time when compared to the status quo assembly method. Hence, the question of technical feasibility is put to rest, in the sense that we have demonstrated evidence that we can do it and do it very well.

    From the standpoint of High Lift's primary product, large integrated centralized storage facilities, we have not proven the centralized concept to be obsolete, but we have provided testimony that it is not appropriate for all applications. Our experience with the decentralized methods and our lessons learned on the High Lifter product line are expected to open new product opportunities, where we can apply our engineering/integration expertise. Hence, the loss of our demonstration model is not expected to be a major problem—and most likely will be a wake-up call for our product development efforts.

Table 17.5   Cash flow model—do nothing versus visual decentralized system alternative (High Lift Case).

Parameters of Difference	Time Frame—End of Year					
	0	1	2	3	4	5
Facilities modifications	$ (425,000)	0	0	0	0	0
New system/equipment	(900,000)	0	0	0	0	0
New system/equipment operations/maintenance	0	(250,000)	(250,000)	(250,000)	(250,000)	(250,000)
Vendor certification	0	(85,000)	(15,000)	(15,000)	(15,000)	(15,000)
Personnel training	0	(45,000)	(12,000)	(12,000)	(12,000)	(12,000)
Inventory switch-out	(265,000)	0	0	0	0	0
Inventory carrying cost savings	0	150,000	150,000	150,000	150,000	150,000
Salvage (current system)	0	850,000	0	0	0	0
Current system operation/ maintenance savings	0	300,000	300,000	300,000	300,000	300,000
Obsolete/damaged parts savings	0	1,000,000	1,000,000	1,000,000	1,000,000	1,000,000
Expediting savings (in house)	0	112,500	112,500	112,500	112,500	112,500
Expediting savings (inbound)	0	202,500	202,500	202,500	202,500	202,500
Line stoppage/off-line assembly savings	0	365,500	365,500	365,500	365,500	365,500
Total	$(1,590,000)	$2,600,500	$1,853,500	$1,853,500	$1,853,500	$1,853,500
Present value (15%)	$(1,590,000)	$2,261,304	$1,401,512	$1,218,706	$1,059,745	$ 921,517
Net present value (Year 0*)	$ 5,272,785					
MARR discount rate	0.15					
Payback analysis:						
Cumulative cash flow	$(1,590,000)	$1,010,500	$2,864,000	$4,717,500	$6,571,000	$8,424,500
Payback period	Less than 1 year					

*Year 0 is defined as the present time.
Other critical parameters:
  Inventory turnover: Inventory turnover is expected to be about 50% faster.
  Risk: Savings proportional to estimates above have been documented in the benchmarked automobile assembly plant.
NOTE: This alternative is attractive because of the annual savings, but it requires a substantial initial investment.

PLAN

All process improvement alternatives require change in our processes/subprocesses—our plan reflects these changes. Regardless of the magnitude of change described in our plan, we review our processes/subprocesses as a whole. Table 17.6 provides details.

From our previous activities, we have a reasonably complete process layout. We update our layout to reflect our improvements. Now, we review the improved layout from the whole-process perspective to assure compatibility with our production system. Next, attention is focused on process/subprocess targets, specifications, and control points.

We clearly visualize and describe the outputs, transformations, and inputs of our improved process, express them in our plan. Then, we concentrate on team structure. **We address both team size and composition needs in our plan.** These adjustments require some level of reassignment and retraining, sometimes extensive, but usually far less extreme than in process redefinition.

Table 17.6 Details related to plan.

Action	Description
Update process/subprocess details	In a whole-process context, review the process/subprocess layouts and add detail where necessary to capture changes in process/subprocess requirements. Develop updated outputs and inputs, goals, objectives, targets, and specifications for the chosen alternative.
Update process team structure	From process definition information, update operating team composition and structure to reflect improvement changes in the process/subprocess.
Update critical control points and process variable targets	Using process definition information and new cause-effect analyses, locate and update critical process control levers and variables and reestablish targets. Some level of experimentation may be required.
Update critical metrics	From process definition information, update critical process metrics for compatibility with improvement changes.
Update PVC model	Update the previous PVC model with respect to process improvement changes.

We update our critical control points by examining cause-effect relationships in the improved process/subprocess plan. Previous cause-effect information, generated during the improvement-redefinition decision analysis, serves as a starting point. We ultimately seek to identify the critical controllable variables acting in the improved process. These variables establish critical control points, which we note on our improved process layout. Our proactive control points usually lie within subprocess transformations and inputs, while reactive control points typically lie at the subprocess outputs. **Each control point will have at least one critical metric that is subject to some form of process control—monitoring and adjustment.**

Finally, **once we have our improved process/subprocess layout, control points, and metrics revisions completed, we update our PVC model.** Here, we see most of the changes at the subprocess and variables levels. Most of the information for updates is already available. Our objective is to depict the interrelationships between what happens upstream at the variables level and what ultimately happens downstream at the production system level. **The PVC model depiction serves to connect the technical world of operations to the business world.**

HIGH LIFT CASE—LINE SUPPORT IMPROVEMENT PLAN   Once we commit to an alternative, | **Example 17.3**
we develop a detailed plan—a plan that we can move into the implementation phase readily. In this case, we have selected the visual production/decentralized alternative. This alternative is described conceptually and depicted in Figures 17.1 and 17.2. At this point, we add detail.

Our plan for the visual, decentralized alternative is summarized in Figure 17.3. This summarized plan includes our impacted subprocess purposes, which need no modification. The plan includes a flow diagram of the impacted areas. Here, we see rather large changes in our subprocesses, e.g., in the "hows." Some impact is expected in our line stocking customers' subprocesses, e.g., assembly. People from these subprocesses are on the improvement team, verbalizing their needs, demands, and expectations. For example, the racks and bins are expected to bring assembly time down, i.e., produce a shorter cycle time, in addition to reducing shortage impacts. This expectation is reasonable since our materials, assemblies, parts, and supplies will be staged for faster movement into the assembly area. In addition, experimentation relative to this expectation is complete; see the Assembly Improvement Case in Chapter 5, Examples 5.2 to 5.5. Empirical results indicate that the technology is both feasible and productive.

Procurement subprocess purpose: To locate and coordinate capable vendors, produce accurate delivery and production schedules, and maintain an adequate inventory.
Receiving subprocess purpose: To accept, unload, and verify contents of arriving items in a timely manner.
Storage subprocess purpose: To accurately and rapidly store, protect, and retrieve all items upon request.
Stocking subprocess purpose: To deliver the right item, at the right place, at the right time, all the time.

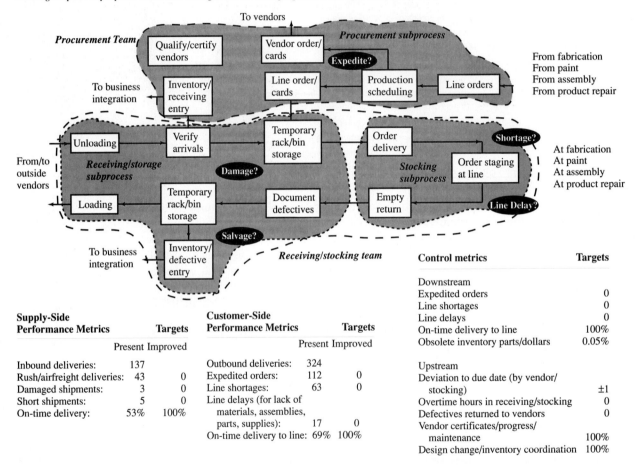

Supply-Side Performance Metrics

	Present	Improved
	Targets	
Inbound deliveries:	137	
Rush/airfreight deliveries:	43	0
Damaged shipments:	3	0
Short shipments:	5	0
On-time delivery:	53%	100%

Customer-Side Performance Metrics

	Present	Improved
	Targets	
Outbound deliveries:	324	
Expedited orders:	112	0
Line shortages:	63	0
Line delays (for lack of materials, assemblies, parts, supplies):	17	0
On-time delivery to line:	69%	100%

Control metrics	**Targets**
Downstream
Expedited orders	0
Line shortages	0
Line delays	0
On-time delivery to line	100%
Obsolete inventory parts/dollars	0.05%
Upstream
Deviation to due date (by vendor/ stocking)	±1
Overtime hours in receiving/stocking	0
Defectives returned to vendors	0
Vendor certificates/progress/ maintenance	100%
Design change/inventory coordination	100%

Figure 17.3 Visual alternative—improved subprocess and team plans, summarized (High Lift Case).

The flow diagram depicts a changed procurement subprocess, driven by line orders, as opposed to a centralized computer algorithm. The receiving/storage subprocess is characterized as an integrated subprocess with an in-out flow, using racked and binned materials, assemblies, parts, and supplies. Stocking is integrated with receiving/storage, with handoffs at the rack/bin level.

Once we develop our improvement plan, we test our plan to identify potential problems and bottlenecks. Tools such as the fault tree analysis (FTA) and failure mode and effects analysis (FMEA) are helpful; see Chapter 20 for details. As an illustration, we know that an unintended line shutdown is clearly an undesirable event. Hence, we can develop an FTA around such an event, considering our improvement plan. Figure 17.4 depicts a simplified FTA for our overall improvement plan. Here, we see four basic events that can lead to an unintended line shutdown: a production error or defect, a part or supply shortage, a part or supply failure at the line, and a line equipment failure. Each of these events is or can be developed down to whatever

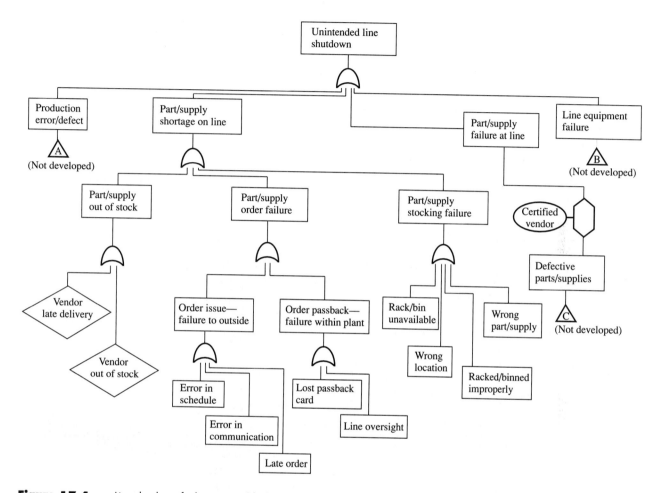

Figure 17.4   Line shutdown fault tree, simplified (High Lift Case).

level of detail is necessary for understanding. Then prevention or at least mitigation can be addressed. If necessary, we modify our plan.

For a broad mistakeproofing tool, we apply the FMEA or the failure mode and effects criticality analysis (FMECA). We develop a simplified FMECA focused on the two-bin system we intend to use for small or commodity parts and supplies. The FMECA in Table 17.7 examines the components of the two-bin system, i.e., the rack, the bin, and the order card. Here, we see a very broad view of possible failure modes and possible effects. We focus on the preventive measures column in order to consider possible countermeasures. From an FMECA, we can seed an FTA in order to study a failure event in more detail. For example, we can select a failure effect or event from the FMECA for our top event and proceed to build a fault tree.

The plan, as developed, has been mistakeproofed as far as possible by resorting to visual stocking methods. This approach is expected to help prevent shortages in real time before they become critical. With the improvement, storage is placed on line. Additionally, storage requirements in terms of inventory value have been substantially reduced, allowing a faster turnover as well as less carrying cost for the inventory itself.

More discipline on the part of vendors is required, as they must deliver more frequently with smaller orders. However, they are included in the information loop regarding High Lift production/product delivery

Table 17.7   Two-bin line stocking alternative FMECA, (High Lift Case).

Functional or Equipment Identification	Functional or Equipment Purpose	Interfaces	Failure Modes	Failure Mechanisms	Failure Detection	Failure Compensation	Failure Effects	Risk Level	Preventive Measures
1.0 Rack	To hold/display bins	Near production line and operators	1.1 Back bin missing	Stocking error	Human observation	None	Possible line shortage/shutdown	II-B	Open rack construction, proper viewing angle, bin detection switch/light
2.0 Bin	To hold/display parts/supplies	Fits in rack	2.1 Bin collapses	Sidewall failure	Human observation	None	Time to pick up contents, damage to contents	III-B	Reinforce sides/corners
			2.2 Drop/spill bin contents	Slips—can't get a grip	Human observation	None	Time to pick up contents, damage to contents, foot injury	III-B	Design handles into bin sides, strength compatible with content weight
			2.3 Drop part	Lack of clearance between rack and bin	Human observation	None	Excessive time to pick/remove part, hand injury	III-A	Design proper clearance between rack and bin opening
3.0 Order card	To allow for reorder notification	Fits in rack pocket	3.1 Lost card	Wind, knocked out by worker	Human observation	See empty bin in rack for prolonged period of time	Replacement bin will not arrive on time, possible line storage/shutdown	III-B	Design clip holder, consider electronic ordering
			3.2 Defaced card	Tear, rip, or smudge	Bar code will not read	Might be able to read code numbers manually	Slow reorder/restock	III-A	Plastic card, consider electronic ordering

schedules for "heads up" information. In order to provide a more disciplined delivery, vendor training and certification are a part of the plan.

The plan provides supply-side and customer-side metrics, and targets that are very aggressive when compared to the operations under the status quo procurement, receiving, storage, and stocking subprocesses. However, these targets appear to be realistic on the basis of the benchmarking information available. Control metrics are also developed and summarized in Figure 17.3. Here, we see both downstream and upstream control metrics and their targets. Once again, these control targets are aggressive but realistic in terms of benchmarks already in existence in other industries. In the interest of brevity, a modified PVC model is not shown. The PVC model structure reflects the new process flows and updated cause-effect relationships; see Figure 3.4 for a general format.

17.3 PROCESS CHANGE

The knowledge and skills necessary for effective and efficient implementation action are different from those necessary for questioning and analysis. Personal involvement in and ownership of the process improvement effort are critical for implementation and subsequent operations. Provided we include people who are involved with and impacted by the improvement, we have already laid the foundation for success in implementation. But, **no matter how well we have prepared our plan and people for implementation, risks and anxieties remain.** Our plan is analogous to a map, marked with a destination, point of departure, and route. Our implementation is analogous to the actual trip. Once we leave our point of departure, things are never the same, and people's lives and livelihoods are impacted.

RESOURCES

Process improvement plans are usually simpler and narrower than process definition or redefinition plans, as described in Section 3, Chapters 7 and 8. Nevertheless, details remain critical. **Once the plan is more or less complete, we review our plan and determine the exact nature, quantity, and location of needed resources;** see Table 17.8.

Table 17.8   Details related to resources.

Actions	Description
Review/preplay the process improvement plan	Working with the process improvement plan, carefully review/preview it in a whole-process context within the production system as a whole.
	Supply any missing details to complete the plan. Address additional buy-ins from other processes/subprocesses.
Identify resources and sources	Identify exact resources and amounts needed to fulfill the process improvement plan. Obtain bids and quotes.
Identify critical components	Locate critical process components that are likely to need special attention as to source and/or timeliness.

Table 17.9   Force field analysis—line shortage improvement (High Lift Case).

Driving Forces (associated with offsetting restraining forces)	Restraining Forces (associated with restraining the improvement)
People Issues	
Improvement will simplify operations.	Everyone is comfortable with the central storage concept and operations.
No involuntary reductions in force will be necessary.	Employees know their present assignments.
Training will be thoroughly planned (all impacted personnel have a chance to participate).	Improvement requires new assignments/responsibilities.
High Lift improvements are likely to improve vendor's operations and their ability to work with other customers.	Many vendors see enhanced service as impossible.
Technical/Business Issues	
Current facilities are a sunk cost (irrelevant to the improvement)—we should look at overall savings.	Huge amount of capital is tied up in central storage facilities.
Capital requirements are large, but payoff is rapid.	A large capital investment is needed for improvement.
Stockouts and line shutdowns are due more to a lack of logistical discipline than reduced inventories.	Lower inventories are (falsely) associated with stockouts/line shutdowns.
New product concepts will be acquired due to the improvement—the current product appears dated.	Abandoning our central storage facility looks bad for our current product line.

Improvement plans require resources ranging from personnel, to training, to equipment, to tooling, and so on. **We identify each needed resource and its acquisition source and lead time.** Critical resources are noted and receive special attention.

In the improvement plan review we typically focus on process integration—the linkages to and impacts on the whole process. These linkages may cut across other processes and subprocesses. What looks good with respect to producing gains or closing gaps in one process or subprocess may wreak havoc in another process or subprocess. Here, **involvement of people in linked processes and subprocesses is critical.** We preplay our plan as best we can in order to identify these areas of primary and secondary impact. We prepare to both assess and act on these impacts.

The review and preplay may be very simple in limited improvement cases involving modular processes/subprocesses. In other cases involving integrated processes/subprocesses, the review and preplay may be rather widespread and complicated. Hopefully, most of the linkages have been identified previously in the analysis—but just in case, adjustments are an order of magnitude easier to deal with before implementation is initiated.

In order to enhance our chances of a successful implementation, we pay particular attention to the changes that are required, both in terms of people and in terms of processes. The force field analysis tool is useful to identify our restraining forces and our driving forces, so that we can reduce the restraining forces as far as possible. Table 17.9 illustrates a force field analysis for our improvement in the High Lift line shortage implementation situation. Here, we have focused on restraining forces first, and then identified driving forces that can offset our identified restraining forces. **Our strategy is to address the restraining forces as far as possible in the implementation planning stage.** See Chapter 20 for force field details.

Table 17.10   Details related to schedule.

Actions	Description
Identify activities and milestones	Given the process improvement plan, identify/list all implementation activities that are necessary to put the new process in place (e.g., tear-outs, procurements, installations). Identify critical implementation activities and project milestones.
Time-sequence activities	Sequence the implementation activities identified above in terms of order of execution, with estimated time to completion for each activity.
Time-sequence resources to activities	Associate/align resources with activities, so that resource inputs are time-sequenced with activities.
Construct the implementation schedule	Piece together the activities, timing, and resource inputs to form a total implementation plan on a calendar-time basis.
Identify critical implementation metrics: Physical Economic Timeliness Customer service	Set up meaningful implementation execution metrics in the appropriate categories to monitor the implementation phase.

SCHEDULE

Depending on the extent of the process/subprocess improvement at hand, scheduling ranges from extremely simple to reasonably complex. **A detailed schedule is helpful for two reasons. First, it helps produce a smooth implementation.** And **second, our team gains valuable experience in preparing for implementation.** This experience or growth effect can be taken into more extensive improvement and redefinition efforts in the future. Section 6, specifically Chapter 18, provides details on leadership and creativity growth elements.

Details related to scheduling are discussed in Table 17.10. These details are identical to those in our process definition/redefinition discussion. However, the extent of time and effort in scheduling is usually considerably less for process improvement.

Our initial task in scheduling is to identify activities and events. These activities and events correspond to the process improvement plan. Usually, activity lists are used to summarize activities and events. **An activity is an action of some type that requires a period of time to accomplish. An event happens at a point in time,** e.g., the beginning or ending point of an activity. **Our critical events represent milestones—points at which we reassess our progress.**

Next, we time-sequence our activities and events, noting our milestones in this sequence. This sequencing effort includes developing time estimates for each activity. Given a start date, we can build a schedule in calendar time. Depending on the complexity of the process improvement plan, this schedule may utilize one or more of several tools, e.g., a Gantt chart or the critical path method (CPM); see Chapter 20 for more details. Typically, scheduling aids as complex as the CPM are not required for process improvement, but may prove to supply experience so that when more complicated improvements or redefinitions are undertaken, we are ready to deal with them. Here, just as in process definition/redefinition, **activity sequences occur in three ways: sequential, parallel, and coupled.**

At this point, we have a complete schedule that includes activities, durations, events, resources, and resource order points. Finally, **we identify critical implementation-based metrics to measure our effectiveness and efficiency.** Here, our aim is to meet or better our plan, as scheduled. In evaluating our implementation, we consider physical, economic, timeliness, and customer service performance parameters associated with implementation activities.

During the course of implementation, we use our scheduling tools to monitor our progress. We use them to help us follow our progress by "filling in" finished activities and then updating our Gantt chart or CPM network. Updating allows us to project a new or updated finish date and review our milestone events.

Example 17.4 | **HIGH LIFT CASE—LINE SUPPORT IMPROVEMENT IMPLEMENTATION SCHEDULE**   To implement our improvement plan, we have identified a set of activities, displayed in Table 17.11. Each activity is provided a symbol and an estimated duration. The ordered sequence can be established from the predecessor column.

Using the critical path method (CPM) we developed a schedule network for our activities. Our schedule network is presented in Figure 17.5. Details to support the network development are provided in Chapter 20. Here, we see each activity, its estimated duration, and its ordered sequence, moving from left to right. We have marked the critical path with a heavy line, and marked the milestone–event–related activities with bold circles—each milestone event occurs at the end of the marked activity. From our estimated times, we project a 93-day improvement project. In other words, we expect to see full-scale production using our improved line support method in 93 days. Once we decide on a calendar start date, we can then project a finish date for our project.

Table 17.11   Line support improvement process activities, sequences, and durations (High Lift Case).

Activity description	Activity symbol	Predecessor	Duration, days
Explain change/plan to affected areas	A		2
Identify rackable/binnable items	B	A	7
Design racks/bins/storage facility modifications	C	B	21
Build/test racks/bins*	D*	C	14
Identify/inform affected vendors	E	A	4
Prepare High Lift and vendor training/certification materials	F	A	15
Gain vendors' cooperation	G	E	8
Certify/train vendors*	H*	G, F	10
Review/modify High Lift team needs	I	F	2
Train High Lift people*	J*	I	5
Modify staging facilities	K	C	12
Modify in/out facilities	L	C	14
Develop procurement scheduling/card system*	M*	F	25
Rack/bin existing bulk inventory*	N*	D, K, L	14
Stage racks/bins to line	O	N	5
Remove/salvage old storage area	P	O	20
Limited-scale operation, test/tune/mistakeproof*	Q*	H, J, M, O	30
Full-scale operations	R	Q	—

| *Indicates milestone activities; milestone occurs at the end of the marked activity.

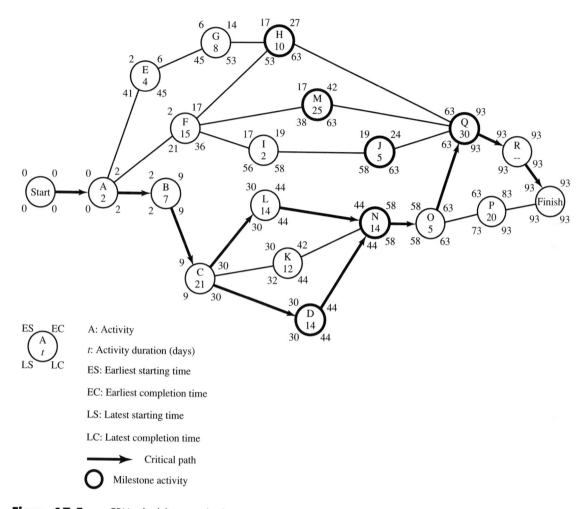

Figure 17.5 CPM schedule network—line support improvement implementation (High Lift Case).

The earliest starting time (ES), earliest completion time (EC), latest starting time (LS), and latest completion time (LC) are noted for each activity in our network. The critical path, i.e., the shortest possible time for completion, provided our time estimates hold, is marked. This path requires a total of 93 days for completion. We see that explaining the project, rack-related activities, in-out facilities area modification, current inventory racking and staging, and limited-scale production start-up are all on the critical path. If any of these critical path activities is delayed, our project will be delayed. If any activity runs short or long in terms of duration, then we update our network and produce an updated finish time, in addition to reassessing our critical path.

Further project scheduling statistics are presented in Table 17.12. Here, we have calculated and displayed the total slack (TS) and free slack (FS) statistics, as well as noted our critical activities. We have also marked our milestone activities with a * symbol, indicating that we intend to review our progress at the completion of these activities. In order to update our CPM network as our implementation plan is executed, when

Table 17.12   Line support improvement process scheduling details (High Lift Case).

Activity description	Activity symbol	Duration, days	ES	EC	LS	LC	TS	FS	Critical?
Explain change/plan to affected areas	A	2	0	2	0	2	0	0	Yes
Identify rackable/binnable items	B	7	2	9	2	9	0	0	Yes
Design racks/bins/storage facility modifications	C	21	9	30	9	30	0	0	Yes
Build/test racks/bins*	D*	14	30	44	30	44	0	0	Yes
Identify/inform affected vendors	E	4	2	6	41	45	39	0	No
Prepare High Lift and vendor training/certification materials	F	15	2	17	21	36	19	0	No
Gain vendors' cooperation	G	8	6	14	45	53	39	3	No
Certify/train vendors*	H*	10	17	27	53	63	36	36	No
Review/modify High Lift team needs	I	2	17	19	56	58	39	0	No
Train High Lift people*	J*	5	19	24	58	63	39	39	No
Modify staging facilities	K	12	30	42	32	44	2	2	No
Modify in/out facilities	L	14	30	44	30	44	0	0	Yes
Develop procurement scheduling/card system*	M*	25	17	42	38	63	21	21	No
Rack/bin existing bulk inventory*	N*	14	44	58	44	58	0	0	Yes
Stage racks/bins to line	O	5	58	63	58	63	0	0	Yes
Remove/salvage old storage area	P	20	63	83	73	93	10	10	No
Limited-scale operation, test/tune/mistakeproof*	Q*	30	63	93	63	93	0	0	Yes
Full-scale operations	R		93	93	93	93	0	0	Yes

*Indicates milestone activities; milestone occurs at the end of the marked activity.

we finish an activity, we place its true time in our network and make new forward and backward passes to update our ES, EC, LS, LC, TS, and FS statistics. See Chapter 20 for details on making these calculations.

As a substitute for, or in addition to, our CPM schedule network, we can develop a Gantt chart. Here, we associate each activity with a horizontal bar. These bars are stacked and labeled as to the activity they represent. Viewing a Gantt chart from left to right, we see activity flows and durations. A properly scaled Gantt chart provides the ES and EC; however, the LS, LC, TS, FS, and critical path are not obvious, as they are in the CPM network analysis.

Using either the Gantt chart or the CPM network, we need an activity table, such as Table 17.11, as a starting point. Figure 17.6 depicts our line shortage improvement activities on a Gantt chart. Here, we see a scaled time line, with each activity superimposed on it. We have labeled each activity and provided the time durations.

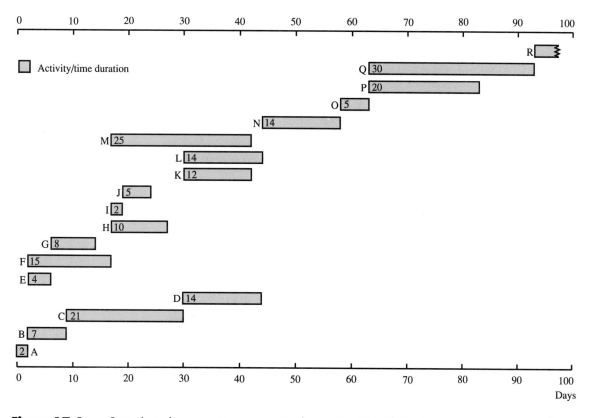

Figure 17.6   Gantt chart—line support improvement implementation (High Lift Case).

As can be seen by comparing the CPM network with the Gantt chart, the CPM network is more detailed, and generally more suited for larger improvement projects. But, on the other hand, the simplicity of the Gantt chart, in its time-scaled bars, helps to communicate the relative length of project activities graphically. The Gantt chart is updated as the project progresses.

ACTION

Our action plan in process improvement is essentially the same as that for process definition/redefinition. However, the scope is usually narrower. Details are provided in Table 17.13.

We follow the process implementation plan schedule, updating the plans and schedule when necessary. This plan may include activities that test parts of the improved process, and perhaps interfaces with other processes, as they are installed. Eventually, the whole improved process/subprocess will be ready to check out and test. Some redesign and rework may be necessary.

The next phase is one of limited-scale operation. Here, we examine operations and results with respect to our targets and requirements. We assess physical, economic, timeliness, and customer service performance aspects of the improvement. **During limited-scale operations we are looking for bottlenecks and the limitations that create them.**

Table 17.13   Details related to action—limited-scale.

Action	Description
Execute scheduled plan	Install the process improvement in accordance with the implementation schedule.
Test	Check out process changes piece by piece, as they are installed.
Start limited-scale operations	Restart operations on a limited scale and observe closely for bottlenecks and related limitations.

Assess improved process:
  Physical
  Economic
  Timeliness
  Customer service
	Collect and display data on critical metrics and critical process variables. Include designed, on-line experiments aimed at optimizing process results when necessary.

Mistakeproof and tune improved process	Address actual and potential process bottlenecks and limitations that appear in limited-scale operations.

Table 17.14   Details related to action—full-scale.

Action	Description
Commence full-scale operations	Move tuned and mistakeproofed improved process on line.

Assess results:
  Physical
  Economic
  Timeliness
  Customer service
	Collect, analyze, and display data on critical process metrics.

Stabilize process/subprocess	Apply process control technology (see Section 4).
Improve process	Apply process improvement technology as new opportunities present themselves (revisit Section 5).
Redefine process	Redefine process as new opportunities and technologies emerge (see Section 3).

Mistakeproofing and bottleneck resolution in limited-scale operations are approached in a variety of ways. Approaches may be reactive in nature, where we isolate a limitation, trace it to a root cause, and resolve it. Or they may be proactive in nature, where we identify a process/subprocess/sub-subprocess leverage point, isolate the lever, and exploit it. In either case, **we look for opportunities to make changes, such that mistakes in full-scale process operations are difficult, if not impossible.** Initiatives and tools such as mistakeproofing, failure mode and effects analysis, fault tree analysis, and cause-effect analysis may be helpful here.

Finally, **we fine-tune our improvement.** Here, we make any necessary adjustments to targets regarding equipment settings or process variables before we commence full-scale operations. In complex process improvements we use designed experiments and resulting process models to help tune or optimize the process.

By the time we commence full-scale operations, see Table 17.14, **our improved process and product metrics should be identified, and sensors and data collection systems in place, so that we can assess on-line results.** Tools such as runs charts help us to assess how our results compare to our targets. Our process control leverage points and their sensors provide us with data useful to assess process stability. Here, **we use process control technology,** see Section 4, Chapters 9 to 15, **to target and stabilize our process through monitoring and adjustments.**

As process operations experience is gained, we are able to identify additional opportunities for process improvement—within the current process definition. To pursue these opportunities, we repeat our process improvement cycle elements. **Eventually, we redefine our process** using our Section 3 elements and subelements.

REVIEW AND DISCOVERY EXERCISES

REVIEW

17.1. What is a feasible alternative? What constitutes feasibility? How is feasibility assessed? Explain.

17.2. How are evaluations of improvement alternatives structured? What bases of comparison exist? Explain.

17.3. Why is it absolutely essential that we use the same time horizon for all alternatives assessed with the net present value method? Explain.

17.4. Why is it virtually impossible to be totally quantitative in our alternative evaluations in process improvement? Explain.

17.5. What advantages do we gain by reviewing our selected alternative in a whole-process context? Explain.

17.6. How do we update our team structure relative to the chosen process improvement alternative? Explain.

17.7. How do we update our control points and process variable targets with respect to the chosen process improvement alternative? Explain.

17.8. How the content and extent of knowledge and skills differ between planning and implementation? Explain.

17.9. What differences exist in the level of detail in resource consideration between the planning and implementation phases of process improvement? Explain.

17.10. What is a milestone in the context of an implementation plan? Explain.

17.11. Why are metrics associated with implementation important? Explain.

17.12. What is a critical path? Explain.

17.13. Explain what is accomplished on a forward pass in a CPM network. A backward pass? See Chapter 20 for details.

17.14. How do we know when to switch from limited-scale to full-scale action in process improvement? Explain.

17.15. What is accomplished in mistakeproofing and tuning a process? Explain.

DISCOVERY

17.16. Why do we refrain from critical assessment until we have all ideas verbalized? Explain.

17.17. Why do we use both critics and champions when assessing the merit of ideas? Explain.

17.18. Why do we focus on input/output impacts in evaluating feasible alternatives? Explain.

17.19. How does "personal ownership" relate to successful process improvement implementation?

17.20. Seamless integration is a term that refers to a smooth continuum of activity and action involving planning and implementation. How can or should seamless integration be approached in process improvement? Explain.

17.21. In the High Lift CPM network, shown in Figure 17.5, what would be the result if activity K was extended by 1 day? By 2 days? By 3 days? Address both project duration and critical path.

See Chapter 16 Review and Discovery Exercises for project ideas.

PROCESS-BASED TRANSFORMATIONS, INITIATIVES, AND TOOLS

The purpose of Section 6 is to describe the elements of process-based transformations and overview associated initiatives and tools.

Chapter 18: Process-Based Transformations

The purpose of Chapter 18 is to identify and address critical elements of transformations to process-based organizations—elements include structure, creativity, and leadership.

Chapter 19: Process-Compatible Initiatives

The purpose of Chapter 19 is to introduce several initiatives that are useful in process-based transformations and operations.

Chapter 20: Process-Compatible Tools

The purpose of Chapter 20 is to introduce several modeling and analysis tools useful in process-based transformations and operations.

chapter

18

PROCESS-BASED TRANSFORMATIONS

18.0   INQUIRY

1. How does organizational structure impact process-based transformation?
2. What does a process-based structure look like?
3. What is creativity? Why is it critical?
4. What is leadership? Why is it critical?

18.1   INTRODUCTION

Transformation from classical department-based organizations to process-based organizations requires change—stimulated by observation, thought, and action. For the most part, preceding chapters have described observation, thought, and action components from a process-technical standpoint. This chapter focuses on these same components from an organizational standpoint. Our discussion of implementation here lays the groundwork for action necessary to install and use the process-based organization. These actions are always taken by people; people-oriented concepts are described herein.

Our overview of process-based transformations and operations, depicted in Figure 18.1, **identifies three critical elements:** (1) **organization,** (2) **creativity,** and (3) **leadership.** The underlying reason why we focus on organization, creativity, and leadership together is that creativity and leadership abilities are widely distributed in our population and therefore in our workplaces. And, organizational structure provides the primary means to tap these creative and leadership abilities.

The first element, organization, is a structural lever that we use to more readily unlock leadership and creative abilities and behaviors that will lead to competitive advantage in our processes. The second and third elements, creativity and leadership, are leverage elements that we use to enhance organizational effectiveness and efficiency in order to gain competitive advantage—by offering our customers a higher benefits to burdens ratio than our competitors.

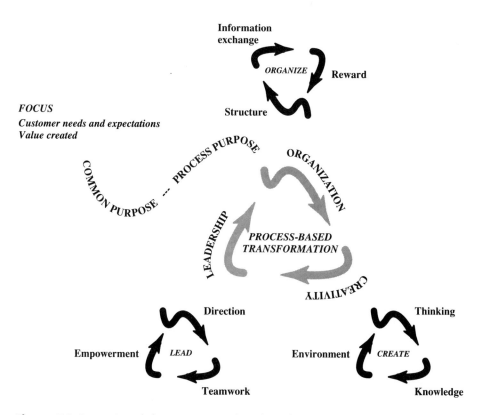

Figure 18.1 Critical elements in process-based transformations and operations.

Each major element is in turn influenced by three subelements, relative to process-based transformations. Each of the three major elements has been studied and documented by thousands of scholars and practitioners, who have provided a plethora of descriptions ranging from highly detailed scientific studies to anecdotal accounts. These studies and accounts have generated scores of basic models and conjectures, many of which are at odds with one another. A lack of agreement is understandable since we are dealing with complex topics in the context of a complex mix of people, constituting a complex mix of organizations, each with complex visions and missions. Our discussions will tend toward the common ground.

18.2 ORGANIZATION

The ability of a cooperative effort to fully tap and utilize both its human and physical resources, as well as its ability to interact with customers, is affected by organization. For example, the manner in which we organize ourselves has a great deal to say about how we expect to and ultimately will interact with each other. Hence, our organization impacts our effectiveness, i.e., quality, and our efficiency, i.e., productivity.

Table 18.1   Traditional versus process-based organization.

Element	Traditional		Process-Based	
	Nature	Focus	Nature	Focus
Organizational structure	Pyramid	Functional departments	Flat	Process teams
Action channels	Specialized by functions	Orders	Whole process	Purposes
Information exchange	Closed	Controlled flow	Open	Shared
Reward structure	Tangible, intangible	Individual	Tangible, intangible	Team

Our goal in process-based organization is to help unlock and develop leadership and creative potentials in everyone—focused on our vision, mission, core values, goals, and objectives. We stress whole-process, self-directed teams as opposed to traditional command and control organizational structures. Table 18.1 provides a summary comparison of traditional and process-based organizational elements.

STRUCTURE AND CHANNELS

The nature of gains from our process definition/redefinition, control, and improvement sequence was originally displayed in Figure 3.2. In this abstract depiction, we see process definition yielding both realizable and unrealizable gains. Then, we see process control gradually add to the realizable gains by targeting and stabilizing the process. Next, we see both realizable as well as unrealizable gains generated by process improvement (within the bounds of the original process definition). Ideally, we see this cycle repeating, e.g., through redefinition, control, and improvement. Our depiction shows gains at all levels. However, losses are possible. We minimize the possibilities of losses by systematic and thorough process definition/redefinition, control, and improvement.

Ambiguous issue identification and resolution render any organization ineffective and inefficient. Regardless of the process technology and team composition, organizational structure facilitates issue identification and corresponding action. **Issue identification and action channels allow us to integrate definition/redefinition, control, and improvement into our process structure, so as to encourage and focus leadership and creativity.**

Action channels are depicted in Figure 18.2. Here, we see a simplified cross section of the organization composed of one process and the business integration process, along with their respective teams, together. Here, the business integration team consists mostly of members drawn from other process teams serving alignment, coordination, and communication functions within the organization.

Three distinct process-issue action channels exist: (1) definition/redefinition, (2) control, and (3) improvement. For each channel, we have labeled the relevant cycles, as described in Section 3 (definition/redefinition cycle), Section 4 (control cycle), and Section 5 (improvement cycle). Hence, Figure 18.2 provides a road map for process-based operations.

Process-based organizational structures are reasonably flat, composed of teams. They are approached in a holistic manner, e.g., as described in Table 18.1. The actual size and composition of teams depends on the nature of the process technology selected and production system scale.

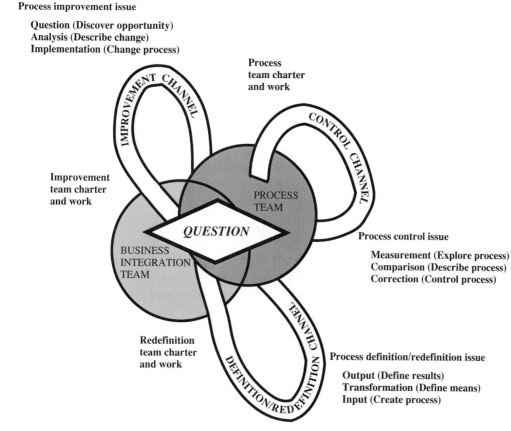

Figure 18.2   Process definition/redefinition, control, improvement issue action channels.

The team structure contains three basic team types: the process team, the definition/ redefinition team, and the improvement team. Questioning what we do, why we do it, how we do it, when we do it, where we do it, and who does it is central to the process basis. **Questioning within any process team leads toward an issue action channel.** The nature of the questioning directs cooperative efforts toward each specific channel. For example, questions about process stability move toward control channels, and questions regarding changes in products and/or processes tend toward improvement channels or redefinition channels.

Process teams are formed on the basis of current process needs. Process teams are necessary so long as the process purpose is relevant. Process definition/redefinition and improvement needs surface in response to questions regarding the current process. Typically, definition/ redefinition and improvement teams have a shorter lifetime than process teams. For example, when the new definition/redefinition or improvement is implemented, these teams disband. All teams change in size and personnel as needs change.

Control channels are navigated by process teams. The issues within control cycles include three elements: (1) measurement, (2) comparison, and (3) correction. In Figure 18.2 we see how the control cycle fits into the organization. The process team "owns" the control cycle and executes it through the control channel.

Definition/redefinition channels are navigated by definition/redefinition teams. Figure 18.2 puts the definition/redefinition cycle into an organizational perspective. The cycle contains three elements: (1) output, (2) transformation, and (3) input. These teams are chartered by the business integration team, which is composed mostly of people from the process teams; many people belong to several teams, e.g., process teams and the business integration team. Redefinition teams are chartered for specific strategic purposes and exist until the purpose is accomplished or abandoned. Then, the team charter is retired, and the members disband and return to their process teams and/or other definition/redefinition and improvement teams.

Improvement channels are navigated by improvement teams. Charters for improvement teams may originate in the process or business integration teams, depending on the scope of tactical change addressed. The broader the scope, the more likely it is that the business integration team issues the charter. Here again, the improvement is purpose-driven. When the team purpose is accomplished or abandoned, the team charter is retired and the team is disbanded. Members then devote their time to their process team and other redefinition or improvement teams.

Figure 18.2 traces the improvement cycle through its elements: (1) question, (2) analysis, and (3) implementation. Questioning leads to a decision option for moving to the redefinition channel or to the improvement channel. The decisive element here is the scope of attention—strategic (big change possibilities) or tactical (incremental change possibilities). Strategic possibilities move into the redefinition channel and tactical possibilities move into the improvement channel. Switches may be made later on, but they require extra time and effort.

As time progresses, we grow in process effectiveness and efficiency by repeatedly questioning, understanding, and moving relevant process issues through action channels. Our process-based organizational structure facilitates this growth.

INFORMATION EXCHANGE AND ARCHIVES

In moving from a traditional organization, characterized by tightly controlled flows of information, **toward a process-based organization, we pursue a more open and shared form of information exchange.** This open format is necessary because the process-based organization—through teamwork and empowerment—decentralizes the responsibility-authority structure, which in turn requires decisions to be decentralized, from a few managers to the process teams. **One critical element in decision making is the need for relevant and timely information.**

In terms of information flow within and between teams, we consider both the information itself, the presentation format, and the forum of presentation. In the process value chain (PVC) model, we see a progression from overall business summary information at one end to technical, i.e., physical-based data at the other. In the course of process definition, redefinition, control, and improvement, **we identify relevant metrics. Each metric is defined/developed as to its** (1) **purpose,** (2) **units of measure,** (3) **source and sensor,** and (4) **display and linkage.**

With regard to purpose, **each metric should make a contribution in establishing our position in terms of physical, economic, timeliness, or customer service performance.** Both upstream (proactive) as well as downstream (reactive) metrics are important. Units of measure should favor proactivity when possible. Accuracy and precision as well as timeliness are issues in source and sensor selection. In terms of display or expression, we should consider units that convey our position as effectively and efficiently as possible. One wasted minute in a group multiplies by the group size and can quickly lead to gross inefficiencies.

We develop our metrics and information exchange so that our teams have timely information support. **Our metrics support team decision making in three ways:** (1) **to spot trends,** (2) **to compare alternatives,** and/or (3) **to calibrate results.** Specifically, process teams require support for day-to-day activities as well as process-questioning activities. Definition/redefinition and improvement teams need information support for calibrating results of current process operations with targets and benchmarks. Data are necessary to weigh against the anticipated results from proposed new and improved processes, in order to make comparisons to assess gap-closing and gain-capturing prospects.

In many cases, team-based information exchange takes place in team meetings; hence presentation formats must be developed for group viewing, e.g., plots for screen projection, with oral description. In most cases information presented in the business integration team flows to the process teams either directly or in a summarized form. Hence, group information flow formats are necessary. The point here is that **inefficiencies in grasping and understanding data-information in a group setting multiplies as the group size expands.**

Frequently, definition, redefinition, control, and improved team endeavors require specialized information, over and above the routine information gathered and distributed through the information system. In other words, our questioning of processes—input, conversion, and output—naturally leads from observation to quantitative measurements. We address this need on an exception basis. We use our four elements of information collection for these exceptional needs, as well as our routine needs: (1) purpose, (2) units of measure, (3) source and sensor, and (4) display.

Process control endeavors require data to characterize each process variable's behavior in order to initiate and calibrate process control models. In the course of operation, process control records, e.g., numerical and descriptive logs, are useful in process performance assessment as well as in process questioning.

The archiving of information is invaluable for process redefinition and improvement, as well as process control, for two primary reasons: (1) **people have a finite capacity for remembering details** and (2) **team membership is dynamic,** with new team members needing to be able to tap old experiences. Numerical information serves as a historic benchmark for future reference. Descriptive cause-effect information serves as a knowledge basis for operations as well as further improvements.

Metric-based data-information is typically held in files. Cause-effect–based information is captured in report files and on storyboards. A storyboard is a display of words, pictures, graphs, and so on, that summarizes improvement or redefinition activities and results. Storyboards are described in more detail in Table 18.2. In general, the storyboard communicates what was done and provides recognition of those who did it.

EXPERIENCE LOST CASE—LID SEAL T^3 (pronounced T-three) is a manufacturer of high-tech equipment serving the semiconductor manufacturing industry. The company started with three partners, essentially working out of a garage facility. Over the past 15 years, they have grown from a small start-up company to a manufacturer with about 300 employees, operating several facilities. Their initial product provided customers with superior semiconductor wafer washing performance. Other follow-on products primarily improved on the initial product, working better, faster, and cheaper.

Example 18.1

Table 18.2   Storyboard structure.

Element	Description
Charter/empowerment	Team title, members, leaders, sponsor, boundaries, date
Purpose/expectations	A clear, concise description of the intent and results expected
Present situation	A summary of the present process/subprocess under study—data, cause-effect, equipment, and so on
Alternatives considered	A summarized description of feasible alternatives considered
Alternative selected	A semidetailed description of the alternative selected, along with the selection criteria/evaluation used
Implementation and results	A summarized description of the installation of, and results from, the redefined or improved process

T^3 purchased a quality training package from a major consultant, sending several of its key people to the consultant's training facility in order to train their trainers. These trained internal people then developed a training package for the entire company. This package followed the consultant's philosophy but contained cases specific to T^3. The trainers, T^3 employees, were very excited and enthusiastic about their training package and the cases therein. They had trained most of the employees in the larger facilities and had done an effective job of communicating their basic message—a message of teamwork, working together to solve production problems and to lower the cost of nonconformance.

During training at one remote, but critical, site housing the research and development group, the training was proceeding as usual. About 20 employees were present, one of whom was the vice president of research and development. This individual was one of the original founders, a very outspoken and pragmatic individual. As the training unfolded, one company case involving a washer lid seal, i.e., a leakage correction problem/solution, was cited as a recent team accomplishment. Several details as to the technical and team nature of the solution were provided to the trainees.

About four-fifths of the way through the case, the attending vice president began to turn somewhat pale, and then somewhat red-faced. Next, he stood up and said, "What's happened to our company? We solved this washer seal problem 10 years ago. Now, we have just solved it again. Don't people talk to each other in this organization? We use to solve problems like this at lunch, on the back of paper bags. How is it that we can't remember what we did in this organization and have to keep reinventing our solutions? Can we figure out how to keep our people aware of our past failures and successes, so that we can keep growing without the need for reinventing solutions?"

REWARD STRUCTURE—AWARDS AND RECOGNITION

All organizations have some sort of reward structure, consisting of two general reward components: (1) **tangible** and (2) **intangible.** Tangible awards include salary, wages, vacation, gain sharing, profit sharing, cash incentives, prizes, and so on. Intangible awards include recognition— public and private acknowledgment—through verbal, written, and graphical mediums, e.g., acknowledgment for efforts/results in ceremonies, letters, and plaques.

Typically, tangible and intangible awards are provided together. In general, people have devised and described a plethora of reward possibilities. The issue of concern here is not the nature of awards, but how awards can be structured and awarded in a manner that will encourage team-based leadership and creativity, aimed at competitive advantage.

Rewarding is fundamental to all cooperative efforts, including process-based organizations as well as traditional organizations. In general, organizations require common purpose, personal ability and willingness, and interpersonal communication for initiation, as well as organizational effectiveness and efficiency for sustained operations, e.g., as described in Chapter 1. Specifically, rewarding focuses on willingness, with the objective of generating and sustaining personal arousal and energy in order to accomplish common purpose. **Rewarding is used in an organizational sense to encourage, align, and direct behavior toward common purpose.**

Reward impact is influenced by ceremonial content. Recognition (an intangible award) and tangible awards are usually transmitted to people via communication formats, e.g., through ceremonial channels. Ceremonies hold a high degree of affective content—emotional feeling, versus cognitive or logical content. Hence, the ceremony itself, as it impacts the emotions of individuals, both the rewarded as well as the observers, is critical. For example, relatively inexpensive awards presented in an effective manner many times do more to cultivate personal willingness than more expensive awards presented in a less effective manner. We design both the award and the presentation format for maximum impact on willingness, considering the whole organization.

The coordination of tangible- and intangible-based rewarding is challenging. It ties together fundamental cooperative effort basics, i.e., purpose and willingness. Any misalignment in rewarding will cut into organizational effectiveness or efficiency. Awards and rewarding are interpreted on an individual basis, in a reasonably open forum. Interpretation involves both the individual receiving the recognition or award and those observing all or at least part of the transaction. **We generate an effect in the persons being rewarded as well as the observers.** The organization is one beneficiary of the net impact. The net impact may be complementary or divisive; i.e., both the rewarded and the observers gain positively, or the rewarded perceive a gain, but the observers perceive a loss.

Organizational rewarding, either tangible or intangible, **essentially represents an exchange between contributors and the organization,** with the purpose of recognizing positive/productive behavior. One party furnishes, or has previously furnished, the behavior, and the other party furnishes the tangible award and/or recognition. This basic exchange seems appropriate in general; however, the exact nature of the award and the manner in which it is exchanged are subject to debate.

In general, we have a wide variety of rewarding options available to us. Because of restrictions on resources and restrictions in the form of procedural rules or laws, some organizations have more options than others. Table 18.3 displays several reward options available.

From Table 18.3, it is obvious we do not lack in vehicles for reward. The challenges occur in implementation. Balancing individual awards with team awards, scaling awards to efforts and results in an equitable fashion, and identifying appropriate people for discretionary awards and formal recognition are formidable challenges.

The literature is rich with both scientific studies as well as anecdotal accounts of rewarding both individuals and groups. Knouse [1] provides a review of rewards relevant to total quality management initiatives. General agreement that rewarding plays a significant role in organizational success and failure is clear. However, specific agreement on a best rewarding structure has not and likely will not appear. Organizations are unique in that they exist in a dynamic environment, both

Table 18.3   Selected reward options and mediums.

Benefits
    Health care
    Retirement
    Life insurance
    Disability insurance
Bonuses
Clothing (caps, shirts, etc.)
Commemorative articles (jewelry, pens, cups, etc.)
Compensation (hourly/salary)
Gatherings (picnics, dinners, banquets, etc.)
Gain sharing
Incentives
New responsibilities and/or promotion
Prizes
Profit sharing
Vacation
Verbal recognition (public and private)
Written recognition
    Letters
    Plaques and pictures
    News releases

---

physically and socially. **Leaders distinguish themselves and their organizations by their creativity in rewarding followers.** The literature offers several basic principles that are useful for guidance in creating or adapting reward structures for team-based organization. These principles are described in Table 18.4.

    Our rewarding principles are divided into two areas: (1) **the award itself** and (2) **the rewarding process.** We first consider the physical essence of the award itself, as to the type of behavior it is intended to encourage. The potential of the award and/or recognition to encourage effective team building and teamwork—leadership, creativity, learning new skills, producing better products and processes, and so forth—is of major concern. **If the award and/or recognition does not have the potential to produce gains in the identified areas, then it should be abandoned.** The determination of effectiveness hinges on the association of the identified behavior desired and the award and/or recognition in terms of personal needs, as perceived by each individual. This association or alignment presents a very challenging endeavor. It requires that we identify behaviors that will be productive to the organization, as well as understand the needs structures of the individuals involved, and then properly associate these elements with the award and/or recognition.

    Effectiveness in the rewarding process involves understanding relevancy of the award or recognition, and how it is associated with the recipient and peers. Here, simplicity aids understanding. Selection involving peer participation that is seen as meaningful and fair is helpful in building credibility that the rewarding process found an appropriate recipient. In many cases, the presentation experience itself, through the emotional appeal of the ceremony, is more valuable than the actual tangible award presented. For example, a letter of recognition may cost $5 to produce but be considered invaluable by the recipient in the context of the entire reward experience. The

Table 18.4 Principles for rewarding in team-based endeavors.

Principle	Description
The Award Itself	
Physical essence	Awards should carry appropriate symbolism in components—cash, gain share, letters, plaques.
Team building	Awards should encourage the team as a whole toward efforts and results.
Individual value	Awards individually and collectively should be valued by individuals contributing to the teams—awards should significantly address one or more human needs (physiological, safety-security, social, esteem, or self-actualization).
The Rewarding Process	
Understanding	A clear understanding of reward requirements and acceptance of the requirements as relevant should exist—simple formats are easier to understand.
Selection	Recipients should be viewed as worthy—peer involvement in selection is appropriate.
Presentation	The ceremony is critical—appealing to appropriate emotions. The presentation should be dignified and appropriate, conducted through private or public forums.

ceremony represents the consummation of exchange. **A significant portion of the reward experience or reward value is coupled with the ceremony.** Personal interpretation is always relevant, since ceremony appeals to the higher-order human needs, especially the esteem need.

AWARD ALIENATION CASE—WHO'S INTERESTED? A large technical organization of about 1500 people developed a tradition of awarding their high performers gold medallions along with cash awards of several thousand dollars. These medallions were displayed on red, white, and blue ribbons, and presented a very attractive appearance. The criteria for receiving this award were spelled out in some detail. Criteria, as described, were based primarily on performance and contribution to the organization.

Example 18.2

About five awards and checks were presented each year, usually in April or May, typically about midafternoon. The awards ceremony was heavily advertised internally within the organization. Most people were aware of both the time and place and were encouraged to attend. The president typically made the awards in person. The setting for the presentation was typically a large room with a bare stage and comfortable seats. Typically, the president would say a few words about each recipient, which she typically had written out on several note cards.

The nominations were made and forwarded by managers. Each manager could essentially designate a person from his or her area for these awards. The awardees were typically notified to attend, and usually did attend unless they were scheduled for out-of-town travel. If they were out of town, the award would be mailed to them. In either case, they felt good about the award—they cashed their checks and proudly wore their medallions at formal company gatherings.

After the last ceremony, the president expressed her disappointment to the managers after the meeting. She stated that only 50 people in attendance, including the managers and president, hardly impressed her. Several suggestions were made that perhaps the medallions were not large enough, and that next year they should consider providing refreshments. Other suggestions included pushing people harder to attend, perhaps making attendance mandatory.

In the "rank and file" group of attendees, exiting the meeting room, summaries of the event were also being voiced. One person stated that he was surprised to see Henry receive a medallion, but guessed it was probably his turn. Another person remarked that she could just as well have stayed back in her area and caught up on the day's work.

18.3 CREATIVITY

Creativity is the only means we have available to define, redefine, and improve our processes. Hence, its importance is obvious. However, its essence is not so obvious. Researchers have studied creativity for decades. Typically, definitions include or infer newness and usefulness in the creation itself. Classical studies of creativity tend to identify individuals who have produced new and useful things, and then attempt to study these individuals' personalities. The focus is usually on creative behavior, i.e., the production of new and useful things, and creative ability, i.e., the capacity to produce concepts and ideas that lead to new and useful things. Recent studies of creativity take a systems perspective.

Creativity is defined in the context of a system; see Csikszentmihalyi [2] :

Creativity is any act, idea, or product that changes an existing domain, or that transforms an existing domain into a new one.

A creative person is someone whose thoughts or actions change a domain or establish a new domain.

In this context, **the system within which creativity occurs consists of three elements:** (1) **the domain,** (2) **the field,** and (3) **the individual. The domain consists of a set of symbolic rules and procedures,** e.g., mathematics, engineering, business, music, and psychology. **Domains** and subdomains, i.e., subdivisions of domains, **are nested in what is termed a culture,** i.e., the symbolic knowledge shared by a particular society or by humanity as a whole.

The field, according to Csikszentmihalyi [2], **is defined as all of the people who serve as gatekeepers to the domain.** For example, the field for the domain of engineering consists of all people who have influence over accepting or rejecting the act, idea, or product offered.

The third and final component is the person acting in a creative capacity. **Creativity occurs when a person, using the symbols of a given domain such as music, engineering, or business, produces a new idea or concept and when the novel idea or concept is selected by the appropriate field for inclusion into the relevant domain, or when the concept or idea establishes a new domain.**

Process definition, redefinition, control, and improvement are compatible with this systems view of creativity. In our process-based context, the domain includes our process or subprocess and relevant knowledge bases. A team member represents the person with the novel idea or concept. The field is composed of the decision makers who decide the fate of the idea and the environment in which these decision makers function.

Csikszentmihalyi [2] justifies the systems model of creativity over more restrictive "trait" models as follows:

Perhaps the most important implication of the systems model is that the level of creativity in a given place at a given time does not depend only on the amount of individual creativity. It depends just as much on how well suited the respective domains and fields are to the recognition and diffusion of novel ideas. This can make a great deal of practical difference to efforts for enhancing creativity. Today many American corporations spend a great deal of money and time trying to increase the originality of their employees, hoping thereby to get a competitive edge in the marketplace. But such programs make no difference unless management also learns to recognize the valuable ideas among the many novel ones, and then finds ways of implementing them.

The point of our subsequent discussion of creativity is to understand and cultivate this creative system so that we can address our vision, mission, core values, and process purposes in a competitive manner. We divide our discussion into three parts: (1) creative thinking (associated with the person), (2) knowledge base (associated with the domain), and (3) environment (associated with the field).

CREATIVE THINKING

Creative thinking has received a great deal of attention over the course of several decades. Several schools of thought and many different models have been proposed. Dacey [3] provides a summary of schools of thought and models that seem plausible. **Two primary schools of thought regarding the formation of creative personality have been active:** (1) **the psychoanalytic theory** and (2) **the humanistic theory.**

Dacey [3] describes the psychoanalytic theories in terms of personal struggle and strife:

In general, psychoanalytic theorists see creativity as the result of overcoming some problem, usually one that began in childhood. The creative person is viewed as someone who has had a traumatic experience, which he or she dealt with by allowing conscious and unconscious ideas to mingle into an innovative resolution of the trauma. The creative act is seen as transforming an unhealthy psychic state into a healthy one.

In contrast, Dacey [3] describes the humanistic theories as driven by the human needs structure:

Unlike most of the psychoanalytic theories, the humanist theories see creativity as a result of a high degree of psychological health. They give a much smaller role to unconscious drives and compensation for deficits in the personality, and much more credit to positive, self-fulfilling tendencies. They also see creativity as developing throughout life, rather than as being restricted to the first five years.

Because of the complexity of the human mind and resulting thinking processes, it is understandable that several theories have emerged to explain them. Each theory adds several insights to our understanding in bringing perspectives to light. However, no one theory appears to be totally adequate.

Early studies of creativity focused on characteristics of individuals acting in a creative capacity and to some degree on their creations. Jackson and Messick [4] connected both people and creations together in a "trait" model. Their model is paraphrased and annotated in Table 18.5. In this model, each row represents a specific categorical association of person and product (creation). Each column describes a set of personality or product (creation) traits or characteristics, respectively.

Table 18.5   Personal and product characteristics associated with creativity.

Personal Characteristics	Product Characteristics	
Tolerance of incongruity or ambiguity (ability to avoid discomfort with unknown or unfamiliar ideas, functions and forms, people, places, etc.)	Unusualness (refers to the novelty of creation)	Surprise (refers to the initial reaction of the gatekeepers of the domain or knowledge base—apparent differences from expectations)
Intuition and analysis (ability to sense opportunity and respond logically)	Appropriateness (usefulness of concept or idea)	Satisfaction (valuable—benefits outweigh burdens)
Open-mindedness (ability to reserve judgment on new concepts and ideas)	Transformation (new formulations in concepts and ideas—totally new or new combinations of old concepts and ideas)	Stimulation (excitement and energy in response to product)
Reflection and spontaneity (ability to gain insight from methodical pondering in combination with flashes of brilliance)	Condensation (efficiency of expression—simplicity of function, form, or fit)	Savoring (captivating—able to hold attention)

The information and comments in Table 18.5 provide an overview of characteristics of creative people and products. The intellectual (personal) traits have been studied in more detail than the product traits—with the tolerance of incongruity or ambiguity being cited as a major determinant of an individual's ability to behave creatively, see Dacey [3]. These general characteristics lack specific details that we can use to understand and assist us in affecting creative behavior; and creative problem solving.

Creative Problem Solving   Many models have been constructed to describe the process of creative problem solving. Dacey [3] summarizes several models and concludes that there is not a great deal of agreement between these models; but taken as a whole they offer some insight to sequential activities. **Classical descriptions of the creative process typically include a sequential description.** A typical description follows:

1. **Perception of need.**   A trigger to energize and direct our action.
2. **Preparation.**   Assimilating information relevant to the perceived need.
3. **Incubation.**   A period of apparent relaxation of conscious effort following intense preparation.
4. **Insight**.   Awareness of new concepts or ideas resulting from either spontaneous insight or systematic reflection.
5. **Verification.**   Test and confirmation regarding new concepts or ideas and communication of the concept or idea to other people.

This sequential description provides a general overview of the creative process; however, it lacks details important to enhance both effectiveness and efficiency in the creative process. Hence, we will describe several more detailed theories that are applicable to our process-based organization.

Guilford's Theory   Guilford [5] describes what he terms the structure of the intellect. This structure contains five mental operations:

1. **Cognition and filtering.**   Cognition refers to discovery and recognition of some part of our physical or social environment. Filtering refers to limiting the nature and amount of the stimuli we encounter in our physical and social environments, i.e., forming our perceptions.

2. **Memory.**   Memory refers to retention of what is discovered and recognized.

3. **Convergent thinking.**   Convergent thinking moves toward or results in a unique outcome, i.e., a single right or wrong answer.

4. **Divergent thinking.**   Divergent thinking moves in different directions or results in multiple outcomes, i.e., several right answers.

5. **Evaluation.**   Evaluation addresses the accuracy, precision, applicability, or appropriateness of information.

Although all operations are essential, Guilford stresses filtering as it may act to block relevant perceptions, i.e., willingness to take risks, aversion to choice, premature judgment. He also stresses, first, the role of divergent thinking in generating a wide range of ideas and, second, the role of convergent thinking in identifying the most appropriate idea and course of action.

DeBono's Theory   DeBono [6] describes the concept of lateral thinking, as opposed to vertical thinking. In this context, lateral thinking addresses a problem or opportunity in a "sideways" manner whereby we enter the endeavor at different entry points. Here, we cut across patterns in our information system. In other words, **lateral thinking seeks to discover multiple thought paths that may be relevant to the problem or opportunity at hand.**

In vertical thinking, we pursue a pattern in a vertical manner. Here, subsequent steps or thoughts are logical extensions of prior steps or thoughts. In other words, vertical thinking is sequential—single path—thinking about our problem or opportunity.

COIN CASE—DEBONO   [6, pp. 58, 59]   **Example 18.3**

A five-year-old in Australia, called Johnny, is offered a choice of two coins by his friends. There is the $1 coin and the smaller $2 coin. He is told he can take one of the coins and keep it. He picks the larger $1 coin. His friends consider him very stupid for not knowing that the smaller coin is twice as valuable. Whenever they want to make a fool of Johnny they offer him the usual choice of coins. He always takes the $1 coin and never seems to learn.

One day an adult who observes this transaction takes Johnny to one side and advises him that the smaller coin is actually worth more than the bigger coin—even though it may seem otherwise.

Johnny listens politely then says, "Yes, I know that. But how often would they have offered me the choice if I had taken the $2 coin the first time?"

Table 18.6   DeBono's lateral thinking hats (perspectives).

Hats	Purpose	Initiation
White hat—data and information	Calls for group to focus exclusively on data and information assimilation	What information do we have? What information do we need? Where can we obtain information?
Red hat—feelings, intuition, hunches, and emotions	Calls for speculation from the group	I feel that . . . I like/don't like . . .
Black hat—pessimistic perspective	Calls for logical negative—critical assessment	It probably won't work because . . . We have never done this before. Last time we tried this it failed. If we fail, then, . . .
Yellow hat—optimistic perspective	Calls for logical positive—optimistic assessment from the group	It is feasible to . . . We can make this work by . . . The benefits are . . . If we succeed, then, . . .
Green hat—creative effort	Calls for creative thinking efforts to stimulate new thoughts—concepts, ideas	How can we address our purpose? Are there new ways of . . . ? Is there another way?
Blue hat—thinking process control	Calls for organizing or discipline in the thinking process—facilitation of the thinking process	Let's switch to the . . . hat. Let's reevaluate/review our purpose/priorities. Let's shift our focus.

DeBono describes lateral thinking in the context of group activities. DeBono [6] illustrates the application of his theory with a thinking hat analogy, consisting of six hats: white hat (information thinking), red hat (intuition and feeling), black hat (caution and the logical negative), yellow hat (logical positive), green hat (creative effort and creative thinking), and blue hat (control of the thinking process). Table 18.6 provides an overview of these six fundamental thinking modes.

The objective of the six thinking hats is to avoid what DeBono describes as argumentative and adversarial thinking, and in its place develop a cooperative application of the subject at hand. In conjunction with the six hats, DeBono [6] sets out several basic creative thinking process concepts/tools. Several of these concepts/tools are described in Table 18.7. These lateral thinking concepts/tools are developed to keep the thinking process moving in a lateral fashion, as opposed to a vertical fashion or stall mode.

Breakthrough Thinking   In contrast to DeBono's hats, which stress shifting roles in creative thinking, other models stress the context of thinking with respect to the subject matter at hand. **Nadler and Hibino [7] propose a seven-point breakthrough thinking structure.** Their seven principles are summarized in Table 18.8.

The diverse nature of the breakthrough thinking principles provides the means for "reality" checks in creative thinking insofar as focused process thinking is concerned. Hence, the seven principles are a valuable asset in our process-based definition/redefinition, control, and improvement activities.

Table 18.7 DeBono's lateral thinking concepts/tools (selected).

Concept/Tool	Purpose	Initiation
Focus—a target or theme for thought	Identifies and communicates a direction for creative thinking to flow	Our general area-oriented focus is . . . Our specific purpose-oriented focus is . . .
Creative challenge—declaration of need for new perspectives	Stimulates uniqueness of thoughts—concepts, ideas—or calls existing concepts, ideas, methods to question	Given these new technologies, what can we do? Why are we doing things this way?
Creative pause—a deliberate interruption in the thinking flow	Allows opportunity for new thought channels—challenges thinkers to pursue new thoughts—concepts, ideas	Are we missing anything here? Think about this . . . Is there another possibility?
Concept fan—logic tree that links ideas to concepts, to direction, to purpose	Systematically generates alternatives (concepts, ideas) that span a wide variety of thought channels	Agree on purpose; identify all directions or relevant technologies; for each direction or technology, identify all relevant concepts; for each concept generate all relevant ideas. Assemble hierarchical tree or fan.

Table 18.8 Nadler's and Hibino's seven principles of breakthrough thinking.

The uniqueness principle: Whatever the apparent similarities, each problem is unique, and each part of a solution (setting it up, writing a report, installing a solution, etc.) requires an approach that dwells on its own contextual needs.

The purposes principle: Focusing on purposes and their own larger purposes helps strip away nonessential aspects to avoid working on the wrong problem.

The solution-after-next principle: Innovation can be stimulated and solutions made more effective by working backward from an ideal target solution for the future.

The systems principle: Every problem is part of a larger system. Understanding the elements and dimensions of a system framework lets you determine in advance the complexities you must incorporate in implementing your solution.

The limited information collection principle: Knowing too much about a problem initially can prevent you from seeing some excellent alternative solutions.

The people design principle: The people who will carry out and use a solution must work together in developing the solution with breakthrough thinking. The proposed solution should include only the minimal, critical details, so that the users of the solution can have some flexibility in applying it.

The betterment time line principle: A sequence of purpose-directed solutions is a bridge to a better future.

Common Threads Creativity as a concept has been examined by many scholars and described in a variety of ways, usually labeled theories. The majority of the studies have focused on the individual's thought processes, since all creative behavior initiates in the human mind. The more useful theories—useful in the sense of cultivating creative ability and behavior—see sequences and patterns in creative thinking that are impacted by the manner in which we traverse and iterate through these sequences, before offering and assessing our alternatives.

Several critical points emerge as bottlenecks to upgrading or enhancing both creative ability and behavior at both the individual and group levels:

1. Our **ability to perceive need or opportunity** for innovation in our physical or social environments.

2. Our **ability to tolerate ambiguity** with unknown and unfamiliar thoughts—concepts and ideas.

3. Our **ability to channel our thoughts in a divergent or lateral manner,** rather than a convergent or vertical manner.

4. Our **ability to persist** in systematically evaluating the benefits and burdens inherent in our alternatives, seeking ways to upgrade benefits and downgrade burdens.

5. Our **aversion to risk** in testing and implementing innovative concepts and ideas.

6. Our **ability to focus on purpose** and visualize both short- and long-term results and outcomes.

KNOWLEDGE BASE—DOMAIN

Creative thinking does not occur in a knowledge vacuum. **Creative thinking is driven by knowledge.** Essentially, **two general types of knowledge are relevant:** (1) **functional knowledge** and (2) **thinking/analytical knowledge.**

Process-specific functional knowledge varies topically by product line, i.e., by industry, and functionally, i.e., by purpose within an organization. Consider, for example, electronics, food, automotive, agricultural industries or domains: each has unique technologies that must be understood before we can function effectively within the particular domain. In addition, we see specific functional domains, e.g., accounting, engineering, and personnel.

Substantial knowledge bases have emerged to deal with product and functional domains. When new domains emerge, e.g., through creative behaviors, new knowledge bases follow. Hence, our domain knowledge bases are constantly expanding.

The second general type of knowledge we utilize in process-based organizations involves thinking and analytical methods. This knowledge represents the "tooling" used to build and expand product, process, and functional knowledge bases or domains. For example, just as physical holding devices help us hold parts while we machine them or weld them, thinking and analytical tools help us to produce product- and process-related knowledge faster and more efficiently than we could otherwise. Just as a vise can hold a wide variety of parts, a C-E exercise and diagram can be used to assess cause-effect relationships in a wide variety of product/process knowledge domains.

On the one hand, we see highly specific knowledge bases or domains that we desire to expand, and on the other hand, we see very general thinking and analytical knowledge bases that we can use in our efforts of expansion. Table 18.9 lists several functional domains as well as selected thinking and analytical domains. Essentially the thinking and analytical knowledge or domains are available to people seeking to expand the product, process, and functional domains. Some level of understanding and mastery in both domains is essential to obtain positive results. Misunderstanding and misapplication can lead to negative results in both technical and social environments.

In practice, the focus is on expansion of our product, process, and functional domains as fast as possible. However, we also seek to expand our thinking and analytical domains in order to expedite expansion of our product, process, and functional domains. In either case, we utilize processes for this expansion, hence, the justification for our process-based organization.

Table 18.9   Selected knowledge bases (domains) relevant to process-based organizations.

Product-oriented (SIC codes):
  Agriculture, forestry, and fishing
  Mining
  Construction
  Manufacturing
  Transport, communications, utilities
  Wholesale trade
  Retail trade
  Finance, insurance, real estate
  Services
  Public administration

Function-oriented:
  Accounting
  Engineering
  Finance
  Management information systems
  Marketing
  Packaging
  Personnel
  Public relations
  Sales
  Training

Thinking/analytical initiatives:
  Benchmarking
  Concurrent engineering
  Continuous improvement
  Cycle time reduction
  Fifth Discipline
  Function-value analysis
  ISO 9000
  Mistakeproofing
  Quality awards
  Quality function deployment
  Reengineering
  Robust design
  Six sigma
  Theory of Constraints
  Total quality management

Tools:
  Activity/sequence list
  Break-even analysis
  Capability analysis
  Cash flow analysis
  Cause-effect diagram
  Check sheet
  Control chart
  Correlation/autocorrelation analysis
  Critical path method
  Experimental design
  Failure mode and effects analysis
  Fault tree analysis
  Flowchart
  Force field analysis
  Gantt chart
  Histogram
  Matrix diagram
  Pareto diagram
  Process value chain analysis
  Relations diagram
  Root cause analysis
  Runs chart
  Scatter diagram
  Stratification analysis

Product, process, and functional knowledge bases, collectively, are extremely large and diverse—well beyond our scope in this book. The thinking and analytical knowledge bases or domains used to expand the product, process, and functional domains are more contained. Brief descriptions and reference materials for selected initiatives and analytical methods or tools are contained in Chapters 19 and 20, respectively.

ENVIRONMENT—FIELD

The environment or theater (field) associated with creativity serves as a gatekeeper for both opportunity and results. Two basic cultures are involved in process-based organizations: (1) the business culture and (2) the technical or scientific culture. These two cultures are described by Kolarik [8]:

Business culture. Business culture is based on authority derived from the organization—it is characterized by submission to the larger needs of the organization and is motivated by a desire to operate an organization and to identify with its success. The business culture looks to society and the business world as its reference point.

Technical and scientific culture. The technical-scientific culture derives its authority from knowledge about nature and is motivated by curiosity and the desire for mastery over and knowledge about technical and scientific problems. It looks to nature instead of business and society and regards scientific method and discipline as its reference points.

The combination of the business and technical cultures forms the environment in which most creative efforts live or die. In general, we would like an environment or gatekeeper that encourages productive efforts and discourages counterproductive efforts regarding our vision, mission, and process purposes in both the short and long term. However, creative efforts and outcomes are seldom obvious. Persistence in forging successful outcomes out of initially marginal or failed efforts is the more usual case.

Within our business and technical cultures we see both social environments and physical environments that influence creative attitudes and behaviors. Throughout history, we have seen **creative endeavors take place when the right people, with the right knowledge, meet at the right place, with the right focus.** In all cases, the environment or field played a supporting role to one degree or another. History typically does not record the "near misses" where most of the ingredients were present, but the environment was so hostile as to kill the creative initiative.

The environment may encourage or discourage efforts toward possible gains anywhere along the need-thought-idea-decision-test-implementation spectrum. **On the basis of both research and experience, we can point to several environment- or field-related influences that help to foster creative behavior.** Table 18.10 summarizes several of these environmental attributes and relates them to elements of our discussion in this chapter. These attributes are not independent of each other and tend to interact and interface with both the knowledge domain and people seeking to behave in a creative manner. Some relate to leadership and some to management activities. Each is highly case specific, requiring flexibility in organization along with persistence and patience.

18.4 LEADERSHIP

The concept of leadership and followership has intrigued people for centuries regarding political, religious, benevolent, and commercial organizations. **Three historical perspectives describing leadership can be identified: (1) the trait period, (2) the behavioral period,** and (3) **the situational-contingency period.** The trait period (pre-1940) focused on associating a personal trait, such as family lineage, intelligence, physical appearance, physical endurance, and so on, with leadership ability.

Table 18.10   Critical attributes of environments that foster creative behavior.

Exposure to need, through observation of current operations and technologies—direction related.

Free access to knowledge-based resources, information and basic facilities for experimentation—knowledge base related.

Opportunity for interaction between significant persons in complementary or competitive roles—teamwork related.

Interest in *pursuit of divergent views,* lateral thinking capacity and tolerance for ambiguity and chaos in thoughts and ideas—creative ability/thinking related.

Existence of significant intrinsic and extrinsic *recognition and rewards* to address human needs—reward oriented.

Risk bearing capacity and adventuresome sprit, excitement and stimulation from challenge and change—leadership related.

Balanced outlook, *short- and long-term vision* in picturing results and outcomes from creative alternatives, and metering resources to creative endeavors—direction related.

Localized *autonomy in decision paths* within the creative endeavor—empowerment related.

Equitable *metrification of outcomes and efforts* to *calibrate* results to purpose and expectations—direction related.
   Persistence and patience, understanding the ebb and flow of creative endeavors, snatching success from the jaws of failure—leadership related.

The behavior period—approximately 1940–1960—focused on the behavior of an individual acting as a leader. This school of thought produced the concept of leadership styles. The leadership behavior or style where the leader makes all or most decisions regarding group activities was labeled authoritarian. Behaviors where people acting as leaders delegated some decision making to others, namely subordinates, were labeled democratic or participative. Leadership behaviors that effectively abdicated most or all decisions to others were labeled laissez-faire. Another manifestation of the latter part of the behavioral period was the concept of task-oriented versus people-oriented leaders. This concept led into the situational leadership theories.

Both the trait and behavior concepts focused almost exclusively on the person acting as a leader. For the most part, they ignored or at least discounted the followers and circumstances surrounding the acts of leadership. In general, they took a rather static view of attitudes and behaviors of people acting as leaders. These concepts, over time, were seen by many as inadequate to explain leadership. Situational and contingency leadership concepts have developed from about 1960 to the present. **Situational leadership concepts expanded the field of view beyond the person acting as a leader, to include followers and circumstances surrounding the acts of leadership.** Here, leadership is viewed in a dynamic light, whereby influences of subordinates and subordinate responses, tasks, organizational policy, superiors and superior responses, peers and peer responses, as well as the characteristics of the person acting as a leader, all play some part in influencing leadership behavior.

Many definitions of leadership have been proposed; two contemporary definitions are captured below:

Leadership is the process of influencing a group toward the achievement of a goal; Reitz [9].

Leadership is a process of giving purpose [meaningful direction] to collective effort, and causing willing effort to be expended to achieve purpose; Jacobs and Jaques [10].

Several observations can be made regarding leadership definitions. First, **leadership is defined in a process context.** Second, **leadership is purpose-related,** in that some vision of accomplishment is shared and pursued. Third, **there is a collectivism about leadership,** where a group of people is engaged in the pursuit. A fourth, but not obvious, observation made by Jacobs and Jaques [10] is that **decision discretion must exist in the leadership process.** Otherwise, the opportunity to exercise leadership does not exist.

Jacobs and Jaques [10] point out that leadership is not a "thing," and it is not necessarily useful to talk of "leaders" as either persons or role incumbents. Rather, **we should view leadership as an influence process that is a type of role behavior**. For example, we might recognize someone for acting or playing a critical leadership role in a given situation; i.e., we can recognize effective leadership only after the fact.

Leadership role behavior can be displayed by any member of a formal or informal organization. This behavior can be directed toward subordinates, toward peers, or toward superiors. For example, consensus building is a leadership behavior.

In the leadership process we see people acting as leaders and followers through their role behaviors. For any given endeavor, these roles may be played out by several people. Here, members of the group have a perception of purpose, i.e., a goal, and are focused on goal attainment in a collective sense. And they are individually committed, i.e., willing, to expend energy and effort toward goal accomplishment. At this point, **we see leadership as a dynamic process, involving a group and a decision context, expressed through role behaviors**.

In implementing and operating process-based organizations, we stress whole-process teams working together. Three leadership-related subelements have previously been identified as critical in process-based organizations: direction, teamwork, and empowerment; see Figure 18.1. Each of these subelements is involved in traditional organizations, but most are concentrated within top management and not typically shared with operational levels. The structure of process-based organizations requires us to rethink the nature and distribution of direction, teamwork, and empowerment. **We must be prepared to encourage and accept new mindsets and acquire new knowledge and skills regarding direction, teamwork, and empowerment.**

DIRECTION

The directional subelement in leadership is dependent on common purpose. We address common purpose through vision, mission, core values, process purposes, goals, objectives, targets, and specifications. Purpose hierarchies exist to delineate and express purpose.

The concept of a purpose hierarchy was introduced in Section 1. In our previous discussions the criticality of expression of purpose was emphasized. In the context of leadership in this section, the expression remains critical; however, the acceptance of purpose by people within the organization becomes our focus. Enthusiastic acceptance of purpose impacts personal willingness and generates the energy required to address common purpose. In other words, **both people acting as leaders and those acting as followers must accept common purpose**.

The process of leadership welds people to purposes through a combination of logical/cognitive as well as emotional/affective appeals; it involves people communicating with people. This leadership-followership interchange or communication takes place in a situational context. Hence, the appeal by the person acting as a leader and the response of people acting as followers are affected by a host of variables related to the physical and social environments.

In summary, acts of leadership and followership are initiated with the acceptance of common purpose, i.e., meaningful direction that leads to involvement. **People involved in leadership and followership are addressing their human needs: physiological, safety and security, social, esteem, and self-actualization.** Involved, willing people eagerly accept burdens and exchange their personal resources, e.g., talent, time, and/or energy, for the prospects of benefits to fulfill their human needs.

TEAMWORK

Transformation to a process-based organization requires three fundamental team types: (1) process or operations teams, (2) definition/redefinition teams, and (3) improvement teams. Process teams are long-standing teams that deal with day-to-day activities. Definition/ redefinition teams are ad hoc teams that deal with strategic opportunities. Improvement teams are ad hoc teams that deal with tactical opportunities. **Teamwork involves people working together toward a common purpose.** People as individuals become involved in teams. Personal interactions among these individuals must take place to conduct the team "business." The more effective and efficient these interactions, the more effective and efficient the team becomes. Hence, we examine several basic subelements in this interchange: trustworthiness, communication, meetings, team interaction roles, and conflicts.

Trustworthiness As a prerequisite for teamwork, individuals must possess some reasonable level of trustworthiness. Otherwise, group function toward common purpose will not take place.

The concept of trust is very broad and abstract. Descriptions usually include faith, honesty, integrity, and so forth, regarding a person in a holistic sense. In our context, we will describe trustworthiness in much narrower terms. **Trustworthiness is an obligation toward common purpose and team welfare.** This definition views trustworthiness as relative to a specific group or team and team purpose. Hence, an individual may be trustworthy in one endeavor and not in another.

By examining this definition, it is apparent that trustworthiness and personal willingness are closely connected. For example, our level of willingness to contribute to the achievement of common purpose is driven by, or a function of, our strength of obligation toward our team—our belief in the abilities and forthcoming behaviors of team members. In other words, **obligation to produce behavior in the best interest of the team and its purpose constitutes trustworthiness in a team member.**

Communication Burgoon et al. [11] define **human communication as a dynamic and ongoing process whereby people create shared meaning through the sending and receiving of messages in commonly understood "codes."** Interpersonal communication can be divided into three major types: (1) written, (2) spoken or verbal, and (3) unwritten or nonverbal.

Written and verbal communications, either formal or informal, constitute a literal form of communication. Because of the contractual nature of business, the written word is typically considered more forceful than the spoken word, even when, at face value, the meaning is the same. Spoken and written words are essential in communicating purpose, facilitating the pursuit of purpose, and recognizing the accomplishment of purpose. We see brevity as an important attribute of effective statements.

Interpersonal face-to-face communications in a social setting consist of both verbal and non-verbal channels. **Research findings estimate that nearly two-thirds of meaning is derived from nonverbal cues**, according to Burgoon et al. [11]. Beebe and Masterson [12] cite three reasons for the importance of nonverbal communications:

1. People in groups spend more time communicating nonverbally than verbally, since only one member speaks at a time while the others watch the speaker and their group peers.

2. People tend to believe nonverbal cues more than verbal communications. That is, "Actions speak louder than words."

3. People communicate emotions and feelings primarily by nonverbal cues—voice tone and facial expression carry a large share of emotional content.

Nonverbal communication is received through the five human senses, sight, hearing, and touch being the most important. Seven major areas of nonverbal communication have received a good deal of attention and appear to play critical roles in communication processes, depending on the situation at hand; see Burgoon et al. [11]:

Kinesics or body language. Body movement, facial activity, and gaze or eye behavior.

Vocalics. Vocal activity such as tone, pitch, and intensity.

Physical appearance. Dress and adornment, as well as physical attributes, e.g., clothing, stature, grooming.

Haptics. Touch and the use of touching, e.g., shaking hands.

Proxemics. Use of space and spatial arrangements, e.g., workplace or office layout.

Chronemics. Timing and the use of time cues, e.g., looking at one's watch during a conversation, arriving late at a meeting.

Artifacts. Environmental surroundings and objects or possessions displayed or used in the proximity of the communication activity or by the communicator, e.g., workplace or office decor, personal and company vehicles, meeting rooms.

When taken together, the written, spoken, and nonverbal communication channels may produce one of three general effects:

1. **Reinforcement and emphasis.** The message is strengthened, e.g., saying "good morning" and smiling.

2. **Complement.** The message is completed, e.g., introducing yourself and shaking hands as a gesture of friendship.

3. **Contradiction.** The communication channels are inconsistent, e.g., saying you feel great with a grimace of pain on your face.

Meetings Because of the distributed decision nature of a process-based organization, it is critical to allow for the open exchange of information, e.g., data, questions, opinions, ideas. In other words, we prepare for effective and efficient group encounters. **Small group meetings serve as vehicles for these open exchanges.**

Meetings, by definition, involve two or more people and interpersonal communications. Communications are critical. They allow us to conduct our business of exchanging information

Table 18.11 Meeting guidelines.

Issue	Meeting Tool	Description
Before		
Preparation/planning	Agenda—spell out purpose and business flow, start and finish time, location	Each meeting should have a purpose, worthy of taking time away from other endeavors.
During		
Time management	Time contracts—encourage participants to be brief	All participants are expected to express themselves in a clear, concise, and efficient manner.
Participation	Facilitation—recognition to speak, taking turns	Each member is encouraged to contribute or participate in an orderly and respectful manner.
Agreement/consensus	Multivoting—to move toward consensus	All members are encouraged to question and assess/communicate their understanding or lack of understanding, e.g., thumb up—agree, thumb sideways—need more information, thumb down—disagree.
Action/commitment	Action items—what, when, who?	Agreement on what, when, and who, with respect to action or inaction—acknowledgement of responsibility and authority for action.
Behavior norms	Rules/procedures for personal and business conduct	Expectations of business conduct and personal conduct—provisions for violations, e.g., time contract violations.
After		
Action/follow-up	Minutes—notes	Documented record of meeting highlights for later referral—documented record of action items, results and lessons learned.

effectively; however, we usually need to add a directional element to our encounter in order to conduct our business effectively and efficiently. **We use meeting guidelines to help us get and stay on track.** Table 18.11 presents meeting guidelines—relevant to before, during, and after the meeting.

We plan our meeting and express our plan with a meeting agenda. The agenda allows people time to think before the group encounter takes place. It also allows each participant to add necessary items to the meeting by adding the items to the agenda. Hence, before the meeting starts, we will have a purpose, i.e., we know what we will be doing, and perhaps be able to have a reasonable expectation of the outcome.

During the meeting we work the agenda. We stress open exchanges, within a set of parliamentary procedures or behavioral norms. We work, through group consensus, toward action items with assigned and accepted responsibilities and authorities. After the meeting, we summarize our progress and action items. We execute our planned actions and observe our results. Finally, we assess our results and communicate our lessons learned.

Teaming Roles   **Teaming is fundamental for process-based organizations.** In the context of complex organizations such as basic production, manufacturing, and service organizations, we see teaming as an intricate form of cooperative effort, open to wide participation on the part of all members.

Viewing teams in the context of the whole organization, rather than as an isolated part of the organization, we see ties back to both leadership activities and management activities. Here, as previously, we describe leadership and management; see Kolarik [8]:

Leadership activities deal with vision and the subsequent defining and delineating of purpose as well as cultivating belief in purpose and motivating people to act accordingly.

Management activities entail providing, metering, and monitoring the resources necessary to sustain the action sparked by leadership.

Both leadership- and management-related elements are necessary for effective and efficient teamwork.

Six specific roles involved with effective and efficient teaming are shown in Table 18.12. Here, **we recognize several teaming roles labeled as team sponsor, lead, facilitator, coordinator, scribe, and member**. It is important to point out that these roles are intended to be taken on by people. These roles may be passed around, or reside with specific individuals, where one individual can play one or more roles during the course of a meeting and/or during the team's lifetime.

These teaming roles relate to all team endeavors, including meetings. The sponsor establishes the legitimacy of the team. This role ties the team to process purpose within the organization as a whole. The sponsor role establishes a linkage to management and the resources management meters out for team activities and rewards. The sponsor defines the general expectation of results or outcomes—establishes general direction.

The lead role details the sponsor's expectations and provides active leadership toward explicit team purpose, all within the sponsor's general direction. The lead role nurtures the team through both coaching and communication, with both the sponsor and the team members. The team lead role resides within the team, whereas the team sponsor role resides primarily outside the team as a contact to the organization as a whole. The facilitator, coordinator, and scribe roles support the functional aspects of the team in terms of team interactions, logistics, and documentation, respectively.

The team member role is a followership role in pursuing team purpose. The team member role is one of helping to establish details of purpose, e.g., goals, objectives, targets, as well as the accomplishment of purpose. Team members' contribute collectively to determine actual results.

Conflicts   **Team-based activities naturally produce some level of conflict.** Focused people and the intensity of their energy levels tend to produce diversity in opinions and outlooks regarding problem diagnosis, alternatives, analysis, plans, and implementation. **Diversity in opinions and perspectives is healthy and ultimately allows us to create synergy in team effectiveness. Diverse opinions and perspectives sometimes create conflict** that can be damaging, if it grows out of check. The opposite of diversity is sameness. **"Groupthink" is a term applied to the situation where a group or team strives to avoid any form of conflict and reaches a consensus without critically testing, analyzing, and evaluating thoughts relative to the problem at hand;** see Beebe and Masterson [12]. Groupthink typically results when one or a few members dominate the group and the remainder play passive roles.

Table 18.12   Teaming roles and descriptions.

Sponsor—direct linkage to management
  Lead/responsibility in challenging, chartering, and empowering team
  Lead/responsibility in providing team resources and expectations
  Lead/responsibility in visualizing the results, setting direction, and defining broad boundaries
  Lead/responsibility for formal recognition, rewarding
  Ultimately responsible/accountable for team success *and* generating (allowing for) team success

Team lead—initiator and promoter (salesperson, advocate) of team purpose to sponsor and team members
  Lead/responsibility in choosing team members
  Lead/responsibility in crystallizing a clear problem/opportunity/process purpose
  Lead/responsibility in communicating challenge and expectations
  Lead/responsibility for coaching, and encouraging team work and efforts
  Lead/responsibility for informal, on-the-spot recognition
  Lead/responsibility in adjustments or updating of purpose (e.g., any change in direction) as new facts and figures are
    exposed
  Lead/responsibility in implementation (if carried out by the team)
  Responsible/accountable for team performance, selling initial purpose, selling initiative to sponsor, establishing
    boundaries and scope for team activities, selling results to sponsor

Facilitator—group interaction catalyst
  Lead/responsibility in reinforcing challenge, purpose, and expectations
  Lead/responsibility in eliciting contributions from members
  Lead/responsibility in preventing, defusing conflicts
  Lead/responsibility in capturing, organizing, summarizing thoughts and ideas
  Lead/responsibility in moving group toward consensus, decisions
  Responsible/accountable for group dynamics

Coordinator—organization
  Lead/responsibility in establishing the agenda and time contracts
  Lead/responsibility in working out the logistics and scheduling
  Responsible/accountable for best meeting time, facilities and equipment

Scribe—documentation
  Lead/responsibility in recording and documenting team activities
  Responsible/accountable for producing accurate permanent record of work

Team member—technical, business, knowledge, expertise, implementation
  Lead/responsibility in day-to-day team operations
  Lead/responsibility in generating thoughts, ideas, creative expressions
  Lead/responsibility in evaluating, problem-solving, decision-making participation
  Responsible/accountable for team success and individual conduct

---

Three categories of conflict have been identified by Beebe and Masterson [12]: (1) **pseudo conflict**, (2) **simple conflict**, and (3) **ego conflict**. Pseudo conflict is a perceived (by team members) conflict resulting primarily from misunderstanding some part of the issues/perspectives being examined. Its resolution is relatively easy, provided it is recognized and exposed. Simple conflict is a real conflict in that the issues/perspectives being examined are in conflict. It is typically of a technical or professional nature, as opposed to a personal nature. Ego conflict is a serious category of conflict where personalities are involved.

Pseudo or simple conflicts can escalate into ego conflicts quickly if not checked. The facilitation role in teaming seeks to keep conflicts within healthy and productive territory—to reduce the possibilities of groupthink and ego conflict, as well as manage or control pseudo

Table 18.13   Types of small group conflict.

	Conflict Type		
	Pseudo Conflict	**Simple Conflict**	**Ego Conflict**
Source of conflict	Misunderstanding of the issues/perspectives	Individual disagreement over the issues/perspectives	Defense of ego—interpersonal conflicts
Suggestions for managing conflict	Ask for clarification of perceptions regarding the issues/perspectives Be supportive, listen, question, paraphrase	Clarify perceptions State, restate the facts Seek, cite common ground in issues/perspectives Stress facts, logical connections Look for alternatives or compromise positions Call a time-out/break	Set baseline rules at outset of meeting—clearly explain and gain consensus for conduct Remove/avoid judgmental content in discussion—stress logical connections/ facts Remain calm, don't panic Call a time-out/break

conflict and simple conflict. Table 18.13 provides a brief description of conflict and conflict resolution useful in teaming.

EMPOWERMENT

Empowerment refers to the sharing of power—responsibility and authority—to act on behalf of the organization. Power to act resides somewhere in the organization—at the supervisory and higher levels of management in traditional organizations, in a graded manner. In the team-based organization, decisions and implementation/action are dispersed within the teams. In short, **transformation from a traditional organization to a process-based organization requires a redistribution or realignment of empowerment.**

In general, **effective empowerment matches the expectation of team results—accomplishment of common purpose—to the level of responsibility and authority necessary for accomplishment.** The more we expect in the way of results from our teams, the more likely it is that the team will require a high level of empowerment. Here, **trust in a team, through trustworthiness of members, along with individual abilities in technical knowledge, skill, and experience, are critical.** If we are limited in personal abilities, technical knowledge, skill, or experience, or in teaming and communication skills in general, we are headed for difficulties, which may render team empowerment and teams in general ineffective and inefficient. **A "readiness" level in trust as well as teaming and communication skills exists such that we can effectively and efficiently redistribute power and provide meaningful and useful empowerment.**

Readiness levels are altered through education and training. Once readiness is above a threshold level, transformational success odds are reasonable, but not 100%. Responsibility and authority boundary movements always create anxiety and concern to those impacted—managers, operators, customers, stakeholders. **Commitment to and persistence in process-based empowerment lead to growth beyond the threshold level, eventually allowing high levels of organizational performance, along with high comfort levels.**

Review and Discovery Exercises

Review

18.1. Compare and contrast the nature and focus of a traditional and a process-based organization in terms of organizational structure.

18.2. How are definition/redefinition, control, and improvement action channels designed into an organization structure? How are they maintained? Explain.

18.3. Why is there a necessity to archive data and information? Explain.

18.4. How do metrics support and serve team decision making? Explain.

18.5. How and where does information exchange take place? How can we make information exchange more effective and efficient? Explain.

18.6. What constitutes an award? A reward structure? Explain.

18.7. Who is impacted by awards? How are they impacted? Explain.

18.8. What role does creativity play in process performance? Explain.

18.9. What is divergent thinking? What is lateral thinking? What is vertical thinking? Explain the differences.

18.10. What role does the knowledge base/domain play in creativity? Explain.

18.11. What role does the environment/field play in creativity? Explain.

18.12. How is leadership defined? What common ground is found in leadership definitions? Explain.

18.13. What team types do we see in process-based organizations? What is the nature of each? Explain.

18.14. What purpose do meetings serve? Explain.

18.15. What forms of conflict tend to crop up in team/group encounters? Explain the source and nature of each.

18.16. What is empowerment and how does it work? Explain.

Discovery

18.17. Critique the Experience Lost Case—Lid Seal (Example 18.1) with respect to data/information exchange and archiving.

18.18. Why do people respond to awards? How does rewarding impact/influence cooperative efforts? Explain.

18.19. How do team-based awards/rewarding differ from individual-based awards/rewarding? Explain.

18.20. Critique the Award Alienation Case—Who's Interested (Example 18.2) with respect to what is right/wrong with the rewarding process described in the case.

18.21. Compare and contrast DeBono's lateral thinking hats, concepts, and tools and Nadler's principles of breakthrough thinking. Address purpose and content.

18.22. Why is conflict (at some level) inevitable in team/group encounters? Is it positive or negative regarding group performance? Explain.

REFERENCES

1. Knouse, S. B., *The Reward and Recognition Process in Total Quality Management,* Milwaukee: ASQC Quality Press, 1995.

2. Csikszentmihalyi, M., *Creativity: Flow and the Psychology of Discovery and Invention,* New York: Harper Collins, 1996.

3. Dacey, J. S., *Fundamentals of Creative Thinking*, Lexington, MA: Lexington Books, 1989.

4. Jackson, P. W., and S. Messick, "The Person, The Product, and The Response: Conceptual Problems in the Assessment of Creativity," *Journal of Personality*, vol. 33, no. 3, pp. 309–329, 1965.

5. Guilford, J.P., *The Nature of Human Intelligence*, New York: McGraw-Hill, 1967.

6. DeBono, E., *Serious Creativity: Using the Power of Lateral Thinking to Create New Ideas,* New York: Harper Collins, 1992.

7. Nadler, G., and S. Hibino, *Breakthrough Thinking*, 2nd ed., Rocklin, CA: Prima Publishing, 1994.

8. Kolarik, W. J., *Creating Quality: Concepts, Systems, Strategies, and Tools,* New York: McGraw-Hill, 1995.

9. Reitz, H. J., *Behavior in Organizations*, Homewood, IL: Irwin, 1977.

10. Jacobs, T. O., and E. Jaques, "Military Executive Leadership," in *Measures of Leadership*, pp. 281–295, edited by Clark, K. E., and M.B. Clark, West Orange, NJ: Leadership Library of America, 1990.

11. Burgoon, J. K., D. B. Bullen, and W. G. Woodall, *Nonverbal Communication*, New York: Harper and Row, 1989.

12. Beebe, S. A., and J. T. Masterson, *Communicating in Small Groups*, 3rd ed., Glenview, IL: Harper Collins, 1989.

chapter

19

PROCESS-COMPATIBLE INITIATIVES

19.0 INQUIRY

1. What constitutes a quality-productivity initiative?
2. How are initiatives formulated?
3. How are initiatives directed?
4. What initiatives have proven their worth?

19.1 INTRODUCTION

Quality-productivity initiatives represent organized efforts, wrapped around central themes, which are introduced into the production system in hopes that gains will be made in either quality or productivity, or both. Typically, initiatives are rather broad and can be implemented throughout the production system, whereas tools are more focused. Initiatives typically make use of tools. For example, total quality management (TQM), as an initiative, is encompassing as to both scope and content and is touted as universally applicable to all organizations, regardless of product or industry. TQM typically utilizes a number of tools in its practice. Other initiatives apply to specific things that need attention. For example, benchmarking initiatives address what other production systems or other parts of the same production system are doing— and the results that they are realizing.

Initiatives typically involve many people and demand attitude and behavior changes, sometimes radical changes. Initiatives are usually formulated in an ad hoc manner within a production system in response to a challenge of some sort, e.g., as a counteraction to low business performance. Then, once observable results are apparent, the initiative tends to be described and generalized and presented to the outside world. This description and generalization is usually positioned around a central theme, with several basic or fundamental concepts or steps described. Although the initiative was likely developed and tested in the context of a reasonably complex organizational setting, descriptions to the outside are generic in the sense that they usually do not consider specific organizational conditions and constraints, e.g., competitive threats and leadership.

Initiatives usually are unsuccessful unless they possess strong champions—people who are willing to put forth the leadership and energy to make the initiative work. Typically, at the original point of development of the initiative, natural champions designed and implemented the things that eventually grew into the initiative, as formally described later. Usually, this same level of commitment of leadership, energy, and resources is necessary, but not necessarily sufficient, to "duplicate" the original results in another production system, or even within other areas of the same production system.

The purpose of this chapter is to briefly describe several selected initiatives for our readers. We must remember that initiatives come and go. Terms like "the flavor of the month" are sometimes used to describe initiatives. **The initiatives we feature all have virtues that will supersede their initiative-label lifetime.** Hence, we can be confident that the technical essence and results that can be produced from these initiatives are long-standing, whereas the specific initiative name, as labeled, may have a finite life. It is interesting to note that technical features of newer initiatives typically encompass many technical features of older "dead" initiatives. Furthermore, high-performance production systems are typically practicing initiative-like concepts without their people knowing or caring about specific initiative labels—they know that **results are the critical issue, not initiatives and their associated slogans and buzzwords.**

19.2 BENCHMARKING

PURPOSE

The purpose of benchmarking is to learn from others—seek out, study, and emulate the best practices associated with high performance/results—so as to enhance or better our own performance.

DESCRIPTION

Benchmarking, although nothing really new, was positioned as an organizational initiative at Xerox. D. T. Kerns (CEO, Xerox) defined benchmarking as

> Benchmarking is the continuous process of measuring products, services, and practices against the toughest competitor or those companies recognized as industry leaders.

Robert Camp [1, 2] defines benchmarking as

> Benchmarking is the search for and implementation of best practices.

Benchmarking encompasses four issues: (1) analyze the operation, (2) know the competition and industry leaders, (3) incorporate the best of the best, and (4) gain superiority.

Obviously, from the definitions and descriptions above, we have seen the essence of benchmarking in sports, as well as in competitive studies of industries, for centuries. What is new is the refinement and systematic pursuit of emulation practices, e.g., shamelessly stealing and creative swiping. The formal scope of benchmarking includes products, processes, and metrics.

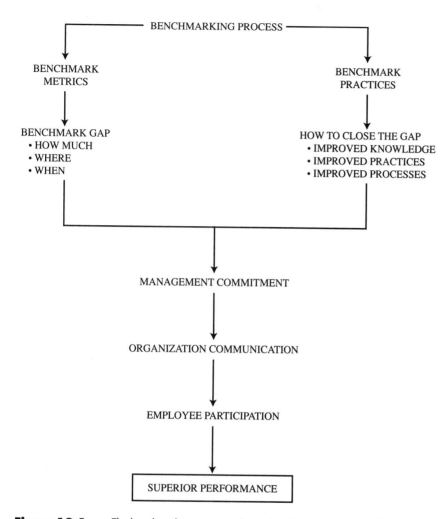

Figure 19.1   The benchmarking process. (Reproduced, with permission, from Camp, R. C., *Benchmarking*, Milwaukee: ASQC Press, 1989, p. 5.)

Camp cites four types of benchmarking: internal, competitive, functional, and generic. Internal benchmarking focuses on best practices within our own organization. Competitive benchmarking provides a comparison between direct competitors. Functional benchmarking refers to comparisons of methods across organizations executing the same basic functions, outside our industry. Generic process benchmarking focuses on innovative work processes in general, wherever they occur.

The benchmarking initiative focuses on two basic issues: (1) best practices and (2) metrics or measurement. Figure 19.1 depicts benchmarking issues where we see a practices channel and a metrics channel. We recognize performance gaps and address them with improvement plans. Management commitment, communication, and employee participation are all critical elements in a benchmarking initiative.

Table 19.1   Phases in benchmarking.

Step	Description
Preplanning	Assess and understand customer needs and business results/outcomes desired.
Planning	Identify what is to be benchmarked. Identify comparative organizations. Determine data collection method and collect data.
Analysis	Determine current performance gap. Project future performance levels/goals.
Integration	Communicate findings and gain acceptance. Establish functional goals.
Action	Develop action plans. Implement specific actions and monitor progress. Recalibrate benchmarks.
Maturity	Attain a leadership position. Integrate benchmarking practices into processes.

Camp [1, 2] describes six phases in benchmarking: (1) preplanning, (2) planning, (3) analysis, (4) integration, (5) action, and (6) maturity. Table 19.1 expands each of these phases. Preplanning involves understanding customer needs and the business results sought. The purpose of this preparatory step is to assure that the specific benchmarking focus is aligned with both customer requirements and business outcomes, before actual benchmarking activities are launched. Planning involves identifying the focal process, comparative organizations, and data needs. Analysis involves data collection, the assessment of performance gaps, and projections and targets for future performance. Integration involves communicating findings, establishing functional goals, and gaining acceptance thereof. Action consists of developing action plans, implementation, and monitoring and measuring progress, as well as recalibrating the benchmarks when necessary. Maturity refers to establishing and maintaining the permanency of benchmarking in the corporate culture.

Benchmarking is a broad initiative. Watson [3] describes the evolution of benchmarking in terms of generations. He cites reverse engineering as the first generation. Here, we see essentially a rote copying strategy. The second generation is termed competitive benchmarking, which focuses on direct competitors. In the third generation he cites process benchmarking, where processes common to different industries are assessed for best practices. The fourth generation is termed strategic benchmarking. Here, the focus is on the strategies that a competitor or noncompetitor uses to guide its organization. The fourth level is used to feed process reengineering initiatives. A futuristic fifth level is cited as global benchmarking. Here, the focus is international in scope and deals with trade, cultural, and business process distinctions among companies. In all cases, the driving force is profit-oriented, as addressed through three parameters: (1) quality beyond that of competitors, (2) technology before that of competitors, and (3) costs below those of competitors.

NOTES

Benchmarking provides a realistic view of our organization in comparison to others, namely competitors, based on measurement rather than speculation. This view helps to communicate and critique our performance, providing a justification for redefinition and improvement efforts. Benchmarking is always relevant to process definition/redefinition, control, and improvement.

19.3 CONCURRENT ENGINEERING

PURPOSE

The purpose of concurrent engineering is to focus a multidisciplinary task force on product and process definition and design so as to shorten lead time and result in fewer changes beyond the definition and design phases.

DESCRIPTION

Concurrent engineering, also known as parallel engineering or simultaneous engineering, initiatives combine a team approach with specialized disciplines. Concurrent engineering is project-centered, with the task force seen as permanent in the sense that it typically remains in force for the duration of the project. Allowances are made to bring new members in as necessary. A typical task force will consist of people with expertise in product design, manufacturing, marketing, purchasing, and finance, as well as principal vendors.

The goal of concurrent engineering is to cut time to market as well as the cost of programs, while bringing improvements in product design, quality, ease of manufacture, and performance in the field; see Hartley [4]. Vital elements involve (1) a multidisciplinary task force, (2) product definition in the customer's terms, (3) parameter design to ensure optimization of quality, (4) design for manufacture, and (5) the simultaneous development of the product, the manufacturing equipment and processes, quality control, and marketing.

Concurrent engineering is driven by the fact that it is faster and less expensive to make changes early in the definition, design, and production sequence than later in the sequence. In concurrent engineering more time and resources are spent in defining the product than in traditional methods of engineering. In other words, resources are front-loaded to assure a more effective and efficient overall project. Hartley [4] cites reductions of 75% in modifications, 50% in work in process, and 33% in production process time in one fighter aircraft project. Other gains are cited in reduced concept-to-production time, life-cycle cost reductions, reduced assembly time, and reduced parts counts.

NOTES

Concurrent engineering is similar to the quality function deployment (QFD) initiative and serves essentially the same purpose. To some degree QFD is more open-ended in the sense that it is readily applicable to services as well as hardgoods, whereas concurrent engineering is typically associated with complex hardgood products.

19.4 CONTINUOUS IMPROVEMENT

PURPOSE

The purpose of continuous improvement is to provide an environment that encourages incremental change in the workplace though collaborative efforts of managers and workers.

DESCRIPTION

Continuous improvement essentially redistributes responsibility, authority, and decision making, regarding tactical or limited-scale improvements, down to the operations level—it essentially empowers the work force. Typically, this redistribution is accomplished through small work or improvement teams, which are focused on problems and challenges at the operations level. Contemporary continuous improvement initiatives vary in both scope and degree of empowerment.

The roots of continuous improvement as practiced today are in the Japanese Kaizen concept. This concept implies ongoing improvements involving everyone—top management, managers, and workers; see Imai [5]. Table 19.2 summarizes the classical Japanese concept of Kaizen. Here, the roles in Kaizen are defined for top management, middle management, supervisors, and work-

Table 19.2   Hierarchy of Kaizen involvement.

Top Management	Middle Management	Supervisors	Workers
Be determined to introduce Kaizen as a corporate strategy	Deploy and implement Kaizen goals as directed by top management through policy deployment and cross-functional management	Use Kaizen in functional roles and provide Kaizen suggestions	Engage in Kaizen through the suggestions system and small-group activities
Provide support and direction for Kaizen by allocating resources	Use Kaizen in functional capabilities	Formulate plans for Kaizen and provide guidance to workers	Practice discipline in the workshop
Establish policy for Kaizen and cross-functional goals	Establish, maintain, and upgrade standards	Improve communication with workers and sustain high morale	Engage in continuous self-development to become better problem solvers
Realize Kaizen goals through policy deployment and audits	Make employees Kaizen-conscious through intensive training programs	Support small-group activities (such as quality circles) and the individual suggestion system	Enhance skills and job performance expertise with cross-education
Build systems, procedures, and structures conducive to Kaizen	Help employees develop skills and tools for problem solving	Introduce discipline in the workshop	

SOURCE: Reprinted with permission, copyright 1986 by Masaaki Imai, from the book *Kaizen*, New York: McGraw-Hill, p. 8.

ers. This flow-down moves operational decisions down to the operations level. The objective is to tap the workers' creative and leadership abilities, within clear-cut boundaries that encompass tactical, as opposed to strategic, decisions.

The Japan Human Relations Association [6] describes Kaizen Teian as a means to develop abilities, solve problems, and devise measures that deal with causes of problems. Three levels of activities are recognized: (1) perceptiveness, (2) idea development, and (3) decision/implementation/effect. Perceptiveness refers to the discovery of a problem. Idea development addresses creative thinking, applied to potential improvements. Decision/implementation/effect refers to converting creative thinking, ideas, into action and results.

The Kaizen concept positions itself as an incremental gain concept as opposed to a concept focused on strategic innovation. Table 19.3 compares the Kaizen concept to the strategic innovation concept. This comparison is critical in positioning Kaizen (continuous improvement) in the workplace, relative to initiatives such as reengineering, described later in this chapter. Here, we see a long-term commitment to small tactical improvements that accumulate in effect over time to yield significant impact on the overall organization. We see a process focus rather than a product focus. We also see a general involvement element coupled to intensive work force training in basic tools, as described in Chapter 20.

The Kaizen concept is tied to the Deming plan-do-check (study)-act (PDCA) cycle. This cycle is typically held up as a structure for all continuous improvement efforts. Empowerment is a major component of continuous improvement. In many cases, work force preparation is necessary in order for empowerment to become effective. Preparation involves training in interpersonal communications, meetings, and team building skills.

Table 19.3   Features of Kaizen and innovation.

	Kaizen	Innovation
Effect	Long-term and long-lasting but undramatic	Short-term but dramatic
Pace	Small steps	Big steps
Time frame	Continuous and incremental	Intermittent and non-incremental
Change	Gradual and constant	Abrupt and volatile
Involvement	Everybody	Select few "champions"
Approach	Collectivism, group efforts, systems approach	Rugged individualism, individual ideas and efforts
Mode	Maintenance and improvement	Scrap and rebuild
Spark	Conventional know-how and state of the art	Technological breakthroughs, new inventions, new theories
Practical requirements	Requires little investment but great effort to maintain it	Requires large investment but little effort to maintain it
Effort orientation	People	Technology
Evaluation criteria	Process and efforts for better results	Results for profits
Advantage	Works well in slow-growth economy	Better suited to fast-growth economy

SOURCE: Reprinted with permission, copyright 1986 by Masaaki Imai, from the book *Kaizen*, New York: McGraw-Hill, p. 24.

NOTES

Continuous improvement is a process-based initiative, compatible with a process-based organization. It is most relevant in terms of process improvement and process control, as described in Sections 5 and 4, respectively. The continuous improvement initiative is generally compatible with and complementary to the total quality management (TQM) initiative described later in this chapter.

19.5   CYCLE TIME/WASTE REDUCTION

PURPOSE

The purpose of cycle time/waste reduction is to lower the time and resources required for any and all activities within the confines of a production system, from the identification of customer needs to the receipt of customer payments.

DESCRIPTION

Cycle time/waste reduction is based on the concept that we have two types of activities in production systems: (1) value-added activities and (2) non-value-added activities. The objective in its simplest form is to eliminate non-value-added activities.

This concept can be addressed at several levels within an organization. For example, traditional industrial engineering takes a work analysis/task approach; see, e.g., Barnes [7]. Figure 19.2 overviews the task-based approach. Here, we see detailed methods applied to specific tasks. Through sequences of eliminate, combine, change sequence, and simplify, we can make substantial gains in time/waste reduction within specific tasks. In addition, we see standardization, training, and incentives in this approach.

On the other hand, Northey and Southway [8] describe the concept of business flow and its relationship to essential activities in an organizationwide approach. Figure 19.3 depicts an organizationwide view of typical push and pull systems. In the pull system, a just-in-time concept, we see what is termed a linear business flow process. The purpose of the linear flow is to avoid impediments involving both production processes as well as support services. For example, tool changeovers, rework, parts shortages, travel time, downtime, and so on all add unnecessarily to cycle time. Support services such as sales order changes, engineering changes, clerical delays, and so on add unnecessarily to business cycle time.

The essence of cycle time reduction/waste elimination is the evaluation of process activities, with the objective of eliminating non-value-added activities. Several steps are suggested in order to reduce cycle times:

1. Identify the process/activity to improve.
2. Establish a baseline cycle time/resource level.
3. Propose process improvements.

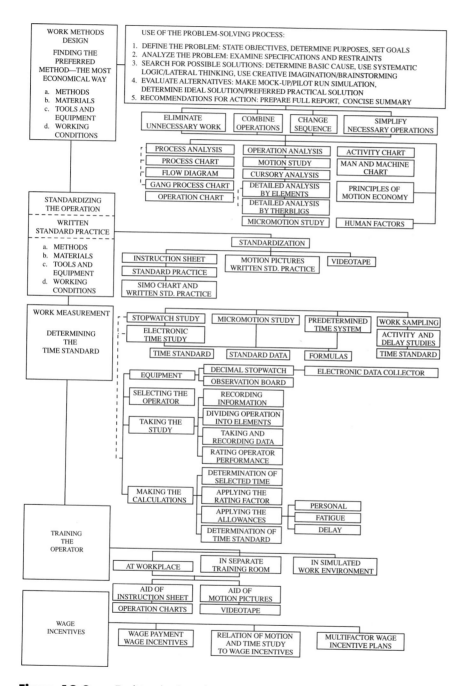

Figure 19.2 Traditional industrial engineering tools overview. (Reproduced, with permission, from Barnes, R. M., *Motion and Time Study: Design and Measurement of Work*, 7th ed., New York: Wiley, 1980, p. 9.)

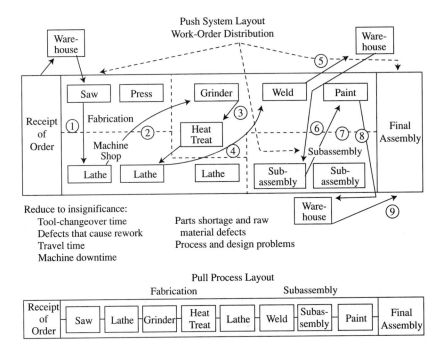

Figure 19.3   Pull versus push system layouts. (Reproduced, with permission, from Northey, P., and N. Southway, *Cycle Time Management: The Fast Track to Time-Based Productivity Improvement*, Portland, OR: Productivity Press, 1993, p. 36.)

4. Evaluate the proposed improvements.

5. Implement the selected improvement.

6. Assess the effectiveness of the implemented improvement.

The manner in which we carry out cycle time reduction initiatives has changed over the years. We have moved from using managers and specialists to perform all of the above steps, to the point that we now train and empower interdisciplinary teams to execute cycle time reduction initiatives.

NOTES

Experience has shown that cycle time/waste reduction can be achieved without sacrificing product quality. Classical industrial engineering has focused on best work methods for several decades; see Barnes [7]. Northey and Southway [8] claim that as many as 90% of existing activities are nonessential and can be eliminated. They point out that an internal process focus can be used to find these nonessential activities and ultimately shorten cycle times. Major benefits include savings in working capital requirements.

Cycle time reduction initiatives address timeliness and cost performance. They vary in labels from cycle time reduction to waste elimination. They represent a targeted focus, as opposed to a broad physical, economic, timeliness, and customer service focus. Hence, they are excellent initiatives to embed within broader initiatives, so that more comprehensive gains can be made. For example, cycle time reduction/waste elimination should be a part of process definition/redefinition, control, and especially improvement efforts. But, in and of themselves, these initiatives sometimes lose the perspectives of interrelationships to other processes/subprocesses—and may, as a result, solve one problem while unintentionally creating other problems.

19.6 FIFTH DISCIPLINE

PURPOSE

The purpose of the Fifth Discipline is to describe the concept of a "learning" organization and encourage transformations toward "learning" organizations.

DESCRIPTION

The Fifth Discipline (see Senge [9]) is a combination of systems thinking along with personal mastery, mental modeling, shared vision, and team learning. These five disciplines, combined together, constitute the essential elements for building the learning organization. The general goal is to create an organization that learns faster than competing organizations. This building process focuses on how we think and how we interact. Each of these five elements plays a critical role in the building process. Each element is described in terms of essence and practices.

The essences of the systems thinking element is holism and interconnectedness. Practices include system archetypes and simulation. Here, cause-effect circles are used to model interrelationships within the whole. Dynamic time concepts regarding changes in both the environments and organization are stressed, as opposed to the analytical concepts of point-in-time snapshots.

The essences of personal mastery include being, generativeness, and connectedness. Personal mastery is described in terms of high levels of proficiency—continually clarifying and deepening personal vision, focusing energy, developing patience, and viewing reality objectively. Practices are described in terms of clarifying personal vision, holding creative tension in terms of focusing on the result and seeing current reality, and making choices.

Mental models focus on ingrained assumptions, generalizations, and images that influence how we understand the world and how we take action. Their essences are the love of truth and openness. Practices include distinguishing data from abstractions based on data, in terms of noticing our jumps from observation to generalization, exposing the "left-hand column" in articulating what we normally do not say, balancing inquiry and advocacy, and testing assumptions.

Shared vision has two essences, commonality of purpose and partnerships. Shared vision develops shared pictures of the future, strong enough to foster genuine commitment and enrollment within the organization. Practices consist of the visioning process, of sharing personal visions, listening to others, and allowing freedom of choice, as well as the acknowledgment of current reality.

The essences of team learning include collective intelligence and alignment of individual energy. Here, the focus is on teams, versus the individual, as the fundamental learning unit within an organization. Dialog and discipline are stressed to create a free flow of meaning through the group and to provide developmental paths for acquiring skills and competencies, respectively. Practices include suspending assumptions, acting as colleagues, and effectively dealing with conflict and defensiveness, as well as gaining experience in team endeavors.

NOTES

The Fifth Discipline provides several basics for group/team work that are similar to, but different from, other initiatives, as well as a holistic perspective from which to view the production system.

19.7   FUNCTION-VALUE ANALYSIS

PURPOSE

The purpose of value or function analysis is to provide higher levels of physical performance in products/processes at reduced cost.

DESCRIPTION

Value is defined as

$$\text{Value} = \frac{\text{worth}}{\text{cost}}$$

Here, in general, worth is associated with customer benefits and cost is associated with customer burdens. Value analysis (VA) attempts to add value by simultaneously increasing the numerator and decreasing the denominator.

A variety of VA practices exists. Kolarik [10] breaks the practices into what are termed high-level and detailed-level analyses. At the high level, four basic inputs are considered: (1) customer needs, (2) product features, (3) cost estimates for product features, and (4) information on competing product features and costs. At this level of analysis, the primary objective is to relate product features or parts and their costs with functions expressed in the form of a verb followed by a noun, e.g., provide protection. Overall, VA attempts to serve customer needs by including essential features while excluding unessential features, thus reducing cost.

The detailed-level analyses are much more comprehensive than the high-level analyses. Detailed VA is usually described in three phases: (1) information and analysis, (2) creativity and synthesis, and (3) evaluation and development. Each phase is typically broken down in more detail; see Mudge [11] and Fowler [12]. VA is applied to simple devices as well as to complex systems. In either case, we see both technical and economic performance addressed, as well as an implementation path for improvements.

The concept of a function analysis system technique (FAST) is introduced in detailed VA. In the FAST, a customer perspective is taken in order to identify "functives," two-word terms consisting of a verb and a noun, e.g., inject plastic, filter dust. We develop a list of functives and stratify the list into primary functives and support functives. Once we produce this functive hierarchy, we associate or allocate costs to each functive.

At this point we have the equivalent to a high-level VA. Next, we add detail by assessing customer demand for each functive. Here, surveys and/or focus groups are used to obtain customer opinions. These opinions are applied against each functive in order to determine how critical the functive is to the customer. Critical functives are accorded high priority. Functives that are not identified as important to customers are subject to elimination. From critical and important functives, we develop a set of target functives for further analysis.

The creativity and synthesis phase encourages creative thinking regarding meeting the target functives. Creative thinking produces thoughts and ideas that are recorded for later synthesis. Through synthesis, creative thoughts are refined and scrutinized. In the evaluation and development phase, feasibility of the synthesized ideas is assessed. Feasibility is considered both in technical terms and in economic terms.

Notes

VA initiatives are typically associated with products rather than processes; however, the concept in its general form is applicable to processes; see Akiyama [13]. VA is most applicable to product/process improvement efforts, as opposed to process definition/redefinition and process control.

19.8   ISO 9000

Purpose

The purpose of ISO 9000 (ASQC/ISO/ASQC Q9000) is to provide an international standard for designing and implementing quality-related systems and activities in order to facilitate supplier-customer relationships.

Description

The ANSI/ASQC 9000 standard package [14] is composed of several components: (1) ANSI/ISO/ASQC A8402-1994 Quality Management and Quality Assurance—Vocabulary, (2) ANSI/ASQC Q9000-1-1994, Quality Management and Quality Assurance Standards—Guidelines for Selections and Use, and (3) ANSI/ISO/ASQC Q9001, Q9002, Q9003, and Q9004, quality management and quality assurance standards.

The vocabulary document, ANSI/ISO/ASQC A8402, establishes definitions for terms used in the actual standard documents. ANSI/ASQC Q9000-1 serves as a master applications guideline for selection and use of the quality management and quality assurance standards, ANSI/ISO/ASQC Q9001, Q9002, Q9003, and Q9004.

ANSI/ISO/ASQC Q9001 is applicable when conformance to specified requirements must be assured by a supplier to a purchaser during several operational stages. ANSI/ISO/ASQC Q9002 is applicable when the contractual arrangement calls for conformance to specified requirements during production and installation. ANSI/ISO/ASQC Q9003 is more limited and relevant for contractual agreements at the final inspection-test stage. Each standard is composed of a number of requirements. These requirement topics are presented in Table 19.4.

Table 19.4   ANSI/ISO/ASQC Q9001, Q9002, Q9003 quality systems requirements summary.

Requirement	Q9001	Q9002	Q9003
Management responsibility	++	++	+
Quality system	++	++	+
Contract review	++	++	++
Design control	++		
Document and data control	++	++	++
Purchasing	++	++	
Customer-supplied product	++	++	++
Product identification and traceability	++	++	+
Process control	++	++	
Inspection and testing	++	++	+
Control of inspection, measuring, and test equipment	++	++	++
Inspection and test status	++	++	++
Control of nonconforming product	++	++	+
Corrective and preventive action	++	++	+
Handling, storage, packaging, preservation, and delivery	++	++	++
Control of quality records	++	++	+
Internal quality audits	++	++	+
Training	++	++	+
Servicing	++	++	
Statistical techniques	++	++	+

++: Comprehensive requirement; +: less comprehensive requirement.

Here, we can see a difference in the level of detail in each of the three standards. For example, in ANSI/ISO/ASQC Q9001 we see the most detail because it applies to the situation where the supplier is involved in product design. For example, Q9001 applies to the situation where a supplier must demonstrate capability to design and supply product to a customer. Q9002 excludes the design requirement but encompasses production processes. For example, Q9002 applies to the situation where a supplier must demonstrate capability to produce, install, and service a product. On the other hand, Q9003 is relevant when the customer is not directly involved with either the design or production processes.

The ANSI/ISO/ASQC Q9004 series (Q9004-1, Q9004-2, Q9004-3, Q9004-4) focuses on quality management and quality system elements. It sets guidelines for internal quality management systems, regarding various categories of products as well as for quality improvement. The elements of ISO 9004 are summarized in Table 19.5. Here, we see a variety of foci, including issues specific to various types of products and quality improvement in general.

NOTES

The ANSI/ISO/ASQC 9000 standards have become a critical consideration for suppliers. They are compatible with process-based organization and encourage process-oriented activities.

Table 19.5   ISO 9004 quality management and quality system elements summary.

Guideline	Comments
Q9004-1	Presents broad quality system elements and discussions relative to the development and implementation of a comprehensive in-house quality system, focused on customer satisfaction.
Q9004-2	Extends Q9000 through Q9004-1 concepts to service activities, providing service-specific concepts and characteristics.
Q9004-3	Extends Q9000 through Q9004-1 concepts to processed materials, considering additional concepts and characteristics.
Q9004-4	Presents an overview of several quality improvement tools.

19.9   MISTAKEPROOFING (POKA-YOKE)

PURPOSE

The purpose of mistakeproofing is to proactively ensure that events will happen as planned, as well as to prevent unintended events from happening.

DESCRIPTION

Mistakeproofing as a formal strategy is not new, and it has been practiced for decades. However, several recent initiatives have been developed and introduced in both products and processes. We will focus on Shingo's manufacturing initiatives—general initiatives aimed at preventing human errors—and strategies used for mistakeproofing in general.

Mistakeproofing in manufacturing processes was introduced by Shingo [15] as a systematic initiative to eliminate defects. Shingo's mistakeproofing concept is a human- or machine-sensor-based series of 100% source inspections, self-checks, or successive checks to detect abnormalities when or as they occur and to correct them on the current unit of production as well as systemwide. Shingo's concept is practiced through what is termed the zero quality control system, which consists of four principles:

1. Use source inspection—the application of control functions at the stages where defects originate.

2. Always use 100% source inspections (rather than sampling inspections).

3. Minimize the time to carry out corrective action when abnormalities appear.

4. Set up poka-yoke (mistakeproofing) devices, such as sensors and monitors, according to product and process requirements.

Shingo pioneered the concept of developing physical and electromechanical devices for manufacturing, whereby mistakes and defects could be minimized, if not totally eliminated, to address zero defects. Shingo [15] developed what are termed warning class source inspection devices, as well as control class source inspection devices. These poka-yoke devices tend to utilize guides of various types, error detection alarms, limit switches, counters, and/or checklists; Shimbun [16].

Other mistakeproofing initiatives involve preventing human errors. According to Sanders and McCormick [17], many mistakes are ultimately the result of human errors: (1) errors of omission, or failure to act; (2) errors of commission, or incorrect action; (3) extraneous acts, or inappropriate action; (4) sequence errors, or an incorrect action sequence; and (5) timing errors, or improper timing of action. Other mistakes may be attributable to acts of nature and physical phenomena. Many mistakes result from a combination of human and physical factors. Norman [18] provides four general system design guidelines to minimize human errors:

1. System state (the current state of a system) feedback (to human operators) should be clear and available.

2. Different classes of actions should have different command sequences.

3. Actions should be reversible as far as possible, and high-consequence actions, which could lead to high-consequence mistakes, should be difficult to accomplish.

4. System command structures should be consistent, e.g., consistent operator response demands, control levers, and push buttons, throughout the system.

Many mistakes, errors, and failures charged to human product users/customers and nature are ultimately product/process definition- and design-related. Kolarik [10] addresses four primary strategies in mistakeproofing as described in Table 19.6: (1) elimination, (2) prevention, (3) detection, and (4) loss control.

The elimination strategy is the most desirable, provided an effective alternative technology or design can be discovered or developed. In this case, the "mistake" is removed or completely designed out of the product or process. For example, fiber-optic technology virtually eliminates electromagnetic interference or noise; a robot may remove a person from the possibility of a hazardous exposure. Caution must be exercised, however, in that new technology may introduce the opportunity for new and perhaps unexpected mistakes.

A wide variety of preventive opportunities may exist. Prevention avoids mistakes but does not eliminate mistake possibilities. Typically, these opportunities are related to design, retrofit, and

Table 19.6   Mistakeproofing strategies.

Strategy and Focus	Example
Elimination:	
Technology (selection/development)	Alternative materials, production processes
Design out (hazards/threats)	Mechanization, automation
Prevention:	
Design in (redundancy/barriers/safeguards)	Controls, shields, guards
Methods (operating procedures)	Training, operating displays, labels
Detection:	
Warning (predictive/corrective)	Warning displays, gauges, horns, lights, odors
Fail-safe (shutdown/minor damage)	Mechanical, electrical fuses, emergency shutdown
Loss control:	
Containment (system damage)	Personal protection and restraints, crush zones, contained releases
Isolation (system loss)	Single-system incident, harmless environmental releases
Catastrophe management (system and extended loss)	Multiple-system incident, harmful environmental releases

operating procedures and methods. For example, machine guards, access barriers, warning labels, product operating instructions and training, production methods training, and so forth help to prevent mistakes. Medicine and food tamperproof production processes and product packaging help to prevent unauthorized product access. Electrical box lockouts provide mistakeproofing for maintenance activities. Smaller gasoline filler holes and nozzle diameters were used for unleaded versus leaded gasoline, and so on.

Detection relies on both proactive and reactive strategies. Detection requires a sensor-based technology whereby critical measures are identified, monitored, and transformed into either advanced warning and corrective action or fail-safe shutdown action. The measurement, detection, and action may be of a closed-loop or open-loop configuration. Action taken may be based on predicted problems or problems in progress. Obviously, more opportunity for correction or recovery is afforded by early warnings.

Loss control is primarily a reactive strategy. It seeks to manage or limit losses. Loss control strategies are the last line of mistakeproofing. Seat belts and air bags in automobiles, dikes around tanks, levies along rivers, and so on provide examples of loss control, given that an unfortunate event or act of nature takes place. In some cases, especially in the case of acts of nature, such as floods, loss control may be the only strategy feasible. In many cases, loss control is used as a redundant strategy, in addition to other higher-order strategies.

NOTES

Effective product and process mistakeproofing strategies go a long way toward producing customer satisfaction or at least preventing customer dissatisfaction. Mistakeproofing is always an integral part of process control. As such it is always relevant to process definition/redefinition and improvement efforts in terms of their relationship to process control in operations.

19.10 QUALITY AWARDS

PURPOSE

The purpose of quality/performance-related awards is to encourage/recognize significant results in improved products/processes—with respect to the well-being of customers, employees, and stakeholders—as well as to promote sharing of strategies/actions that led to the results.

DESCRIPTION

Most quality/performance awards are offered to encourage higher levels of performance with regard to broad criteria, usually set out as a result of a perceived challenge of one form or another. Most quality-related awards in the United States are patterned after the Malcolm Baldrige National Quality Award [19]. This award was created by Public Law 100-107, enacted in 1987. The Baldrige Award has evolved over the past decade. Currently, it encompasses seven categories, using a point system of scoring, as described in Table 19.7. The criteria within each category are described in detail; see Malcolm Baldrige Quality Award Criteria for Performance Excellence [19].

Table 19.7   Malcolm Baldrige National Quality Award categories (1998)

1. Leadership		110
1.1 Leadership System	80	
1.2 Company Responsibility and Citizenship	30	
2. Strategic Planning		80
2.1 Strategy Development Process	40	
2.2 Company Strategy	40	
3. Customer and Market Focus		80
3.1 Customer and Market Knowledge	40	
3.2 Customer Satisfaction and Relationship Enhancement	40	
4. Information and Analysis		80
4.1 Selection and Use of Information and Data	25	
4.2 Selection and Use of Comparative Information and Data	15	
4.3 Analysis and Review of Company Performance	40	
5. Human Resource Focus		100
5.1 Work Systems	40	
5.2 Employee Education, Training, and Development	30	
5.3 Employee Well-Being and Satisfaction	30	
6. Process Management		100
6.1 Management of Product and Service Processes	60	
6.2 Management of Support Processes	20	
6.3 Management of Supplier and Partnering Processes	20	
7. Business Results		450
7.1 Customer Satisfaction Results	125	
7.2 Financial and Market Results	125	
7.3 Human Resource Results	50	
7.4 Supplier and Partner Results	25	
7.5 Company-Specific Results	125	
Total points		1000

All seven of the Baldrige Award Criteria are linked together to represent a system perspective in terms of strategy and structure. Figure 19.4 depicts the current Baldrige system perspective. Here, all seven categories are viewed, together, in terms of a customer- and market-focused concept. We see three distinct clusters of categories. First, on the left side, we see a strategic focus in terms of strategy and planning. Below, we see a linkage to information and analysis thereof. Then, on the right, we see the action/results component that is the product of the first two clusters.

Initially, applicants develop their applications in open-ended descriptions and make their submission. Applications are examined by several examiners, and detailed evaluation reports are developed. The award categories are examined individually and reviewed as a whole, and a composite score is provided. A consensus process is followed, with the most competitive applicants afforded a site visit. Final selection is made and announced by the president of the United States.

Each applicant is furnished a detailed set of comments regarding its performance as expressed in its application. Award winners are expected to share their philosophies, strategies, and experiences with others. Hence, a lasting effect of the award is obtained, both within the applicant's organization as well as in the community in general.

Strategy and Action Plans

Strategy and Action Plans are the set of customer and market focused company-level requirements, derived from short- and long-term strategic planning, that must be done well for the company's strategy to succeed. Strategy and Action Plans guide overall resource decisions and drive the alignment of measures for all work units to ensure customer satisfaction and market success.

System

The system is comprised of the six Baldrige Categories in the center of the figure that define the organization, its operations, and its results.

Leadership (Category 1), Strategic Planning (Category 2), and Customer and Market Focus (Category 3) represent the leadership triad. These Categories are placed together to emphasize the importance of a leadership focus on strategy and customers. Senior leaders must set company direction and seek future opportunities for the company. If the leadership is not focused on customers, the company as a whole will lack that focus.

Human Resource Focus (Category 5), Process Management (Category 6), and Business Results (Category 7) represent the results triad. A company's employees and its key processes accomplish the work of the organization that yields its business results.

All company actions point toward Business Results—a composite of customer, financial, and non-financial performance results, including human resource results and public responsibility.

The large arrow in the center of the framework links the leadership triad to the results triad, a linkage critical to company success. Furthermore, the arrow indicates the central relationship between Leadership (Category 1) and Business Results (Category 7). Leadership must keep its eyes on the business results and must learn from them to drive improvement.

Information and Analysis

Information and Analysis (Category 4) is critical to the effective management of the company and to a fact-based system for improving company performance and competitiveness. Information and analysis serve as a foundation for the performance management system.

BALDRIGE CRITERIA FOR PERFORMANCE EXCELLENCE FRAMEWORK
A Systems Perspective

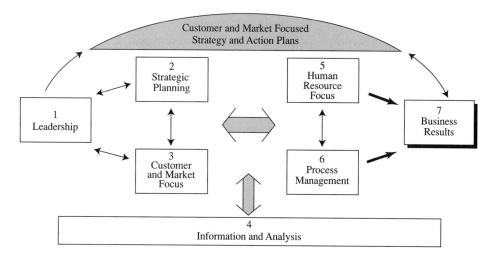

Figure 19.4   Malcolm Baldrige National Quality Award Criteria framework. (Reproduced from "Malcolm Baldrige National Quality Award—1998 Criteria," Washington, DC: U.S. Department of Commerce, 1997, p. 43.)

NOTES

The Baldrige Award encourages process thinking and process enhancement. For every applicant, thousands of Award Criteria booklets are distributed. Many organizations use these criteria for self-examinations and to aid in organizational enhancement on an internal basis. The Baldrige Award is totally compatible with a process-based organization structure.

19.11   QUALITY FUNCTION DEPLOYMENT

PURPOSE

The purpose of quality function deployment (QFD) is to ensure that customer needs, demands, and expectations are identified and integrated into product/process quality plans and requirements.

DESCRIPTION

QFD is a semigraphical aid in linking customer needs, demands, and expectations to product and production design that helps us in several ways:

1. To identify and prioritize true quality characteristics (expressed in customer language).
2. To identify and relate true quality characteristics to substitute quality characteristics (expressed in technical language).
3. To address competitive edge and product selling points.
4. To express high-level goals, objectives, targets, and requirements for both product and production design.
5. To identify technical and business-related bottlenecks as early as possible.

Quality function deployment represents a planning shell for systematic product and production definition, design, and development. Formats differ in scope, content, organization, and level of detail. Common ground exists in the context of matrix diagrams (see Chapter 20), whereby we assess relationships between customer needs, demands, and expectations, and technical characteristics, all in the context of our competitors' products and processes. Several matrices are developed, each cascading the level of details from general to specific; see Figure 19.5. Essentially, we move from customer requirements to product, to production, to production/quality control characteristics. This developmental cascade is systematic and tailored to our product and production needs.

Typically, the first matrix is developed around product planning, relating the customer's true quality characteristics to fundamental substitute quality characteristics or technical design goals, objectives, and/or requirements. Then, the development format tends to vary because of the nature of the product, production priorities, and bottlenecks identified. Hence, the standard format depicted in Figure 19.5 is a flexible guide rather than a fixed format. Figure 7.4 provides a simplified example of a product definition matrix. For details regarding QFD formats, see Akao [20], Day [21], and Kolarik [10].

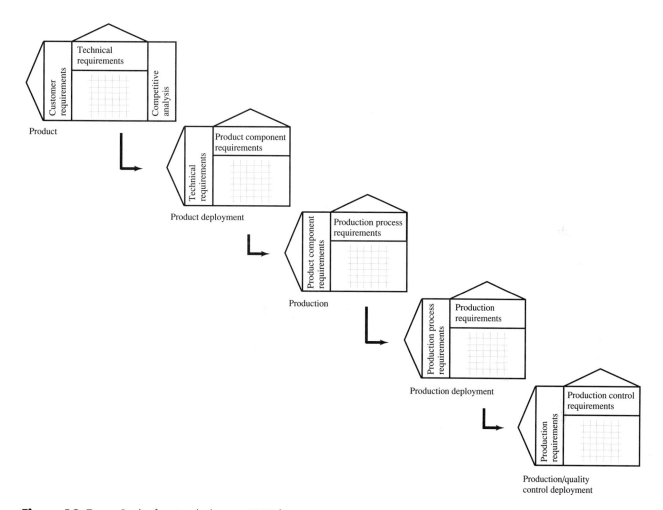

Figure 19.5 Quality function deployment (QFD) format.

QFD is similar to concurrent engineering in many respects, but generally is more flexible in applications to both goods and services. It is implemented through a multidisciplinary team, composed of people with expertise in marketing, sales, product design, process design, production, purchasing, and so on. Applications of its use include both simple and complex products and processes, e.g., shipbuilding, automobiles, airplanes.

NOTES

QFD is extremely useful in process definition or redefinition initiatives, where customer interests are involved. QFD is open-ended in the sense that we must exercise judgment in the extent and detail to which we expand our analysis.

19.12　REENGINEERING

PURPOSE

The purpose of reengineering is to fundamentally rethink and radically redesign business processes to achieve dramatic improvements in critical measures of performance such as cost, quality, service, and speed.

DESCRIPTION

Several forces have been recognized as driving the reengineering initiative: (1) customers, (2) competition, and (3) change; see Hammer and Champy [22]. Customers now have the upper hand in telling and demanding what they want, when they want it, how they want it, and what they will pay for it. Customers demand that their unique and particular needs be addressed by their suppliers. Competitors have taken niche marketing positions, whereby products are tailored for customers more so than in the past. One superior performer, e.g., a freewheeling start-up, from the standpoint of a producer, raises customer expectations and thereby intensifies customer demands. This loop of producer step-outs and elevated customer expectations creates a spiral that places yesterday's exceptional performances, once very competitive, at a disadvantage today.

Business process reengineering is based on radical and dramatic changes or shifts in business process thinking. The objective is to start over, with the expectation of producing a quantum leap in performance. This initiative is process-based, rather than task- or departmental-based, requiring a new process definition/design within an organization. Typically, the reengineered business process is structured around a new technology of some type, e.g., new information technology or new communication technology.

Hammer and Champy [22] cite several commonalties and recurring themes or characteristics that are encountered in reengineering initiatives:

1. Several jobs are combined into one, yielding compressed responsibility.
2. Worker-made decisions, yielding vertical compression in the organization.
3. The steps in the process are performed in a natural order.
4. Processes have multiple versions—tailored to needs, as opposed to one process for several needs.
5. Work is performed where it makes the most sense—work is shifted upstream or downstream in the process.
6. Checks and controls are reduced according to value added.
7. Reconciliation is minimized—external contact points are reduced.
8. A case manager provides a single point of contact; empowered case managers settle customer inquiries and complaints.
9. Hybrid centralized/decentralized operations are prevalent; advantages of both centralization and decentralization are combined in the same process.

According to Hammer and Champy [22], successful reengineering leads to several common results. Results from reengineering include work unit changes from functional departments to process teams. Job focus changes from simple tasks to multidimensional work. People are empowered, rather than controlled. Job preparation changes from training to education. The focus for performance measures changes from activity to results. Advancement criteria change from past performance to ability. Values change from protective to productive. Organizational structure changes from hierarchical to flat. Executives change from scorekeepers to leaders.

NOTES

Hammer and Champy claim that from 50 to 70% of redefinition efforts do not produce the dramatic results expected. They cite many common errors in reengineering:

1. Attempts to fix a process instead of changing it.
2. A lack of focus on business processes.
3. Ignoring everything except process design—everything associated with the process must be refashioned in order to support the business system.
4. A failure to consider people's values and beliefs.
5. A willingness to settle for minor results.
6. Giving up too soon.
7. Placing prior constraints on the definition of the problem and the scope of the reengineering effort.
8. Allowing existing corporate cultures and management attitudes to prevent reengineering from getting started.
9. Attempting to make reengineering happen from the bottom up, rather than from the top down.
10. Assigning someone who doesn't understand reengineering to lead the effort.
11. Limiting the resources devoted to reengineering.
12. Burying reengineering in the middle of the corporate agenda.
13. Dissipating energy across a great many reengineering projects.
14. Attempting to reengineer when the CEO is 2 years from retirement.
15. Failing to distinguish reengineering from other business improvement programs.
16. Concentrating exclusively on design.
17. Trying to make reengineering happen without making anybody unhappy.
18. Pulling back when people resist making reengineering changes.
19. Dragging the effort out longer than necessary.

Reengineering is a process-based initiative of a radical nature. It is similar to our concept of process definition/redefinition; but it is executed in an isolated context. We propose that an organization view all three process options—definition/redefinition, control, and improvement—together as a whole, rather than as disjoint parts; see Sections 3, 4, and 5.

19.13   ROBUST DESIGN

PURPOSE

The purpose of robust design is to improve the performance of a product by minimizing the effect of the causes of variation, without eliminating the causes.

DESCRIPTION

Taguchi [23] defines the concept of quality as follows:

> Quality is the loss a product causes to society after being shipped, other than any losses caused by its intrinsic functions.

This definition implies that quality can be approached from a standpoint of minimizing losses, as opposed to the traditional approach of maximizing benefits. Taguchi's definition of quality and the concept of loss minimization lead to robust design. Robust design consists of a conceptual framework of three design levels: (1) system design, (2) parameter design, and (3) tolerance design. Within these design levels, quantitative methods based on loss functions are developed.

System Design Level   System design focuses on relevant product or process technologies and approaches. System-level issues may include the total system view of the way the customer views and uses the product, or the producer intends to build the product, or both. It may also reach within the product to include technical issues in subsystems, components, materials, production process technologies, assembly strategies, maintainability, and so forth. For example, in order to build a shaft for a golf club, we might select a composite material, rather than steel, in order to reduce shaft weight and provide a variety of "stiffness" options for our customers. This product choice is made at a high level. It then dictates high-level production process design alternatives. For example, the process technology to produce composite golf club shafts is distinctly different from that required to produce steel shafts.

Parameter Design Level   Parameter design is a secondary design level, within or below the system design level. Parameter design focuses on determining a "best" level, or target, for the design parameters identified and selected at the system design level. The point is to meet the performance target with the least expensive materials and processes and to produce a robust product—one that is on target and insensitive to variation.

Parameter design can be thought of as a process of optimizing the functional design with respect to both performance and cost. Taguchi's approach to parameter design concentrates on designed experiments and specialized signal-to-noise ratio (SNR) measures. Whether or not we use Taguchi's SNR measures or other means, a sound and efficient experimental design program is mandatory to elicit an effective parameter design. Systematic experimentation, valid sampling techniques, and valid analyses, involving both location and dispersion measures, are required. For example, at the parameter design level, we determine the best parameter levels or targets—specific materials, temperatures, pressures, additives, and so forth—that take the least amount of time and create the least expense. The goal is to gain product-process performance, economic, and timeliness advantages in the marketplace.

Tolerance Design Level   Tolerance design is the final step, where the product/process tolerances or specifications are set. The parameter design step sets the targets for the process parameters. The tolerance design step is a logical extension of parameter design to the point of a complete specification or requirement, whereby we can judge a single item as conforming or not conforming to specifications. In general, narrow tolerances should be given only to parameters where variation will create critical performance problems. The point is to recognize that production costs and time demands escalate at nonlinear rates, as tolerances are tightened. For example, a tight hole tolerance in the hosel in the golf club head might call for a reaming operation in addition to a drilling operation, whereas the reaming operation may be eliminated if performance can be assured with a relaxed tolerance, superior tooling, different materials, or a different processing method.

Robustness   Effective and efficient processes produce performance to target—smaller, bigger, nominal—with minimal variation or dispersion. Phadke [24] describes the Taguchi parameter design philosophy (labeled robust design here) as an experiment-based technique:

> The fundamental principle of robust design is to improve the quality of a product by minimizing the effect of the causes of variation without eliminating the causes. This is achieved by optimizing the product and process designs to make the performance minimally sensitive to the various causes of variation, a process called "parameter design."
>
> Robust design uses many concepts from statistical experimental design and adds a new dimension to it by explicitly addressing two major concerns faced by all product and process designers:
> *a.* How to reduce economically the variation of a product's function in the customer's environment.
> *b.* How to ensure that decisions found optimum during laboratory experiments will prove to be so in manufacturing and in customer environments.

NOTES

The concept of robust design is virtually unchallenged for its ability to guide design initiatives. These concepts are universally useful in process definition/redefinition, control, and improvement regarding all products. However, the quantification of the concepts, using the Taguchi loss functions, has been challenged in terms of its effectiveness and efficiency. Both proponents and antagonists exist; see Taguchi [23, 25], Phadke [24], Ross [26], and Roy [27], for proponents, and Nair et al. [28], Montgomery [29], and Pignatiello and Ramberg [30] for antagonists. The key to the argument is to assess and understand the underlying model assumptions, and then, on a case-by-case basis, weigh these assumptions against the physical process characteristics and the corresponding process purpose.

19.14   SIX SIGMA

PURPOSE

The purpose of 6 sigma is to systematically improve process performance, to the point of producing 3.4 of fewer defects per million opportunities.

DESCRIPTION

Many high-tech production processes expose their product to literally thousands of opportunities for defects. With such a broad exposure, final process yields can be very low when $\pm$ 3 sigma, i.e., the natural tolerance, quality levels exist. Figure 19.6 depicts statistical performance levels, in terms of defective parts per million, for various processes, including a 6-sigma process. Here, we can see that plus or minus 3-sigma processes produce 2,700 and 66,810 defective parts per million, under unshifted/centered conditions and plus or minus 1.5-sigma shifted conditions, respectively. In summary, classical quality expectations are not adequate for complex processes/ products.

Motorola [31] developed the 6-sigma initiative as an outgrowth of its manufacturing experience. The initiative helped the company win the Malcolm Baldrige National Quality Award in 1988. The 6-sigma initiative is characterized by a general process improvement protocol consisting of six steps:

1. Identify your product.
2. Identify your customer.
3. Identify your needs in producing your product for your customer.
4. Define your processes.
5. Mistakeproof your processes and eliminate waste.
6. Continuously improve your processes.

In the manufacturing environment, the six general steps are refined and stated in a more explicit fashion:

1. Identify product characteristics that will satisfy your customer.
2. Classify the characteristics as to criticality.
3. Determine if the classified characteristics are controlled by part and/or process.
4. Determine the maximum allowable tolerance for each classified characteristic.
5. Determine the process variation for each classified characteristic.
6. Change the design of product, process, or both to achieve a 6-sigma process capability, $C_p \geq 2.0$.

The concept of process capability, C_p, is described in Chapter 11.

NOTES

The 6-sigma initiative is used much like a zero-defects initiative. Both focus on every product/process meeting every product/process requirement, all of the time. The 6-sigma initiative shares many common attributes with the continuous improvement initiative, but differs in its quantitative target parameter. This difference focuses on product/process metrics and emphasizes a correspondence between the design requirements, as expressed through substitute quality characteristics, and production results, as expressed through location and dispersion metrics regarding any one specific characteristic.

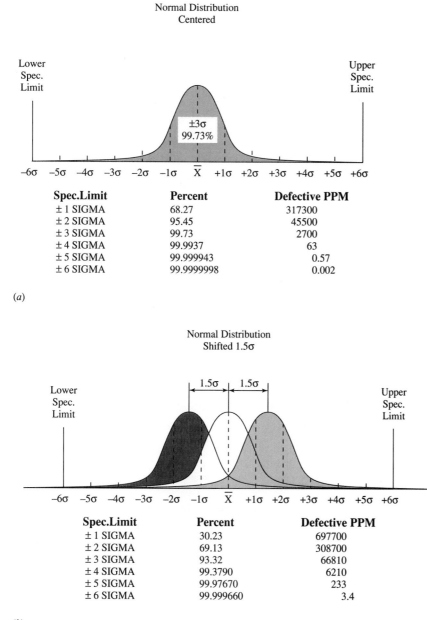

Normal Distribution
Centered

Spec.Limit	Percent	Defective PPM
± 1 SIGMA	68.27	317300
± 2 SIGMA	95.45	45500
± 3 SIGMA	99.73	2700
± 4 SIGMA	99.9937	63
± 5 SIGMA	99.999943	0.57
± 6 SIGMA	99.9999998	0.002

(a)

Normal Distribution
Shifted 1.5σ

Spec.Limit	Percent	Defective PPM
± 1 SIGMA	30.23	697700
± 2 SIGMA	69.13	308700
± 3 SIGMA	93.32	66810
± 4 SIGMA	99.3790	6210
± 5 SIGMA	99.97670	233
± 6 SIGMA	99.999660	3.4

(b)

Figure 19.6   Six-sigma process concept. (a) True 6-sigma process. (b) six-sigma process with a ±1.5-sigma shift. (Reproduced with permission from "The Motorola Guide to Statistical Process Control," Phoenix, AZ: Motorola Semiconductor Products Sector, 1989, p. 5.)

19.15 THEORY OF CONSTRAINTS

PURPOSE

The purpose of the Theory of Constraints (TOC) is to expose and address the limiting factor in a production system, such that the constraint's negative impact on production can be eliminated.

DESCRIPTION

Because of the complexity of production systems and traditional means of measuring perform-ance, constraints many times go unnoticed and untreated. Goldratt [32, 33] developed what he terms the Theory of Constraints. This initiative focuses on locating constraints and breaking them down so that they are no longer the constraining factor.

The primary premise in TOC is that we can readily identify our overall goal, e.g. profits. This goal then drives the decisions we make and everything we do as a result of the decision. Dettmer [34] summarizes the basic principles underlying TOC. These principles are listed in Table 19.8.

The TOC principles are practiced through a sequence of tree or relationship diagrams. The five logical tree tools are described in Table 19.9. Current reality trees are used to describe and model the present situation. Here, we map out the cause-effect relationships that exist and move

Table 19.8   Summarized TOC principles.

Principles
Systems thinking is preferable to analytical thinking.
An optimal system solution deteriorates over time as the system's environment changes.
The system optimum is not the sum of the local optima (of the parts).
Systems are analogous to chains—each has a weakest link.
Strengthening of any link in a chain, other than the weakest link, does nothing to improve the strength of the chain as a whole.
Knowing what to change requires an understanding of the system's current reality and its goal, and the difference between the two.
Most of the undesirable effects within a system are caused by a few core problems.
Core problems manifest themselves through undesirable effects (UDEs), the causes of which are never readily apparent.
Elimination of UDEs provides a false sense of security, while solution of a core problem simultaneously eliminates all resulting UDEs.
Core problems are usually perpetuated by a hidden or underlying conflict; resolution of a core problem typically requires challenging the assumptions underlying the conflict and invalidating at least one.
System constraints can be of a physical or a policy nature.
Inertia is the worst enemy of a process of ongoing improvement—solutions tend to assume a mass of their own that resists further change.
Ideas are not solutions.

Table 19.9 TOC tree structure.

Stage of Change	Tree Tool	Comments
What to change	Current reality tree	Relates/links UDEs to intermediate causes to root causes to core problems
What to change to	Conflict resolution diagram	Relates/links objectives to requirements to prerequisites
	Future reality tree	Relates/links desired effects to intermediate effects to injections, e.g., a potential/suggested action
How to cause the change	Prerequisite tree	Relates/links objective to intermediate objectives to obstacles
	Transition tree	Relates/links objective to intermediate effects to specific action

toward the discovery of the core problem. The conflict resolution diagram is used to resolve hidden conflicts that usually perpetuate chronic problems. The future reality tree serves two purposes. First, we use it to model and verify that we can break the constraint with our proposed action. And second, we use it to help identify any new undesirable effects (UDEs) that may develop as a result of our action. The prerequisite tree is used to help implement a potential action in that it maps out ways to overcome obstacles to implementation. The transition tree provides a detailed set of instructions for implementing a course of action.

NOTES

The TOC is compatible with general systems theory, as described in Chapter 2. It makes use of relationship diagrams, tailored to relate UDEs to core problems and their resolution. TOC is compatible with process-based organizations, especially in the redefinition and improvement elements.

19.16 TOTAL QUALITY MANAGEMENT

PURPOSE

The purpose of total quality management (total quality control) is to generate, coordinate, and sustain quality improvement efforts that focus on customers, continuous improvement, and involvement of workers and managers.

DESCRIPTION

Total quality management (TQM) and total quality control (TQC) begin with philosophies, as described by Deming [35], Juran [36, 37], Feigenbaum [38], Ishikawa [39], and others; see Kolarik [10] for details. These philosophies involve people, products, and processes in a systems context. Specifically, they relate to a participative work culture, where customers, suppliers, workers, managers, and functional specialists are viewed as an integrated whole with one mission, rather than separate entities, each with specific missions. Kolarik [10] describes several of these philosophies and divides them into two basic groups: (1) people-based and (2) production-based. Table 19.10 summarizes four of these philosophies. Additional TQM-related bases include ANSI/ISO/ASQC Q9000 [14] and quality awards, e.g., the Malcolm Baldrige National Quality Award Criteria [19].

In general, sound quality philosophies contain several common focal points; see Kolarik [10]:

1. Definition and delineation of purpose in the context of the organization.

2. Commitment to a common purpose.

3. Leadership through words, action, and role models.

4. Employee empowerment and creative license.

5. Organizationwide teamwork and communications.

6. Education and training to prepare people to lead and create.

7. Customer and supplier relationships—partnerships to define, produce, and deliver customer satisfaction.

8. Metrics and measurement to gauge progress.

9. Recognition for both efforts and results in the creation of quality.

TQM/TQC organizational implementation varies a great deal. Resulting quality systems are diverse, but share several common elements. First, implementation structures involve three basic organizational elements: (1) quality councils, i.e., high-level teams that deal with strategic direction and decisions regarding quality issues; (2) work-unit teams, i.e., teams that deal with day-to-day production activities; and (3) cross-functional teams, i.e., teams that are chartered in order to address improvement and problem issues.

Several fundamentals underlie TQM/TQC. Empowerment and teamwork are used to redistribute responsibility, authority, and decision making down to the operations and cross-functional team levels. Interpersonal communications are critical to allow for coordination between empowered work teams. Education and training in fundamental communications, meeting skills, and basic quality and improvement tools are part of TQM/TQC initiatives; see Kolarik [10] for details.

NOTES

TQM/TQC are very broad; they stretch from philosophy, to workplace culture, to organizational structure, to redistribution of responsibility, authority, and decision making. They impact people throughout the production system, including customers and suppliers. Hence, TQM/TQC are difficult to describe and even more difficult to implement. About one in three TQM/TQC initiatives is cited as being successful; see Brown et al. [40].

Table 19.10 Widely used and publicized total quality philosophies.

Deming	Ishikawa	Crosby	Juran
1. Create and publish to all employees a statement of the aims and purposes of the company or other organization. The management must demonstrate constantly its commitment to this statement.	1. Quality first—not short-term profits first.	Four quality absolutes:	1. Quality planning: Determine who the customers are. Determine the needs of the customers. Develop product features that respond to customers' needs. Develop processes able to produce the product features. Transfer the plans to the operating forces.
2. Learn the new philosophy, top management and everybody.	2. Consumer orientation—not producer orientation (think from the standpoint of the other party).	1. Quality is defined as conformance to requirements, not goodness or elegance.	
3. Understand the purpose of inspection, for improvement of processes and reduction of cost.	3. The next process is your customer—breaking down the barrier of sectionalism.	2. The system for causing quality is prevention, not appraisal.	
4. End the practice of awarding business on the basis of price tag alone.	4. Using facts and data to make presentations—utilization of statistical methods.	3. The performance standard must be zero defects, not "that's close enough."	2. Quality control: Evaluate actual product performance. Compare actual performance to product goals. Act on the differences.
5. Improve constantly and forever the system of production and service.	5. Respect for humanity as a management philosophy—full participatory management.	4. The measurement of quality is the price of nonconformance, not indexes.	
6. Institute training.	6. Cross-functional management (by divisions and functions).	Fourteen steps:	3. Quality improvement: Establish the infrastructure. Identify the improvement projects. Establish project teams. Provide the teams with resources, training, and motivation to Diagnose the causes. Stimulate remedies. Establish controls to hold the gains.
7. Teach and institute leadership.		1. Management commitment.	
8. Drive out fear. Create trust. Create a climate for innovation.		2. Quality improvement team.	
9. Optimize toward the aims and purposes of the company the efforts of teams, groups, staff areas.		3. Quality measurement.	
10. Eliminate exhortations for the work force.		4. Cost of quality evaluation.	
11a. Eliminate numerical quotas for production. Instead, learn and institute methods for improvement.		5. Quality awareness.	
11b. Eliminate Management by Objective. Instead, learn the capabilities of processes, and how to improve them.		6. Corrective action.	
12. Remove barriers that rob people of pride of workmanship.		7. Ad hoc committee for the zero defects program.	
13. Encourage education and self-improvement for everyone.		8. Supervisor training.	
14. Take action to accomplish the transformation.		9. Zero defects day.	
		10. Goal setting.	
		11. Error-cause removal.	
		12. Recognition.	
		13. Quality councils.	
		14. Do it over again.	

SOURCE: Reproduced, with permission, from Kolarik, W. J., *Creating Quality: Concepts, Systems, Strategies, and Tools*, New York: McGraw-Hill, 1995, p. 773.

D ISCOVERY E XERCISES

19.1. Why is there a tendency for people within organizations to focus so tightly on initiatives that they lose sight of common purpose and see initiatives as the common purpose? Explain.

19.2. Select an initiative and study its essence—its purpose, its structure, its implementation, its advocates, its adversaries, and so on.

19.3. Select an initiative and demonstrate its features in the context of a problem or opportunity.

R EFERENCES

1. Camp, R. C., *Benchmarking*, Milwaukee: ASQC Press, 1989.

2. Camp, R. C., *Business Process Benchmarking*, Milwaukee: ASQC Press, 1995.

3. Watson, G. H., *Strategic Benchmarking*, New York: Wiley, 1993.

4. Hartley, J. R., *Concurrent Engineering*, Portland, OR: Productivity Press, 1992.

5. Imai, M., *Kaizen*, New York: McGraw-Hill, 1986.

6. *Kaizen Teian 2*, edited by Japan Human Relations Association, Portland, OR: Productivity Press, 1992.

7. Barnes, R. M., *Motion and Time Study: Design and Measurement of Work,* 7th ed., New York: Wiley, 1980.

8. Northey, P., and N. Southway, *Cycle Time Management: The Fast Track to Time-Based Productivity Improvement*, Portland, OR: Productivity Press, 1993.

9. Senge, P. M., *The Fifth Discipline*, New York: Currency Doubleday, 1994.

10. Kolarik, W. J., *Creating Quality: Concepts, Systems, Strategies, and Tools*, New York: McGraw-Hill, 1995.

11. Mudge, A. E., *Value Engineering: A Systematic Approach*, New York: McGraw-Hill, 1971.

12. Fowler, T. C., *Value Analysis in Design*, New York: Van Nostrand Reinhold, 1990.

13. Akiyama, K., *Function Analysis: Systematic Improvement of Quality and Performance*, Portland, OR: Productivity Press, 1991.

14. ANSI/ISO/ASQC A8402-1994, Q9000-1-1994, Q9001-1994, Q9002-1994, Q9003-1994, Q9004-1-1994, Q9004-2-1991, Q9004-3-1993, Q9004-4-1993, Milwaukee: American Society for Quality.

15. Shingo, S., *Zero Quality Control: Source Inspection and the Poka-Yoke System*, Portland, OR: Productivity Press, 1985.

16. Shimbun, N. K., ed., *Poka-Yoke: Improving Product Quality by Preventing Defects*, Portland, OR: Productivity Press, 1988.

17. Sanders, M. S., and E. J. McCormick, *Human Factors in Engineering and Design*, New York: McGraw-Hill, 1987.

18. Norman, D. A., "Steps Toward a Cognitive Engineering: Design Rules Based on Analysis of Human Error," from "Five Papers on Human Machine Interaction," ONR-8205 (AD-A116031, DTIC), Washington, DC: 1982.

19. "Malcolm Baldrige National Quality Award—1998 Award Criteria," Washington, DC: U.S. Department of Commerce, 1997.

20. Akao, Y., ed., *Quality Function Deployment, Integrating Customer Requirements into Product Design*, Portland, OR: Productivity Press, 1990.

21. Day, R. G., *Quality Function Deployment, Linking a Company with Its Customers*, Milwaukee: ASQC Press, 1993.

22. Hammer, M., and J. Champy, *Reengineering the Corporation*, New York: Harper Collins, 1993.

23. Taguchi, G., *Introduction to Quality Engineering: Designing Quality into Products and Processes*, New York: Kraus International Publications, 1986.

24. Phadke, M. S., *Quality Engineering Using Robust Design*, Englewood Cliffs, NJ: Prentice Hall, 1989.

25. Taguchi, G., *System of Experimental Design*, vols. 1 and 2, New York: Kraus International Publications, 1987.

26. Ross, P. J., *Taguchi Techniques for Quality Engineering*, New York: McGraw-Hill, 1988.

27. Roy, R., *A Primer on the Taguchi Method*, New York: Van Nostrand Reinhold, 1990.

28. Nair, V. N., et al., "Taguchi's Parameter Design: A Panel Discussion," *Technometrics*, vol. 34, no. 2, pp. 127–161, 1992.

29. Montgomery, D. C., *Design and Analysis of Experiments*, 4th ed., New York: Wiley, 1997.

30. Pignatiello, J. J., and J. S. Ramberg, "Top Ten Triumphs and Tragedies of Genichi Taguchi," *Quality Engineering*, vol. 4, no. 2, pp. 221–225, 1991–1992.

31. "The Motorola Guide to Statistical Process Control," Phoenix, AZ: Motorola Semiconductor Products Sector, 1989.

32. Goldratt, E. M., *The Goal*, 2nd ed., Great Barrington, MA: North River Press, 1992.

33. Goldratt, E. M., *It's Not Luck*, Great Barrington, MA: North River Press, 1994.

34. Dettmer, H. W., *Goldratt's Theory of Constraints: A System Approach to Continuous Improvement*, Milwaukee: ASQC Press, 1997.

35. Deming, W. E., *Out Of The Crisis*, Cambridge, MA: MIT Center for Advanced Engineering Studies, 1986.

36. Juran, J. M., *Juran on Leadership for Quality*, New York: Free Press, 1989.

37. Juran, J. M., *Juran on Quality by Design*, New York: Free Press, 1992.

38. Feigenbaum, A. V., *Total Quality Control*, 3rd ed., New York: McGraw-Hill, 1983.

39. Ishikawa, K., *What Is Total Quality Control? The Japanese Way*, Englewood Cliffs, NJ: Prentice-Hall, 1985.

40. Brown, M. G., K. E. Hitchcock, and M. L. Willard, *Why TQM Fails and What to Do About It*, New York: Irwin, 1994.

chapter

20

PROCESS-COMPATIBLE TOOLS

20.0 INQUIRY

1. What constitutes a quality-productivity analysis tool?
2. Why do we have process-based tools?
3. How do we select an appropriate tool?
4. What tools have proven their worth?

20.1 INTRODUCTION

In quality-productivity transformation work, we use tools to help us discover, understand, describe, organize, analyze, and communicate something about our physical, economic, or social environments to ourselves and others. We use these tools to help us make better decisions regarding competitive edge in our production systems. **Tools help us to be more systematic and thorough in our assessments.**

Over the years, people have experimented with many possible ways to discover, describe, organize, analyze, and communicate information. The net result of these experimental endeavors is a set of process-based tools. **We see tools that deal with concepts in a qualitative fashion, and tools that deal with facts and figures in a quantitative fashion.** Typically, we begin with qualitative concepts and evolve qualitative models. Eventually, we move toward quantitative models and analyses. Usually, quantitative models require more simplifying assumptions than do qualitative models. Sometimes no realistic quantitative model exists, mostly because of limitations in quantitative methods and a lack of understanding of physical, economic, or social cause-effect relationships.

If we can find an appropriate tool, off the shelf, to help us accomplish our goals and objectives, then there is no need to invent a new tool. However, if we cannot find an appropriate tool, then we tend to press ahead and develop or evolve a "new" tool of some sort. In order to gain familiarity with current tools, we provide brief descriptions of several useful tools.

20.2 ACTIVITY/SEQUENCE LIST

PURPOSE

The purpose of an activity/sequence list is to identify and organize a set of activities as to sequence and duration.

DESCRIPTION

In general, each activity involved with an endeavor will be sequential, parallel, or coupled to other activities. Sequential activities require that a predecessor activity be completed before its successor can begin. Parallel activities can be undertaken and executed simultaneously. Coupled activities are executed together, and hence their progression is linked together in some manner. The activity/sequence list addresses these relationships. First, each activity in the list is uniquely identified. Then, the sequence as to predecessor and successor activities is established. Finally, we estimate a duration for each activity. Example activity/sequence lists appear in Table 8.11 and Table 17.11.

NOTES

An activity list is an indispensable tool in project scheduling. It allows us to organize a complex project in workable chunks, i.e., activities. Activity/sequence lists provide input to primary scheduling tools; see the discussions of the critical path method network (CPM) and the Gantt chart tools later in this chapter (20.10 and 20.16).

20.3 BREAK-EVEN ANALYSIS

PURPOSE

The purpose of break-even analysis is to estimate the number of product units we must sell and total dollars we must generate in order for total revenues to equal total costs.

DESCRIPTION

Break-even analysis is usually performed under the assumption of linear relationships in both costs and revenues, setting costs equal to revenues. Essentially, we develop a total cost function as the sum of fixed costs and variable costs. Fixed costs represent the costs that do not tend to vary with production level, e.g., facilities and equipment. Variable costs are costs that tend to vary in direct proportion to product volume, e.g., materials, supplies, and energy. Total revenue is

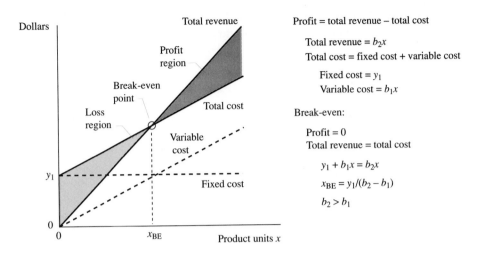

Figure 20.1   Break-even analysis graph and numeric solution.

expressed in terms of volume times dollars per unit sold, a simple linear relationship. We set our total cost function of fixed plus variable costs equal to our total revenue function and solve for the unknown product volume. Figure 20.1 illustrates the cost/revenue nature of a simple linear break-even analysis.

NOTES

Break-even analysis, although perhaps an oversimplification of the financial nature of a production system, allows a rough, first-cut approximation of the sales volume we must generate in order to yield financial success. The slopes of the total cost and revenue curves also provide an approximation as to the sensitivity of profit (or loss) to volume.

20.4   CAPABILITY ANALYSIS

PURPOSE

The purpose of a capability analysis is to link product design–related specifications, which are based on our customers' needs, to process-related results, which are related to our production processes.

DESCRIPTION

Process capability analyses provide a quantitative means of relating what we want to happen in our process with our abilities to make it happen. We use what are termed process capability indices to quantify capability. The two most popular indices are (1) the C_p index, an inherent or potential measure of capability, and (2) the C_{pk} index, a realized or actual measure of capability. See Chapter 11 (11.10, Process Capability) for details.

NOTES

Customer demands always drive our processes. We use capability analyses to characterize our process/production in light of what customers demand, as expressed through substitute quality characteristics.

20.5 CASH-FLOW ANALYSIS

PURPOSE

The purpose of a cash-flow analysis is to build a meaningful economic model of our process/subprocess so that we can study/quantify economic aspects relevant to the decisions at hand and/or project economic activity into the future.

DESCRIPTION

Cash-flow analyses are based on estimated cash flows, both inflows and outflows. First, we develop the scope of our analysis, i.e., define relevant economic categories of interest for our analysis. Then, we determine a time horizon for our analysis. Next, we estimate our inflows and outflows in each time period. Traditionally, we use an end-of-period basis, e.g., quarterly, annually, for our cash flows. At this point we have a cash-flow profile, based on estimates.

Two basic options exist for modeling time in our cash-flow analyses. First, we can use the payback method where the time value of money is ignored. Here, we simply accumulate our cash flows and note the point where the cumulated cash flow crosses from negative to positive. This point is, by definition, the payback or payout period. This method will not work for pure cost analyses or situations where we fail to recover our costs over the time horizon of interest.

Second, we can take the time value of money, i.e., compounding, into consideration in our model. Here, we can use methods such as the net present value (NPV), equivalent annual value (EAV), or internal rate of return (IRR). In the NPV method we assume a minimum attractive rate of return (MARR) in order to discount our future cash flows. The MARR represents a benchmark interest or earnings potential we use to judge the merit of our alternative. We assume our capital resources can earn the MARR, and hence justify it as a benchmark.

The net present value is calculated as

$$PV_n = \text{(total cash flow entry for period } n) [(1 + i)^{-n}] \qquad \textbf{[20.1]}$$

$$NPV = \sum_{\text{all } n} PV_n \qquad \textbf{[20.2]}$$

where PV_n = present value associated with the cash flow from time period n

NPV = net present value at the present time, sometimes called year zero

i = minimum attractive rate of return (MARR)

n = time period index

If our time horizons differ, we can develop an equivalent annual value (EAV), whereby we take our net present value and spread it over our time horizon, starting at the end of year 1, using our MARR value.

$$EAV = NPV \left[\frac{i(1 + i)^n}{(1 + i)^n - 1} \right]$$

[20.3]

where EAV = equivalent annual value, i.e., at the end of each year

NPV = net present value at the present time, sometimes called year zero

i = MARR

n = time period index

The internal rate of return (IRR) is defined as the interest rate at which our NPV or EAV is equal to zero. Hence, we determine the IRR by searching over the interest rate i in our NPV or EAV model. The IRR method will not work on a pure cost project, i.e., all costs and no revenues. Examples of cash-flow models appear in Chapters 8 and 17. Details regarding economic cash-flow models, including the incorporation of income tax effects, are addressed by Park [1] and other authors of engineering economy books.

NOTES

Cash-flow models are critical in establishing the economic merit of any production system regarding the process/subprocess decisions therein. We typically use the best estimates available at the time of analysis. Hence, our model reflects what we think will happen. We can run best-case and worst-case scenarios to assess the sensitivity of our alternatives to what we perceive to be optimistic and pessimistic estimates, respectively.

20.6 CAUSE-EFFECT DIAGRAM

PURPOSE

The purpose of a cause-effect diagram is to aid in discovering cause and effect by providing a systematic picture of effects and causes.

DESCRIPTION

A cause is a fundamental condition or stimulus that ultimately creates an effect or result of some type. Cause-effect analyses are essentially systematic inquiries into potential causes, given an effect of interest, or consequently a systematic inquiry as to potential effects resulting from given causes. Ishikawa [2] developed the concept of the cause-effect (C-E) diagram—also known as an Ishikawa diagram or a fishbone diagram—as one of the seven indispensable tools for quality improvement.

Causes

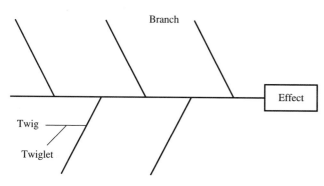

Figure 20.2 Cause-effect diagram format.

Cause-effect diagrams consist of an effect located on the right-hand side of the diagram and a series of causes stratified and structured along branches and twigs on the left-hand side. Figure 20.2 provides a basic structural layout for a C-E diagram. Development of a C-E diagram typically involves brainstorming potential causes relative to a given effect, and then stratifying or clustering them into categories. Major categories are represented by branches, while subcategories are represented by twigs. An example of a process analysis C-E diagram appears in Figure 8.5, while Figure 4.4 depicts a cause enumeration C-E diagram.

Three basic C-E diagram formats exist: (1) cause enumeration, (2) dispersion analysis, and (3) process analysis. The cause enumeration C-E diagram is the most common. The objective is to identify as many potential causes for a given effect as possible, without regard to the strength of any particular cause. The dispersion analysis C-E diagram looks like the enumeration C-E diagram; however, the process of development is more systematic in that we pursue lines of questioning associated with identified branches, in order to develop cause details. The process analysis C-E diagram tends to follow process and subprocess lines in its branch structure. See Ishikawa [3] and Kolarik [4] for more details on C-E diagrams and analyses.

NOTES

The cause-effect diagram is a basic tool that can and should be mastered by every member of an organization. Cause-effect analysis is fundamental to understanding processes. Cause-effect diagramming is a good starting point for process control and improvement exercises. Cause and effect diagrams typically break down to fairly specific variables—both controllable and uncontrollable—as we see on the left side of the process value chain; see Figures 3.5 and 8.8. Broad conceptual causes and effects are more easily depicted on a relations diagram. See the relations diagram discussion (20.21) later in this chapter.

20.7   CHECK SHEET

PURPOSE

The purpose of a check sheet is to classify, stratify, and tally observed data in the form of occurrences of events or outcomes, so that we can obtain a feel for relative frequency and dispersion of the events or outcomes.

DESCRIPTION

The check sheet is one of Ishikawa's seven indispensable quality improvement tools. Two basic types of check sheets exist, a tabular check sheet and a pictorial check sheet. A tabular check sheet contains several categories or strata associated with some classification of interest, e.g., time of day, category of defect, equipment unit. A pictorial check sheet provides a graphic cue as to a product/process profile that serves as the basis for tallying our observations. For example, we could use a process flow diagram and mark/tally subprocesses or processing units where defects were generated, allowing stacks of marks to develop over time. For products, a product pictorial might include an outline or profile of the product, with defects marked or tallied at the point of occurrence.

We assess our graphic as to where the highest concentration of defects occurs simply by looking for the highest frequency stacks. Figure 12.9 depicts a simple tabular check sheet used in conjunction with a Pareto analysis. Other examples of check sheets are described by Ishikawa [3] and Kolarik [4].

NOTES

Although the check sheet is a simple tool, it is an indispensable aid in helping to collect data regarding both products and processes. It is extremely useful in initial data collection and monitoring for process improvement efforts.

20.8   CONTROL CHART

PURPOSE

The purpose of a control chart is to monitor a process metric in order to expose the presence of special cause influences operating in the process.

DESCRIPTION

Process control charts make up a significant portion of what is termed statistical quality control/statistical process control (SQC/SPC). The process control chart is one of the seven fundamental or indispensable tools identified by Ishikawa [3]. Descriptions of both attributes-based and variables-based SPC are introduced in Chapter 9 and described in detail in Chapters 10 to 13.

NOTES

Process control charts add objectivity to what might otherwise be total subjectivity in process monitoring. They work extremely well when the underlying assumptions of normality and independence in our process data stream can be satisfied. They can be misleading and frustrating when these assumptions do not hold. See Chapters 9 and 13 for discussions regarding proper application.

20.9 CORRELATION/AUTOCORRELATION ANALYSIS

PURPOSE

The purpose of correlation and autocorrelation analyses is to quantitatively assess mathematical relationships between and within data sets, respectively.

DESCRIPTION

Correlation is used to assess the mathematical relationship between two variables, as expressed through empirical data. Autocorrelation is used to assess the relationship across time or order of production within a single data stream. Both correlation and autocorrelation analyses are described in the context of process characterization in Chapter 4. Walpole, Myers, and Myers [5] and other basic probability and statistics texts provide details regarding correlation in general. Autocorrelation is described by Newbold and Bos [6] and other authors of time series analysis texts.

NOTES

We utilize correlation and autocorrelation analyses in order to provide guidance as to appropriate control strategies. We typically use computer aids to calculate correlation coefficients as well as to produce autocorrelation correlograms.

20.10 CRITICAL PATH METHOD (CPM)

PURPOSE

The purpose of a CPM network is to organize and display a sequenced schedule of project activities/events with regard to starting and finishing estimates, both for the whole project and for individual activities.

DESCRIPTION

In order to develop a CPM network for a project, we first identify activities and events. An activity is something that requires action of some type over a time duration. An event happens at a point in time, e.g., the beginning or ending point of an activity. Our critical events represent milestones—points at which we reassess our progress. The activity/sequence list is a helpful tool to summarize activities, sequences, and time estimates.

A CPM network diagram is depicted in Figure 8.9. Here, we have taken the symbol, sequence, and duration information from an activity/sequence list in Tables 8.11 and 8.12. The network flows from left to right in a time sequence. Each activity is represented by a node, i.e., a circle. Within each circle we list the activity's symbol and its estimated time duration. Other information developed includes earliest start time (ES), earliest completion time (EC), latest start time (LS), and latest completion time (LC). These estimates are provided for each node/activity on the network.

The critical path is defined as the path that determines the minimum completion time for the entire project. The critical path is usually depicted by boldface arrows. If a delay occurs on any activity on the critical path, then the project duration will be increased. Hence, we watch the activities on the critical path very carefully with respect to time duration violations.

The ES and EC estimates are developed on a forward pass through the network of activities and durations. We develop the network using a start event and a finish event. We usually start at time zero and finish at the shortest time possible, considering our time and duration estimates. On the forward pass we begin at the start node and develop our ES_j estimates for each node. Usually, we assume the $ES_{\text{start node}}$ is equal to zero. However, we could assume some positive value. Then, we develop ES_j estimates for each activity as we move from left to right through the network. Each ES_j is equal to the maximum of the EC_i estimates taken from the set of all immediate predecessor activities. Each EC_j is estimated by summing its ES_j and its duration, t_j. The $EC_{\text{finish node}}$ is equal to the maximum of the EC_i estimates taken from the set of all immediate predecessor activities.

The LC and LS estimates are developed on a backward pass through the network of activities. Starting at the finish node, we set the $LC_{\text{finish node}}$ equal to the $EC_{\text{finish node}}$. We set the $LS_{\text{finish node}}$ equal to the $LC_{\text{finish node}}$. We estimate LC_j as the minimum of the LS_i estimates taken from the set of all immediate successor activities. Each LS_j is equal to its LC_j minus its activity duration t_j.

Tables, such as Tables 8.11 and 8.12, are constructed in order to both facilitate our network development and summarize our results. We usually repeat our activity descriptions, symbols, and durations. We list our ES, EC, LS, and LC estimates, which match those in our CPM network, e.g., Figure 8.9. Additionally, we include total slack (TS) and free slack (FS) estimates. In the CPM network method, we define total slack as the amount of time activity j may be delayed from its earliest starting time without delaying the latest completion time of the project.

$$TS_j = LC_j - EC_j = LS_j - ES_j \qquad \textbf{[20.4]}$$

Whenever the TS_j equals zero, we have a critical path activity.

Free slack is defined as the amount of time activity j may be delayed from its earliest starting time, without delaying the starting time of any of its immediate successor activities.

$$FS_j = \min [(ES_{i=1} - EC_j), (ES_{i=2} - EC_j), \ldots, (ES_{i=\text{last successor activity}} - EC_j)] \qquad \textbf{[20.5]}$$

where i corresponds to the index for all successor activities, $i = 1, 2, \ldots,$ last successor for activity j.

We can use updated CPM graphics and tables to update our plan as activities are completed. Additionally, we can project changes in subsequent activity estimates. Here, we use the same basic rules that we used to develop the initial CPM network, but begin at the end of the completed event. Hence, we can generate updated ES, EC, LS, and LC estimates for the remaining activities, and redevelop our slack estimates as well. We can also determine if our critical path has changed as a result of our changes.

NOTES

Implementation of large-scale process definition/redefinition, control, and improvement efforts usually involves an intricate set of activities and events. The CPM network scheduling tool is useful for organizing and managing projects—the more complex the project, the more aid CPM provides. Alternatives for less complex projects include Gantt charts, described later in this chapter (20.16). Additional details pertaining to project planning and implementation are available in project management texts such as Badiru and Pulat [7].

20.11 EXPERIMENTAL DESIGN

PURPOSE

The purpose of experimental design is to provide a systematic plan of investigation and analysis, based on established statistical principles, so that the interpretation of the observations can be defended as to technical relevance.

DESCRIPTION

In general, a designed or planned experiment is conducted using controlled physical conditions, along with a random ordering of treatments, i.e., factor levels or factor-level combinations. Planned experimentation consists of six basic steps:

1. Establish purpose.
2. Identify the variables.
3. Design the experiment.
4. Execute the experiment.
5. Analyze the results.
6. Interpret and communicate the analysis.

Most applications of experimental design are of a classical nature. An introduction to designed experiments is provided in Chapter 5. Chapter 6 introduces response surface modeling. A wide variety of texts explaining both fundamentals and advanced topics associated with experimental design is available; see Montgomery [8] and Box, Hunter, and Hunter [9]. Regression approaches are available in sources such as Neter, Wasserman, and Kutner [10], Draper and Smith [11], and Box and Draper [12]. Specialized approaches such as Taguchi experiments [also see the robust design discussion (19.13) in Chapter 19] are described by Taguchi [13] and [14], Phadke [15], Roy [16], and Ross [17]. Kolarik [4] provides an introduction to each of these approaches.

NOTES

The efficiency of designed experiments provides both timeliness and economic advantages in process exploration and characterization. The effectiveness of designed experiments allows us to learn more about the variables that underlie our processes, and thus supports process-related decisions.

20.12   FAILURE MODE AND EFFECTS ANALYSIS (FMEA)

PURPOSE

The purpose of failure mode and effects analysis is to aid in developing product and process action plans that address elimination, prevention, or at least mitigation of failures and the undesirable results they produce.

DESCRIPTION

FMEA consists of a qualitative assessment of a product or process regarding single-point failure—where a single failure in a system can cause system failure—and the unwanted consequences they might produce. The FMEA format consists of a columnar description of a product or process on either a functional or part-by-part basis. A basic FMEA consists of a set of eight columns:

1. Function/equipment/process identification.   Identify and list the specific functions or process/product parts to be analyzed. Break out one row in the FMEA for each.

2. Function/equipment/process purpose: Describe each identified and listed item (above) in enough detail to communicate its functional essence.

3. Interfaces: Identify other functions/equipment/processes/subprocesses which interface with the named function/equipment/process.

4. Failure mode.   Identify all potential failure modes, e.g., unwanted results, for the associated item, row by row.

5. Failure mechanism.   Identify all possible failure mechanisms associated with each item's failure modes. Here, a failure mechanism is a physical process that ultimately leads to the failure mode or unwanted result. Considerations for initiating failure mechanisms include environment, application, configuration, and operating method.

6. Failure detection.   Identify the means of failure detection. Detection means vary from early warning sensors and alarms, to observation of the failure mode, to observation of the unwanted failure effects themselves.

7. Failure compensation.   Identify any means available to compensate for the failure. Possibilities include redundancy, standby systems, and so on.

8. Failure effects.   Describe the expected immediate and long-term results that failure may produce. These effects include potential chains of consequences of the failure mode. For exam-

Table 20.1   Risk levels.

Hazard Categories	Consequences (People, Property, Environment)
Category I—catastrophic	Death, loss of facility/equipment/surroundings, severe environmental damage
Category II—critical	Severe injury/illness, major damage to facility/equipment/surroundings, major environmental damage
Category III—marginal	Minor injury/illness, minor damage to facility/equipment/surroundings, minor environmental damage
Category IV—negligible	No injury/illness, little or no damage to facility/equipment/surroundings, little or no environmental damage

Likelihood	Probability of Occurrence Range
A—likely	P(annual occurrence) $\geq 10^{-2}$
B—unlikely	$10^{-2} > P$(annual occurrence) $\geq 10^{-4}$
C—extremely unlikely	$10^{-4} > P$(annual occurrence) $\geq 10^{-6}$
D—rare	P(annual occurrence) $< 10^{-6}$

ple, the failure of a latch might lead to the opening of a door, which might lead to someone falling out of a moving vehicle, which might lead to an injury or death.

8. **Preventive measures.**   Produce a list of countermeasures available that may be pursued in order to counter the failure mode and its effects. This list of countermeasures should be detailed enough to serve as the basis for a prevention plan.

As an addition to the FMEA, we can develop a criticality analysis, and produce a failure mode and effects criticality analysis (FMECA). Here, we assess risk, where risk is a combination of both probability of occurrence and possible consequences. Table 20.1 describes several levels of both probability of occurrence and consequences. We can map these two dimensions into a risk matrix as shown in Figure 20.3. With this matrix, we associate the upper right-hand portion with probability/consequence combinations that merit careful scrutiny. Hence, the risk associated with any given item/row in the FMECA helps to determine the level of resources that should be applied to pursue failure mode prevention. The FMECA is typically an enhanced FMEA. The FMECA typically has one additional column, titled risk level, inserted between the failure effects and preventive measures columns of a FMEA. An example FMECA appears in Table 17.7. See Kolarik [4] for further details and discussions of FMEA and FMECA.

NOTES

FMEA is a fundamental tool, useful in improving reliability, maintainability, safety, and survivability of products and processes. It encourages systematic evaluation of a product or process, the postulation of single-point failures, possible effects, possible counteractions, and the documentation thereof.

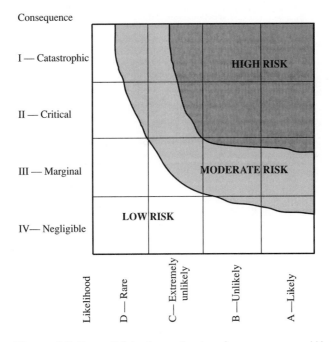

Figure 20.3   Risk levels as a function of consequences and likelihood.

20.13   FAULT TREE ANALYSIS

PURPOSE

The purpose of a fault tree analysis (FTA) is to depict logical pathways from sets of basic causal events to a single undesirable result or top event.

DESCRIPTION

The FTA is a logic tree that consists of a top event and every possible (conceivable) means, or combination of events, that could lead to the top event. We use symbols to link or depict relationships between events in the FTA. Figure 20.4 depicts several generic symbols used in an FTA. Several steps are involved in the development of the FTA:

1. Identify the top event. The top event is an undesirable event that we are motivated to prevent.
2. Identify the second-level events. Second-level events represent events that could lead to the top event.
3. Develop logical relationships between the top and next level events. Use logic gates, e.g., AND or OR gates, to connect the second-level events to the top event.
4. Identify and link lower-level events. Develop the logic tree down to the lowest level desired.
5. Quantify the FTA (optional). Develop estimates of probability occurrence for the events in the FTA and then develop a probability statement and estimate for the top event.

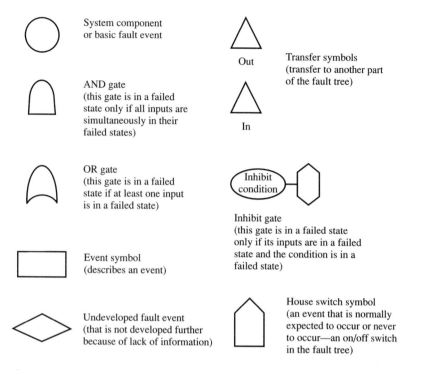

Figure 20.4 Selected fault tree analysis symbols and definitions.

A simplified FTA appears in Figure 17.4. A fault tree does not contain all possible failure modes or all possible fault events that could cause system failure. However, an FTA is capable of considering and modeling human error as well as hardware and software failures and acts of nature. It finds widespread usage in the fields of reliability, safety, and risk analysis. See Kolarik [4] for more details.

NOTES

The FTA is a more focused tool than the FMEA. FMEA is sometimes used to help determine the top event in an FTA. FTA works well for independent events, but common cause is difficult to model, especially in terms of quantification.

20.14 FLOWCHART

PURPOSE

The purpose of a flowchart is to provide an annotated graphical depiction of a process flow in terms of input, transformation, and output.

DESCRIPTION

A flowchart depicts process flow by using a sequence of symbols and words to represent process flow components, all connected with directional lines to indicate flow paths. A wide variety of processes are charted, and hence a wide variety of symbols are used. In some cases, simple box or rectangular symbols are used that are self-descriptive or annotated near the symbol. In other cases, the symbols are iconic in the sense that the symbol shape is indicative of the process element. Usually, a legend is provided to define specialized symbols. For example, Figure 20.5 depicts several specialized symbols used in detailed process flow mapping regarding production processes. Typically, the more focused the flowchart, the more specialized the symbols.

Operation. An operation consists of an activity that changes or transforms an input (material, supply, in-process product, and so forth). An operation may be performed by a person or a machine.

Transportation. A transportation consists of the physical movement of an input (material, supply, in-process product, and so forth) when the movement is not an integral part of an operation (e.g., the movement of a part held in a machine necessary to plane a surface is not a transportation, whereas the movement of a part from one machine to another is a transportation).

Delay. A delay is created when an input is waiting for the next planned activity. During a delay, no activity is taking place with regard to the input (material, supply, in-process product, and so forth).

Storage. A storage is created when a material, supply, in-process product, or finished product is placed somewhere such that authorization must be obtained to remove it from its location (e.g., placed in a stockoom or warehouse, etc.).

Source inspection. A source inspection is created through a self-check or a successive check of an input at the point of origination of possible defects (e.g., as activities are performed).

SPC charting point. Critical points in the production process are identified with an SPC charting point to indicate that a statistical process control chart is positioned to track and help control (monitor) the production quality of the input.

Sorting inspection. A sorting inspection is created when an inspection or test is performed for the purpose of examining input or final product before it is turned over to the next internal customer or the external customer. At the point of sorting inspection, the input or product is declared as either acceptable or unacceptable as judged against a product or process specification. The sorting inspection point is in essence a formal decision point.

Figure 20.5   Process flowchart symbols. (Reproduced, with permission, from Kolarik, W. J., *Creating Quality: Concepts, Systems, Strategies, and Tools*, New York: McGraw-Hill, 1995, p. 206.

We chart process flows to help us understand how processes actually work, or how they are expected to work. Process flow mapping usually involves several steps:

1. *Flowchart purpose.*   Initially, we clearly state the purpose for our charting efforts. This purpose will dictate the level of detail we need in our flowchart.

2. *Boundary definition.*   We determine the starting and ending points for the flowchart effort, relative to purpose and necessary observations.

3. *Process observation.*   Provided the process is in operation, direct process observation and experience are necessary to understand the process flow. We may also observe and chart processes in other organizations through benchmarking activities.

4. *Gross process flow.*   Initially, we develop and chart a process overview, depicting the production system or process in terms of major components, i.e., processes or subprocesses, respectively.

5. *Detailed process flow.*   Once we have obtained and captured the general essence of the process flow, we focus on details, cascading the level of detail down to the point that it is compatible with our purpose. Details are sequenced to represent the order and position that they occupy in the actual process.

6. *Validity and completeness checks.*   As we move from level to level in our flowcharting, we examine our chart for validity and completeness. Validity checks typically involve chart review as to accuracy of inputs, transformation, output, and sequence. Completeness refers to level of detail with the target process as well as interactions with other processes.

Process flowcharting is performed by teams and individuals—operators, technicians, engineers, specialists, and/or managers. Diverse perspectives are gained through process flowcharting when an interdisciplinary team is involved with the charting. Flowcharts are used throughout this text; specifically, see Chapters 7, 8, 16, and 17. See Kolarik [4] for general details, Barnes [18] for specialized charting techniques relative to classical industrial engineering, and Hughes [19] for automatic process control–related flowcharting basics.

NOTES

Flowcharts are universally applicable as an aid in understanding and modeling processes. The flowchart is the most widely used tool in process definition/redefinition and improvement and is an indispensable tool in process control.

20.15 FORCE FIELD ANALYSIS

PURPOSE

The purpose of force field analysis is to identify both driving and restraining forces associated with an activity.

DESCRIPTION

Force field analysis is based on Lewin's concept and principle of force fields, as applied to organizational activities. This principle states that you can move a problem by either increasing the intensity of the driving forces behind it or by reducing the restraining forces that resist the driving.

Force field analysis consists of developing two columns, one containing driving forces and the other containing restraining forces. Analogies are made to physical systems. For example, if we desire to remove a tree stump, we can increase the force against the stump to the point we can push it out directly, or we can dig a cellar behind the stump, removing dirt and cutting roots, to allow for smaller pushing forces to pry the stump out of the ground. In terms of organizational changes, we can use disincentives and penalties of several types to push a change ahead, or we can use incentives and rewards to reduce the amount of force required to allow the change to take hold. For an example force field analysis table, see Table 17.9.

NOTES

Force field analysis is very useful in implementation planning. It allows us to do a reality check on our plan regarding process definition/redefinition, control, and improvement. Our objective is to gain acceptance and support for the process plan and thereby help assure positive results for our customers and our organization as a whole.

20.16   GANTT CHART

PURPOSE

The purpose of a Gantt chart is to organize and display a sequenced schedule of project activities and events in a graphical format.

DESCRIPTION

A Gantt chart is essentially a horizontal bar chart, where the bars, representing activities, are arranged in sequential (chronological) order. We develop Gantt charts from activity and sequence lists. Usually, we place the earliest activities at the lower left-hand positions and the later activities in the upper right-hand positions. A Gantt chart is illustrated in Figure 17.6. Several variations of Gantt charts are discussed by Badiru and Pulat [7].

NOTES

Gantt charts predate CPM networks by several decades. They are simpler, but offer less information than CPM networks in terms of timeliness details regarding project management. However, their simplicity adds to their ability to rapidly communicate the basic sequence of activities. They are adequate for managing relatively simple sequences of project activities, and useful in depicting overall flows in any given project's implementation plan.

20.17 HISTOGRAM

PURPOSE

The purpose of a histogram is to provide a graphical depiction of both location and dispersion in a univariate data set.

DESCRIPTION

A histogram essentially depicts a data set by assigning each observation in the data set to one of several cells or predefined categories and then depicting associated cell counts and relative frequencies. Typically a minimum of 20 to 30 observations is necessary to produce a meaningful histogram. Histograms are constructed in several steps:

1. Determine the range of the data. Here we calculate the difference between the largest and smallest values.
2. Determine the number of cells or categories. Usually we develop from 5 to 15 cells.
3. Determine cell midpoints and boundaries. Cells of equal width are usually defined; however, cells can vary in width.
4. Place each observation in a cell. Each observation must fall in only one cell. A check sheet format is useful to classify each observation.
5. Display the cells. The frequency and/or relative frequency of each cell determine the cell's relative height in vertical histograms.

Computer aids are available to develop histograms. Modern spreadsheets as well as dedicated statistical analysis packages can be used to produce and display histograms. Other descriptive statistical analysis tools used to assess location and dispersion include stem and leaf plots and box plots; see Walpole, Myers, and Myers [5] for details.

NOTES

Histograms are fundamental to data exploration. They are universally useful in quantitative data analysis to assess location and dispersion. They are not capable of capturing the time sequence associated with data collection. If the time sequence is important, we use the runs chart tool, described later in the chapter (20.23).

20.18 MATRIX DIAGRAM

PURPOSE

The purpose of a matrix diagram is to help identify, organize, display, and communicate relationships between two or more sets of characteristics or factors.

DESCRIPTION

A matrix diagram usually relates two sets of characteristics or factors. The typical layout is a two-dimensional matrix with the vertical dimension used to lay out one set of factors and the horizontal dimension used to lay out the other set. We typically identify and document relationships within each set and between the two sets.

One application of the matrix diagram is to relate customer needs, demands, and expectations in the customer's language to technical characteristics of the product/process, expressed in the producer's language. Figure 7.4 illustrates this particular application of a matrix diagram. This illustration contains the within-set relationships in the triangular appendages at the left side and the top of the matrix. Here, we use + and − symbols to represent positive and negative relationships, respectively. We use other symbols such as the bull's-eye, open circle, and triangle to represent very strong, strong, and weak relationships between characteristics of the two sets, respectively. In this particular matrix diagram, we have included customer needs, demands, and expectations, technical definition characteristics, and competitor characteristics together.

Matrix diagrams differ in scope and detail, as well as layout format. See Mizuno [20] for details regarding the matrix diagram in general and Akao [21] and Day [22] for specific QFD applications in particular.

NOTES

The concept of a matrix diagram is simple; essentially we develop them to help us to relate sets of factors, usually in a qualitative fashion. The actual development of a matrix diagram is rather involved in terms of both level of detail and completeness. For example, customers talk in broad terms regarding their needs, demands, and expectations, while producers must focus on technical details in terms of product/process requirements and definition. In this example, two different languages are involved, and the translation from customer to technical language is not typically unique. The quantitative counterpart to matrix diagrams is multivariate statistical analysis.

20.19 PARETO ANALYSIS

PURPOSE

The purpose of a Pareto analysis is to systematically stratify and rank causes or results associated with past performance so as to help us visualize the maldistribution thereof.

DESCRIPTION

Vilfredo Pareto, a nineteenth-century Italian economist, observed that about 20 percent of the population controlled about 80 percent of the wealth in Italy. The result of this observation and its generalization leads to what is commonly referred to as the Pareto principle, also known as the 80-20 rule. The split may not be exactly 80-20; however, this principle of maldistribution is observed in all endeavors. For example, about 20 percent of our sales force produces about 80 percent of our sales. Or, about 20 percent of our equipment produces about 80 percent of our downtime.

The Pareto principle, applied to quality work, suggests that the majority of quality losses are maldistributed in such a way that a "vital few" quality defects or problems always constitute a high percentage of the total quality losses. The same argument can be put forth regarding productivity losses. Hence, we reason that if we can identify our primary loss points in terms of category and/or source, we can place improvement efforts at these points and enhance our performance accordingly. This concept is compatible with that of seeking process levers and leverage, as described in previous chapters. The Pareto analysis helps us to identify our leverage points.

Using the Pareto analysis, we stratify and rank process characteristics. Our characteristics fall into one of two categories: (1) results or (2) causes. Regardless of the category, our analysis procedure is the same:

1. Determine the general result or cause to be analyzed, an appropriate unit of measure, and a meaningful analysis period, e.g., a week, month, year.

2. Collect and stratify the associated data. A check sheet may serve as an appropriate tool to help categorize the data. Usually 5 to 10 categories, including an "all others" category, are sufficient.

3. Quantify the Pareto analysis data. Tally the total observations in each category and determine relative percentages.

4. Develop the Pareto diagram. Plot each category along the horizontal axis and its associated relative frequency on the vertical axis, along with a cumulative probability across the bars on the plot.

5. Interpret the Pareto diagram. Because of the 80-20 rule, we typically interpret the few categories that produce roughly 80 percent of the cumulative result or cause total.

Once our Pareto diagram is completed and we have identified our major contributors, we assess our process redefinition, improvement, and control options. In other words, we use our Pareto analysis to identify problems and opportunities for future efforts. An example Pareto analysis and diagram are illustrated in Figure 12.9. For more detailed discussion of Pareto analysis and its association with cause-effect analysis, see Kolarik [4] and Ishikawa [3].

NOTES

The Pareto analysis is simply a means to help us organize our inquiries into process effectiveness and efficiency. It is simple in concept as well as in application. However, in its simplicity, we gain an appreciation for systematic process assessment based on measurement—as opposed to guessing and trial and error—whereby we can objectively support process definition/redefinition, control, and improvement decisions.

20.20 PROCESS VALUE CHAIN ANALYSIS

PURPOSE

The purpose of a process value chain (PVC) analysis is to depict a sequence of cause-to-effect and effect-to-cause relationships between business results and outcomes and basic physical, economic, and social variables.

DESCRIPTION

PVC analysis links basic variables with business results so that value-added process sequences are clearly depicted. This linkage is not precise in the sense that each basic variable has its own natural/technical unit of measure, e.g., length, pressure, volume, composition, while business results are expressed in their own units or unitless ratios, e.g., production units, percent conformance, scrap rate, cost, revenue. Hence, PVCs have discontinuities, where unitary incompatibility presents gaps and challenges. The point is to link variables related to specific process decisions and process control points to business results and vice versa as best we can. Hence, understanding as to cause-effect and time lags in moving from cause to effect become more obvious for all concerned, e.g., operators, engineers, managers.

A generic PVC is depicted in Figure 3.5 with a partial example provided in Figure 8.8. Here, we see basic outputs on the right-hand side and basic inputs on the left-hand side. Processes and subprocesses are depicted in the middle. We develop a PVC working from one of several starting points; we may start somewhere in our outputs, e.g., business results, and work toward inputs, e.g., basic variables. Or, we may start somewhere in our inputs and work toward our outputs. The focus is to understand how our processes work and how they add value as they operate. We use the PVC to locate definition/redefinition, control, and improvement points and opportunities.

A working knowledge of processes and their subprocesses is essential for the development of a PVC. Typically, our process flow diagrams provide a good deal of information. Here, we see processes/subprocesses as input/transformation/output sequences. C-E analysis and diagrams are used to develop PVC details. The PVC is usually a pragmatic and summarized combination of flow diagrams and C-E branches, as illustrated in Figure 8.8.

The entire PVC spans all of our processes, extending back to both controllable and uncontrollable variables, and forward to general business results—producing a line of sight between the two. We are usually unable to develop precise numeric linkage models, but we do expect to describe basic causal relationships. In addition, we address time lags between cause stimulus and effect impact, in order to understand timeliness aspects in causal relationships.

NOTES

The PVC diagram connects the business world to the technical world through a logical, sequential linkage that cascades up and down all processes and their respective subprocesses. PVC diagrams are useful for operators to see how operational decisions in the technical world ultimately impact business results. They are useful for managers and leaders to clearly see that business targets are met through a sequence of operational decisions.

20.21 RELATIONS DIAGRAM

PURPOSE

The purpose of a relations (or relationship) diagram is to support qualitative thinking, discovery, and communication for basic, high-level, conceptual cause-effect thinking.

DESCRIPTION

The relations diagram is a graphical aid in basic problem-opportunity/cause-effect discovery and relationship determination and expression, which helps us to

1. Identify and relate basic causal factors.
2. Express basic causal sequences.
3. Introduce assertions and assess or project resulting effects.
4. Communicate critical relationships in terms of basic cause and effect.

A relations diagram typically contains open-ended statements that include observations and assertions. These diagrams typically contain facts and figures, as well as speculation as to relationships and results.

Relations diagramming is best addressed in a team environment, so as to capture a diversity of perspectives regarding both effects and causes. Typical starting points include effects, both undesirable as well as desirable. Logical development from these effects back to potential causes is common to most relations diagramming efforts. Clustering and sequencing of causes are common to all relations diagrams. Boxes, circles, ovals, loops, and directional arrows are used to depict cause-to-effect flows.

Since the effects are usually a complex result of some type, e.g., low business performance, and include physical, economic, and social variables, fundamental relationships are not usually obvious. This complexity makes relationship diagramming more speculative than traditional C-E diagramming discussed earlier. Figures 7.3 and 16.6 depict simple relations diagrams. See Mizuno [20] or Kolarik [4] for specific details.

Relationships underlie all systems and systems analysis. General systems theory uses the concept of relations diagrams; see Chapter 2 for details. The Theory of Constraints, described in Chapter 19, and the analyses therein, rely on relationslike diagrams. These diagrams are referred to as trees—current reality trees, future reality trees, prerequisite trees, transition trees; see Dettmer [23] for more details.

NOTES

Relationships and the understanding thereof are fundamental to process mastery. Relations diagrams are fundamental tools useful in graphically displaying and communicating relationships. The concept is simple: Provide a summarized description of what is happening or is likely to happen. The application is complex because many related causes and effects, as well as assertive impacts, are complex and not at all obvious. From relations depictions, we develop strategies and plans, and assess potential outcomes.

20.22 ROOT CAUSE ANALYSIS

PURPOSE

The purpose of root cause analysis is to provide a systematic means to determine the basic causes that underlie an observed problem/symptom.

DESCRIPTION

Root cause analysis uses disciplined and systematic thinking to logically work from a defined problem or symptom to apparent causes to root/basic causes. This systematic thinking sequence requires logical reasoning and some form of summary and depiction of this reasoning. A variety of basic, structured tools are available for root cause analysis:

1. Basic cause-effect analysis—an open-ended descriptive tool. See the discussion of cause-effect diagrams (20.6).

2. General logic tree diagrams—event-to-cause pathways. See the discussion of fault tree analysis (20.13). In addition to a fault tree, which is built from an undesirable event, a goal tree can be built from a desirable event, result, or outcome, using essentially the same methods used in fault tree construction. In goal tree analysis, we seek root causes that are associated with successful outcomes; see Kolarik [4] for details.

3. Basic relations analysis—conceptual cause-effect analysis. See the discussion of relations diagrams (20.21).

4. Other root cause analysis tools. A number of root cause analysis tools have been described over the years. Wilson, Dell, and Anderson [24] describe several classes of tools and provide brief descriptions.

NOTES

The determination of root causes is fundamental to process understanding, from both reactive and proactive viewpoints. Several of our previously described tools, as listed above, are useful in root cause analysis.

20.23   RUNS CHART

PURPOSE

The purpose of a runs chart or plot is to display a time- or order-sequenced data set so as to expose shifts, trends, or any other patterns that may be reflected by the data.

DESCRIPTION

A runs chart/plot typically charts a univariate metric across time or order of production. Several steps are involved:

1. Determine a metric of interest, an appropriate sensor, and a means of data collection. Record each response value and its associated time or order coordinate together.

2. Develop a set of axes. Typically the horizontal axis is used for time or order sequence and the vertical axis for the response value.

3. Plot the response points. Typically the responses are plotted across the axes and connected with a solid line.

4. Assess the runs chart or plot for patterns—shifts, trends, and so on.

Runs charts are supported by both computer-aided statistical analysis packages and spreadsheets. Several runs charts are displayed in Chapters 4, 9, and 10.

NOTES

The runs chart or plot is a fundamental tool for data exploration. It is a good first step in assessing time- or order-related stability or instability.

20.24 SCATTER DIAGRAM

PURPOSE

The purpose of a scatter diagram or plot is to provide a graphical display of the numerical relationship between two or more variables.

DESCRIPTION

A scatter diagram or plot graphically depicts numerical relationships between variables. The development of a scatter diagram or plot involves several steps:

1. Identify the variables of interest. Identify the metric, sensor, and data collection means.

2. Collect the data in multivariate sets. Each set represents one observation and contains one data point for each variable. For example, for a bivariate set, say X_1 and X_2, we collect data pairs.

3. Develop a set of axes. Each variable occupies one axis. Two variables require two axes; three variables require three axes.

4. Plot the data observations on the axes. Typically, the time order or sequence is not important or noted on the plot.

5. Interpret the plot. We assess relationships in terms of correlation, looking for increasing or decreasing patterns in the plots. For example, as in Figure 20.6, a positive correlation is seen when large values of one variable correspond to large values of another variable. Or, negative correlation occurs when large values of one variable correspond to small values of another variable. No correlation implies a scattered arrangement of points without any noticeable shape or direction.

Scatter diagrams provide a graphic picture of quantitative relationships. The correlation tool described previously provides a counterpart quantitative measure of numerical (linear) association. Patterns that are not of a linear nature, e.g., an arc of some type, are readily apparent on a scatter diagram or plot, whereas they are not picked up with a linear correlation metric. For further details regarding scatter diagrams see Kolarik [4] and Ishikawa [3].

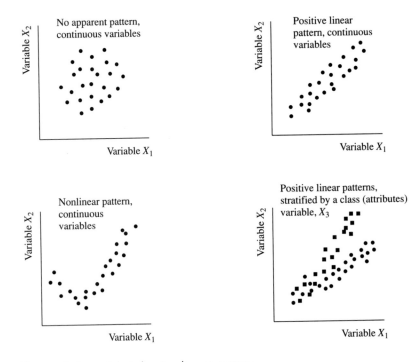

Figure 20.6   Selected scatter diagram patterns.

NOTES

Scatter diagrams and plots are one of the seven fundamental quality improvement tools described by Ishikawa. They are simple to develop, provided the data have been collected properly—in sets. They support data exploration prior to formal numerical modeling. Scatter diagrams or plots are supported by computer-aided statistical packages as well as spreadsheets.

20.25   STRATIFICATION ANALYSIS

PURPOSE

The purpose of stratification analysis is to divide a data set according to some criteria so that a more meaningful interpretation of the data can be exposed.

DESCRIPTION

Stratification is used to mine data for meaningful clues toward understanding process activities and events. Using the stratification concept, we break down or sort a large database so that we can find meaningful strata or subsets and/or produce meaningful summaries. Strata may be developed along product lines, process lines, time lines, geographical lines, and so forth. Stratification is another of Ishikawa's seven fundamental quality improvement tools.

Data collection must be disciplined and documented in order to accommodate stratification. We must record data in multivariate sets, including values of covariates. A covariate is a variable that is not necessarily controllable, but measurable. For example, if we are measuring revenues, covariates can include sales region, salesperson, time of sale, and so on. Database software as well as computer-aided statistical analysis and spreadsheet packages have stratification capabilities. See Kolarik [4] and Ishikawa [3] for more details and examples of stratification.

NOTES

Stratification is a basic data analysis tool, always useful to some degree when we assess a multivariate data set.

DISCOVERY EXERCISES

20.1. Why is there a tendency for people within organizations to focus so tightly on tools that they lose sight of common purpose and see tools as the common purpose? Explain.

20.2. Select a tool and demonstrate its features in the context of a problem or opportunity.

REFERENCES

1. Park, C. S., *Contemporary Engineering Economics*, 2nd ed., Menlo Park, CA: Addison-Wesley, 1997.

2. Ishikawa, K., *What is Total Quality Control? The Japanese Way*, Englewood Cliffs, NJ: Prentice-Hall, 1985.

3. Ishikawa, K., *Guide to Quality Control*, Tokyo: Asian Productivity Organization, 1982.

4. Kolarik, W. J., *Creating Quality: Concepts, Systems, Strategies, and Tools*, New York: McGraw-Hill, 1995.

5. Walpole, R. E., R. H. Myers, and S. L. Myers, *Probability and Statistics for Engineers and Scientists*, 6th ed., Upper Saddle River, NJ: Prentice Hall, 1998.

6. Newbold, P., and T. Bos, *Introductory Business and Economic Forecasting*, 2nd ed., Cincinnati, OH: South-Western Publishing, 1994.

7. Badiru, A. B., and P. S. Pulat, *Comprehensive Project Management*, Englewood Cliffs, NJ: Prentice Hall, 1995.

8. Montgomery, D. C., *Design and Analysis of Experiments*, 4th ed., New York: Wiley, 1997.

9. Box, G. E. P., W. G. Hunter, and J. S. Hunter, *Statistics for Experimenters*, New York: Wiley, 1978.

10. Neter, J., W. Wasserman, and M. H. Kutner, *Applied Linear Statistical Models*, 3rd ed., Homewood, IL: Irwin, 1990.

11. Draper, N. R., and H. Smith, *Applied Regression Analysis*, 2nd ed., New York: Wiley, 1981.

12. Box, G. E. P., and N. R. Draper, *Empirical Model-Building and Response Surfaces*, New York: Wiley, 1987.

13. Taguchi, G., *Introduction to Quality Engineering: Designing Quality into Products and Processes*, New York: Kraus International Publications, 1986.

14. Taguchi, G., *System of Experimental Design*, vols. 1 and 2, New York: Kraus International Publications, 1987.

15. Phadke, M. S., *Quality Engineering Using Robust Design*, Englewood Cliffs, NJ: Prentice Hall, 1989.

16. Roy, R., *A Primer on the Taguchi Method*, New York: Van Nostrand Reinhold, 1990.

17. Ross, P. J., *Taguchi Techniques for Quality Engineering*, New York: McGraw-Hill, 1988.

18. Barnes, R. M., *Motion and Time Study Design and Measurement of Work*, 7th ed., New York: Wiley, 1980.

19. Hughes, T. A., *Measurement and Control Basics*, 2nd ed., Research Triangle Park, NC: Instrument Society of America, 1995.

20. Mizuno, S., *Management for Quality Improvement*, Portland, OR: Productivity Press, 1988.

21. Akao, Y., ed., *Quality Function Deployment*, Portland, OR: Productivity Press, 1990.

22. Day, R. G., *Quality Function Deployment: Linking a Company with Its Customers*, Milwaukee: ASQC Press, 1993.

23. Dettmer, H. W., *Goldratt's Theory of Constraints: A System Approach to Continuous Improvement*, Milwaukee: ASQC Press, 1997.

24. Wilson, P. F., L. D. Dell, and G. F. Anderson, *Root Cause Analysis*, Milwaukee: ASQC Press, 1993.

7

PROCESS CASES—
DESCRIPTIONS AND DATA

The purpose of Section 7 is to provide a set of case descriptions, including data sets, to support review and discovery exercises in Sections 2 and 4.

VII.1　INTRODUCTION

Section 7 consists of several cases used to support process-related exercises in Sections 2 and 4. Both attributes and variables data appear, along with brief process descriptions. The cases range from relatively simple to complex; accordingly, models ranging from simple to complex are required. The following cases are provided:

Case VII.1　AA Fiberglass
Case VII.2　Apple Core—Dehydration
Case VII.3　Apple Dehydration Exploration
Case VII.4　Back-of-the-Moon—Mining
Case VII.5　Big City Waterworks
Case VII.6　Big Dog—Dog Food Packaging
Case VII.7　Bushings International—Machining
Case VII.8　Door-to-Door—Pizza Delivery
Case VII.9　Downtown Bakery—Bread Dough
Case VII.10　Downtown Bakery—pH Measurement
Case VII.11　Fix-Up—Automobile Repair
Case VII.12　Hard-Shell Aquaculture

Case VII.13   Health Assist—Service

Case VII.14   High-Precision—Collar Machining

Case VII.15   High-Precision—Collar Measurement

Case VII.16   Link-Lock Chain

Case VII.17   LNG—Natural Gas Liquefaction

Case VII.18   M-Stick Manufacturing

Case VII.19   Night Hauler Trucking

Case VII.20   PCB—Printed Circuit Boards

Case VII.21   Punch-Out—Sheet Metal Fabrication

Case VII.22   Re-Use—Recycling

Case VII.23   Re-Use—Sensor Precision

Case VII.24   Silver Bird—Baggage

Case VII.25   Snappy—Plastic Injection Molding

Case VII.26   Squeaky Clean Laundry

Case VII.27   Rainbow—Paint Coating

Case VII.28   Sure-Stick Adhesive

Case VII.29   TexRosa—Salsa

Case VII.30   Tough-Skin—Sheet Metal Welding

VII.2   Data Extensions

Case-specific data sets may be extended or modified in order to make the case more interesting and challenging. For example, we can introduce process shifts, once the SPC chart parameters are calculated, to demonstrate how a shift impacts our chart. In this case, we can generate additional data using the relationship in Equation (VII.1).

$$RND_i = \mu + (\sigma \, SND_i) \qquad \textbf{[VII.1]}$$

where RND_i = the ith pseudo-random number "drawn" from a normal distribution

μ = the mean of the normal distribution

σ = the standard deviation of the normal distribution

SND_i = the ith pseudo random number "drawn" from a standard normal distribution, having $\mu = 0$ and $\sigma = 1$

SND_i values are available in Section 8, Table VIII.8, or they can be generated by external means using a computer-aided statistical package or, in some cases, spreadsheet software.

Example VII.1

Using the standard normal deviates in Table VIII.8, develop a set of normally distributed pseudo-random numbers with a mean of 5 units and a standard deviation of 2 units.

Solution

We use the first five numbers (pseudo-random numbers sampled from a standard normal distribution) that appear in the upper left-hand section of Table VIII.8, located in Section 8. These five SND_is are processed through Equation (VII.1) as

$$RND_1 = 5 + [2(0.47671)] = 5.95342$$

$$RND_2 = 5 + [2(-1.81764)] = 1.36472$$

$$RND_3 = 5 + [2(1.13274)] = 7.26548$$

$$RND_4 = 5 + [2(-0.06503)] = 4.86994$$

$$RND_5 = 5 + [2(-0.12146)] = 4.75708$$

If we have a reasonably large set of pseudo-random numbers, we can characterize their statistical properties using the methods described in Chapter 4.

VII.3 CASES

CASE VII.1: AA FIBERGLASS

AA Fiberglass Inc. manufactures fiberglass parts. Because of both quality and productivity concerns, AA needs to learn more about gel time responses regarding several process input variables. They have designed an in-house experiment to measure the effect of three factors on the gel time of polyester fiberglass resin. Gel time is the time interval when fiberglass resin is workable. According to the manufacturer of the polyester resin, gel time may be affected by ambient humidity, ambient temperature, and the amount of catalyst used.

The engineering team of AA designed and conducted a FAT-CRD experiment to assess these effects. Two replications were run in the experiment. The data collected appear below, where the response is measured in minutes. AA has large parts, where long gel times are essential—bigger is better. They have small parts where smaller is better. And they have many parts where a gel time of 1 hour is about right—a nominal-is-best response. The purpose of the experiment is to find "best" or adequate factor levels for each of the three gel time situations. [DM]

	Temperature (B)							
	60°F				80°F			
	Catalyst (C)				Catalyst (C)			
Humidity (A)	1%		2%		1%		2%	
30%	50	65	12	18	22	35	6	12
70%	75	90	20	25	38	60	15	36

Response is in minutes

CASE VII.2:   APPLE CORE—DEHYDRATION

A fruit dehydration process cores and slices raw fruit, which is subsequently moved through a drying tunnel. The sliced fruit drops onto a conveyor belt running through the drying tunnel. There are three slicers feeding a single conveyor that passes through the drying tunnel. In general, it is difficult to balance the conveyor load, as it is difficult to load all slicers predictably back upstream. Conveyor load is measured in a percentage of design throughput capacity.

The output target is 10% moisture by weight. High-moisture product does not possess a suitable shelf life. Product that is too dry is both undesirable from the customer's perspective and expensive, i.e., requires additional energy, from the producer's perspective. Any product with a moisture content of over 13% is recycled through the dryer. Any product under 7% moisture must be taken off line and blended with standard-moisture product.

The moisture of the entering fruit is more or less constant, as the fruit is saturated with a treatment liquid (to preserve the fruit's natural color) as it hits the conveyor, with excess liquid recaptured under the stainless steel belt. Data consisting of tunnel temperature, throughput load percent, and exit moisture content were collected from the process and appear below and on the next page.

Sample Hour	Temp., °F	Throughput, % Capacity	Moisture, %	Sample Hour	Temp., °F	Throughput, % Capacity	Moisture, %
1	385	102	11.31	31	452	76	7.34
2	384	98	12.45	32	448	82	7.63
3	382	110	13.67	33	453	79	8.32
4	390	115	14.47	34	452	83	8.78
5	382	112	13.60	35	449	87	9.35
6	388	117	13.98	36	446	90	9.00
7	391	121	13.53	37	443	91	8.86
8	387	116	13.62	38	447	88	8.41
9	393	113	13.42	39	450	92	8.03
10	384	109	12.93	40	453	95	7.63
11	395	108	13.21	41	456	88	6.81
12	392	106	12.48	42	452	89	7.34
13	397	103	11.87	43	449	94	8.98
14	398	104	12.23	44	450	96	9.36
15	403	100	11.37	45	447	101	8.83
16	404	98	11.19	46	449	97	9.00
17	411	99	11.00	47	446	99	9.45
18	420	101	10.89	48	443	101	9.87
19	423	97	10.74	49	440	108	10.20
20	437	96	10.26	50	445	110	9.83
21	435	98	10.89	51	443	107	9.36
22	441	102	10.31	52	449	105	9.21
23	437	97	9.76	53	446	103	8.39
24	429	95	9.89	54	449	98	8.52
25	435	91	9.13	55	452	97	8.89
26	447	87	8.69	56	455	102	9.21
27	448	85	8.42	57	451	104	9.45
28	450	81	8.56	58	447	107	9.37
29	449	79	8.21	59	449	102	9.01
30	450	82	7.98	60	444	98	9.59

(continued)

(concluded)

Sample Hour	Temp., °F	Throughput, % Capacity	Moisture, %	Sample Hour	Temp., °F	Throughput, % Capacity	Moisture, %
61	441	95	9.14	81	437	103	10.21
62	440	93	8.63	82	442	101	10.07
63	445	96	8.45	83	451	98	9.42
64	443	98	8.92	84	443	97	9.31
65	448	101	9.37	85	448	100	9.73
66	452	108	9.87	86	446	110	10.26
67	447	110	10.03	87	442	106	10.11
68	453	104	9.67	88	440	95	9.45
69	448	102	9.46	89	438	91	9.21
70	443	98	8.47	90	437	96	9.34
71	439	96	7.23	91	434	98	9.87
72	432	87	6.48	92	438	102	10.29
73	438	85	6.32	93	444	108	11.42
74	431	89	7.34	94	439	115	12.78
75	426	93	8.56	95	443	118	13.85
76	437	95	8.91	96	441	117	14.00
77	432	94	8.63	97	438	112	12.60
78	441	98	9.26	98	445	109	11.23
79	438	100	9.13	99	443	101	9.67
80	443	97	9.98	100	451	98	9.43

CASE VII.3: APPLE DEHYDRATION EXPLORATION

Apple Core has experienced erratic dehydration performance in their production operations, as described in Case VII.2. The purpose of this experimental sequence is to seek out suitable dehydration alternatives. Here, one strategy calls for slicing the apple flesh in smaller cross sections so as to expedite dehydration. Cross-section dimensions are usually specified by customers; hence, operating strategies at various cross-sectional dimensions are desired. Furthermore, a cross section of 0.25 inches is as small as any customer has specified and the smallest size that can be placed on the operating line belt; smaller sections have a tendency to fall through or attach to the belt mesh. Nevertheless, small sections can be accommodated in the laboratory and are included in the experiment.

In order to seek out the best performance area for operations, the company has developed a strategy of response surface analysis. First, a limited experiment using a 2^2 FAT-CRD with a center point is designed to obtain a first-order response surface. Then, a follow-on central composite design is developed for the construction of a second-order response surface. Isolating a feasible operating region, given a customer's cross-sectional requirements, is the primary concern.

It is desired to dry the fruit to a target level of 10% moisture, by weight, in a 2-hour drying period. The 2-hour drying period is selected because faster dehydration is believed to produce a product that is too hard on the surface. Two factors are studied, temperature and cross-sectional dimensions. The experiment consists of preparing Granny Smith apples according to the cross-sectional dimensions and drying them in a laboratory oven under simulated "bulk volume" production conditions. The ultimate purpose is to develop a model whereby we can take a customer's cross-sectional requirement and select a drying temperature to meet the 10% moisture requirement. The results of the first experiment are shown below.

Observation	Initial Cross Section, Inches	Temp., °F	Coded Cross Section	Coded Temp.	Response, % Moisture
1	0.375	300	−1	−1	10
2	0.625	300	1	−1	20
3	0.375	400	−1	1	5
4	0.625	400	1	1	8
5	0.500	350	0	0	7
6	0.500	350	0	0	9
7	0.500	350	0	0	10
8	0.500	350	0	0	8

The results of the second experiment appear below.

Initial Cross Section, Inches	Temp., °F	Coded Cross Section	Coded Temp.	Response, % Moisture
0.250	250	−1	−1	11
0.500	250	1	−1	13
0.250	350	−1	1	6
0.500	350	1	1	9
0.198	300	−1.414	0	7
0.552	300	1.414	0	16
0.375	229.3	0	−1.414	15
0.375	370.7	0	1.414	5
0.375	300	0	0	8
0.375	300	0	0	11
0.375	300	0	0	12
0.375	300	0	0	10
0.375	300	0	0	9

CASE VII.4: BACK-OF-THE-MOON—MINING

Back-of-the-Moon (BOM) is a futuristic company that has set up an automated production process on the back side of the moon. Their product line consists of bags of moon rocks. This product has found many uses both as a medicinal and therapeutic aid as well as a novelty product (on Earth). Production process specifications call for a mass of 1 kilogram per unit or bag (bagged and measured on the moon). Each bag contains several rocks.

Current production processes involve digging, crushing, sifting, sorting, and bagging. The automated production process calls for each bag to contain one fairly large rock; robotic manipulators add smaller rocks to the bag so as to bring the mass up to or over the 1-kilogram mass specification. Data collected on the moon were transmitted back to Earth and appear on the next page. Several arguments at BOM headquarters on Earth have focused on package size and process modeling and stability. One point of view is that a 1-kilogram mass is too large to effectively handle and ship; the opposing argument is that it is about the right size for customers to carry home in the back seat of their automobile.

Sample	Mass, Kilograms	Sample	Mass, Kilograms
1	1.205	51	1.811
2	1.356	52	1.398
3	1.342	53	1.553
4	1.804	54	1.259
5	1.937	55	1.050
6	1.794	56	1.826
7	1.700	57	1.797
8	1.068	58	1.622
9	1.003	59	1.575
10	1.048	60	1.404
11	1.066	61	1.714
12	1.025	62	1.384
13	1.047	63	1.121
14	1.341	64	1.444
15	1.302	65	1.967
16	1.388	66	1.353
17	1.713	67	1.218
18	1.612	68	1.527
19	1.490	69	1.117
20	1.756	70	1.062
21	1.219	71	1.499
22	1.854	72	1.314
23	1.858	73	1.253
24	1.122	74	1.655
25	1.686	75	1.374
26	1.486	76	1.738
27	1.691	77	1.397
28	1.621	78	1.664
29	1.595	79	1.563
30	1.089	80	1.335
31	1.040	81	1.667
32	1.832	82	1.663
33	1.748	83	1.493
34	1.722	84	1.911
35	1.819	85	1.382
36	1.412	86	1.137
37	1.730	87	1.170
38	1.789	88	1.160
39	1.343	89	1.287
40	1.468	90	1.218
41	1.131	91	1.428
42	1.067	92	1.788
43	1.404	93	1.225
44	1.302	94	1.996
45	1.140	95	1.795
46	1.370	96	1.405
47	1.684	97	1.459
48	1.391	98	1.963
49	1.979	99	1.940
50	1.762	100	1.498

Case VII.5:   Big City Waterworks

A wastewater treatment plant at Big City utilizes hydrogen peroxide at various temperatures to destroy organic contaminants in the incoming water. It is of interest to the manager of the facility to determine if and how the treatment is influenced by operating temperatures and peroxide flow rates. She selects three peroxide flow rates and three temperatures. She then designs and conducts a FAT-CRD experiment, using fixed effects factors, on contaminated water drawn from a lagoon. In each case, she measures the percentage removal of organics as the response—a bigger-is-better response. The data obtained are shown below. The numbers in parentheses indicate the order in which data were obtained. [MR]

Hydrogen Peroxide Flow Rate	Temperature, °C			
	10°C	**50°C**	**70°C**	**90°C**
50 ml/h	16.7 (14)*	25.1 (10)	36.1 (12)	32.3 (6)
	24.3 (1)	29.3 (24)	29.3 (3)	36.8 (22)
100 ml/h	29.8 (15)	35.0 (23)	45.3 (11)	52.1 (4)
	21.4 (21)	28.7 (19)	51.6 (16)	57.5 (8)
150 ml/h	31.3 (13)	50.1 (17)	60.3 (20)	63.2 (5)
	25.7 (9)	56.3 (7)	57.2 (2)	58.8 (18)

*Response is in percent removal; numbers in parentheses indicate experimental order.

Case VII.6:   Big Dog—Dog Food Packaging

A dry dog food manufacturer produces a low-fat, high-energy dog food primarily for working dogs. The dog food contains a proprietary mix of proteins and carbohydrates, along with fast-release sugars and added fiber. The dog food is formed in the shape of small bones and is dried to a uniform moisture of 12% before bagging.

The purpose of the bagging subprocess is to fill each bag to a full measure, labeled at 20 pounds. It is critical to avoid underfilled bags. An overfill is acceptable, but not desirable, since product is given away free with an overfill. The process utilizes a volume release metering system, triggered by a manual setpoint; i.e., an operator sets and adjusts the setpoint. Once the setpoint is "hit," an actuator is triggered and shuts off the material flow. Finally, the bags are indexed forward, with the filled bag sewed and moved to a packing area. Targets and subgrouped data, for a subgroup size of four, are provided on the next page. Bag net weight targets are $\mu_0 = 20.2$ lb and $\sigma_0 = 0.035$ lb.

	Observation			
Subgroup	1	2	3	4
1	20.181	20.166	20.253	20.167
2	20.159	20.125	20.143	20.220
3	20.211	20.223	20.209	20.187
4	20.213	20.193	20.241	20.197
5	20.216	20.227	20.245	20.230
6	20.194	20.206	20.181	20.236
7	20.223	20.284	20.227	20.222
8	20.239	20.234	20.216	20.215
9	20.183	20.203	20.218	20.200
10	20.182	20.207	20.220	20.196
11	20.160	20.210	20.234	20.258
12	20.198	20.204	20.209	20.198
13	20.206	20.240	20.231	20.178
14	20.215	20.193	20.212	20.201
15	20.184	20.263	20.218	20.219
16	20.207	20.258	20.120	20.261
17	20.165	20.243	20.211	20.242
18	20.264	20.266	20.144	20.212
19	20.124	20.191	20.171	20.173
20	20.209	20.203	20.250	20.127
21	20.217	20.188	20.224	20.177
22	20.202	20.171	20.202	20.185
23	20.233	20.197	20.139	20.229
24	20.173	20.192	20.193	20.209
25	20.174	20.150	20.250	20.200
26	20.204	20.251	20.180	20.188
27	20.199	20.207	20.183	20.170
28	20.191	20.248	20.261	20.232
29	20.253	20.177	20.211	20.177
30	20.201	20.181	20.184	20.239

Response is in pounds.

CASE VII.7:   BUSHINGS INTERNATIONAL—MACHINING

A bushing is manufactured from a proprietary brass material impregnated with a dry lubricant. A machining operation is performed on a tubular section of stock, whereby the inside diameter is machined. The inside diameter is a critical dimension. The part specifications and targets are provided below:

Inside diameter specifications: 2.500 + 0.020, −0.000 inches

Inside diameter targets: $\mu_0 = 2.510$ inches, $\sigma_0 = 0.002$ inches

Subgrouped data, for a subgroup size of two, were collected from the machining subprocess.

	Observation	
Subgroup	**1**	**2**
1	2.507	2.510
2	2.511	2.510
3	2.509	2.507
4	2.512	2.511
5	2.510	2.511
6	2.506	2.509
7	2.509	2.508
8	2.510	2.512
9	2.510	2.509
10	2.509	2.509
11	2.509	2.507
12	2.509	2.507
13	2.510	2.510
14	2.511	2.507
15	2.510	2.512
16	2.510	2.514
17	2.512	2.509
18	2.508	2.514
19	2.508	2.508
20	2.509	2.512
21	2.510	2.512
22	2.511	2.512
23	2.509	2.509
24	2.509	2.510
25	2.510	2.509
26	2.508	2.506
27	2.512	2.510
28	2.512	2.506
29	2.512	2.513
30	2.512	2.511

Response is in inches.

CASE VII.8: DOOR-TO-DOOR—PIZZA DELIVERY

The Door-to-Door pizza organization delivers pizza from its food service facility to the customer's door. Customers like to receive their pizza hot, so that they can begin eating immediately, rather than warming it after it arrives. Hence, the delivery process is critical. Door-to-Door has set targets for the in-box temperatures, as shown below. However, the delivered temperature has been difficult to deal with effectively.

In-box temperature targets: $\mu_0 = 350°F$, $\sigma_0 = 5°F$

Door-to-Door has been experimenting with an "improved" delivery subprocess—new technology to contain and insulate the product during delivery. The ideal situation is to deliver pizza to each customer with an out-of-box temperature equal to the in-box temperature. Data have been collected over the late summer and fall seasons, using both the existing and the "improved" subprocesses. Data are displayed on the following page.

Sample	In-the-Box Temp., °F	Travel Time, Minutes	Ambient Temp., °F	Out-of-Box Temp., °F
Existing Subprocess				
1	347.7	10.1	89.2	175.3
2	343.6	9.4	74.2	132.3
3	349.3	9.5	69.3	145.7
4	351.2	15.3	61.8	106.5
5	354.9	9.3	58.3	152.4
6	358.9	11.1	87.3	169.0
7	345.7	9.6	87.1	153.9
8	352.4	8.8	67.7	168.3
9	357.7	10.1	63.1	138.9
10	361.3	9.6	61.0	156.3
11	347.3	9.0	88.0	163.7
12	346.6	7.5	78.3	175.4
13	347.6	7.6	69.0	178.1
14	343.9	9.1	63.5	154.2
15	343.9	8.8	62.7	146.8
16	347.1	10.1	55.6	164.7
17	342.2	12.1	53.2	112.3
18	348.9	10.2	37.2	123.6
19	342.3	10.3	34.8	85.2
20	344.6	9.3	34.6	116.5
21	348.3	11.5	27.1	89.4
22	351.2	15.0	25.8	78.2
23	359.4	10.4	24.9	115.4
24	348.5	10.9	21.4	112.9
25	349.4	10.3	15.8	98.1
Experimental "Improved" Subprocess (under test)				
1	349.9	11.2	88.0	161.7
2	350.8	10.0	78.1	165.3
3	357.5	11.9	69.9	173.8
4	354.7	10.5	54.0	159.5
5	350.9	9.7	87.3	178.2
6	356.1	9.1	68.4	179.8
7	345.6	10.7	62.5	156.0
8	359.4	10.7	54.3	161.0
9	345.8	7.5	54.0	183.4
10	344.9	9.4	48.1	175.2
11	345.5	10.4	78.2	165.0
12	356.9	9.6	62.7	182.9
13	348.0	9.2	46.2	156.9
14	353.9	13.8	43.6	147.4
15	347.3	10.7	42.6	163.1
16	340.9	6.4	47.1	174.0
17	351.5	11.5	42.9	172.6
18	352.1	9.9	38.9	181.5
19	353.5	11.7	35.5	168.3
20	350.4	10.5	35.8	162.6
21	349.7	10.7	28.3	174.2
22	343.8	11.6	25.2	176.3
23	352.8	10.8	21.7	174.3
24	350.2	10.2	18.4	162.1
25	358.4	10.8	12.2	156.5

CASE VII.9:   DOWNTOWN BAKERY—BREAD DOUGH

Depending on the specific bread product, several fundamental ingredients are mixed and blended to form dough. The base dough characteristics come together to determine taste, aroma, and texture. In a bread-making subprocess, the dough pH is a critical metric. A pH of 7.8 works best, with pH between 7.6 and 8.4 considered acceptable. Additionally, a dough temperature of 98°F is best, with an acceptable temperature range of 89 to 102°F. Fifty subgroups with 5 observations each are measured in the dough-making subprocess; each physical sample taken is measured for both temperature and pH. The targets and data are displayed below, and on the next page.

Temperature targets: $\mu_0 = 98$, $\sigma_0 = 0.8$

pH targets: $\mu_0 = 7.8$, $\sigma_0 = 0.1$

[MS]

Bread pH data.

Subgroup	Observation					Subgroup	Observation				
	1	2	3	4	5		1	2	3	4	5
1	7.92	8.19	8.00	7.91	8.00	26	8.00	8.05	8.11	8.10	8.11
2	8.04	7.94	7.90	7.96	7.90	27	8.12	8.02	8.01	8.04	8.01
3	8.07	7.90	7.88	8.00	7.88	28	8.06	7.92	8.02	8.02	8.02
4	8.00	7.89	8.06	8.07	8.06	29	8.02	7.88	7.99	8.02	7.99
5	7.91	7.99	8.01	8.03	8.01	30	8.16	7.82	7.90	7.88	7.90
6	8.01	8.15	8.18	7.93	8.18	31	8.11	7.90	8.13	8.12	8.13
7	7.94	8.04	7.89	7.94	7.89	32	7.98	8.05	7.99	8.14	7.99
8	7.93	8.02	7.95	7.97	7.95	33	8.08	7.96	7.89	8.29	7.89
9	8.00	7.91	7.91	8.10	7.91	34	7.94	8.01	8.00	8.12	8.00
10	7.92	8.00	7.89	7.93	7.89	35	7.92	8.14	8.11	7.95	8.11
11	7.98	7.90	7.87	8.08	7.87	36	8.03	7.83	8.05	7.98	8.05
12	8.00	8.00	7.94	7.91	7.94	37	8.23	7.68	8.11	8.05	8.11
13	8.04	7.99	7.92	8.02	7.92	38	7.95	7.98	7.90	8.08	7.90
14	7.97	8.10	8.09	8.07	8.09	39	8.05	7.94	7.92	7.98	7.92
15	7.97	7.92	8.03	8.23	8.03	40	7.98	7.94	7.94	7.78	7.94
16	8.01	8.09	8.01	7.99	8.01	41	8.13	7.89	8.21	7.99	8.21
17	7.88	8.07	8.04	7.80	8.04	42	7.77	7.81	8.05	8.01	8.05
18	7.96	7.81	8.00	8.02	8.00	43	8.13	8.03	7.91	7.91	7.91
19	8.06	7.89	7.87	7.72	7.87	44	8.04	8.14	7.87	7.90	7.87
20	8.03	7.71	8.01	8.22	8.01	45	7.94	8.02	8.05	7.95	8.05
21	7.92	7.98	7.96	7.80	7.96	46	8.06	8.05	7.87	8.01	7.87
22	8.13	8.03	7.91	7.92	7.91	47	8.11	8.02	8.24	8.24	8.24
23	7.92	8.12	8.02	8.10	8.02	48	8.15	7.90	7.99	8.05	7.99
24	8.18	7.96	8.05	8.07	8.05	49	7.86	7.99	7.88	7.90	7.88
25	7.95	7.91	8.05	8.03	8.05	50	8.03	8.00	7.97	7.94	7.97

Bread dough temperature data, °F.

Subgroup	Observation					Subgroup	Observation				
	1	2	3	4	5		1	2	3	4	5
1	97.75	97.66	98.29	97.42	98.53	26	97.38	96.67	98.66	97.75	98.07
2	97.64	97.79	98.12	98.60	98.68	27	98.56	98.81	97.29	98.73	97.57
3	97.40	97.48	98.14	98.49	98.27	28	98.05	98.93	99.03	97.92	98.14
4	98.32	98.82	97.12	96.75	97.61	29	98.32	97.94	97.40	97.90	98.26
5	98.64	99.15	98.90	98.90	97.75	30	98.65	99.15	98.11	97.59	97.38
6	98.38	98.28	99.03	96.37	97.64	31	98.78	97.62	97.35	95.84	98.56
7	97.21	98.30	100.17	100.56	97.40	32	97.42	99.12	97.24	99.23	98.05
8	97.46	98.75	98.89	97.64	98.32	33	98.66	97.63	98.39	96.74	98.32
9	97.06	97.58	97.62	99.02	98.64	34	98.62	97.59	97.97	97.27	98.65
10	96.70	98.25	97.84	98.89	98.38	35	97.64	96.88	98.69	97.83	98.78
11	98.56	98.08	98.41	97.99	97.21	36	97.89	98.89	99.24	96.08	97.42
12	98.20	97.78	98.57	97.84	97.46	37	98.63	97.75	98.15	99.00	98.66
13	97.23	97.76	97.88	98.46	97.06	38	98.53	98.01	98.01	98.40	98.62
14	98.79	98.26	96.32	97.23	96.70	39	96.62	98.89	97.55	98.37	97.64
15	97.84	98.57	97.95	98.25	98.56	40	98.96	98.30	98.00	98.41	97.89
16	97.19	98.63	98.07	97.76	98.20	41	97.17	98.44	97.10	98.82	98.63
17	99.08	98.00	97.31	97.74	97.23	42	98.65	97.86	99.41	97.81	98.53
18	98.70	98.25	97.27	96.87	98.79	43	97.44	96.85	98.70	98.65	96.62
19	96.79	97.53	97.62	98.38	97.84	44	98.66	96.45	97.57	98.06	98.96
20	98.46	98.04	98.10	98.44	97.19	45	97.29	99.68	98.30	97.78	97.17
21	96.49	97.54	99.78	97.62	99.08	46	99.03	99.26	98.46	97.75	98.65
22	98.07	97.91	98.34	97.20	98.70	47	97.40	99.23	97.58	97.97	97.66
23	97.57	97.74	97.25	98.30	96.79	48	98.11	98.17	98.79	98.08	97.79
24	98.14	99.62	97.57	98.33	98.46	49	97.35	98.79	98.31	98.00	97.48
25	98.26	98.63	97.44	98.18	96.49	50	97.24	97.34	98.03	96.79	98.82

CASE VII.10: DOWNTOWN BAKERY—pH MEASUREMENT

In the bread production system (see Case VII.9), the pH of the product is a critical metric. In order to measure the pH of the final product, a dough sample is pulled—a 10-gram cross section is cut and then mixed and blended with 100 ml of distilled water in a beaker. Then, a pH probe is inserted in the beaker and the pH is read.

Three different operators prepare samples and perform the measurements (with the same pH meter) throughout a two-shift-per-day production run. There is some question as to the precision of the resulting pH measurements. In order to assess the precision of the pH measurements, each operator was asked to prepare a 10-gram sample, from a common 100-gram dough mass, and independently measure the sample two times. The purpose of the experiment is to assess the variation introduced by technicians as well as the pH meter. Data are displayed on the next page.

Dough Sample	Technician 1 Observations, pH		Technician 2 Observations, pH		Technician 3 Observations, pH	
	1	2	1	2	1	2
1	7.998	7.969	8.035	7.947	7.969	7.905
2	7.979	7.991	8.073	8.021	7.852	7.899
3	8.061	8.018	8.116	8.114	8.008	8.041
4	7.898	7.895	7.933	8.003	7.857	7.876
5	8.033	8.065	8.067	8.104	8.019	7.978
6	7.968	8.041	8.016	8.003	7.911	8.009
7	7.966	7.923	8.021	8.014	7.931	7.916
8	7.850	7.886	7.900	7.910	7.766	7.825
9	8.050	8.050	8.020	8.060	7.957	7.964
10	8.058	8.084	8.212	8.208	8.073	8.109
11	7.950	8.033	7.967	8.102	7.923	7.873
12	7.894	7.910	7.971	8.055	7.797	7.810
13	8.039	8.055	8.041	8.008	7.973	8.010
14	8.044	8.088	8.144	8.147	7.881	7.784
15	8.024	7.963	8.056	8.137	7.909	7.977
16	7.967	7.906	7.964	7.972	7.932	7.942
17	8.097	8.192	8.162	8.142	8.083	8.074
18	7.990	7.958	8.091	8.124	7.941	7.994
19	7.987	8.019	8.023	7.986	7.961	7.934
20	7.946	7.858	7.905	7.969	7.927	7.957
21	7.712	7.710	7.765	7.791	7.624	7.604
22	8.164	8.148	8.254	8.207	8.112	8.121
23	7.832	7.821	7.811	7.864	7.825	7.810
24	7.902	7.872	7.985	8.038	7.802	7.730
25	7.978	7.927	8.010	8.058	7.860	7.866
26	7.744	7.740	7.793	7.770	7.728	7.740
27	8.134	8.182	8.280	8.271	8.015	8.050
28	8.053	8.064	8.110	8.181	7.937	7.982
29	8.049	7.950	8.120	8.082	7.966	7.984
30	8.054	8.067	8.062	8.036	8.021	7.995

CASE VII.11:   FIX-UP—AUTOMOBILE REPAIR

An automobile repair shop is concerned about customer satisfaction in terms of the entire experience customers receive from the repair shop. In order to quantify the customer experience, several critical service characteristics have been identified:

Complaints of a failure to fix the vehicle.

Delay beyond the promised pickup time.

Complaints of damage to the inside/outside of the vehicle during repair.

A level of zero defects, with no dissatisfied customers is the long-term goal. In order to address this goal, statistics have been collected over the past few months regarding these critical characteristics.

Sample Day	Sample Size, Vehicles	Number of Items Not Fixed Properly	Number of Delays from Promised Completion Times	Number of Damaged Items in Repair	Dissatisfied Customers
1	15	2	2	0	4
2	23	3	3	1	5
3	17	1	2	0	3
4	27	2	3	1	6
5	18	1	1	1	3
6	16	1	1	0	2
7	25	3	3	1	7
8	19	2	2	1	5
9	17	1	2	0	3
10	16	1	1	0	1
11	23	0	2	1	3
12	29	2	3	2	5
13	11	0	1	1	2
14	15	1	2	1	4
15	31	3	4	1	6
16	17	1	2	0	3
17	21	1	3	1	4
18	25	2	3	1	6
19	19	1	2	0	2
20	27	2	3	1	6
21	18	1	2	1	3
22	24	2	3	1	5
23	21	1	2	0	3
24	17	0	1	0	1
25	31	3	4	1	8
26	23	1	2	2	4
27	26	2	3	0	5
28	18	1	2	1	4
29	15	1	1	0	2
30	19	0	2	0	2

CASE VII.12: HARD-SHELL AQUACULTURE

In the last few years many entrepreneurs, including Hard-Shell Inc., have entered the field of aquaculture. Aquaculture is defined as the regulation and cultivation of water plants and animals for human consumption. In order to learn more about the aquaculture practices best suited to raising oysters in captivity, Hard-Shell has designed a three-factor experiment. The three factors considered for this experiment are bed density, salinity, and water flow position.

Response Variable	Measurement Technique
Average weight of oyster meat	Laboratory scale (ounces)
Factors under Study	**Levels**
F: Water flow position	1: near water inlet, position 1 2: middle position 2 3: middle position 3 4: near water outlet, position 4
D: Bed density	600, 300 spat per tray (a spat is a young oyster)
S: Salinity	10 ppt, 15 ppt

This experiment was conducted for 12 months in order to gather the necessary data. Four different troughs were prepared to simulate the possible conditions under which oysters could be grown. The troughs were loaded with 16 trays (each trough contained four trays). Eight trays were loaded with 600 spat (young oysters) per tray, and 8 trays were loaded with 300 spat per tray. Spat were assigned at random to the trays. The trays were laid in the troughs. Bay water was pumped over the trays at a controlled rate. The troughs were numbered and arranged 1 through 4, with number 1 being located at the water inlet point.

At the end of the experiment oysters were selected at random from each tray, within each trough. The oysters were then measured umbo to bill, and the oyster meat was carefully removed and weighed—a bigger-is-better response. Two oysters were randomly removed from each tray and identified with the corresponding tray and trough numbers. The weight of the oyster meat and the relationship to the best combinations of growing conditions were the focus of this experiment. A listing of the variables and the data is shown below. [DM]

Water Flow Position	Low-Density Population (300 Spat)		High-Density Population (600 Spat)	
	Salinity (10 ppt)	Salinity (15 ppt)	Salinity (10 ppt)	Salinity (15 ppt)
1	2.6 2.5	2.9 2.7	2.8 2.9	3.0 2.9
2	2.5 2.4	2.6 2.5	2.8 2.7	2.8 2.8
3	2.3 2.3	2.5 2.3	2.4 2.2	2.4 2.2
4	1.5 1.6	2.1 2.0	1.6 1.4	1.8 1.6

CASE **VII.13:**   **H**EALTH **A**SSIST—**S**ERVICE

A managed health care organization pays for health-related services rendered within their network of health care providers. Once services are rendered, qualified costs for services are reimbursed to customers, on a case-by-case basis. Either hard copy or electronic submissions are made by customers; then the organization works from these submissions and pays claims. Several defects are encountered:

Wrong form/format.

Incomplete details on submission.

Service items not covered in plan.

One form/format is used for each claim. Several details and service items appear in each claim. The goal is to develop a zero defect service process, resulting in no dissatisfied customers. Over the course of a 30-day period, statistics have been logged regarding defects.

Sample, Day	Sample Size, Claims	Wrong Form/Format	Number of Incomplete Details	Number of Service Items Not Covered	Dissatisfied Customers
1	346	7	17	12	28
2	223	4	10	6	16
3	198	3	8	5	13
4	311	7	14	9	24
5	274	6	12	6	20
6	157	3	7	3	11
7	248	5	11	6	18
8	312	6	17	8	26
9	348	8	19	14	32
10	273	4	12	9	19
11	283	6	14	5	22
12	327	7	13	12	24
13	350	7	19	7	29
14	176	3	8	3	12
15	231	4	10	8	16
16	211	4	9	4	14
17	317	7	14	13	25
18	298	6	13	11	22
19	190	3	9	3	13
20	267	5	14	10	22
21	233	5	12	8	19
22	257	4	10	7	17
23	308	7	17	9	28
24	222	4	10	2	16
25	389	9	21	15	35
26	264	5	13	7	19
27	187	3	9	0	12
28	271	4	14	13	21
29	260	5	12	11	19
30	344	8	16	14	28

CASE VII.14: HIGH-PRECISION—COLLAR MACHINING

The inside diameter of a steel collar is machined to a target of 50 mm. The roundness of the diameter is an important characteristic. After machining, the inside diameter is probed with a coordinate measuring machine (CMM) in two locations 90° apart. Subgroups of two collars are used.

Here, the target is critical, as is the roundness of the inside diameter. Specifications on the inside diameter call for each of the two measurements to be compared with a specification of 50 mm ± 0.02 mm. Targets are set as

Diameter targets: $\mu_0 = 50$ mm, $\sigma_0 = 0.015$ mm

Data from 50 subgroups of 2 collars each are displayed below.

	Collar 1 Location		Collar 2 Location			Collar 1 Location		Collar 2 Location	
Subgroup	1	2	1	2	Subgroup	1	2	1	2
1	49.9961	49.9903	49.9916	49.9936	26	49.9953	49.9871	50.0004	49.9962
2	50.0054	50.0058	49.9916	49.9981	27	49.9961	49.9901	49.9927	50.0008
3	50.0055	50.0076	50.0016	49.9974	28	49.9968	49.9917	50.0008	50.0010
4	49.9956	49.9946	49.9893	49.9873	29	49.9993	50.0045	50.0006	49.9976
5	49.9897	49.9825	49.9875	49.9937	30	49.9880	49.9869	49.9881	49.9835
6	49.9992	49.9967	49.9821	49.9759	31	50.0188	50.0240	50.0071	50.0093
7	49.9890	49.9855	49.9979	50.0009	32	50.0043	50.0065	50.0049	50.0017
8	50.0146	50.0204	49.9858	49.9969	33	49.9957	49.9941	49.9957	49.9908
9	50.0044	50.0044	49.9969	49.9981	34	50.0021	49.9993	49.9879	49.9847
10	49.9879	49.9752	49.9924	49.9915	35	50.0030	50.0029	50.0030	50.0099
11	49.9989	49.9981	49.9965	49.9938	36	50.0100	50.0169	50.0067	50.0068
12	49.9941	49.9893	50.0125	50.0156	37	49.9929	49.9901	50.0035	50.0006
13	49.9863	49.9840	49.9927	49.9910	38	50.0110	50.0172	50.0005	50.0023
14	50.0128	50.0145	50.0040	50.0121	39	50.0010	49.9995	49.9994	50.0004
15	50.0009	49.9950	49.9867	49.9833	40	49.9918	49.9891	49.9904	49.9842
16	49.9883	49.9871	50.0185	50.0181	41	49.9926	49.9877	50.0021	49.9993
17	50.0076	50.0104	50.0085	50.0093	42	49.9886	49.9862	50.0072	50.0112
18	50.0092	50.0182	49.9929	50.0012	43	50.0052	50.0138	49.9891	49.9879
19	49.9861	49.9803	50.0240	50.0231	44	50.0120	50.0231	50.0037	50.0039
20	49.9979	49.9972	50.0152	50.0054	45	49.9879	49.9888	49.9834	49.9845
21	49.9967	49.9915	49.9771	49.9695	46	50.0065	50.0142	50.0099	50.0115
22	49.9984	49.9952	49.9958	49.9824	47	50.0082	50.0161	49.9993	49.9988
23	50.0009	50.0026	49.9975	49.9988	48	49.9850	49.9792	49.9955	49.9943
24	50.0044	50.0096	49.9934	49.9966	49	49.9965	49.9983	50.0076	50.0071
25	50.0039	50.0102	50.0054	50.0089	50	49.9886	49.9819	49.9979	50.0015

Response is measured in mm.

CASE VII.15:   HIGH-PRECISION—COLLAR MEASUREMENT

A gauge study was initiated regarding Case VII.14, High-Precision—Collar Machining. The same coordinate measuring machine (CMM) was used by two different operators. Operator 1 is a first-shift operator with several years' experience, while operator 2 is a second-shift operator with about 6 months' experience. A total of 30 collars were drawn from production at random. Each operator was asked to measure each collar at two positions, as described in Case VII.14. Then, each operator was asked to repeat the two measurements on each collar. The purpose of this experiment is to estimate the components of variation associated with the operators and the CMM, and assess their compatibility with the measurement needs of the process/product. Data are displayed below.

Collar No.	Operator 1				Operator 2			
	Position 1 Observations, mm		Position 2 Observations, mm		Position 1 Observations, mm		Position 2 Observations, mm	
	1	2	1	2	1	2	1	2
1	50.0000	49.9976	49.9984	49.9958	50.0104	50.0142	50.0103	50.0076
2	50.0106	50.0098	50.0107	50.0076	50.0073	50.0082	50.0063	50.0045
3	50.0055	49.9998	50.0114	50.0129	50.0082	50.0092	50.0113	50.0081
4	50.0110	50.0078	50.0130	50.0134	50.0061	50.0086	50.0010	49.9967
5	49.9899	49.9907	49.9928	49.9974	49.9949	49.9935	49.9966	49.9984
6	49.9924	49.9942	49.9915	49.9886	49.9907	49.9915	49.9890	49.9897
7	49.9937	49.9938	49.9861	49.9847	49.9884	49.9887	49.9867	49.9842
8	50.0211	50.0223	50.0108	50.0085	50.0226	50.0219	50.0151	50.0177
9	50.0052	50.0068	50.0164	50.0136	50.0104	50.0096	50.0129	50.0124
10	49.9914	49.9895	49.9998	49.9966	49.9900	49.9909	49.9929	49.9902
11	49.9871	49.9871	49.9953	49.9939	49.9781	49.9800	49.9756	49.9792
12	50.0051	50.0073	50.0063	50.0043	50.0055	50.0076	50.0002	50.0025
13	49.9868	49.9843	49.9921	49.9943	49.9908	49.9908	49.9928	49.9887
14	50.0240	50.0207	50.0197	50.0205	50.0169	50.0178	50.0192	50.0207
15	49.9991	50.0008	49.9946	49.9950	50.0024	50.0008	50.0036	49.9986
16	49.9884	49.9850	49.9865	49.9874	49.9866	49.9867	49.9849	49.9852
17	49.9967	49.9934	49.9965	49.9964	49.9966	49.9950	50.0014	50.0000
18	50.0001	49.9985	50.0038	50.0006	50.0097	50.0094	50.0092	50.0096
19	50.0145	50.0153	50.0158	50.0162	50.0152	50.0144	50.0146	50.0154
20	49.9961	49.9950	49.9924	49.9913	49.9982	50.0036	49.9955	49.9934
21	50.0032	50.0037	50.0000	49.9978	49.9989	50.0010	49.9845	49.9864
22	50.0019	49.9972	50.0001	50.0005	50.0040	49.9995	50.0122	50.0123
23	49.9972	49.9982	50.0021	50.0033	49.9895	49.9922	49.9811	49.9822
24	49.9932	49.9923	49.9896	49.9909	49.9929	49.9960	49.9880	49.9902
25	49.9977	49.9946	50.0018	50.0046	49.9986	49.9984	49.9975	50.0001
26	49.9852	49.9847	49.9808	49.9811	49.9888	49.9926	49.9760	49.9741
27	50.0066	50.0043	50.0077	50.0082	50.0058	50.0046	50.0125	50.0147
28	49.9923	49.9897	49.9960	49.9957	49.9938	49.9975	49.9964	49.9985
29	50.0075	50.0054	50.0188	50.0164	50.0041	50.0029	50.0066	50.0054
30	50.0006	50.0004	50.0002	50.0035	49.9962	49.9948	49.9989	49.9985

CASE VII.16: LINK-LOCK CHAIN

Link-Lock is a progressive manufacturer of roller chains. They have developed an experimental purpose regarding investigating the effect of four specific press operating speeds on the pitch of link plates. The press speeds chosen were 90 sfm, 120 sfm, 140 sfm, and 160 sfm. For this experiment they used standard tooling and the same press for all observations. Their nominal-is-best responses were recorded in deviations from the target value in thousandths of an inch. Hence, a response of exactly zero is best. [MM]

	Observation			
Operating Speed	**1**	**2**	**3**	**4**
90	0.35	0.15	−0.46	−0.53
120	0.54	−0.25	−0.11	−0.05
140	0.98	0.68	0.65	0.14
160	0.97	0.70	0.85	0.25

Response is deviation from target, measured in thousandths of an inch.

Case VII.17:   LNG—Natural Gas Liquefaction

A natural gas liquefying process takes feedstock in the form of natural gas and liquefies it. The liquefaction subprocess involves chilling and pressurizing the gas until it reaches a liquid state. The chiller temperature is a critical process metric that determines the throughput of the process. In general, the lower the temperature, the higher the throughput. However, if the temperature is pushed too low, a freeze-up occurs, creating a flow interruption and requiring expensive maintenance. At temperatures below −40°F the risk of producing a freeze-up is considered high. Furthermore, the ambient temperature tends to affect the chiller temperature as well as the entering feedstock temperature. For example, on a hot day, the feedstock temperature increases, whereas on a cold day, the feedstock temperature decreases. Data were collected from the chiller at the beginning of each hour and are displayed below and on the next page. All temperatures are measured in degrees Fahrenheit.

Sample, Hour	Ambient Temp., °F	Chiller Temp., °F	Sample, Hour	Ambient Temp., °F	Chiller Temp., °F
1	74.0	−35.77	26	94.8	−33.50
2	74.8	−36.16	27	95.0	−34.15
3	75.0	−35.22	28	95.4	−33.20
4	75.7	−35.40	29	95.0	−34.08
5	76.0	−35.60	30	95.3	−33.00
6	77.0	−35.88	31	96.1	−33.16
7	79.0	−35.42	32	95.9	−33.63
8	81.0	−35.64	33	96.0	−33.21
9	82.0	−36.00	34	96.3	−32.54
10	83.0	−35.30	35	96.5	−34.02
11	84.0	−35.64	36	97.1	−32.34
12	84.6	−35.70	37	97.0	−32.25
13	85.0	−35.39	38	97.8	−33.27
14	86.0	−34.58	39	98.1	−33.70
15	87.6	−35.33	40	98.3	−33.33
16	89.0	−35.56	41	98.0	−32.38
17	90.0	−35.18	42	98.4	−32.90
18	89.9	−35.05	43	98.9	−32.76
19	90.5	−34.78	44	98.8	−31.94
20	91.0	−34.94	45	99.0	−33.03
21	92.0	−35.53	46	98.5	−32.63
22	92.3	−35.23	47	98.0	−32.19
23	93.0	−34.66	48	97.3	−32.48
24	93.7	−34.72	49	97.0	−32.73
25	94.0	−33.96	50	96.0	−32.91

(continued)

(concluded)

Sample, Hour	Ambient Temp., °F	Chiller Temp., °F	Sample, Hour	Ambient Temp., °F	Chiller Temp., °F
51	95.0	−32.93	76	76.6	−34.97
52	94.7	−33.54	77	76.9	−35.96
53	94.0	−33.62	78	75.0	−35.86
54	92.0	−33.96	79	75.4	−36.42
55	93.0	−34.58	80	75.2	−34.98
56	90.0	−33.98	81	75.7	−35.86
57	87.0	−34.39	82	74.0	−35.66
58	86.4	−34.35	83	73.8	−35.56
59	86.0	−35.32	84	74.2	−35.62
60	85.9	−34.92	85	74.3	−34.99
61	85.0	−35.33	86	74.0	−35.93
62	84.9	−34.37	87	73.9	−35.57
63	84.0	−35.32	88	74.4	−34.87
64	83.6	−34.92	89	73.9	−35.65
65	83.0	−34.43	90	74.0	−34.84
66	82.0	−34.87	91	74.1	−36.09
67	81.5	−35.70	92	73.8	−35.40
68	80.8	−35.62	93	73.6	−35.06
69	80.0	−35.38	94	73.0	−35.75
70	79.4	−34.79	95	73.4	−35.02
71	79.0	−35.22	96	74.1	−35.62
72	78.3	−36.04	97	73.9	−35.34
73	78.0	−35.16	98	74.2	−34.57
74	76.0	−34.77	99	73.9	−35.05
75	76.4	−34.94	100	73.2	−35.47

CASE VII.18:   M-STICK MANUFACTURING

At M-Stick Plant I, the lengths of meter sticks are measured with an automatic measuring machine. M-Stick wants to estimate the variation associated with the population of recorded measurements of their product. They also want to estimate the proportion of variation attributable to the product and that attributable to the measuring machine, in order to assess whether or not the measuring machine has adequate precision relative to their product.

They have designed a CRD, random effects, experiment where 20 meter sticks are randomly sampled from production output and carefully measured with the one measuring machine available in the plant. Each one of the meter sticks is independently measured 2 times, yielding two replications. The data in millimeters appear below and on the next page.

Meter Stick	Observation, mm 1	2
1	1000.158	1000.157
2	1000.003	1000.006
3	999.722	999.720
4	1000.303	1000.305
5	1000.216	1000.217
6	999.636	999.638
7	1000.153	1000.153

(continued)

(concluded)

	Observation, mm	
Meter Stick	**1**	**2**
8	1000.383	1000.381
9	999.754	999.757
10	1000.237	1000.238
11	999.968	999.968
12	1000.073	1000.075
13	1000.186	1000.184
14	999.948	999.945
15	999.868	999.869
16	999.951	999.952
17	999.900	999.903
18	1000.387	1000.387
19	999.961	999.962
20	999.813	999.812

At M-Stick Plant II, a different measurement process is used to measure the meter sticks. This plant has a purpose similar to that of M-Stick Plant I; however, they use a manual measuring device operated by machine attendants. Here, the objective is to isolate the operator as well as the measuring device portions of the variation. From a practical point of view, the objective is to compare the operator and measuring machine portions of variation, relative to the product metrics, and determine if operator training and/or machine upgrades are needed.

M-Stick Plant II has designed a two-factor random effects CRD experiment to address their needs. Once again, a total of 20 meter sticks were selected at random. Then, 3 operators were selected at random. The same measuring device was used by all operators to measure each meter stick 2 times. The measuring device was carefully calibrated before the experiment. The data collected (in millimeters) are shown below.

	Operator 1 Observations, mm		Operator 2 Observations, mm		Operator 3 Observations, mm	
Meter Stick	**1**	**2**	**1**	**2**	**1**	**2**
1	1000.16	1000.18	1000.08	999.99	1000.25	1000.28
2	1000.00	999.89	999.97	999.93	1000.11	1000.09
3	999.74	999.84	999.73	999.76	999.83	999.87
4	1000.29	1000.25	1000.28	1000.29	1000.33	1000.31
5	1000.21	1000.20	1000.12	1000.08	1000.27	1000.29
6	999.74	999.81	999.75	999.71	999.79	999.80
7	1000.16	1000.05	1000.03	1000.08	1000.21	1000.23
8	1000.38	1000.39	1000.33	1000.30	1000.44	1000.42
9	999.85	999.90	999.87	999.83	999.87	999.87
10	1000.24	1000.18	1000.21	1000.20	1000.21	1000.23
11	999.94	999.92	999.91	999.92	999.99	999.96
12	1000.07	1000.08	1000.08	1000.02	1000.12	1000.11
13	1000.19	1000.19	1000.14	1000.11	1000.23	1000.25
14	999.99	1000.02	999.97	999.93	1000.13	1000.11
15	999.87	999.82	999.85	999.81	999.93	999.93
16	999.95	999.91	999.98	999.94	999.98	999.99
17	999.90	999.87	999.84	999.81	999.97	999.96
18	1000.40	1000.47	1000.31	1000.34	1000.41	1000.41
19	999.96	999.96	999.93	999.95	999.95	999.97
20	999.79	999.82	999.72	999.68	999.89	999.90

CASE VII.19: NIGHT HAULER TRUCKING

The Night Hauler Trucking Company is considering the use of a new diesel fuel additive that is claimed to effectively capture more usable energy from diesel fuel by enhancing the combustion process. The manufacturer of the additive recommends using 150 ml of additive per 10 gallons of diesel fuel. The Night Haulers are not sure that they want to use the additive, but are convinced that they should give it a try. Hence, they have set up an experiment to determine what amount, if any, of the additive should be used in their truck fleet.

In order to test the effectiveness of the additive, a team of Night Haulers has devised a CRD experiment using diesel additive levels as fixed effects treatment levels. They have selected 3 similar light trucks and tested the mileage of each truck with 0, 50, 100, and 150 ml of additive per 10 gallons of diesel. Since the Night Haulers are not capable of developing detailed measurements in combustion engineering, they have decided to use a substitute process performance measure, miles per gallon of diesel fuel, in their field operations. The response data collected are shown below. [JM]

	Observation, mpg		
Additive Level, ml/10 gal	1	2	3
0	13.5 (8)*	14.6 (3)	15.6 (12)
50	17.8 (5)	18.8 (2)	18.2 (10)
100	16.8 (7)	20.6 (9)	22.8 (4)
150	18.9 (1)	17.7 (11)	23.9 (6)

*Numbers in parentheses indicate experimental ordering.

CASE VII.20: PCB—PRINTED CIRCUIT BOARDS

A precision automatic indexing and feeding mechanism is coordinated with a computer numerical control (CNC) drilling machine in a printed circuit board production process. The objective is to produce a number of holes at specified x, y coordinates. The indexing and feeding mechanism indexes a part into place, and then the CNC drilling head locates the first hole and drills it. Then, it locates the next hole and drills it. Finally, the workpiece is indexed on down the line.

Deviation from target is a critical dimension. A deviation of 0, 0, for x and y, respectively, is best; hence

x and y deviation targets: $\mu_0 = 0$ thousandths, $\sigma_0 = 1.2$ thousandths of an inch

Data collected from the process for two holes on each board are displayed on the following page. Deviations are expressed in thousandths of an inch.

| | Hole | | | | | | Hole | | | |
| | 1 | | 2 | | | | 1 | | 2 | |
Sample	*x* dev	*y* dev	*x* dev	*y* dev	Sample		*x* dev	*y* dev	*x* dev	*y* dev
1	−1.484	0.705	−0.878	0.715	26		1.212	−0.532	0.047	−0.286
2	−1.424	−1.253	−0.071	−0.316	27		1.019	−0.945	1.521	−0.870
3	0.258	−0.964	0.341	−0.412	28		−0.437	1.228	−0.444	0.465
4	0.464	0.857	1.051	0.157	29		−1.238	−0.372	−0.130	−0.855
5	−1.278	−1.077	−0.625	−1.297	30		1.013	0.967	−0.550	1.781
6	1.282	0.992	1.198	2.132	31		1.103	0.287	0.235	1.191
7	0.590	0.598	1.417	0.370	32		0.609	−0.565	1.711	0.329
8	0.588	1.303	1.564	0.309	33		−0.435	0.798	−0.833	2.469
9	−0.519	−1.692	−0.335	−0.200	34		−0.652	−0.109	−0.005	−0.221
10	−0.339	−0.549	−0.323	−0.032	35		0.339	0.480	1.105	0.878
11	0.222	0.794	0.026	0.032	36		0.917	−0.545	−1.328	0.171
12	2.179	0.982	1.252	0.258	37		−1.783	1.407	−1.553	0.474
13	−0.827	−0.126	−0.569	−0.483	38		−0.817	−0.507	−0.856	−0.307
14	0.537	0.329	−0.964	0.861	39		−1.010	−0.828	−0.316	−0.579
15	0.422	−1.293	0.594	−0.488	40		−0.326	−0.553	−1.057	−1.254
16	−1.130	0.452	−0.294	0.006	41		1.245	1.541	0.549	1.050
17	1.375	0.484	0.154	1.244	42		0.380	0.789	0.979	0.738
18	−0.183	0.197	−1.034	1.558	43		−0.563	0.285	−0.600	0.026
19	1.098	−0.513	0.379	−2.770	44		0.432	0.434	0.517	1.345
20	−2.080	0.323	−1.709	1.833	45		−1.570	0.120	−1.120	0.794
21	−0.241	0.703	−2.180	0.951	46		−0.667	1.061	−0.691	1.111
22	−0.068	−0.852	−0.341	−0.961	47		0.299	0.271	−0.103	0.598
23	0.501	−0.139	0.878	−0.198	48		−0.779	1.262	−0.133	−0.352
24	0.852	−0.732	0.891	−0.800	49		−0.311	−0.955	0.633	−1.674
25	−1.020	0.679	−0.614	0.183	50		−0.794	−1.065	0.408	−0.588

Response is deviation from target, in thousandths of an inch.

CASE VII.21:   PUNCH-OUT—SHEET METAL FABRICATION

A production system specializes in sheet metal fabrication. A wide variety of parts are fabricated for assembly. One simple part involves shearing a small flat blank from coil stock. The material is fed to a stop and then the shear is engaged to cut the length dimension. Next, the sheared strip is moved to a second shear and placed against the stop; the second shear is engaged, and the cut to width is produced. The sheared-off metal is recycled. The part is then moved to the next operation, a break operation for a bend. The sheet metal part specifications and targets are as follows:

Length specifications: 6 ± 0.05 inches

Width specifications: 4 ± 0.05 inches

Length targets: $\mu_0 = 6$ inches, $\sigma_0 = 0.015$ inches

Width targets: $\mu_0 = 4$ inches, $\sigma_0 = 0.015$ inches

Data were collected in subgroups. Each subgroup contains four parts, where each part is measured for both length and width.

Length data, inches.

Subgroup	Observation				Subgroup	Observation			
	1	2	3	4		1	2	3	4
1	6.008	6.027	6.004	5.977	26	5.994	6.007	6.022	5.988
2	5.994	5.993	5.992	5.975	27	6.002	5.995	6.025	6.009
3	6.016	6.005	6.026	5.986	28	5.989	5.982	6.000	5.991
4	5.996	6.012	5.986	5.993	29	5.997	5.984	5.999	5.993
5	6.019	6.017	6.013	5.974	30	6.001	5.995	5.987	6.015
6	5.991	5.998	6.013	6.017	31	5.985	5.992	5.987	5.990
7	5.996	5.969	6.003	5.997	32	5.981	5.981	5.994	6.016
8	5.956	5.982	6.025	5.993	33	5.977	5.985	6.021	5.994
9	5.973	6.015	5.997	6.023	34	6.004	6.028	6.053	6.007
10	6.016	5.997	5.998	6.005	35	5.998	6.006	5.995	5.999
11	6.029	6.011	6.004	5.996	36	6.015	6.008	5.993	6.008
12	5.985	6.019	5.990	6.008	37	5.977	6.001	6.003	6.005
13	6.013	5.993	5.995	6.044	38	5.993	5.994	5.976	5.992
14	5.997	5.993	6.017	6.004	39	6.027	6.018	6.011	5.991
15	6.011	6.002	6.032	6.007	40	5.993	6.002	6.001	6.000
16	5.987	6.017	6.025	5.986	41	5.987	6.000	5.996	6.012
17	5.995	5.997	6.012	5.997	42	5.982	5.987	5.988	6.007
18	5.996	6.038	6.008	6.007	43	6.023	5.992	6.001	6.007
19	5.973	6.012	6.006	6.004	44	6.004	6.003	6.050	5.997
20	6.006	5.991	5.980	5.989	45	6.034	6.014	5.996	6.005
21	5.967	5.997	6.013	6.023	46	5.990	6.012	5.975	6.038
22	6.002	6.028	5.986	5.996	47	5.991	6.021	6.001	5.996
23	6.019	5.989	6.008	6.007	48	5.997	5.973	6.002	5.964
24	5.995	5.999	6.006	6.011	49	6.028	6.011	6.004	6.017
25	5.998	5.985	5.996	6.001	50	5.998	6.005	6.003	6.015

Width data, inches.

Subgroup	Observation				Subgroup	Observation			
	1	2	3	4		1	2	3	4
1	3.983	3.992	3.981	4.023	16	4.019	3.993	3.983	4.011
2	4.007	3.999	4.005	3.992	17	3.998	4.015	4.008	3.990
3	4.001	4.017	3.986	4.004	18	3.992	4.014	4.014	3.993
4	4.012	3.993	4.007	3.981	19	4.003	4.015	3.996	4.015
5	4.049	3.987	4.024	3.985	20	3.979	4.005	4.004	4.006
6	4.035	3.991	3.989	4.006	21	4.028	4.005	3.997	4.002
7	3.980	3.971	3.986	4.008	22	3.994	4.026	4.028	4.022
8	3.982	3.998	4.029	3.983	23	3.980	3.992	3.997	3.984
9	3.991	3.974	4.012	4.005	24	3.990	4.018	3.989	3.984
10	4.001	4.029	3.974	4.019	25	4.000	3.995	4.003	3.995
11	4.016	3.993	4.019	3.993	26	4.016	4.002	4.022	4.003
12	4.006	3.986	4.004	3.979	27	3.995	3.981	4.018	4.012
13	3.987	3.998	4.016	3.976	28	4.003	3.998	3.972	3.981
14	3.991	3.990	4.015	4.004	29	4.004	3.985	4.000	4.003
15	4.013	3.992	3.998	3.994	30	4.007	3.981	4.013	4.028

(continued)

Width data, inches *(concluded)*.

Subgroup	Observation				Subgroup	Observation			
	1	**2**	**3**	**4**		**1**	**2**	**3**	**4**
31	4.004	3.986	4.002	4.006	41	4.013	3.972	3.998	3.995
32	4.004	3.998	3.993	3.986	42	4.005	3.995	4.009	4.004
33	4.012	3.985	3.994	3.987	43	3.993	3.981	3.997	3.973
34	4.011	4.008	3.988	3.985	44	3.977	3.989	3.982	4.017
35	4.008	4.018	3.989	3.972	45	3.991	4.007	4.012	4.013
36	4.026	3.982	3.985	3.995	46	4.031	4.010	4.015	4.011
37	3.978	3.984	3.993	3.995	47	4.003	3.997	3.998	3.983
38	4.013	4.005	3.979	3.984	48	4.006	4.002	4.019	3.978
39	3.984	4.006	3.982	3.996	49	4.012	4.012	4.001	3.979
40	4.029	4.000	3.994	3.980	50	4.001	3.998	3.993	3.947

CASE VII.22:   RE-USE—RECYCLING

A recycling process shreds waste paper. The shredder uses blades to sever the feedstock. These blades wear and require sharpening or replacing periodically. Since blade wear is difficult to measure directly, the recycler uses shredder energy demand to meter blade wear. As the blade edges become duller, the energy demand increases.

The feedstock is manually inspected before entering the shredder, but occasionally a foreign object, such as a rock or piece of metal, is encountered. In some cases, the foreign object creates blade damage that must be repaired. At some point, the machine is taken out of service and the worn or damaged blades removed and replaced with sharp blades. The replacement process is both expensive and demanding.

The energy requirement is easy and cheap to monitor and appears to provide a good indication of when to change blades. These questions arise: What constitutes a stable process? How can we best determine when blade maintenance or replacement is necessary? Currently, when the energy hits about 350 energy units with a load of about 7 mass units, the recycler considers the blades to be marginal. The following data were collected under normal operating conditions:

Sample, Hour	Throughput, Mass Units	Energy Demand, Energy Units	Sample, Hour	Throughput, Mass Units	Energy Demand, Energy Units
1	9.7	106.447	14	12.9	99.240
2	12.1	99.078	15	11.6	110.831
3	11.4	91.267	16	11.2	98.535
4	10.3	90.114	17	12.7	97.990
5	11.8	104.645	18	11.3	104.580
6	11.3	114.378	19	10.9	102.142
7	12.4	87.984	20	10.7	120.329
8	11.9	94.770	21	12.2	128.297
9	13.1	95.204	22	11.6	113.539
10	12.7	100.652	23	11.5	113.669
11	10.8	103.991	24	12.5	124.445
12	11.7	106.580	25	12.3	114.228
13	12.8	100.314	26	11.4	119.465

(continued)

(concluded)

Sample, Hour	Throughput, Mass Units	Energy Demand, Energy Units	Sample, Hour	Throughput, Mass Units	Energy Demand, Energy Units
27	12.5	115.793	64	11.1	209.014
28	12.8	129.162	65	10.9	217.811
29	11.3	126.771	66	9.7	199.567
30	11.8	128.278	67	9.9	213.724
31	11.7	123.030	68	10.1	225.122
32	11.3	128.806	69	9.5	240.139
33	12.6	133.467	70	8.9	229.843
34	10.4	138.900	71	9.7	250.434
35	11.7	131.191	72	8.8	228.031
36	10.9	132.454	73	9.4	238.176
37	12.7	128.393	74	8.6	226.109
38	11.5	149.461	75	9.3	256.187
39	10.9	150.968	76	8.6	246.975
40	10.5	132.318	77	8.5	244.152
41	12.4	127.813	78	9.1	257.130
42	11.0	139.329	79	8.8	260.265
43	11.2	146.517	80	9.0	254.173
44	10.8	139.308	81	8.4	251.569
45	10.2	153.987	82	8.7	263.158
46	11.5	160.683	83	7.6	267.906
47	11.0	167.568	84	7.9	273.626
48	10.8	167.280	85	7.5	277.654
49	10.4	161.022	86	8.9	280.700
50	10.4	174.039	87	7.7	290.072
51	9.9	173.994	88	8.5	310.794
52	10.0	178.565	89	7.6	312.479
53	10.3	169.168	90	8.2	308.329
54	11.1	180.335	91	7.4	321.598
55	10.7	168.701	92	8.1	343.562
56	11.5	192.279	93	7.9	342.937
57	10.6	188.940	94	6.8	375.892
58	11.0	168.142	95	7.1	360.271
59	9.8	210.042	96	7.2	385.947
60	10.2	190.630	97	6.4	392.730
61	10.6	190.651	98	6.9	410.693
62	10.3	201.389	99	7.3	400.683
63	10.0	213.290	100	6.6	432.819

CASE VII.23: RE-USE—SENSOR PRECISION

An automatic energy sensor is used to measure the energy usage level for the shredding sub-process described in Case VII.22. The precision of the sampler is an unknown at this point. In order to assess the sampler precision (measurement error) in the instrumentation, the sampler was set to produce three measurements within a 10-second interval. The purpose of this experiment is to quantify the sampler precision and to assess if it is small relative to the magnitude of the energy measurement. The data collected are shown on the next page.

	Observation, Energy Units		
Sample	**1**	**2**	**3**
1	86.85	85.65	86.75
2	115.37	114.88	114.32
3	106.16	109.70	109.22
4	119.99	120.25	114.88
5	122.46	120.33	124.13
6	105.32	96.26	103.72
7	119.09	118.08	117.37
8	119.76	123.81	112.24
9	108.59	109.66	111.11
10	112.83	115.53	115.74
11	126.58	128.45	124.08
12	121.29	119.80	115.99
13	127.15	132.50	129.13
14	121.96	125.77	124.18
15	145.87	143.61	147.04
16	132.77	132.60	131.13
17	143.16	144.83	148.00
18	143.98	135.39	143.47
19	169.67	161.43	169.02
20	148.64	151.19	145.92
21	153.08	150.67	138.66
22	167.94	171.21	176.14
23	150.83	155.83	142.44
24	172.75	178.95	167.85
25	189.74	188.76	188.63
26	188.85	197.65	176.06
27	194.93	197.58	201.63
28	171.82	173.94	174.48
29	205.95	203.50	208.40
30	235.27	235.90	237.98
31	206.84	207.32	212.31
32	232.00	229.58	230.72
33	237.03	236.57	241.38
34	252.43	258.64	252.86
35	238.74	241.26	237.26
36	265.23	271.28	263.56
37	279.52	280.57	279.34
38	274.29	271.36	274.80
39	282.18	287.94	282.18
40	309.64	314.19	301.60
41	296.81	292.93	307.15
42	334.24	338.26	330.94
43	330.97	334.25	333.71
44	339.14	330.78	334.25
45	320.99	327.00	325.96
46	338.23	327.38	336.49
47	331.23	348.28	325.93
48	338.70	336.27	340.18
49	343.84	350.63	348.99
50	354.06	360.00	352.70

CASE VII.24:   SILVER BIRD—BAGGAGE

Customers who check their bags during airline travel expect to receive their bags at the end of their journey, on time and in the same condition as when they checked them at the ticket counter. The baggage handling subprocess consists of several activities. First, the bags are checked, labeled, and moved to a staging area. Then, the bags are loaded in the cargo bay of the appropriate aircraft. As the trip unfolds, bags may be moved from one flight to the next. Once at the final destination, the bags are unloaded and transported to the baggage pickup area.

Silver Bird collected data on checked baggage. Each sample consisted of 1000 bags. The metrics recorded were late arrivals, number of damage points observed, and lost baggage. A failure as defined by the customer is any event whereby a bag (in the same condition as it was checked) is not waiting at the termination of the flight. Silver Bird has set a goal of total customer satisfaction. Data are displayed below and on the next page.

Sample	Number of Late Bags	Number of Damage Points	Number of Lost Bags	Dissatisfied Customers
1	25	3	0	28
2	43	4	1	47
3	31	4	1	35
4	17	1	2	18
5	42	5	1	47
6	33	7	3	40
7	23	4	3	27
8	40	6	1	46
9	37	3	5	40
10	29	2	2	31
11	31	2	4	33
12	23	0	2	23
13	45	3	3	48
14	21	8	2	29
15	34	4	3	38
16	28	1	5	29
17	37	3	4	40
18	19	6	1	25
19	27	3	0	30
20	38	3	0	41
21	24	5	1	29
22	32	4	3	36
23	27	5	3	32
24	17	3	1	20
25	53	5	0	58
26	47	8	1	55
27	36	7	1	42
28	41	3	1	44
29	38	1	0	39
30	44	1	1	45
31	30	5	0	35
32	33	3	1	36
33	26	7	2	33

(continued)

(concluded)

Sample	Number of Late Bags	Number of Damage Points	Number of Lost Bags	Dissatisfied Customers
34	45	6	0	51
35	34	2	2	36
36	31	1	0	32
37	28	6	4	34
38	37	4	3	41
39	29	3	2	32
40	31	2	0	33

CASE VII.25: SNAPPY—PLASTIC INJECTION MOLDING

An injection molding subprocess transforms plastic pellets into molded plastic parts that must snap into an assembly. Two parts are produced for every shot (cycle) of the injection molding equipment. Defects are primarily of two types: (1) short shots, where the mold cavity is not entirely filled and a void exists and (2) flash, locations on the parts where plastic extends beyond the edge of the mold cavity. The general requirements for the subprocess/product are zero short shots and zero flash locations. The following data pertaining to short shots and flashes were collected:

Attributes data (two parts per shot).

Sample	Sample Size, 100s of Shots	Number of Short Shots	Number of Flash Locations	Number of Defective Parts
1	5	3	7	20
2	7	1	5	8
3	4	0	9	15
4	9	4	11	30
5	3	0	2	4
6	4	2	3	10
7	2	0	2	4
8	8	5	12	25
9	7	3	14	30
10	5	2	4	12
11	1	0	0	0
12	7	3	10	26
13	4	4	8	20
14	3	2	5	14
15	5	4	15	25
16	4	3	12	30
17	6	2	15	32
18	4	1	9	20
19	7	4	12	32
20	5	2	9	20
21	5	3	11	28
22	3	1	6	12
23	4	3	9	24

(continued)

Attributes data (two parts per shot) *(concluded)*.

Sample	Sample Size, 100s of Shots	Number of Short Shots	Number of Flash Locations	Number of Defective Parts
24	6	4	13	34
25	3	0	8	16
26	5	2	12	28
27	7	4	16	40
28	4	2	10	20
29	6	4	7	22
30	5	4	11	22

Case VII.26: Squeaky Clean Laundry

The Squeaky Clean Laundry prides itself in stain removal. They have a money-back guarantee for any stain of any type; if they can't remove it, you don't pay. Squeaky Clean personnel have been approached by three vendors, each claiming their spot remover will outclean all others (guaranteed). Furthermore, each vendor has multicolored charts and brochures claiming that third-party tests at "a leading university" prove ($\alpha = 0.05$) that their spot removers are more effective than their competitors'. The products offered are as follows:

Vendor 1—"Spotfree" tested at University X

Vendor 2—"Spotless" tested at University Y

Vendor 3—"Spotaway" tested at University Z

At the present time, Squeaky Clean is using a product furnished by a fourth vendor, called "Nospot." It so happens that all of the three calling vendors' test results, written up in their brochures, include "Spotfree," "Spotless," "Spotaway," and "Nospot."

Squeaky Clean personnel are confused with all of the "scientific" claims. Their purpose is to remove spots. How can/should they interpret the vendors' claims? How can/should they evaluate the spot removers?

Case VII.27: Rainbow—Paint Coating

In response to customer complaints regarding paint deterioration and peeling on automotive products, a manufacturer now offers two basic paint options, (1) standard metallic and (2) rough service metallic, in several colors. The rough service (RS-metallic) is more expensive, but is selling very well. The paint subprocess is an in-line process. The paint is shot through the same equipment, with changeovers in paint between production units, as necessary.

The color does not seem to impact the process, but the standard and rough service paints are different in their flow and coating characteristics. The requirement for the standard metallic paint coat thickness is at least 0.2 mm and for the rough service metallic at least 0.4 mm. Additional thickness beyond the minimum requirement does not adversely affect the product, but does present additional material costs.

Since a physical measurement of paint thickness is necessary, a sample coupon is used in the paint subprocess. It is painted, dried with the part, scraped, and measured for thickness. Data are collected on the paint thickness on the coupons. Paint thickness is measured in millimeters.

Standard paint targets: $\mu_0 = 0.25$ mm, $\sigma_0 = 0.01$ mm

Rough service paint targets: $\mu_0 = 0.50$ mm, $\sigma_0 = 0.02$ mm

Variables data—paint thickness, mm.

Sample	Paint Type	Target	Thickness	Sample	Paint Type	Target	Thickness
1	Metallic	0.2	0.239	43	Metallic	0.2	0.264
2	RS-metallic	0.4	0.475	44	RS-metallic	0.4	0.492
3	RS-metallic	0.4	0.489	45	RS-metallic	0.4	0.506
4	RS-metallic	0.4	0.484	46	RS-metallic	0.4	0.504
5	Metallic	0.2	0.259	47	Metallic	0.2	0.247
6	Metallic	0.2	0.253	48	Metallic	0.2	0.243
7	RS-metallic	0.4	0.503	49	RS-metallic	0.4	0.495
8	Metallic	0.2	0.254	50	RS-metallic	0.4	0.470
9	Metallic	0.2	0.250	51	Metallic	0.2	0.254
10	RS-metallic	0.4	0.474	52	Metallic	0.2	0.263
11	RS-metallic	0.4	0.503	53	RS-metallic	0.4	0.483
12	RS-metallic	0.4	0.492	54	RS-metallic	0.4	0.492
13	Metallic	0.2	0.241	55	RS-metallic	0.4	0.525
14	Metallic	0.2	0.252	56	Metallic	0.2	0.237
15	RS-metallic	0.4	0.510	57	Metallic	0.2	0.256
16	Metallic	0.2	0.255	58	RS-metallic	0.4	0.544
17	RS-metallic	0.4	0.522	59	Metallic	0.2	0.252
18	Metallic	0.2	0.251	60	RS-metallic	0.4	0.496
19	Metallic	0.2	0.252	61	Metallic	0.2	0.245
20	Metallic	0.2	0.249	62	RS-metallic	0.4	0.513
21	RS-metallic	0.4	0.481	63	RS-metallic	0.4	0.493
22	Metallic	0.2	0.263	64	RS-metallic	0.4	0.532
23	Metallic	0.2	0.249	65	RS-metallic	0.4	0.486
24	RS-metallic	0.4	0.477	66	Metallic	0.2	0.249
25	Metallic	0.2	0.250	67	RS-metallic	0.4	0.503
26	RS-metallic	0.4	0.521	68	Metallic	0.2	0.266
27	Metallic	0.2	0.255	69	RS-metallic	0.4	0.496
28	Metallic	0.2	0.261	70	Metallic	0.2	0.230
29	Metallic	0.2	0.240	71	Metallic	0.2	0.235
30	RS-metallic	0.4	0.485	72	RS-metallic	0.4	0.446
31	RS-metallic	0.4	0.487	73	RS-metallic	0.4	0.505
32	RS-metallic	0.4	0.542	74	RS-metallic	0.4	0.513
33	Metallic	0.2	0.255	75	RS-metallic	0.4	0.514
34	Metallic	0.2	0.241	76	Metallic	0.2	0.242
35	RS-metallic	0.4	0.474	77	Metallic	0.2	0.266
36	Metallic	0.2	0.255	78	Metallic	0.2	0.250
38	Metallic	0.2	0.274	79	Metallic	0.2	0.244
37	Metallic	0.2	0.237	80	Metallic	0.2	0.241
39	RS-metallic	0.4	0.498	81	RS-metallic	0.4	0.509
40	RS-metallic	0.4	0.477	82	RS-metallic	0.4	0.487
41	Metallic	0.2	0.247	83	RS-metallic	0.4	0.480
42	Metallic	0.2	0.250	84	Metallic	0.2	0.244

(continued)

Variables data—paint thickness, mm *(concluded).*

Sample	Paint type	Target	Thickness	Sample	Paint type	Target	Thickness
85	Metallic	0.2	0.264	93	Metallic	0.2	0.248
86	RS-metallic	0.4	0.500	94	Metallic	0.2	0.251
87	Metallic	0.2	0.244	95	Metallic	0.2	0.252
88	RS-metallic	0.4	0.507	96	RS-metallic	0.4	0.506
89	RS-metallic	0.4	0.504	97	RS-metallic	0.4	0.498
90	RS-metallic	0.4	0.475	98	RS-metallic	0.4	0.495
91	RS-metallic	0.4	0.489	99	Metallic	0.2	0.249
92	RS-metallic	0.4	0.516	100	Metallic	0.2	0.257

CASE VII.28: SURE-STICK ADHESIVE

The Sure-Stick Adhesive Corporation manufactures several product lines of commercial and industrial adhesives for the automotive industry. A new product line of high-strength industrial adhesives, made exclusively for one of its automotive customers, is being developed and tested.

In trials at the automotive customer's facilities, problems in the shear strength of the new adhesive have surfaced. In fact, the shear strength varied from 600 to 1000 psi; while the customer strength specification requires at least 800 psi. Sure-Stick Corporation also states a guarantee that its minimum shear strength will be at least 900 psi. The automotive customer, who has been a loyal Sure-Stick customer for years, has requested that Sure-Stick address this strength problem. If Sure-Stick cannot "fix" this problem, their automotive customer will consider other adhesive suppliers.

In an attempt to keep one of their most profitable accounts, and to avoid tarnishing their quality reputation, Sure-Stick developed an experimental program. First, Sure-Stick organized a team of engineers, technicians, and operators both from their own company as well as from their customer's facilities. This team's objective is to maximize the shear strength of the adhesive so as to keep the customer satisfied. After a cause-effect analysis, this group produced a list of factors that are most likely affecting the process:

A: Volume of additive

B: Application pressure (psi)

C: Application temperature

The team decided to perform a 3-factor central composite design experiment and develop a response surface so as to isolate the best combination of these factors. The purpose is to assess product strength with regard to customer requirements and manufacturer claims, under "optimal" application conditions. The data collected are shown on the next page. [RM]

Trial	Additive, ml	Additive Coded Units	Application Pressure, Psi	Application Pressure Coded Units	Application Temperature, ºF	Application Temperature Coded Units	Shear Strength, 100 psi
1	20	−1	50	−1	100	−1	6.6
2	40	1	50	−1	100	−1	8.0
3	40	1	50	−1	200	1	7.5
4	20	−1	50	−1	200	1	7.0
5	20	−1	100	1	100	−1	7.2
6	40	1	100	1	100	−1	9.0
7	40	1	100	1	200	1	8.3
8	20	−1	100	1	200	1	6.0
9	30	0	75	0	150	0	10.0
10	30	0	75	0	150	0	11.4
11	30	0	75	0	150	0	11.6
12	30	0	75	0	150	0	9.7
13	30	0	75	0	150	0	10.2
14	30	0	75	0	150	0	9.8
15	13.18	−1.682	75	0	150	0	10.0
16	46.82	1.682	75	0	150	0	6.0
17	30	0	33.0	−1.682	150	0	6.8
18	30	0	117.0	1.682	150	0	6.3
19	30	0	75	0	65.9	−1.682	6.5
20	30	0	75	0	234.1	1.682	8.1

CASE VII.29:　TEXROSA—SALSA

Salsa production involves the combination of several ingredients, including tomatoes, peppers, onions, and spices. The pH of the mixture is a critical metric in salsa processing, as it determines the nature of the taste. A pH of 3.8 to 4.6 is acceptable. The temperature at the time of container filling is also a critical metric. An acceptable range of 80 to 88°C is sufficient to control bacteria.

Data consisting of 50 subgroups of size 5 were collected during normal operations. Each pH, temperature observation data pair comes from a different batch; i.e., a batch mixing process is used. Reasonable targets for the means and standard deviations are

pH targets: $\mu_0 = 4.2$, $\sigma_0 = 0.15$
Temperature targets: $\mu_0 = 84$, $\sigma_0 = 0.5$

[MS]

Salsa pH data.

	Observation						Observation				
Subgroup	1	2	3	4	5	Subgroup	1	2	3	4	5
1	4.25	4.29	4.21	4.26	4.18	8	4.05	4.30	4.09	3.94	4.26
2	4.24	4.07	4.16	4.16	4.41	9	4.21	4.01	4.31	3.95	4.08
3	4.28	4.31	4.07	4.36	4.09	10	4.21	3.88	4.20	4.28	4.49
4	4.12	4.16	4.17	4.31	4.12	11	4.02	4.43	4.11	4.13	4.15
5	4.22	4.14	3.94	4.13	4.01	12	4.28	4.28	4.55	4.30	4.38
6	4.27	4.12	4.13	4.19	4.34	13	4.39	4.22	4.16	4.35	4.22
7	4.18	4.26	4.14	4.25	4.30	14	4.03	4.40	4.28	4.39	4.31

(continued)

Salsa pH data *(concluded)*.

Subgroup	Observation 1	2	3	4	5	Subgroup	Observation 1	2	3	4	5
15	4.22	4.17	4.45	4.17	4.05	33	4.35	4.07	4.23	4.30	4.25
16	3.82	4.17	4.20	4.46	4.24	34	4.11	4.10	4.02	3.95	4.12
17	4.36	4.41	4.27	4.28	4.31	35	4.10	4.16	4.38	4.38	4.15
18	4.48	4.11	4.16	4.26	3.91	36	4.28	4.29	4.41	3.87	4.13
19	4.10	4.36	4.14	4.13	4.41	37	4.10	4.10	4.63	4.71	4.08
20	4.18	4.31	4.20	4.22	4.23	38	4.20	4.21	4.38	4.13	4.26
21	4.34	4.06	3.90	4.21	4.17	39	4.24	4.14	4.12	4.40	4.33
22	4.01	4.32	4.24	4.13	4.36	40	3.97	4.11	4.17	4.38	4.28
23	4.41	4.13	3.79	4.19	4.12	41	4.34	4.14	4.28	4.20	4.04
24	4.30	4.01	4.52	4.39	4.29	42	4.25	4.12	4.31	4.17	4.09
25	4.21	4.33	3.90	4.28	4.14	43	4.02	4.14	4.18	4.29	4.01
26	4.23	4.00	4.08	4.38	4.23	44	4.16	4.30	3.86	4.05	3.94
27	3.98	4.38	4.35	4.23	4.16	45	4.26	4.16	4.19	4.25	4.31
28	4.29	4.25	4.30	4.11	3.98	46	4.07	4.21	4.21	4.15	4.24
29	4.10	4.13	4.24	4.37	4.22	47	4.18	4.01	4.06	4.15	4.05
30	4.31	4.29	4.35	4.34	4.23	48	4.22	4.48	4.05	3.97	4.36
31	4.47	4.47	4.26	4.08	4.31	49	4.26	4.39	4.12	4.28	4.17
32	4.38	4.30	4.22	4.32	4.34	50	4.26	4.22	4.22	4.29	4.04

Salsa temperature data, °C.

Subgroup	Observation 1	2	3	4	5	Subgroup	Observation 1	2	3	4	5
1	84.04	84.02	84.23	83.47	84.46	26	83.51	83.09	83.72	83.98	83.78
2	84.03	83.70	83.50	84.20	83.19	27	84.30	83.90	84.52	84.05	83.86
3	83.40	83.54	83.71	84.22	84.30	28	84.91	84.40	84.21	84.00	83.65
4	84.35	84.22	83.63	84.12	83.00	29	84.41	84.11	84.02	83.20	84.55
5	84.24	83.69	84.44	83.84	84.05	30	83.76	84.27	83.48	85.03	84.77
6	83.79	83.51	83.53	84.48	83.71	31	84.05	84.19	83.39	83.67	84.19
7	83.40	83.68	84.69	83.95	84.09	32	83.86	83.85	84.30	84.27	84.20
8	84.15	84.69	83.60	83.93	84.17	33	83.58	84.53	84.07	83.51	84.50
9	84.33	84.01	84.07	83.73	83.58	34	83.91	84.38	84.91	84.50	83.72
10	84.17	83.71	83.57	82.56	84.37	35	83.13	83.77	83.43	83.83	84.17
11	84.02	84.19	83.49	84.82	84.03	36	83.78	83.98	83.73	83.47	84.06
12	83.97	84.10	84.26	83.16	84.22	37	83.80	84.17	83.55	84.15	83.86
13	83.52	83.38	83.98	83.51	84.43	38	83.65	83.14	83.43	84.51	83.84
14	84.10	83.72	84.46	83.89	84.52	39	84.37	83.18	83.36	83.86	84.17
15	84.36	84.40	84.83	82.72	83.61	40	84.02	84.25	83.72	83.09	84.38
16	83.46	83.88	84.10	84.67	84.44	41	83.71	83.76	83.61	84.04	84.42
17	84.18	84.03	84.01	84.27	84.41	42	85.15	84.33	84.43	84.40	84.00
18	83.17	84.10	83.70	84.24	83.76	43	83.88	84.50	84.17	83.29	84.17
19	84.50	84.16	84.00	84.27	83.93	44	84.28	84.62	84.07	84.33	83.69
20	83.96	83.95	83.40	84.55	84.42	45	84.84	83.90	84.18	83.82	84.02
21	83.78	83.88	84.94	83.87	84.35	46	83.99	84.88	83.99	83.99	83.70
22	84.38	83.95	84.47	84.43	83.08	47	84.23	84.26	83.35	84.96	83.94
23	83.90	84.35	83.71	84.04	84.64	48	83.88	84.21	84.07	84.07	83.83
24	83.60	84.03	84.20	83.85	83.45	49	83.80	83.76	83.79	84.21	85.08
25	84.72	83.79	84.31	83.83	84.43	50	84.01	84.06	83.57	83.58	84.42

CASE VII.30: TOUGH-SKIN—SHEET METAL WELDING

In an automotive part fabrication shop, a sheet metal spot welding subprocess is used to fasten a door panel to a support structure. Here, the voltage and amperage of the welding equipment are set according to the metal piece-part characteristics. The electrodes hinge around the two parts, holding them together. Then, the welder is engaged for a preset time. The current heats the two metals and the force causes them to fuse together. Finally, the electrodes are released and pulled back.

The clamping force at the electrodes varies, but is measurable. The strength specification target for the weld is 35,000 psi. Any weld less than 30,000 psi is considered a defect. Data collected are shown below and on the next page.

Weld strength targets: $\mu_0 = 35$ kpsi, $\sigma_0 = 2$ kpsi

Sample	Clamp Force, psi	Weld Strength, kpsi	Sample	Clamp Force, psi	Weld Strength, kpsi
1	50.2	36.805	36	53.1	36.315
2	46.3	33.929	37	49.3	34.829
3	61.7	30.793	38	50.9	35.908
4	52.8	33.425	39	45.8	32.887
5	45.6	33.880	40	62.4	28.921
6	48.4	35.982	41	53.8	36.148
7	51.9	35.523	42	47.3	33.868
8	54.2	34.097	43	54.2	34.836
9	52.5	37.369	44	52.3	37.411
10	48.1	34.518	45	44.9	32.643
11	65.4	30.892	46	53.8	35.776
12	63.1	30.553	47	47.2	34.513
13	52.9	34.408	48	46.8	34.483
14	49.3	34.285	49	41.9	33.737
15	48.1	35.208	50	54.7	37.745
16	38.4	29.940	51	48.4	37.763
17	42.1	32.611	52	60.3	33.412
18	39.1	32.996	53	52.4	35.483
19	52.2	41.365	54	49.5	35.193
20	48.3	34.759	55	46.3	34.727
21	53.7	34.332	56	49.8	34.994
22	55.9	34.405	57	58.2	33.073
23	54.7	37.242	58	51.9	36.348
24	48.3	36.972	59	58.4	34.210
25	57.8	34.646	60	39.7	31.931
26	44.3	34.863	61	43.9	33.253
27	41.5	33.972	62	42.6	31.848
28	42.1	32.933	63	53.9	35.638
29	52.5	36.444	64	52.7	34.892
30	39.7	31.545	65	48.1	35.373
31	50.4	35.934	66	49.7	35.312
32	58.6	33.517	67	48.2	36.399
33	48.3	34.107	68	63.5	31.920
34	41.8	32.887	69	64.8	32.879
35	53.1	36.803	70	46.5	34.407

(continued)

(concluded)

Sample	Clamp Force, psi	Weld Strength, kpsi	Sample	Clamp Force, psi	Weld Strength, kpsi
71	47.3	34.284	86	54.3	37.317
72	38.3	32.580	87	53.6	37.723
73	51.4	38.367	88	49.1	35.821
74	50.9	37.079	89	52.3	37.579
75	39.0	31.880	90	43.0	34.236
76	57.1	37.697	91	49.4	36.348
77	45.0	32.620	92	48.2	36.307
78	46.2	33.249	93	59.6	31.929
79	52.8	35.517	94	57.7	31.436
80	48.0	34.383	95	48.8	34.757
81	63.2	33.154	96	42.1	32.053
82	50.0	36.255	97	56.3	37.884
83	46.7	33.551	98	59.4	34.100
84	53.3	38.474	99	51.5	36.905
85	38.0	29.663	100	45.0	34.285

STATISTICAL TABLES

Table VIII.1   Cumulative standard normal distribution table

Table VIII.2   *t* distribution table—critical values

Table VIII.3   Chi-squared distribution table—critical values

Table VIII.4   *F* distribution tables—critical values

Table VIII.5   *X*-bar, *R,* and *S* control chart—3-sigma limit constants

Table VIII.6   *X*-bar, *R,* and *S* control chart—probability limit constants

Table VIII.7   EWMA and EWMD control chart limit constants

Table VIII.8   Tabled pseudo-standard normal random numbers

Table VIII.9   Normal probability plotting paper

Table VIII.1   Cumulative standard normal distribution table.

				Areas Under the Normal Curve [$\Phi(z)$]						
z	.00	.01	.02	.03	.04	.05	.06	.07	.08	.09
−3.4	.0003	.0003	.0003	.0003	.0003	.0003	.0003	.0003	.0003	.0002
−3.3	.0005	.0005	.0005	.0004	.0004	.0004	.0004	.0004	.0004	.0003
−3.2	.0007	.0007	.0006	.0006	.0006	.0006	.0006	.0005	.0005	.0005
−3.1	.0010	.0009	.0009	.0009	.0008	.0008	.0008	.0008	.0007	.0007
−3.0	.0013	.0013	.0013	.0012	.0012	.0011	.0011	.0011	.0010	.0010
−2.9	.0019	.0018	.0017	.0017	.0016	.0016	.0015	.0015	.0014	.0014
−2.8	.0026	.0025	.0024	.0023	.0023	.0022	.0021	.0021	.0020	.0019
−2.7	.0035	.0034	.0033	.0032	.0031	.0030	.0029	.0028	.0027	.0026
−2.6	.0047	.0045	.0044	.0043	.0041	.0040	.0039	.0038	.0037	.0036
−2.5	.0062	.0060	.0059	.0057	.0055	.0054	.0052	.0051	.0049	.0048
−2.4	.0082	.0080	.0078	.0075	.0073	.0071	.0069	.0068	.0066	.0064
−2.3	.0107	.0104	.0102	.0099	.0096	.0094	.0091	.0089	.0087	.0084
−2.2	.0139	.0136	.0132	.0129	.0125	.0122	.0119	.0116	.0113	.0110
−2.1	.0179	.0174	.0170	.0166	.0162	.0158	.0154	.0150	.0146	.0143
−2.0	.0228	.0222	.0217	.0212	.0207	.0202	.0197	.0192	.0188	.0183
−1.9	.0287	.0281	.0274	.0268	.0262	.0256	.0250	.0244	.0239	.0233
−1.8	.0359	.0352	.0344	.0336	.0329	.0322	.0314	.0307	.0301	.0294
−1.7	.0446	.0436	.0427	.0418	.0409	.0401	.0392	.0384	.0375	.0367
−1.6	.0548	.0537	.0526	.0516	.0505	.0495	.0485	.0475	.0465	.0455
−1.5	.0668	.0655	.0643	.0630	.0618	.0606	.0594	.0582	.0571	.0559
−1.4	.0808	.0793	.0778	.0764	.0749	.0735	.0722	.0708	.0694	.0681
−1.3	.0968	.0951	.0934	.0918	.0901	.0885	.0869	.0853	.0838	.0823
−1.2	.1151	.1131	.1112	.1093	.1075	.1056	.1038	.1020	.1003	.0985
−1.1	.1357	.1335	.1314	.1292	.1271	.1251	.1230	.1210	.1190	.1170
−1.0	.1587	.1562	.1539	.1515	.1492	.1469	.1446	.1423	.1401	.1379
−0.9	.1841	.1814	.1788	.1762	.1736	.1711	.1685	.1660	.1635	.1611
−0.8	.2119	.2090	.2061	.2033	.2005	.1977	.1949	.1922	.1894	.1867
−0.7	.2420	.2389	.2358	.2327	.2296	.2266	.2236	.2206	.2177	.2148
−0.6	.2743	.2709	.2676	.2643	.2611	.2578	.2546	.2514	.2483	.2451
−0.5	.3085	.3050	.3015	.2981	.2946	.2912	.2877	.2843	.2810	.2776
−0.4	.3446	.3409	.3372	.3336	.3300	.3264	.3228	.3192	.3156	.3121
−0.3	.3821	.3783	.3745	.3707	.3669	.3632	.3594	.3557	.3520	.3483
−0.2	.4207	.4168	.4129	.4090	.4052	.4013	.3974	.3936	.3897	.3859
−0.1	.4602	.4562	.4522	.4483	.4443	.4404	.4364	.4325	.4286	.4247
−0.0	.5000	.4960	.4920	.4880	.4840	.4801	.4761	.4721	.4681	.4641

(continued)

Table VIII.1   Cumulative standard normal distribution table *(concluded)*.

					Areas Under the Normal Curve [$\Phi(z)$]					
z	.00	.01	.02	.03	.04	.05	.06	.07	.08	.09
0.0	.5000	.5040	.5080	.5120	.5160	.5199	.5239	.5279	.5319	.5359
0.1	.5398	.5438	.5478	.5517	.5557	.5596	.5636	.5675	.5714	.5753
0.2	.5793	.5832	.5871	.5910	.5948	.5987	.6026	.6064	.6103	.6141
0.3	.6179	.6217	.6255	.6293	.6331	.6368	.6406	.6443	.6480	.6517
0.4	.6554	.6591	.6628	.6664	.6700	.6736	.6772	.6808	.6844	.6879
0.5	.6915	.6950	.6985	.7019	.7054	.7088	.7123	.7157	.7190	.7224
0.6	.7257	.7291	.7324	.7357	.7389	.7422	.7454	.7486	.7517	.7549
0.7	.7580	.7611	.7642	.7673	.7704	.7734	.7764	.7794	.7823	.7852
0.8	.7881	.7910	.7939	.7967	.7995	.8023	.8051	.8078	.8106	.8133
0.9	.8159	.8186	.8212	.8238	.8264	.8289	.8315	.8340	.8365	.8389
1.0	.8413	.8438	.8461	.8485	.8508	.8531	.8554	.8577	.8599	.8621
1.1	.8643	.8665	.8686	.8708	.8729	.8749	.8770	.8790	.8810	.8830
1.2	.8849	.8869	.8888	.8907	.8925	.8944	.8962	.8980	.8997	.9015
1.3	.9032	.9049	.9066	.9082	.9099	.9115	.9131	.9147	.9162	.9177
1.4	.9192	.9207	.9222	.9236	.9251	.9265	.9278	.9292	.9306	.9319
1.5	.9332	.9345	.9357	.9370	.9382	.9394	.9406	.9418	.9429	.9441
1.6	.9452	.9463	.9474	.9484	.9495	.9505	.9515	.9525	.9535	.9545
1.7	.9554	.9564	.9573	.9582	.9591	.9599	.9608	.9616	.9625	.9633
1.8	.9641	.9649	.9656	.9664	.9671	.9678	.9686	.9693	.9699	.9706
1.9	.9713	.9719	.9726	.9732	.9738	.9744	.9750	.9756	.9761	.9767
2.0	.9772	.9778	.9783	.9788	.9793	.9798	.9803	.9808	.9812	.9817
2.1	.9821	.9826	.9830	.9834	.9838	.9842	.9846	.9850	.9854	.9857
2.2	.9861	.9864	.9868	.9871	.9875	.9878	.9881	.9884	.9887	.9890
2.3	.9893	.9896	.9898	.9901	.9904	.9906	.9909	.9911	.9913	.9916
2.4	.9918	.9920	.9922	.9925	.9927	.9929	.9931	.9932	.9934	.9936
2.5	.9938	.9940	.9941	.9943	.9945	.9946	.9948	.9949	.9951	.9952
2.6	.9953	.9955	.9956	.9957	.9959	.9960	.9961	.9962	.9963	.9964
2.7	.9965	.9966	.9967	.9968	.9969	.9970	.9971	.9972	.9973	.9974
2.8	.9974	.9975	.9976	.9977	.9977	.9978	.9979	.9979	.9980	.9981
2.9	.9981	.9982	.9982	.9983	.9984	.9984	.9985	.9985	.9986	.9986
3.0	.9987	.9987	.9987	.9988	.9988	.9989	.9989	.9989	.9990	.9990
3.1	.9990	.9991	.9991	.9991	.9992	.9992	.9992	.9992	.9993	.9993
3.2	.9993	.9993	.9994	.9994	.9994	.9994	.9994	.9995	.9995	.9995
3.3	.9995	.9995	.9995	.9996	.9996	.9996	.9996	.9996	.9996	.9997
3.4	.9997	.9997	.9997	.9997	.9997	.9997	.9997	.9997	.9997	.9998
3.5	.9997674									
4.0	.9999683									
5.0	.9999997133									
6.0	.9999999990									

Table VIII.2　t distribution table—critical values.

ν	0.40	0.30	0.20	0.15	0.10	0.05	0.025	0.02	0.015	0.01	0.0075	0.005	0.0025	0.0005
1	0.325	0.727	1.376	1.963	3.078	6.314	12.706	15.895	21.205	31.821	42.434	63.657	127.322	636.590
2	0.289	0.617	1.061	1.386	1.886	2.920	4.303	4.849	5.643	6.965	8.073	9.925	14.089	31.598
3	0.277	0.584	0.978	1.250	1.638	2.353	3.182	3.482	3.896	4.541	5.047	5.841	7.453	12.924
4	0.271	0.569	0.941	1.190	1.533	2.132	2.776	2.999	3.298	3.747	4.088	4.604	5.598	8.610
5	0.267	0.559	0.920	1.156	1.476	2.015	2.571	2.757	3.003	3.365	3.634	4.032	4.773	6.869
6	0.265	0.553	0.906	1.134	1.440	1.943	2.447	2.612	2.829	3.143	3.372	3.707	4.317	5.959
7	0.263	0.549	0.896	1.119	1.415	1.895	2.365	2.517	2.715	2.998	3.203	3.499	4.029	5.408
8	0.262	0.546	0.889	1.108	1.397	1.860	2.306	2.449	2.634	2.896	3.085	3.355	3.833	5.041
9	0.261	0.543	0.883	1.100	1.383	1.833	2.262	2.398	2.574	2.821	2.998	3.250	3.690	4.781
10	0.260	0.542	0.879	1.093	1.372	1.812	2.228	2.359	2.527	2.764	2.932	3.169	3.581	4.587
11	0.260	0.540	0.876	1.088	1.363	1.796	2.201	2.328	2.491	2.718	2.879	3.106	3.497	4.437
12	0.259	0.539	0.873	1.083	1.356	1.782	2.179	2.303	2.461	2.681	2.836	3.055	3.428	4.318
13	0.259	0.537	0.870	1.079	1.350	1.771	2.160	2.282	2.436	2.650	2.801	3.012	3.372	4.221
14	0.258	0.537	0.868	1.076	1.345	1.761	2.145	2.264	2.415	2.624	2.771	2.977	3.326	4.140
15	0.258	0.536	0.866	1.074	1.341	1.753	2.131	2.249	2.397	2.602	2.746	2.947	3.286	4.073
16	0.258	0.535	0.865	1.071	1.337	1.746	2.120	2.235	2.382	2.583	2.724	2.921	3.252	4.015
17	0.257	0.534	0.863	1.069	1.333	1.740	2.110	2.224	2.368	2.567	2.706	2.898	3.222	3.965
18	0.257	0.534	0.862	1.067	1.330	1.734	2.101	2.214	2.356	2.552	2.689	2.878	3.197	3.922
19	0.257	0.533	0.861	1.066	1.328	1.729	2.093	2.205	2.346	2.539	2.674	2.861	3.174	3.883
20	0.257	0.533	0.860	1.064	1.325	1.725	2.086	2.197	2.336	2.528	2.661	2.845	3.153	3.849
21	0.257	0.532	0.859	1.063	1.323	1.721	2.080	2.189	2.328	2.518	2.649	2.831	3.135	3.819
22	0.256	0.532	0.858	1.061	1.321	1.717	2.074	2.183	2.320	2.508	2.639	2.819	3.119	3.792
23	0.256	0.532	0.858	1.060	1.319	1.714	2.069	2.177	2.313	2.500	2.629	2.807	3.104	3.768
24	0.256	0.531	0.857	1.059	1.318	1.711	2.064	2.172	2.307	2.492	2.620	2.797	3.091	3.745
25	0.256	0.531	0.856	1.058	1.316	1.708	2.060	2.167	2.301	2.485	2.612	2.787	3.078	3.725
26	0.256	0.531	0.856	1.058	1.315	1.706	2.056	2.162	2.296	2.479	2.605	2.779	3.067	3.707
27	0.256	0.531	0.855	1.057	1.314	1.703	2.052	2.158	2.291	2.473	2.598	2.771	3.057	3.690
28	0.256	0.530	0.855	1.056	1.313	1.701	2.048	2.154	2.286	2.467	2.592	2.763	3.047	3.674
29	0.256	0.530	0.854	1.055	1.311	1.699	2.045	2.150	2.282	2.462	2.586	2.756	3.038	3.659
30	0.256	0.530	0.854	1.055	1.310	1.697	2.042	2.147	2.278	2.457	2.581	2.750	3.030	3.646
40	0.255	0.529	0.851	1.050	1.303	1.684	2.021	2.125	2.250	2.423	2.542	2.704	2.971	3.551
60	0.254	0.527	0.848	1.045	1.296	1.671	2.000	2.099	2.223	2.390	2.504	2.660	2.915	3.460
120	0.254	0.526	0.845	1.041	1.289	1.658	1.980	2.076	2.196	2.358	2.468	2.617	2.860	3.373
∞	0.253	0.524	0.842	1.036	1.282	1.645	1.960	2.054	2.170	2.326	2.432	2.576	2.807	3.291

Table VIII.3   Chi-squared distribution table—critical values.

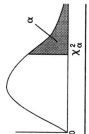

υ = degrees of freedom

υ	.995	.99	.975	.95	.90	.80	.70	.50	.30	.20	.10	.05	.025	.01	.005
1	$.0^3393$	$.0^3157$	$.0^3982$	$.0^3393$	.0158	.0642	.148	.455	1.074	1.642	2.706	3.841	5.024	6.635	7.879
2	.0100	.0201	.0506	.103	.211	.446	.713	1.386	2.408	3.219	4.605	5.991	7.378	9.210	10.597
3	.0717	.115	.216	.352	.584	1.005	1.424	2.366	3.665	4.642	6.251	7.815	9.348	11.345	12.838
4	.207	.297	.484	.711	1.064	1.649	2.195	3.357	4.878	5.989	7.779	9.488	11.143	13.277	14.860
5	.412	.554	.831	1.145	1.610	2.343	3.000	4.351	6.064	7.289	9.236	11.070	12.832	15.086	16.750
6	.676	.872	1.237	1.635	2.204	3.070	3.828	5.348	7.231	8.558	10.645	12.592	14.449	16.812	18.548
7	.989	1.239	1.690	2.167	2.833	3.822	4.671	6.346	8.383	9.803	12.017	14.067	16.013	18.475	20.278
8	1.344	1.646	2.180	2.733	3.490	4.594	5.527	7.344	9.524	11.030	13.362	15.507	17.535	20.090	21.955
9	1.735	2.088	2.700	3.325	4.168	5.380	6.393	8.343	10.656	12.242	14.684	16.919	19.023	21.666	23.589
10	2.156	2.558	3.247	3.940	4.865	6.179	7.267	9.342	11.781	13.442	15.987	18.307	20.483	23.209	25.188
11	2.603	3.053	3.816	4.575	5.578	6.989	8.148	10.341	12.899	14.631	17.275	19.675	21.920	24.725	26.757
12	3.074	3.571	4.404	5.226	6.304	7.807	9.034	11.340	14.011	15.812	18.549	21.026	23.337	26.217	28.300
13	3.565	4.107	5.009	5.892	7.042	8.634	9.926	12.340	15.119	16.985	19.812	22.362	24.736	27.688	29.819
14	4.075	4.660	5.629	6.571	7.790	9.467	10.821	13.339	16.222	18.151	21.064	23.685	26.119	29.141	31.319
15	4.601	5.229	6.262	7.261	8.547	10.307	11.721	14.339	17.322	19.311	22.307	24.996	27.488	30.578	32.801
16	5.142	5.812	6.908	7.962	9.312	11.152	12.624	15.338	18.418	20.465	23.542	26.296	28.845	32.000	34.267
17	5.697	6.408	7.564	8.672	10.085	12.002	13.531	16.338	19.511	21.615	24.769	27.587	30.191	33.409	35.718
18	6.265	7.015	8.231	9.390	10.865	12.857	14.440	17.338	20.601	22.760	25.989	28.869	31.526	34.805	37.156
19	6.844	7.633	8.907	10.117	11.651	13.716	15.352	18.338	21.689	23.900	27.204	30.144	32.852	36.191	38.582
20	7.434	8.260	9.591	10.851	12.443	14.578	16.266	19.337	22.775	25.038	28.412	31.410	34.170	37.566	39.997
21	8.034	8.897	10.283	11.591	13.240	15.445	17.182	20.337	23.858	26.171	29.615	32.671	35.479	38.932	41.401
22	8.643	9.542	10.982	12.338	14.041	16.314	18.101	21.337	24.939	27.301	30.813	33.924	36.781	40.289	42.796
23	9.260	10.196	11.688	13.091	14.848	17.187	19.021	22.337	26.018	28.429	32.007	35.172	38.076	41.638	44.181
24	9.886	10.856	12.401	13.848	15.659	18.062	19.943	23.337	27.096	29.553	33.196	36.415	39.364	42.980	45.558
25	10.520	11.524	13.120	14.611	16.473	18.940	20.867	24.337	28.172	30.675	34.382	37.652	40.646	44.314	46.928
26	11.160	12.198	13.844	15.379	17.292	19.820	21.792	25.336	29.246	31.795	35.563	38.885	41.923	45.642	48.290
27	11.808	12.879	14.573	16.151	18.114	20.703	22.719	26.336	30.319	32.912	36.741	40.113	43.194	46.963	49.645
28	12.461	13.565	15.308	16.928	18.939	21.588	23.647	27.336	31.391	34.027	37.916	41.337	44.461	48.278	50.993
29	13.121	14.256	16.047	17.708	19.768	22.475	24.577	28.336	32.461	35.139	39.087	42.557	45.722	49.588	52.336
30	13.787	14.953	16.791	18.493	20.599	23.364	25.508	29.336	33.530	36.250	40.256	43.773	46.979	50.892	53.672

α

χ_α^2

Table VIII.4   F distribution tables—critical values.

$F_{0.10, v_1, v_2}$; v_1 = degrees of freedom numerator; v_2 = degrees of freedom denominator

v_2 \ v_1	1	2	3	4	5	6	7	8	9	10	12	15	20	24	30	40	60	120	∞
1	39.86	49.50	53.59	55.83	57.24	58.20	58.91	59.44	59.86	60.19	60.71	61.22	61.74	62.00	62.26	62.53	62.79	63.06	63.33
2	8.53	9.00	9.16	9.24	9.29	9.33	9.35	9.37	9.38	9.39	9.41	9.42	9.44	9.45	9.46	9.47	9.47	9.48	9.49
3	5.54	5.46	5.39	5.34	5.31	5.28	5.27	5.25	5.24	5.23	5.22	5.20	5.18	5.18	5.17	5.16	5.15	5.14	5.13
4	4.54	4.32	4.19	4.11	4.05	4.01	3.98	3.95	3.94	3.92	3.90	3.87	3.84	3.83	3.82	3.80	3.79	3.78	3.76
5	4.06	3.78	3.62	3.52	3.45	3.40	3.37	3.34	3.32	3.30	3.27	3.24	3.21	3.19	3.17	3.16	3.14	3.12	3.10
6	3.78	3.46	3.29	3.18	3.11	3.05	3.01	2.98	2.96	2.94	2.90	2.87	2.84	2.82	2.80	2.78	2.76	2.74	2.72
7	3.59	3.26	3.07	2.96	2.88	2.83	2.78	2.75	2.72	2.70	2.67	2.63	2.59	2.58	2.56	2.54	2.51	2.49	2.47
8	3.46	3.11	2.92	2.81	2.73	2.67	2.62	2.59	2.56	2.54	2.50	2.46	2.42	2.40	2.38	2.36	2.34	2.32	2.29
9	3.36	3.01	2.81	2.69	2.61	2.55	2.51	2.47	2.44	2.42	2.38	2.34	2.30	2.28	2.25	2.23	2.21	2.18	2.16
10	3.29	2.92	2.73	2.61	2.52	2.46	2.41	2.38	2.35	2.32	2.28	2.24	2.20	2.18	2.16	2.13	2.11	2.08	2.06
11	3.23	2.86	2.66	2.54	2.45	2.39	2.34	2.30	2.27	2.25	2.21	2.17	2.12	2.10	2.08	2.05	2.03	2.00	1.97
12	3.18	2.81	2.61	2.48	2.39	2.33	2.28	2.24	2.21	2.19	2.15	2.10	2.06	2.04	2.01	1.99	1.96	1.93	1.90
13	3.14	2.76	2.56	2.43	2.35	2.28	2.23	2.20	2.16	2.14	2.10	2.05	2.01	1.98	1.96	1.93	1.90	1.88	1.85
14	3.10	2.73	2.52	2.39	2.31	2.24	2.19	2.15	2.12	2.10	2.05	2.01	1.96	1.94	1.91	1.89	1.86	1.83	1.80
15	3.07	2.70	2.49	2.36	2.27	2.21	2.16	2.12	2.09	2.06	2.02	1.97	1.92	1.90	1.87	1.85	1.82	1.79	1.76
16	3.05	2.67	2.46	2.33	2.24	2.18	2.13	2.09	2.06	2.03	1.99	1.94	1.89	1.87	1.84	1.81	1.78	1.75	1.72
17	3.03	2.64	2.44	2.31	2.22	2.15	2.10	2.06	2.03	2.00	1.96	1.91	1.86	1.84	1.81	1.78	1.75	1.72	1.69
18	3.01	2.62	2.42	2.29	2.20	2.13	2.08	2.04	2.00	1.98	1.93	1.89	1.84	1.81	1.78	1.75	1.72	1.69	1.66
19	2.99	2.61	2.40	2.27	2.18	2.11	2.06	2.02	1.98	1.96	1.91	1.86	1.81	1.79	1.76	1.73	1.70	1.67	1.63
20	2.97	2.59	2.38	2.25	2.16	2.09	2.04	2.00	1.96	1.94	1.89	1.84	1.79	1.77	1.74	1.71	1.68	1.64	1.61
21	2.96	2.57	2.36	2.23	2.14	2.08	2.02	1.98	1.95	1.92	1.87	1.83	1.78	1.75	1.72	1.69	1.66	1.62	1.59
22	2.95	2.56	2.35	2.22	2.13	2.06	2.01	1.97	1.93	1.90	1.86	1.81	1.76	1.73	1.70	1.67	1.64	1.60	1.57
23	2.94	2.55	2.34	2.21	2.11	2.05	1.99	1.95	1.92	1.89	1.84	1.80	1.74	1.72	1.69	1.66	1.62	1.59	1.55
24	2.93	2.54	2.33	2.19	2.10	2.04	1.98	1.94	1.91	1.88	1.83	1.78	1.73	1.70	1.67	1.64	1.61	1.57	1.53
25	2.92	2.53	2.32	2.18	2.09	2.02	1.97	1.93	1.89	1.87	1.82	1.77	1.72	1.69	1.66	1.63	1.59	1.56	1.52
26	2.91	2.52	2.31	2.17	2.08	2.01	1.96	1.92	1.88	1.86	1.81	1.76	1.71	1.68	1.65	1.61	1.58	1.54	1.50
27	2.90	2.51	2.30	2.17	2.07	2.00	1.95	1.91	1.87	1.85	1.80	1.75	1.70	1.67	1.64	1.60	1.57	1.53	1.49
28	2.89	2.50	2.29	2.16	2.06	2.00	1.94	1.90	1.87	1.84	1.79	1.74	1.69	1.66	1.63	1.59	1.56	1.52	1.48
29	2.89	2.50	2.28	2.15	2.06	1.99	1.93	1.89	1.86	1.83	1.78	1.73	1.68	1.65	1.62	1.58	1.55	1.51	1.47
30	2.88	2.49	2.28	2.14	2.05	1.98	1.93	1.88	1.85	1.82	1.77	1.72	1.67	1.64	1.61	1.57	1.54	1.50	1.46
40	2.84	2.44	2.23	2.09	2.00	1.93	1.87	1.83	1.79	1.76	1.71	1.66	1.61	1.57	1.54	1.51	1.47	1.42	1.38
60	2.79	2.39	2.18	2.04	1.95	1.87	1.82	1.77	1.74	1.71	1.66	1.60	1.54	1.51	1.48	1.44	1.40	1.35	1.29
120	2.75	2.35	2.13	1.99	1.90	1.82	1.77	1.72	1.68	1.65	1.60	1.55	1.48	1.45	1.41	1.37	1.32	1.26	1.19
∞	2.71	2.30	2.08	1.94	1.85	1.77	1.72	1.67	1.63	1.60	1.55	1.49	1.42	1.38	1.34	1.30	1.24	1.17	1.00

(continued)

SOURCE: Reproduced from Pearson, E. S., and H. O. Hartley, eds., *Biometrika Tables for Statisticians*, Cambridge: University Press, Vol. 1, pp. 169–175, 1976. Reproduced with permission of the *Biometrika* Trustees.

Table VIII.4   F distribution tables—critical values *(continued)*.

$F_{0.05, \nu_1, \nu_2}$; ν_1 = degrees of freedom numerator; ν_2 = degrees of freedom denominator

$\nu_2 \backslash \nu_1$	1	2	3	4	5	6	7	8	9	10	12	15	20	24	30	40	60	120	∞
1	161.4	199.5	215.7	224.6	230.2	234.0	236.8	238.9	240.5	241.9	243.9	245.9	248.0	249.1	250.1	251.1	252.2	253.3	254.3
2	18.51	19.00	19.16	19.25	19.30	19.33	19.35	19.37	19.38	19.40	19.41	19.43	19.45	19.45	19.46	19.47	19.48	19.49	19.50
3	10.13	9.55	9.28	9.12	9.01	8.94	8.89	8.85	8.81	8.79	8.74	8.70	8.66	8.64	8.62	8.59	8.57	8.55	8.53
4	7.71	6.94	6.59	6.39	6.26	6.16	6.09	6.04	6.00	5.96	5.91	5.86	5.80	5.77	5.75	5.72	5.69	5.66	5.63
5	6.61	5.79	5.41	5.19	5.05	4.95	4.88	4.82	4.77	4.74	4.68	4.62	4.56	4.53	4.50	4.46	4.43	4.40	4.36
6	5.99	5.14	4.76	4.53	4.39	4.28	4.21	4.15	4.10	4.06	4.00	3.94	3.87	3.84	3.81	3.77	3.74	3.70	3.67
7	5.59	4.74	4.35	4.12	3.97	3.87	3.79	3.73	3.68	3.64	3.57	3.51	3.44	3.41	3.38	3.34	3.30	3.27	3.23
8	5.32	4.46	4.07	3.84	3.69	3.58	3.50	3.44	3.39	3.35	3.28	3.22	3.15	3.12	3.08	3.04	3.01	2.97	2.93
9	5.12	4.26	3.86	3.63	3.48	3.37	3.29	3.23	3.18	3.14	3.07	3.01	2.94	2.90	2.86	2.83	2.79	2.75	2.71
10	4.96	4.10	3.71	3.48	3.33	3.22	3.14	3.07	3.02	2.98	2.91	2.85	2.77	2.74	2.70	2.66	2.62	2.58	2.54
11	4.84	3.98	3.59	3.36	3.20	3.09	3.01	2.95	2.90	2.85	2.79	2.72	2.65	2.61	2.57	2.53	2.49	2.45	2.40
12	4.75	3.89	3.49	3.26	3.11	3.00	2.91	2.85	2.80	2.75	2.69	2.62	2.54	2.51	2.47	2.43	2.38	2.34	2.30
13	4.67	3.81	3.41	3.18	3.03	2.92	2.83	2.77	2.71	2.67	2.60	2.53	2.46	2.42	2.38	2.34	2.30	2.25	2.21
14	4.60	3.74	3.34	3.11	2.96	2.85	2.76	2.70	2.65	2.60	2.53	2.46	2.39	2.35	2.31	2.27	2.22	2.18	2.13
15	4.54	3.68	3.29	3.06	2.90	2.79	2.71	2.64	2.59	2.54	2.48	2.40	2.33	2.29	2.25	2.20	2.16	2.11	2.07
16	4.49	3.63	3.24	3.01	2.85	2.74	2.66	2.59	2.54	2.49	2.42	2.35	2.28	2.24	2.19	2.15	2.11	2.06	2.01
17	4.45	3.59	3.20	2.96	2.81	2.70	2.61	2.55	2.49	2.45	2.38	2.31	2.23	2.19	2.15	2.10	2.06	2.01	1.96
18	4.41	3.55	3.16	2.93	2.77	2.66	2.58	2.51	2.46	2.41	2.34	2.27	2.19	2.15	2.11	2.06	2.02	1.97	1.92
19	4.38	3.52	3.13	2.90	2.74	2.63	2.54	2.48	2.42	2.38	2.31	2.23	2.16	2.11	2.07	2.03	1.98	1.93	1.88
20	4.35	3.49	3.10	2.87	2.71	2.60	2.51	2.45	2.39	2.35	2.28	2.20	2.12	2.08	2.04	1.99	1.95	1.90	1.84
21	4.32	3.47	3.07	2.84	2.68	2.57	2.49	2.42	2.37	2.32	2.25	2.18	2.10	2.05	2.01	1.96	1.92	1.87	1.81
22	4.30	3.44	3.05	2.82	2.66	2.55	2.46	2.40	2.34	2.30	2.23	2.15	2.07	2.03	1.98	1.94	1.89	1.84	1.78
23	4.28	3.42	3.03	2.80	2.64	2.53	2.44	2.37	2.32	2.27	2.20	2.13	2.05	2.01	1.96	1.91	1.86	1.81	1.76
24	4.26	3.40	3.01	2.78	2.62	2.51	2.42	2.36	2.30	2.25	2.18	2.11	2.03	1.98	1.94	1.89	1.84	1.79	1.73
25	4.24	3.39	2.99	2.76	2.60	2.49	2.40	2.34	2.28	2.24	2.16	2.09	2.01	1.96	1.92	1.87	1.82	1.77	1.71
26	4.23	3.37	2.98	2.74	2.59	2.47	2.39	2.32	2.27	2.22	2.15	2.07	1.99	1.95	1.90	1.85	1.80	1.75	1.69
27	4.21	3.35	2.96	2.73	2.57	2.46	2.37	2.31	2.25	2.20	2.13	2.06	1.97	1.93	1.88	1.84	1.79	1.73	1.67
28	4.20	3.34	2.95	2.71	2.56	2.45	2.36	2.29	2.24	2.19	2.12	2.04	1.96	1.91	1.87	1.82	1.77	1.71	1.65
29	4.18	3.33	2.93	2.70	2.55	2.43	2.35	2.28	2.22	2.18	2.10	2.03	1.94	1.90	1.85	1.81	1.75	1.70	1.64
30	4.17	3.32	2.92	2.69	2.53	2.42	2.33	2.27	2.21	2.16	2.09	2.01	1.93	1.89	1.84	1.79	1.74	1.68	1.62
40	4.08	3.23	2.84	2.61	2.45	2.34	2.25	2.18	2.12	2.08	2.00	1.92	1.84	1.79	1.74	1.69	1.64	1.58	1.51
60	4.00	3.15	2.76	2.53	2.37	2.25	2.17	2.10	2.04	1.99	1.92	1.84	1.75	1.70	1.65	1.59	1.53	1.47	1.39
120	3.92	3.07	2.68	2.45	2.29	2.17	2.09	2.02	1.96	1.91	1.83	1.75	1.66	1.61	1.55	1.50	1.43	1.35	1.25
∞	3.84	3.00	2.60	2.37	2.21	2.10	2.01	1.94	1.88	1.83	1.75	1.67	1.57	1.52	1.46	1.39	1.32	1.22	1.00

(continued)

Table VIII.4   F distribution tables—critical values *(continued)*.

$F_{0.025, \nu_1, \nu_2}$; ν_1 = degrees of freedom numerator, ν_2 = degrees of freedom denominator

ν_2 \ ν_1	1	2	3	4	5	6	7	8	9	10	12	15	20	24	30	40	60	120	∞
1	647.8	799.5	864.2	899.6	921.8	937.1	948.2	956.7	963.3	968.6	976.7	984.9	993.1	997.2	1001	1006	1010	1014	1018
2	38.51	39.00	39.17	39.25	39.30	39.33	39.36	39.37	39.39	39.40	39.41	39.43	39.45	39.46	39.46	39.47	39.48	39.49	39.50
3	17.44	16.04	15.44	15.10	14.88	14.73	14.62	14.54	14.47	14.42	14.34	14.25	14.17	14.12	14.08	14.04	13.99	13.95	13.90
4	12.22	10.65	9.98	9.60	9.36	9.20	9.07	8.98	8.90	8.84	8.75	8.66	8.56	8.51	8.46	8.41	8.36	8.31	8.26
5	10.01	8.43	7.76	7.39	7.15	6.98	6.85	6.76	6.68	6.62	6.52	6.43	6.33	6.28	6.23	6.18	6.12	6.07	6.02
6	8.81	7.26	6.60	6.23	5.99	5.82	5.70	5.60	5.52	5.46	5.37	5.27	5.17	5.12	5.07	5.01	4.96	4.90	4.85
7	8.07	6.54	5.89	5.52	5.29	5.12	4.99	4.90	4.82	4.76	4.67	4.57	4.47	4.42	4.36	4.31	4.25	4.20	4.14
8	7.57	6.06	5.42	5.05	4.82	4.65	4.53	4.43	4.36	4.30	4.20	4.10	4.00	3.95	3.89	3.84	3.78	3.73	3.67
9	7.21	5.71	5.08	4.72	4.48	4.32	4.20	4.10	4.03	3.96	3.87	3.77	3.67	3.61	3.56	3.51	3.45	3.39	3.33
10	6.94	5.46	4.83	4.47	4.24	4.07	3.95	3.85	3.78	3.72	3.62	3.52	3.42	3.37	3.31	3.26	3.20	3.14	3.08
11	6.72	5.26	4.63	4.28	4.04	3.88	3.76	3.66	3.59	3.53	3.43	3.33	3.23	3.17	3.12	3.06	3.00	2.94	2.88
12	6.55	5.10	4.47	4.12	3.89	3.73	3.61	3.51	3.44	3.37	3.28	3.18	3.07	3.02	2.96	2.91	2.85	2.79	2.72
13	6.41	4.97	4.35	4.00	3.77	3.60	3.48	3.39	3.31	3.25	3.15	3.05	2.95	2.89	2.84	2.78	2.72	2.66	2.60
14	6.30	4.86	4.24	3.89	3.66	3.50	3.38	3.29	3.21	3.15	3.05	2.95	2.84	2.79	2.73	2.67	2.61	2.55	2.49
15	6.20	4.77	4.15	3.80	3.58	3.41	3.29	3.20	3.12	3.06	2.96	2.86	2.76	2.70	2.64	2.59	2.52	2.46	2.40
16	6.12	4.69	4.08	3.73	3.50	3.34	3.22	3.12	3.05	2.99	2.89	2.79	2.68	2.63	2.57	2.51	2.45	2.38	2.32
17	6.04	4.62	4.01	3.66	3.44	3.28	3.16	3.06	2.98	2.92	2.82	2.72	2.62	2.56	2.50	2.44	2.38	2.32	2.25
18	5.98	4.56	3.95	3.61	3.38	3.22	3.10	3.01	2.93	2.87	2.77	2.67	2.56	2.50	2.44	2.38	2.32	2.26	2.19
19	5.92	4.51	3.90	3.56	3.33	3.17	3.05	2.96	2.88	2.82	2.72	2.62	2.51	2.45	2.39	2.33	2.27	2.20	2.13
20	5.87	4.46	3.86	3.51	3.29	3.13	3.01	2.91	2.84	2.77	2.68	2.57	2.46	2.41	2.35	2.29	2.22	2.16	2.09
21	5.83	4.42	3.82	3.48	3.25	3.09	2.97	2.87	2.80	2.73	2.64	2.53	2.42	2.37	2.31	2.25	2.18	2.11	2.04
22	5.79	4.38	3.78	3.44	3.22	3.05	2.93	2.84	2.76	2.70	2.60	2.50	2.39	2.33	2.27	2.21	2.14	2.08	2.00
23	5.75	4.35	3.75	3.41	3.18	3.02	2.90	2.81	2.73	2.67	2.57	2.47	2.36	2.30	2.24	2.18	2.11	2.04	1.97
24	5.72	4.32	3.72	3.38	3.15	2.99	2.87	2.78	2.70	2.64	2.54	2.44	2.33	2.27	2.21	2.15	2.08	2.01	1.94
25	5.69	4.29	3.69	3.35	3.13	2.97	2.85	2.75	2.68	2.61	2.51	2.41	2.30	2.24	2.18	2.12	2.05	1.98	1.91
26	5.66	4.27	3.67	3.33	3.10	2.94	2.82	2.73	2.65	2.59	2.49	2.39	2.28	2.22	2.16	2.09	2.03	1.95	1.88
27	5.63	4.24	3.65	3.31	3.08	2.92	2.80	2.71	2.63	2.57	2.47	2.36	2.25	2.19	2.13	2.07	2.00	1.93	1.85
28	5.61	4.22	3.63	3.29	3.06	2.90	2.78	2.69	2.61	2.55	2.45	2.34	2.23	2.17	2.11	2.05	1.98	1.91	1.83
29	5.59	4.20	3.61	3.27	3.04	2.88	2.76	2.67	2.59	2.53	2.43	2.32	2.21	2.15	2.09	2.03	1.96	1.89	1.81
30	5.57	4.18	3.59	3.25	3.03	2.87	2.75	2.65	2.57	2.51	2.41	2.31	2.20	2.14	2.07	2.01	1.94	1.87	1.79
40	5.42	4.05	3.46	3.13	2.90	2.74	2.62	2.53	2.45	2.39	2.29	2.18	2.07	2.01	1.94	1.88	1.80	1.72	1.64
60	5.29	3.93	3.34	3.01	2.79	2.63	2.51	2.41	2.33	2.27	2.17	2.06	1.94	1.88	1.82	1.74	1.67	1.58	1.48
120	5.15	3.80	3.23	2.89	2.67	2.52	2.39	2.30	2.22	2.16	2.05	1.94	1.82	1.76	1.69	1.61	1.53	1.43	1.31
∞	5.02	3.69	3.12	2.79	2.57	2.41	2.29	2.19	2.11	2.05	1.94	1.83	1.71	1.64	1.57	1.48	1.39	1.27	1.00

(continued)

Table VIII.4   F distribution tables—critical values (concluded).

$F_{0.01, \nu_1, \nu_2}$; ν_1 = degrees of freedom numerator; ν_2 = degrees of freedom denominator

ν_2 \ ν_1	1	2	3	4	5	6	7	8	9	10	12	15	20	24	30	40	60	120	∞
1	4052	4999.5	5403	5625	5764	5859	5928	5981	6022	6056	6106	6157	6209	6235	6261	6287	6313	6339	6366
2	98.50	99.00	99.17	99.25	99.30	99.33	99.36	99.37	99.39	99.40	99.42	99.43	99.45	99.46	99.47	99.47	99.48	99.49	99.50
3	34.12	30.82	29.46	28.71	28.24	27.91	27.67	27.49	27.35	27.23	27.05	26.87	26.69	26.60	26.50	26.41	26.32	26.22	26.13
4	21.20	18.00	16.69	15.98	15.52	15.21	14.98	14.80	14.66	14.55	14.37	14.20	14.02	13.93	13.84	13.75	13.65	13.56	13.46
5	16.26	13.27	12.06	11.39	10.97	10.67	10.46	10.29	10.16	10.05	9.89	9.72	9.55	9.47	9.38	9.29	9.20	9.11	9.02
6	13.75	10.92	9.78	9.15	8.75	8.47	8.26	8.10	7.98	7.87	7.72	7.56	7.40	7.31	7.23	7.14	7.06	6.97	6.88
7	12.25	9.55	8.45	7.85	7.46	7.19	6.99	6.84	6.72	6.62	6.47	6.31	6.16	6.07	5.99	5.91	5.82	5.74	5.65
8	11.26	8.65	7.59	7.01	6.63	6.37	6.18	6.03	5.91	5.81	5.67	5.52	5.36	5.28	5.20	5.12	5.03	4.95	4.86
9	10.56	8.02	6.99	6.42	6.06	5.80	5.61	5.47	5.35	5.26	5.11	4.96	4.81	4.73	4.65	4.57	4.48	4.40	4.31
10	10.04	7.56	6.55	5.99	5.64	5.39	5.20	5.06	4.94	4.85	4.71	4.56	4.41	4.33	4.25	4.17	4.08	4.00	3.91
11	9.65	7.21	6.22	5.67	5.32	5.07	4.89	4.74	4.63	4.54	4.40	4.25	4.10	4.02	3.94	3.86	3.78	3.69	3.60
12	9.33	6.93	5.95	5.41	5.06	4.82	4.64	4.50	4.39	4.30	4.16	4.01	3.86	3.78	3.70	3.62	3.54	3.45	3.36
13	9.07	6.70	5.74	5.21	4.86	4.62	4.44	4.30	4.19	4.10	3.96	3.82	3.66	3.59	3.51	3.43	3.34	3.25	3.17
14	8.86	6.51	5.56	5.04	4.69	4.46	4.28	4.14	4.03	3.94	3.80	3.66	3.51	3.43	3.35	3.27	3.18	3.09	3.00
15	8.68	6.36	5.42	4.89	4.56	4.32	4.14	4.00	3.89	3.80	3.67	3.52	3.37	3.29	3.21	3.13	3.05	2.96	2.87
16	8.53	6.23	5.29	4.77	4.44	4.20	4.03	3.89	3.78	3.69	3.55	3.41	3.26	3.18	3.10	3.02	2.93	2.84	2.75
17	8.40	6.11	5.18	4.67	4.34	4.10	3.93	3.79	3.68	3.59	3.46	3.31	3.16	3.08	3.00	2.92	2.83	2.75	2.65
18	8.29	6.01	5.09	4.58	4.25	4.01	3.84	3.71	3.60	3.51	3.37	3.23	3.08	3.00	2.92	2.84	2.75	2.66	2.57
19	8.18	5.93	5.01	4.50	4.17	3.94	3.77	3.63	3.52	3.43	3.30	3.15	3.00	2.92	2.84	2.76	2.67	2.58	2.49
20	8.10	5.85	4.94	4.43	4.10	3.87	3.70	3.56	3.46	3.37	3.23	3.09	2.94	2.86	2.78	2.69	2.61	2.52	2.42
21	8.02	5.78	4.87	4.37	4.04	3.81	3.64	3.51	3.40	3.31	3.17	3.03	2.88	2.80	2.72	2.64	2.55	2.46	2.36
22	7.95	5.72	4.82	4.31	3.99	3.76	3.59	3.45	3.35	3.26	3.12	2.98	2.83	2.75	2.67	2.58	2.50	2.40	2.31
23	7.88	5.66	4.76	4.26	3.94	3.71	3.54	3.41	3.30	3.21	3.07	2.93	2.78	2.70	2.62	2.54	2.45	2.35	2.26
24	7.82	5.61	4.72	4.22	3.90	3.67	3.50	3.36	3.26	3.17	3.03	2.89	2.74	2.66	2.58	2.49	2.40	2.31	2.21
25	7.77	5.57	4.68	4.18	3.85	3.63	3.46	3.32	3.22	3.13	2.99	2.85	2.70	2.62	2.54	2.45	2.36	2.27	2.17
26	7.72	5.53	4.64	4.14	3.82	3.59	3.42	3.29	3.18	3.09	2.96	2.81	2.66	2.58	2.50	2.42	2.33	2.23	2.13
27	7.68	5.49	4.60	4.11	3.78	3.56	3.39	3.26	3.15	3.06	2.93	2.78	2.63	2.55	2.47	2.38	2.29	2.20	2.10
28	7.64	5.45	4.57	4.07	3.75	3.53	3.36	3.23	3.12	3.03	2.90	2.75	2.60	2.52	2.44	2.35	2.26	2.17	2.06
29	7.60	5.42	4.54	4.04	3.73	3.50	3.33	3.20	3.09	3.00	2.87	2.73	2.57	2.49	2.41	2.33	2.23	2.14	2.03
30	7.56	5.39	4.51	4.02	3.70	3.47	3.30	3.17	3.07	2.98	2.84	2.70	2.55	2.47	2.39	2.30	2.21	2.11	2.01
40	7.31	5.18	4.31	3.83	3.51	3.29	3.12	2.99	2.89	2.80	2.66	2.52	2.37	2.29	2.20	2.11	2.02	1.92	1.80
60	7.08	4.98	4.13	3.65	3.34	3.12	2.95	2.82	2.72	2.63	2.50	2.35	2.20	2.12	2.03	1.94	1.84	1.73	1.60
120	6.85	4.79	3.95	3.48	3.17	2.96	2.79	2.66	2.56	2.47	2.34	2.19	2.03	1.95	1.86	1.76	1.66	1.53	1.38
∞	6.63	4.61	3.78	3.32	3.02	2.80	2.64	2.51	2.41	2.32	2.18	2.04	1.88	1.79	1.70	1.59	1.47	1.32	1.00

Table VIII.5 X-bar, R, and S control chart—3-sigma limit constants.

(Based on Sampling from a Normal Distribution)

Subgroup size n	d_2	d_3	c_4	D_3	D_4	B_3	B_4	A	A_2	A_3
2	1.128	0.8525	0.7979	0.00	3.27	0.00	3.27	2.12	1.88	2.66
3	1.693	0.8884	0.8862	0.00	2.57	0.00	2.57	1.73	1.02	1.95
4	2.059	0.8798	0.9213	0.00	2.28	0.00	2.27	1.50	0.73	1.63
5	2.326	0.8641	0.9400	0.00	2.11	0.00	2.09	1.34	0.58	1.43
6	2.534	0.8480	0.9515	0.00	2.00	0.03	1.97	1.22	0.48	1.29
7	2.704	0.8332	0.9594	0.08	1.92	0.12	1.88	1.13	0.42	1.18
8	2.847	0.8198	0.9650	0.14	1.86	0.19	1.81	1.06	0.37	1.10
9	2.970	0.8078	0.9693	0.18	1.82	0.24	1.76	1.00	0.34	1.03
10	3.078	0.7971	0.9727	0.22	1.78	0.28	1.72	0.95	0.31	0.98
11	3.173	0.7873	0.9754	0.26	1.74	0.32	1.68	0.90	0.29	0.93
12	3.258	0.7785	0.9776	0.28	1.72	0.35	1.65	0.87	0.27	0.89
13	3.336	0.7704	0.9794	0.31	1.69	0.38	1.62	0.83	0.25	0.85
14	3.407	0.7630	0.9810	0.33	1.67	0.41	1.59	0.80	0.24	0.82
15	3.472	0.7562	0.9823	0.35	1.65	0.43	1.57	0.77	0.22	0.79
16	3.532	0.7499	0.9835	0.36	1.64	0.45	1.55	0.75	0.21	0.76
17	3.588	0.7441	0.9845	0.38	1.62	0.47	1.53	0.73	0.20	0.74
18	3.640	0.7386	0.9854	0.39	1.61	0.48	1.52	0.71	0.19	0.72
19	3.689	0.7335	0.9862	0.40	1.60	0.50	1.50	0.69	0.19	0.70
20	3.735	0.7287	0.9869	0.41	1.59	0.51	1.49	0.67	0.18	0.68
30	4.086	0.6926	0.9914	*	*	0.60	1.40	0.55	*	0.55
40	4.322	0.6692	0.9936			0.66	1.34	0.47		0.48
50	4.498	0.6521	0.9949			0.70	1.30	0.42		0.43
60	4.639	0.6389	0.9958			0.72	1.28	0.39		0.39
70	4.755	0.6283	0.9964			0.74	1.26	0.36		0.36
80	4.854	0.6194	0.9968			0.76	1.24	0.34		0.34
90	4.939	0.6118	0.9972			0.77	1.23	0.32		0.32
100	5.015	0.6052	0.9975			0.79	1.21	0.30		0.30

*R chart factors are given only up to $n = 20$; for larger subgroup sizes the S chart should be used.
SOURCE: Adapted with permission from Grant, E. L., and R. S. Leavenworth, *Statistical Quality Control*, 7th ed., New York: McGraw-Hill, pp. 717–720, 1996.

Table VIII.6 X-bar, R, and S control chart—probability limit constants.

Probability-Based X-Bar and R Chart Constants

Subgroup size n	$A_{2,0.001}$	$A_{3,0.999}$	$D_{0.001}$	$D_{0.999}$	$A_{2,0.005}$	$A_{2,0.995}$	$D_{0.005}$	$D_{0.995}$	$A_{2,0.025}$	$A_{2,0.975}$	$D_{0.025}$	$D_{0.975}$
2	1.94	1.94	0.00	4.12	1.61	1.61	0.01	3.52	1.23	1.23	0.04	2.81
3	1.05	1.05	0.04	2.99	0.88	0.88	0.08	2.61	0.67	0.67	0.18	2.17
4	0.75	0.75	0.10	2.58	0.63	0.63	0.17	2.28	0.48	0.48	0.29	1.93
5	0.59	0.59	0.16	2.36	0.50	0.50	0.24	2.10	0.38	0.38	0.37	1.81
6	0.50	0.50	0.21	2.22	0.41	0.41	0.30	1.99	0.32	0.32	0.42	1.72
7	0.43	0.43	0.26	2.12	0.36	0.36	0.34	1.90	0.27	0.27	0.46	1.66
8	0.38	0.38	0.29	2.04	0.32	0.32	0.38	1.84	0.24	0.24	0.50	1.62
9	0.35	0.35	0.33	1.99	0.29	0.29	0.41	1.80	0.22	0.22	0.52	1.58
10	0.32	0.32	0.35	1.94	0.26	0.26	0.43	1.76	0.20	0.20	0.54	1.55

Probability-Based X-Bar and S Chart Constants

Subgroup size n	$A_{3,0.001}$	$A_{3,0.999}$	$B_{0.001}$	$B_{0.999}$	$A_{3,0.005}$	$A_{3,0.995}$	$B_{0.005}$	$B_{0.995}$	$A_{3,0.025}$	$A_{3,0.975}$	$B_{0.025}$	$B_{0.975}$
2	2.74	2.74	0.00	4.14	2.28	2.28	0.00	3.52	1.74	1.74	0.04	2.82
3	2.01	2.01	0.05	2.99	1.68	1.68	0.08	2.60	1.28	1.28	0.18	2.17
4	1.68	1.68	0.10	2.53	1.40	1.40	0.16	2.25	1.06	1.06	0.29	1.92
5	1.47	1.47	0.16	2.29	1.23	1.23	0.23	2.04	0.93	0.93	0.37	1.78
6	1.33	1.33	0.22	2.13	1.10	1.10	0.29	1.92	0.84	0.84	0.43	1.68
7	1.22	1.22	0.26	2.01	1.01	1.01	0.34	1.83	0.77	0.77	0.47	1.63
8	1.13	1.13	0.30	1.93	0.94	0.94	0.38	1.76	0.72	0.72	0.51	1.58
9	1.06	1.06	0.34	1.86	0.89	0.89	0.42	1.70	0.67	0.67	0.54	1.53
10	1.00	1.00	0.37	1.81	0.84	0.84	0.45	1.67	0.64	0.64	0.57	1.49

SOURCE: Adapted with permission from Grant, E. L., and R. S. Leavenworth, *Statistical Quality Control*, 7th ed., New York: McGraw-Hill, p. 381, 1996.

Table VIII.7　　EWMA and EWMD control chart limit constants.

Weighting factor	Equivalent sample size	Means	Standard deviations		
r	n	A^*	D_1^*	D_2^*	d_2^*
0.050	39	0.480	0.514	1.102	0.808
0.100	19	0.688	0.390	1.247	0.819
0.200	9	1.000	0.197	1.486	0.841
0.250	7	1.132	0.109	1.597	0.853
0.286	6	1.225	0.048	1.676	0.862
0.333	5	1.342	0.000	1.780	0.874
0.400	4	1.500	0.000	1.930	0.892
0.500	3	1.732	0.000	2.164	0.921
0.667	2	2.121	0.000	2.596	0.977
0.800		2.449	0.000	2.990	1.030
0.900		2.714	0.000	3.321	1.076
1.000	1	3.000			

SOURCE: Reproduced with permission from Sweet, A. L., "Control Charts Using Coupled Exponentially Weighted Moving Averages," *Transactions of the IIE*, vol. 18, no. 1, pp. 26–33, 1986.

Table VIII.8　　Tabled pseudo-standard normal random numbers. (distributed $\sim N$ ($\mu = 0$, $\sigma = 1$)

0.47671	−0.63157	1.17682	1.38893	0.29935	−0.36990	−1.38133
−1.81764	0.98664	−0.47724	−0.75603	0.44318	−0.43958	−3.52913
1.13274	−0.47155	0.42744	−2.24605	0.27279	1.08531	0.48250
−0.06503	0.59924	−1.64883	1.30748	0.23868	−0.32504	−1.69861
−0.12146	1.15719	−0.29187	−0.35456	0.80306	0.18250	0.25864
−1.78527	−0.34949	−1.02516	−0.19274	0.74288	1.94339	−0.95295
0.27995	0.10390	1.11556	1.06622	0.53846	−0.18519	−0.63666
−0.77037	0.02947	0.32742	1.80489	1.71187	0.25690	0.13542
−0.28073	−2.25984	0.57612	−0.21378	−1.47336	−0.98618	0.07830
1.33213	−0.47535	0.53908	0.43944	0.87327	0.87017	0.36186
−0.82266	−0.88714	1.49105	−1.26902	−1.09260	1.13430	0.82205
−0.87072	0.94592	−1.33734	0.55962	1.91605	0.83984	−1.47412
0.53503	0.14339	0.14455	1.14011	0.83885	0.75693	−0.43518
−0.87180	1.15180	−1.19089	−0.26841	0.32592	−1.11069	1.24766
0.01901	0.96600	2.11745	0.97474	−0.44043	−1.49246	−0.29409
−0.44330	0.67814	−0.77063	−0.27262	−1.55516	0.08973	−0.92432
0.48498	1.25696	1.27885	−0.61958	−0.57276	0.83685	−1.65208
0.27078	−2.47169	−1.14441	1.36950	2.09914	−0.55102	0.32399
−0.62991	−0.00271	−0.23542	−1.72502	0.18031	−0.35410	0.37645
−1.31138	1.47100	0.29224	−0.15858	0.39250	−0.08269	−0.20215

(continued)

Table VIII.8   Tabled pseudo-standard normal random numbers *(continued)*.
(distributed $\sim N$ ($\mu = 0$, $\sigma = 1$))

−0.29397	−0.41734	1.70816	−0.81394	0.77061	1.89595	2.00776
0.31803	1.55201	−0.78720	2.04463	0.05768	2.43915	1.53412
−0.40793	1.94149	−0.25707	−0.51363	0.11788	−0.20433	0.94817
1.06835	0.76433	−1.31800	−1.21051	−1.24440	−1.49616	1.75307
0.96553	1.00749	−1.16989	0.23741	−0.15470	1.14125	−0.84570
−0.90689	0.66249	1.42437	−1.98771	−1.01506	−0.96540	−0.26753
−0.87109	0.83365	0.39018	1.53325	−1.27082	1.07872	−1.35132
−1.99934	−0.38330	0.86691	−0.27925	−0.90759	−0.15311	−0.35032
0.48963	0.92281	−1.37902	0.01905	−0.10005	−2.00573	0.27594
−0.16047	−0.07504	1.39062	0.03640	−1.02716	−2.79012	−1.79301
−2.58217	−1.34700	1.82321	0.48759	0.56465	−1.07733	0.27382
−0.87284	0.72691	−2.06094	−0.28378	1.18091	1.08537	0.12944
−1.59212	−0.94154	1.37182	0.00672	−1.22821	−0.49724	3.16433
−1.38737	1.23570	0.63250	−0.34858	−1.08189	−0.83716	−1.41709
0.17473	−0.71040	0.74947	−1.30692	0.33843	0.24488	1.07147
1.16507	0.58156	1.53238	−0.37844	0.37260	−0.99431	−1.46289
1.43023	1.93126	−0.26095	0.88359	−0.01167	−1.85355	1.23951
0.48559	−0.43282	−1.65439	0.14399	−1.85664	−0.33894	0.77974
0.65470	−1.76289	0.05637	0.80908	−0.30514	−0.33201	−0.62029
1.18528	−1.32500	0.48102	−1.28570	−1.28060	−1.05597	−0.92865
−1.33752	−1.07506	−0.15686	0.32091	−0.73147	0.67558	0.41067
0.62665	−0.86615	−0.38441	2.53346	0.47012	1.06813	0.69642
−0.61632	−0.79291	0.01271	0.54359	0.66926	−0.59139	−0.21235
0.04412	−1.19252	−2.17056	−0.25257	−0.22779	−0.14993	−1.22567
1.60263	−0.45730	0.40670	1.86190	0.11872	−0.99868	0.21332
−1.58043	0.30849	1.52788	−0.41211	0.79218	0.64335	1.36038
0.78892	−0.85921	−0.05265	−1.14521	−0.15867	−0.81497	−1.25836
−0.71441	−0.86211	1.92703	−0.94077	1.48201	−2.29965	0.12787
−0.53046	−2.18511	−0.18933	−0.51900	1.18857	−0.52730	−0.86464
0.82538	−0.11016	1.96364	−1.86201	−1.85296	0.68331	1.74397
−0.54065	−0.41538	−0.39555	0.37638	0.00705	0.47141	0.17480
0.34592	0.41595	0.09323	−0.98873	0.86559	−0.27252	1.85097
−0.28545	−0.51224	0.34789	−0.63056	0.15749	0.94059	0.37386
−0.91667	0.29619	−0.50496	1.80110	−0.46236	1.35905	−0.93862
1.51105	0.47584	−0.00715	1.85356	−0.42587	−0.59658	−0.87669
−1.73765	−0.28764	0.42201	−1.64691	−0.81150	0.67667	−0.09687
1.09100	0.60061	0.52409	−0.00199	−0.73509	2.48864	−0.07095
1.88160	0.47218	−2.16934	−0.03293	−1.01896	−0.07852	−1.70595
−0.96712	−0.42203	0.75606	−0.30334	−0.46192	−0.02807	−0.54189
0.35249	−1.24655	1.59211	−0.37005	−1.40560	1.51940	0.33775

(continued)

Table VIII.8   Tabled pseudo-standard normal random numbers *(concluded)*.
(distributed $\sim N\,(\mu = 0,\ \sigma = 1)$)

0.205990.86116	−0.96530	−1.18356	−1.21906	0.07680	−0.26683	
0.32013	−0.31332	0.16963	−1.21574	−0.42939	−0.48684	1.40092
1.12241	−1.09303	1.18014	0.06198	−0.13097	0.21023	0.29320
0.02863	−0.33323	0.90950	−0.15770	0.61930	0.78937	−0.25414
−0.24430	−1.49912	0.07584	−1.11657	−0.17693	−1.24935	1.33492
−2.46315	0.48054	1.26719	−0.58017	−1.18390	0.07791	−0.09606
1.01627	−0.94174	3.14190	0.21292	0.77692	−0.66749	−0.32351
1.07282	−0.23358	−1.61208	1.13695	1.02769	−0.30758	−1.06518
−0.11808	−0.45955	−0.60773	−1.02224	−0.14825	0.74662	0.13940
−1.77940	1.76685	2.46160	−0.38517	1.28755	0.53529	0.69768

Table VIII.9   Normal probability plotting paper.

Normal probability paper

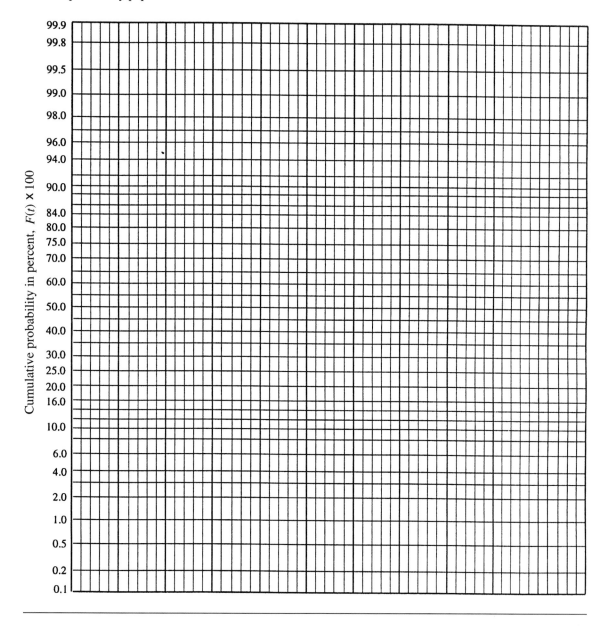

INDEX

CASE

Section 1—Production System and Process Performance Cases

Example 1.1—True and substitute quality characteristics, 6

Example 1.2—Experience and creations of quality, 8

Example 1.3—Productivity ratio calculations, 9–11

Example 1.4—Cost of quality problems, 12

Example 1.5—Aircraft Fleet Support—Strong Department/Weak Processes, 16–18

Example 1.6—Software Development—Strong Processes, 20–21

Example 2.1—Rainmaker—Systems Thinking, 37-40

Example 3.1—Rainmaker—Purpose Hierarchy, 48

Example 3.2—Amplifier Acquisition—Product Variation, 50

Example 3.3—Rainmaker—Variables Identification, 58–59

Example 3.4—Rainmaker—Benchmarks, 59-60

Section 2—Process Characterization, Exploration, and Response Modeling Cases

Example 4.1—Meter Stick Production—Physical Characterization, 71–73

Example 4.2—Meter Stick Blank—Length Data Characterization, 81–84

Example 4.3—Natural Gas Production—Physical and Statistical Characterization, 84–87

Example 5.1—Roll Pressure—Random Effects Model, 96–100

Example 5.2—Assembly Improvement—Fixed Effects Model, 102–105

Example 5.3—Assembly Improvement—Treatment Level Confidence Intervals, 106–107

Example 5.4—Assembly Improvement—Treatment Comparisons, 108–109

Example 5.5—Assembly Improvement—Model Adequacy, 110–113

Example 5.6—Energy Cost/Operations—Two Fixed Effects Factors, 116–123

Example 6.1—Energy Cost/Operations—Response Surface in One Factor, 136–142

Example 6.2—Energy Cost/Operations—Response Surface in Two Factors, 143–145

Example 6.3—Packaging Process—2^f Design with a Center, 147–150

Example 6.4—Packaging Process—Follow-On Experiment and Model, 152–157

Section 3—Process Definition and Redefinition Cases

Example 7.1—Downtown Bakery—Market and Direction, 167–169

Example 7.2—Downtown Bakery—Product Definition, Design, and Development Processes, 169–174

Example 7.3—Downtown Bakery—Production Process Boundaries, Customers, and Suppliers, 177–178

Example 7.4—Downtown Bakery—Production Process Outcomes, 179–180

Example 7.5—Downtown Bakery—Production Process Concepts, 181–185

Example 8.1—Downtown Bakery—Production Process Options, 190–193

Example 8.2—Downtown Bakery—Evaluation of Process Options, 195–198

Example 8.3—Downtown Bakery—Production Process Plan, 199–209

Example 8.4—Downtown Bakery—Production Process Implementation, 212–215

Section 4—Process Control Cases

Example 9.1—Automatic Closed-Loop Feedback Control, 226–227

Example 10.1—Process Monitoring Fundamentals, 236–239

Example 10.2—X-bar and R chart Mechanics, 247–249

625

Example 10.3—Probability Control Limits, 256

Example 10.4—Subgrouped EWMA and EWMD, 261–263

Example 10.5—Tabular CuSum Location Chart for PCB Data, 267–268

Example 10.6—Shift Detection Probabilities, 276–277

Example 11.1—Steel Production, 286–290

Example 11.2—Individuals EWMA and EWMD Charts, 292–294

Example 11.3—Individuals CuSum Chart, 296–297

Example 11.4—Production Source—Corn Flakes for Cattle, 300–302

Example 11.5—Meter Stick Process Capability, 307–309

Example 11.6—Six-Sigma Capability, 309

Example 11.7—Estimation of Variation Inflation, 310–311

Example 11.8—Estimation of Gauge Repeatability, Operator Reproducibility, and P/T Ratio, 312–313

Example 12.1—Golf Club Shaft Production Process, 323–325

Example 12.2—Golf Club Shaft Production Case Revisited, 329–332

Example 12.3—35-mm Camera Body Production Process, 332–334

Example 12.4—Tractor Production Process, 335–338

Example 12.5—35-mm Camera Body Production Process Revisited, 339–340

Example 13.1—Meter Stick Production Process Revisited and Simulated— Appropriate SPC Application, 348–354

Example 13.2—Shredder—Simulated Applications and Misapplications of SPC, 355–359

Example 13.3—Natural Gas Chiller— Simulated Applications and Misapplications of SPC, 360–364

Example 13.4—Panel Assembly Deviation, 372–376

Example 13.5—Panel Assembly Deviation, continued, 377–378

Example 14.1—Salsa Production Process— Basics, 390–392

Example 14.2—Salsa Production Process— Discontinuous Control, 393–397

Example 14.3—Salsa Production Process— Proportional Control, 397–400

Example 14.4—Salsa Production Process— PI Control, 402–404

Example 14.5—Salsa Production Process— PD Control, 407–408

Example 14.6—Salsa Production Process— Transfer Function, 410–411

Example 14.7—Block Diagrams, 412

Example 14.8—Transfer Function and Block Diagram, 414–415

Example 15.1—Photovoltaic System— Adaptive Controller, 429–431

Section 5—Process Analysis and Improvement Cases

Example 16.1—High Lift—Background, 440–443

Example 16.2—High Lift— Receiving/Stocking Subprocess Observation, 445–446

Example 16.3—High Lift—Procurement, Receiving, Storage, and Stocking Subprocesses Improvement Concepts, 448–451

Example 16.4—High Lift—Procurement, Receiving, Storage, and Stocking Subprocesses Enhancement Options, 453–454

Example 17.1—High Lift—Line Support Improvement Alternatives, 458,461

Example 17.2—High Lift—Line Support Improvement Evaluation, 463–466

Example 17.3—High Lift—Line Support Improvement Plan, 467–471

Example 17.4—High Lift—Line Support Improvement Implementation Schedule, 474–477

Section 6—Process-Based Transformation Cases

Example 18.1—Experience Lost—Lid Seal, 487–488

Example 18.2—Award Alienation—Who's Interested? 491–492

Example 18.3—Coin—DeBono, 495

Section 7—Process Cases—Descriptions and Data

Case VII.1 AA Fiberglass, 573

Case VII.2 Apple Core—Dehydration, 574–575

Case VII.3 Apple Dehydration Exploration, 575–576

Case VII.4 Back-of-the-Moon—Mining, 576–577

Case VII.5 Big City Waterworks, 578

Case VII.6 Big Dog—Dog Food Packaging, 578–579

Case VII.7 Bushings International—Machining, 579–580

Case VII.8 Door-to-Door—Pizza Delivery, 580–581

Case VII.9 Downtown Bakery—Bread Dough, 582–583

Case VII.10 Downtown Bakery—pH Measurement, 583–584

Case VII.11 Fix-Up—Automobile Repair, 584

Case VII.12 Hard-Shell Aquaculture, 585–586

Case VII.13 Health Assist—Service, 586–587

Case VII.14 High-Precision—Collar Machining, 587–588

Case VII.15 High-Precision—Collar Measurement, 588–589

Case VII.16 Link-Lock Chain, 589–590

Case VII.17 LNG—Natural Gas Liquefaction, 590–591

Case VII.18 M-Stick Manufacturing, 591–592

Case VII.19 Night Hauler Trucking, 593

Case VII.20 PCB—Printed Circuit Boards, 593–594

Case VII.21 Punch-Out—Sheet Metal Fabrication, 594–596

Case VII.22 Re-Use—Recycling, 596–597

Case VII.23 Re-Use—Sensor Precision, 597

Case VII.24 Silver Bird—Baggage, 599–600

Case VII.25 Snappy—Plastic Injection Molding, 600–601

Case VII.26 Squeaky Clean Laundry, 601

Case VII.27 Rainbow—Paint Coating, 601–603

Case VII.28 Sure-Stick Adhesive, 603–604

Case VII.29 TexRosa—Salsa, 604–605

Case VII.30 Tough-Skin—Sheet Metal Welding, 606–607

INDEX

SUBJECT AND AUTHOR

A

Accuracy, of measurements, 68
Activities, definition of, 211, 473
Activity/sequence list, 545
Activity sequences, 211, 473–474
Akao, Y., 530, 562
Akiyama, K., 523
Alt, F. B., 369, 375, 380
Anderson, C. W., 426
Anderson, G. F., 566
ANSI/ISO/ASQC Q9000, 523–525, 540
Antsaklis, P. J., 422
Asfahl, C. R., 392
Attributes process control charts (*see* Statistical
    process control charting)
AT&T, 252
Autocorrelation, 79–81
    correlograms, 82, 86, 355–356, 360
    measurement of process inertia, 82, 86,
        229, 355–356, 360
Automatic process control (APC), basics
    adaptive control, 422
    automatic versus manual control, discussion
        of, 390–392
    dynamic models, characteristics
        of, 421
    goal, 386
    instability, 387–388
    intelligent control, 422
    linearity, 389–390
    mathematical models, 420–421
    nonlinearity, 389–390
    nonparametric models, 422
    process identification, 420–422
    relationships to human-based control,
        385–386, 422, 426–427
    stability, 387–388
    steady-state response, 387–388
    transient response, 387–388
Automatic process control (APC),
    conventional models
    block diagrams, 411–413
    Boolean theorems and laws, 396
    continuous control action, types of, 397
    controllers, tuning of, 407, 409

Automatic process control (APC), conventional
    models *(cont'd)*
    derivative, continuous control action,
        405–406, 408
    discontinuous/discrete control action,
        392–397
    gain, for derivative control action, 405
    gain, for integral control action, 401
    integral, continuous control action, 401–403
    ladder logic, 395–396
    on/off control action, 392–395
    proportional, continuous control
        action, 397–400
    proportional band, 399
    proportionality constant, 397–399
    proportional plus derivative (PD) control
        action, 405–407
    proportional plus integral (PI) control
        action, 401–402, 404
    proportional plus integral plus derivative
        (PID) control action, 407
    transfer functions, 409–411, 413–415
Automatic process control (APC),
    unconventional models
    adaptive tuning, 429–431
    cascade control, 419
    evolutionary computation, 428
    expert systems and controllers, 426–427
    feedforward control, 419–420
    neural networks
        applications in control, 425–426
        architectures for, 424–425
        feedforward networks, 424–425
        learning in, 425
        nodal operation in, 423–424
        nodal structure in, 423, 425
        recurrent networks, 424
    ratio control, 419
    self-tuning control, 429–431
Average run lengths (ARLs) in SPC, 253, 255,
    277–279, 347–364

B

Badiur, A. B., 553, 560
Barnard, C. I., 14

Barnard, G. A., 264
Barnes, R. M., 518–520, 559
Baxley, R. V., Jr., 221
Beebe, S. A., 504, 507
Benchmarking and benchmark performance, 55, 58–60, 443, 512–515
Bias in measurements, 68
Binomial distribution, 326–327
Blocked experimental models, 124
Bos, T., 81, 551
Boulding, K. E., 26
Box, G. E. P., 100, 107, 124, 127–129, 131, 151, 221, 432, 553
Break-even analysis, 174, 545–546
Brook, D., 296
Brown, M. G., 540
Bullen, D. B., 503–504
Burgoon, J. K., 503–504
Business integration, as a fundamental process, 11, 18–19, 166, 441
Business process reengineering, 532–533

C

Camp, R. C., 512–514
Capability analysis, 546–547
Cash-flow analysis, 547–548
Cause-effect diagram, 72, 200, 548–549
C control chart
    area of opportunity for defects, 335
    concepts and mechanics, 334–337
    inspection unit, definition of, 335
    usage/application, 335
Central limit theorem, 346
Champ, C., 380, 381
Champy, J., 532–533
Chan, L. K., 303, 317
Chang, T., 244, 259, 264, 283, 348
Characterization of processes (*see* Process characterization)
Checkland, P., 25, 27–29
Checkland's general systems thinking model, 28–29, 37–40
Check sheet, 343, 550
Cheng, S. W., 303, 317
Chinnam, R. B., 426
Chi squared location chart, subgrouped data, 368
Christopher, W. F., 11
Closed-loop process structures, 51–53, 224–227
Cochran, W. G., 151

Coefficients of multiple determination, R^2, R^2_{max}, and R^2_{adj}. 135, 138, 141–142, 144
Coleman, D. E., 221
Common and special cause variation
    definition of, 65
    mention of, 74, 221, 227, 231, 235, 248–252, 281, 322, 328, 354
Common purpose, establishment of
    core values, 15, 18, 44, 47–48, 52
    goals, 47–48, 52
    mission, 15, 18, 44, 47–48, 52
    objectives, 47–48, 52
    specifications, 47–48, 52
    targets, 47–48, 52
    vision, 15, 18, 44, 47–48, 52
Concurrent engineering, 515
Conover, W. J., 112
Continuous improvement, 516–518
Control chart, as a tool, 550–551
Cooperative efforts, introduction to
    common purpose, 14–15
    components, 14–15
    effectiveness, 14–15
    efficiency, 14–15
    individual willingness, 14–15
    interpersonal communications, 14–15
Core values (*see* Vision, mission, and core values)
Correlation, 78–79, 87
Correlation/autocorrelation analysis, as a tool, 551
Cox, G. M., 151
C_p index, 303, 304–309, 317–318
C_{pk} index, 303, 304–309, 317–318
C_{pL} index, 305–306
C_{pm} index, 303, 317–318
C_{pmk} index, 303, 317–318
C_{pU} index, 305–306
Creativity, 15, 483, 492–500, (*also see* Process-based transformation, creativity)
Critical path method (CPM), 211–215, 473–477, 551–553
Crosby, P. B., 5, 541
Crosier, R., 381
Csikszentmihalyi, M., 492
Customer benefits and burdens, 3, 7, 175, 189
Customer demands, 6–7, 169–172, 175–180, 190, 193
Customers and customer satisfaction
    benefits and burdens, associated with, 3, 7, 175, 189
    usage processes, related to, 7
    value as a measure of, 4, 6

CuSum control chart
  concepts, individuals data, 294–295
  concepts, subgrouped data, 263–264
  mechanics, individuals data, 295–299
  mechanics, tabular chart, subgrouped
      data, 266–268
  mechanics, V-mask chart, subgrouped
      data, 264–266
  variations, standardized and FIR,
      individuals data, 298–299
Cycle time/waste reduction, 518–521

D

Dacey, J. S., 493–494
Day, R. G., 530, 562
DeBono, E., 495–497
Defect, definition of, 321
Defective, definition of, 321
Dell, L. D., 566
Deming, W. E., 5, 11, 230, 299, 539, 541, 565
Departmental-based organization, concept of,
      15–18, 21
Design/development, as a fundamental process,
      6, 11, 18–19, 166, 441
Dettmer, H. W., 29, 538
DeVor, R. E., 244, 259, 264, 283, 348
Direction, as a subelement of leadership, 483,
      502–503
Dispersion and location, 48–49, 75, 230, 235
Disposal/recycle, as a fundamental process, 6,
      11, 18–19, 166, 441
Distribution/marketing/sales/customer service,
      as a fundamental process, 6, 11, 18–19,
      166, 441
Draper, N. R., 127–129, 131, 133–135, 151,
      157, 553
Duncan, A. J., 245, 256, 264, 273, 348, 432
Dynamic system, definition of, 387–389

E

Ellis, K., 27
Empowerment, 483, 508
Engineered process controls (*see* Automatic
      process control)
Engineering and science
  differences, 25–26
  linkages, 25–26, 37
Eppinger, S. D., 164–165
Equipment-based control models, 222,
      225–228, (*also see* Automatic process
      control)
Equivalent annual value, 201, 547–548

Evans, D. A., 296
Events, definition of, 211, 473
Evolutionary computation, 428
EWMA and EWMD control charts
  concepts, individuals data, 291
  concepts, subgrouped data, 258
  mechanics, individuals data, 291–294
  mechanics, subgrouped data, 259–263
Experimental design, general issues in
  analysis forms
      descriptive, 91
      inferential, 91
      predictive, 92
  experimental error, definition of, 93–94
  experimental unit (eu), definition of, 93–94
  hypotheses, 93
  model adequacy, 109–110
  model assumptions, 95, 109–113
  protocol, 90–91
  purpose, 92, 97, 103, 117, 147
  P value, definition of, 93, 95, 116
  practical significance, 93
  statistical significance, 93
  terms and features, 92–93
Experimental design as a tool, 553–554
Experimental error, definition of, 93–94
Experimental unit (eu), definition of, 93–94
Experiments, designs for response surfaces
  assumptions, 128
  central composite designs, 149–157
      coded data in, 153–154
      general characteristics of, 149, 151
      optimization in, 155, 157
      orthogonal design, 152
      regression ANOVAs in, 154–155
      residuals, 155
      uniform precision design, 152
  2^f with a center point, 146–149
      ANOVA in, 149
      coded data in, 148, 150
      pure error in, 146
      regression ANOVAs in, 150
Experiments, multiple factors, completely
      randomized design (FAT-CRD)
  experimental error, 93, 115–117, 121
  factorial arrangement of treatments
      (FAT), 114
  fixed effects
      ANOVA table for, 116, 121
      expected mean squares for, 116
      hypothesis testing for, 116–122
      interpretation of, 115–122
      treatment mean comparisons, 119–120,
          122–123

Experiments, multiple factors, completely
    randomized design (FAT-CRD) *(cont'd)*
  fixed effects *(cont'd)*
    treatment mean confidence
        intervals, 106
  model, 114, 116–117
  model adequacy, 109–113
  random effects
    ANOVA table for, 116
    components of variation in, 116
    expected mean squares for, 116
    hypothesis testing for, 116
    interpretation of, 115–116
    model, 116
  sum of squares equation, 115–116
Experiments, single factor, completely
    randomized design (CRD)
  experimental error, 93–96, 99, 105–109
  fixed effects
    ANOVA table for, 94–95, 105
    assumptions for, 95
    expected mean squares for, 95
    experimental error in, 93–95,
    hypothesis testing for, 95, 101, 104–105
    interpretation of, 94–95, 100–102,
        104–105
    model, 92–95
    model adequacy, 109–113
    model assumptions in, 95, 109–110
    predictions, of model terms, 102, 105
    treatment mean comparisons
        in, 107–109
    treatment mean confidence intervals
        for, 106–107
  random effects
    ANOVA table for, 94–95, 99
    assumptions for, 95
    components of variation in, 95, 99–100
    expected mean squares for, 95–96, 99
    hypothesis testing for, 95, 99
    interpretation of, 94–96, 99–100
    model, 95–100
    model assumptions, 95, 109–110
    model adequacy, 109–113
  sum of squares equation, 94–95
Expert systems and controllers, 426–427

F

Factorial arrangement of treatments *(see
    Experiments, multiple factors)*
Failure mode and effects analysis (FMEA),
    468–470, 554–555

Fault tree analysis, 468–469, 556–557
Feedback control, nature of, 51–53
Feedforward control, nature of, 51–53
Feigenbaum, A. V., 5, 35, 539
Fifth Discipline, 29, 521–522
Fixed effects model/factor, definition of, 92
Flowchart, as a tool, 557–559
Fogel, D. B., 428
Force field analysis, 472, 559
Fowler, T. C., 522
Franklin, J. A., 426
Function-value analysis, 522–523
Fundamental processes for creating quality in a
        production system
    business integration, 11, 18–19, 166, 441
    design/development, 6, 11, 18–19, 166, 441
    disposal/recycle, 6, 11, 18–19, 166, 441
    distribution/marketing/sales/customer
        service, 6, 11, 18–19, 166, 441
    market/definition, 6, 11, 18–19, 166, 441
    production, 6, 11, 18–19, 166, 441
    use/support, 6, 11, 18–19, 166, 441

G

Gantt chart, 211, 473, 476–477, 560
Gauge studies
    concept of, 309–310
    precision to tolerance (P/T) ratio, 310
    relation to SPC charts, 310–313
General systems theory, 25–30, 37–40, 53–54
Georgiopoulos, M., 425
Gilford, J. P., 495
Goals, definition of, 47–48, 52
Goldberg, D. E., 428
Goldman, S. L., 36
Goldratt, E. M., 29, 538
Goodness of fit, 76–78, 83–84
Grant, E. L., 250, 256, 264, 283, 299, 328–329,
        339, 348
Gupta, M. M., 426–427

H

Haesloop, D., 426
Hammer, M., 532–533
Hartley, J. R., 515
Hassoun, M. H., 425
Hawkins, D. M., 263, 266–267, 296
Haykin, S., 424–425
Hayter, A., 380
Heileman, G. L., 425

Hibino, S., 496–497
Histogram, 75, 561
Hitchcock, K. E., 540
Holt, B. R., 426
Homogeneity of variance, 110, 113
Hoskins, J., 252
Hotelling, H., 381
Hotelling T^2 location chart
    individuals data, 379–380
    subgrouped data, 368–375
Hsiang, T. C., 317
Hughes, T. A., 407, 419, 559
Human- and equipment-based control,
    relationship between, 385–386, 422,
    426–427
Human needs related to quality, 6–7
Hunter, J. S., 100, 107, 124, 294, 553
Hunter, W. G., 100, 107, 124, 553
Hurley, P., 306

I

Imai, M., 516–517
Inappropriate applications of SPC,
    demonstration of, 356–357, 361–362
Individual experience of quality (IQE) model,
    6–7, 175
Industrial engineering, traditional, 518–520
Initiatives in quality/productivity
    characteristics of, 511
    success in, 512
    virtues of, 512
Integrated processes, 165–166, 441, 449, 462
Isermann, R., 409
Ishikawa, K., 5, 11, 35, 539, 541, 548–550,
    563, 567, 569
ISO 9000, 523–525

J

Jackson, P. W., 493–494
Jacobs, T. O., 501–502
Japan Human Relations Association, 517
Jaques, E., 501–502
Johnson, M. L., 303, 318
Jones, D. T., 35
Juran, J. M., 5, 11, 539, 541

K

Kaizen, 516–517
Kane, V. E., 303

Karim, M., 348
Knouse, S. B., 489
Kolarik, W. J., 6, 8, 11, 15, 107, 124, 426, 500,
    506, 522, 526, 530, 539–541, 549–550,
    553, 555, 557–559, 563, 565–567, 569
Kotz, S., 303, 318
Kushler, R. H., 306
Kutner, M. H., 127–129, 157, 553

L

Laszlo, E., 26
Leadership 15, 483, 500–508 (*also see* Process-
    based transformation, leadership issues)
Leadership, differences from management,
    15, 506
Least significant difference (LSD), 107–109,
    119–120, 122–123
Least squares estimation, in regression
    assumptions, 128
    general form, 126–128
Leavenworth, R. S., 250, 256, 264, 283, 299,
    328–329, 339, 348
Location and dispersion, 48–49, 75, 230, 235
Lowry, C., 369, 381
Lu, Y., 426–427
Luceno, A., 221, 432

M

Mader, D., 380
Malcolm Baldrige National Quality Award, 21,
    527–530, 536, 540
Management, differences from leadership,
    15, 506
Market/definition, as a fundamental process, 6,
    11, 18–19, 166, 441
MARR (minimum attractive rate of return),
    195, 197–198
Maslow, A. H., 6, 15
Maslow's human needs hierarchy, 6–7
Mason, R., 379, 380
Masterson, J. T., 504, 507
Matrix diagram, 171–173, 561–562
McCormick, E. J., 526
Measurement scales, classification of
    attributes, 67–68
    interval, 67–68
    nominal, 67–68
    ordinal, 67–68
    ratio, 67–68
    variables, 67–68

Messick, S., 493–494
Metrics, direct and indirect, 64–65
Minimum attractive rate of return (MARR), 195, 197–198
Mission (*see* Vision, mission and core values)
Mistakeproofing, 216, 478, 525–527
Mistakeproof performance, 55–59
Mizuno, S., 562
Models and modeling
    abstract, 26
    analogic, 26
    Checkland's general systems model, 28
    classification of, 26
    iconic, 26
    physical, 26
    roles in engineering and business, 26
Modular processes, 165–166, 441, 449, 462
Montgomery, D. C., 100, 107, 112, 124, 147, 244, 263–264, 266, 348, 358, 369, 381, 369, 381, 432, 535, 553
Motorola, 536–537
Moving average control chart mechanics, 285
Moving range control chart
    concepts and usage, 283–285
    mechanics, 283–291
Mudge, A. E., 522
Mukhopadhyay, S., 426
Multiple factor experiments (*see* Experiments, multiple factors)
Multivariate SPC (*see* Statistical process control charting using variables, multivariate data)
Murray, A. F., 424
Myers, R. H., 76, 79, 100, 110, 127–128, 133–134, 250, 277, 551, 561
Myers, S. L., 76, 79, 100, 110, 127–128, 133–134, 250, 277, 551, 561
Myler, H. R., 425

N

Nadler, G., 496–497
Nagel, R. N., 36
Nair, V. N., 124, 535
Narendra, K. S., 389, 426
Neter, J., 127–129, 157, 553
Net present value analysis
    definition of, 197, 463, 548
    usage of, 195, 197–198, 201, 206–207, 463, 465–466
Neural networks, 422–426
Newbold, P., 81, 551
Nonconformity, defintion of, 321
Nonconforming item, definition of, 321

Normal, Gaussian, distribution, 242
Norman, D. A., 526
Northey, P., 518, 520

O

Oakland, J. S., 256
Objective-based control models (*also see* Statistical process control), 222, 225–228
Objectives, in a purpose hierarchy, 47–48, 52
Ogata, K., 389
Open-loop process structures, 51–53, 224–227

P

Page, E. S., 264
Papadourakis, G. M., 425
Pareto analysis, 341, 343, 562–563
Park, C. S., 548
Parthasarathy, K., 389, 426
Passino, K. M., 422
Patro, S., 426
Payback period, 206–207, 463, 465–466
P control chart
    concepts and mechanics, 322–334
    hypotheses, involved in, 322
    process improvement, relationship to, 324, 330–332
    usage/application, 329
    variable sample sizes, 332–334
Pearn, W. L., 303, 318
Performance assessment in SPC
    appropriate applications, 347–354
    appropriate applications with residuals, 356, 358- 359, 361, 363–364
    inappropriate applications, 356–357, 361–362
Performance measurement parameters
    benchmarks, 55, 58–60
    dispersion, 48–50
    location, 48–50
    mistakeproof, 55–59
    physical, economics/scale, timeliness, customer service, flexibility, 3, 11, 48, 53, 194, 211–213, 462
    robustness, 55–59
Phadke, M. S., 124, 535, 553
Pham, D. T., 428
Physical characterization of processes, 69–73, 84–87
Pignatiello, J., 381, 535
Poisson distribution, 327
Poka-yoke, 525–527

Precision of measurements, 68–69, 309–313

Preiss, K., 36

Prins, J., 380

Probability limits for the *X*-bar, *R*, and *S* control charts, 255–257

Process, basic issues
  capability, criticality of, 224, 303
  close-loop, 51–53, 224–227
  common and special cause variation
    definition of, 65
    mention of, 74, 221, 227, 231, 235, 248–252, 281, 322, 328, 354
  concept, 3–4
  control structures, 51–53, 221–227
  definition/redefinition, control, improvement triad, 3, 28, 37, 43–47
  features and synergy, 43–47
  fundamental processes used to create quality, 6, 11–12, 18–21, 46, 166, 441
  hierarchy, 19, 166, 441
  inertia, 79–86, 227–229, 281, 346–347, 354
  levers/leverage, 65, 69, 72–73, 223
  open-loop, 51–53, 224–225
  performance parameters, 3, 11, 48, 53, 55–60, 194, 211–213
  performance variability, 48–50, 55–57
  purpose, 18–21, 47–48, 63, 446, 458–459, 468
  special and common cause variation
    definition of, 65
    mention of, 74, 221, 227, 231, 235, 248–252, 281, 322, 328, 354
  synergy, 46–47
  value chain (PVC)
    analysis, 563–564
    definition of, 53–55, 199, 201, 204–205
    mention of, 440, 444, 447, 467, 471
  variables, classification of
    controllable, 53–54, 65–67
    input/output, 65–67
    uncontrollable, 53–54, 65–67

Process-based organization, 18–21

Process-based transformation, basics of
  elements and subelements, 482–483
  purpose, 482, 484

Process-based transformation, creativity, 482–483, 492–500
  characteristics, product related, 493–494
  creative thinking, 483, 493–498
    bottlenecks in, 497–498
    breakthrough thinking, 496–497
    characteristics, personal, 493–494
    creative problem solving, sequence in, 494–495

Process-based transformation, creativity, *(cont'd)*
  creative thinking, *(cont'd)*
    DeBono's Theory, 495–496
    divergent thinking, 495
    Gilford's Theory, 495
    lateral thinking, 495–497
    primary schools of thought for, 493
    thinking hats, DeBono's, 496
  criticality of, 492
  definition of, 492
  elements of, 483, 492
  environment—field, associated with creativity, 483, 500–501
    business culture in, 500
    creative behavior and environments, critical attributes of, 500–501
    technical culture in, 500
  knowledge base—domain, associated with creativity, 483, 498–499
    functional knowledge, 498–499
    thinking/analytical knowledge, 498–499
  systems perspective in, 492

Process-based transformation, leadership issues, 483, 500–508
  behavior-related concepts, 501
  definition, 501–502
  direction, as a subelement of leadership, 483, 502–503
    human needs relationships in, 503
    leadership and followership relationships in, 502
    purpose of, 502
  empowerment, 483, 508
    definition of, 508
    readiness level in, 508
    responsibility/authority/decision making, involvement in, 508
  perspectives, historical, 500
  situation-related concepts, 501
  teams/small groups, 502
  teamwork, 483, 503–508
    common purpose relationships in, 503
    communication, definition and forms of, 503–504
    conflict management/resolution in, 508
    conflicts in teaming, 506–508
    conflicts in teaming, sources of, 506
    ego conflict in, 507–508
    meetings, guidelines for, 505
    meetings, purpose of, 504
    nonverbal communication, extent of, 504
    pseudo conflict in, 507–508

Process-based transformation, leadership
    issues, *(cont'd)*
    teamwork, *(cont'd)*
        simple conflict in, 507–508
        teaming roles, definition of, 506–507
        teaming roles, in process-based
            organizations, 506
        team types, in process-based
            organizations, 503
        trustworthiness, definition of and need
            for, 503
    trait-related concepts, 501
Process-based transformation, organizational
    elements and issues, 482–492
    action channels, in general, 484–485
    control action channel, 485–486
    definition/redefinition action
        channel, 486
    improvement channel, 486
    information exchange and archiving, 483,
        486–488
        considerations in, 486–487
        metric definition in, 486–487
        need for, 486–487
        storyboard structure in, 487–488
        team decision support from, 487
    issue identification, 484
    purpose, common and process purpose
        relationships, 483–484
    questioning what we do, 484–485
    reward structures, 483, 488–492
        effectiveness of, 490–492
        leadership, relationship to, 490
        options for, 489–490
        principles of rewarding, 490–491
        purpose of, 489
        tangible and intangible components of,
            488–489
    structural characteristics, 483–486
    team structures, 484–485
    traditional versus process-based,
        comparisons of, 484
Process capability, measurement of
    concept, 303–305
    C_p index, 303, 304–309, 317–318
    C_{pk} index, 303, 304–309, 317–318
    C_{pL} index, 305–306
    C_{pm} index, 303, 317–318
    C_{pmk} index, 303, 317–318
    C_{pU} index, 305–306
    hypothesis test for C_{pk}, 306–309
    interpretation of indices, 306–307
    six sigma capability, 309

Process cash flow models, 201, 206–209,
    465–466
Process characterization
    physical, 69–73, 84–87
    principles analysis, 64–65
    statistical, 73–87, 355, 360
        data collection in, 74–75
        graphical assessment for, 75–76,
            355, 360
        numerical assessment for, 76–87,
            355, 360
    variables analysis, 64–65
Process classification
    integrated, 165–166, 441, 449, 462
    modular, 165–166, 441, 449, 462
Process control, general issues
    adjustment strategies, 221–224
    automatic process control, equipment-
        based (APC)
        model options, conventional and
            unconventional, 228–229
        origin of, 228
        uses for, 228
    common and special cause variation
        definition of, 65
        mention of, 74, 221, 227, 231, 235,
            248–252, 281, 322, 328, 354
    critical elements/subelements, 222–223
    equipment-based, form of, 222, 225–228
    feedback, 224–225
    feedforward, 224–225
    goal, associated with, 223
    growth in calibration, 225–226
    growth in understanding, 225–226
    instability, 223, 230
    introduction to, 3, 28, 37, 43–47
    monitoring strategies, 221–224
    objective-based, form of, 222, 225–228
    relationship between subjective-, objective-,
        and equipment-based models, 227
    special and common cause variation
        definition of, 65
        mention of, 74, 221, 227, 231, 235,
            248–252, 281, 322, 328, 354
    stability, 223–224, 230
    statistical process control (SPC),
        introduction to, 229–232
        failure to detect, type II errors in, 230
        false alarm, type I errors in, 230
        in-control, statistical terminology, 230
        limitations of, 227
        model options, conventional and
            unconventional, 227–229
        origin of, 227

Process control, general issues *(cont'd)*
   statistical process control (SPC),
      introduction to, *(cont'd)*
         out-of-control, statistical
            terminology, 230
         purpose of, 229
         relationship to statistical inference,
            229–230
         statistical errors, type I and type II, in,
            230, 235
      subjective-based, form of, 222, 225–228
Process definition/redefinition, basic issues
   common purpose, relationship to, 164, 169
   customers, relationship to, 164–165, 168
   elements and subelements, 160–162, 164
   general systems theory, relationship
      to, 164
   goal, 164, 181
   initiatives, useful in, 162
   introduction, 3, 28, 37, 43–47
   personnel involved, 163
   philosophies, 163
   tailoring within a production system, 164
   teaming, 163, 167–168, 190
   timing of efforts, 163
   tools for support, 162
Process definition/redefinition, definition of
      expected results, 164–185
   process-level results, 175–185
      boundaries, 175–178, 189
      business results/parameters, 177, 179
      concepts for, 179–185
      customers involved, 175–180, 189, 189,
         193, 199
      evaluation in, 183–185
      outcomes expected, 176–179, 193
      process purpose relationships, 175,
         177–178, 189, 191
      suppliers involved, 175–180, 189,
         193, 199
      synergy in, 181
      targets in, 179
   production system-level results, 164–174
      business results/parameters, 174
      common purpose relationships,
         169, 171
      customers/market, 164, 166, 167–173
      outcomes expected, 169–174
      product definition, 169–173
      targets in, 172–174
Process definition/redefinition, definition of
      means, 188–209
   boundaries, 189
   customers involved, 189, 193, 199

Process definition/redefinition, definition of
      means, *(cont'd)*
   evaluation of situation, 194–209
      alternatives for, 194–198
      cash flow models for, 195, 197–198,
         201, 206–209
      performance parameters for
         comparisons, 194, 196
      relative comparisons of alternatives,
         194–198
      risk assessment in, 195
   options, meaningful, 189–193
      alternatives for, 189, 194–198
      process/subprocess layouts in, 189–193
   outcomes desired, 193
   plans, 198–209
      control and control points in, 198
      metrics in, 199
      process value chain in, 199–201,
         204–205
   process purpose, associated with, 189, 191
   suppliers involved, 189, 193, 199
Process definition/redefinition,
      implementation/process creation,
      210–217
   action involved, 213, 216–217
   control and control points, 216–217
   critical path method, 211–215
   full-scale action, 216–217
   improvement, need for, 217
   limited-scale action, 216
   metrics for implementation, 211–212
   mistakeproofing, 216
   resource acquisition, 210
   scheduling, 210–215
Process improvement, basic issues
   analysis involved in, 438–439
   boundaries, 438
   distinctions from
      definition/redefinition, 439
   elements and subelements, 438–439
   implementation, 438–439
   initiatives and tools useful for, 437–438
   introduction, 3, 28, 37, 43–47
   leverage and improvement, 437, 444
   questioning involved in, 438–439
Process improvement, discovery of opportunity,
      443–454
   concepts relative to improvement, 445,
      446–451
      boundaries in, 449
      cause-effect reasoning in, 448
      customer-side issues in, 446
      description of, 448

Process improvement, discovery of
opportunity, *(cont'd)*
concepts relative to improvement, *(cont'd)*
process levers in, 445, 447
purpose statement for, 439, 446, 453
root causes in, 445, 448, 451
supplier-side issues in, 446
teams involved in, 443, 449
vision, mission, and core values in, 444
observation relative to improvement,
443–446
performance gains from, 444, 447
performance gaps in, 443–444, 447
proactivity in, 443, 445, 447–448
reactivity in, 443, 445, 447–448
options relative to improvement, 451–454
assessment parameters in, 452
improvement option, 451–453
purpose statement for, 453
pursue other opportunities,
abandonment option, 451–453
redefinition option, 451–453
Process improvement, exploration for
change, 456–471
alternatives, formulation of, 457–461
facts and figures in, 457
feasible alternative generation, 457–461
idea generation, 457–461
layout formulation, 457
purpose and purpose linkages in,
457–459
evaluation of alternatives, 462–466
baseline for comparison, 462–464
cash flow models, 465–466
evaluation formats, 462, 465–466
MARR, 463, 465–466
net present value analysis, 463,
465–466
payback period, 463, 465–466
involvement of people, 457
plan, formulation of, 466–471
control points, 467–468
risk assessment/analysis, 468–470
updating of process-related details,
467–468
Process improvement, implementation of
change, 471–479
action involved, 477–479
control, 479
critical path method, 473–477
full-scale action, 478–479
limited-scale action, 477–478
mistakeproofing, 478
redefinition, eventual need for, 478–479

Process improvement, implementation of
change, *(cont'd)*
resource acquisition, 471–472
risks and anxieties, associated with,
471–472
scheduling, 473–477
Process inertia, 79–86, 227–229, 281,
346–347, 354
Process levers/leverage, 65, 69, 72–73, 223
Process logs, 74, 341–342
Process purpose
definition of, 18–21, 47–48
mention of, 63, 446, 458–459, 468
Process sensors, 68–69
Process synergy through definition/redefinition,
control, and improvement, 43–47, 63,
87, 437
Process value chain (PVC)
definition of, 53–55, 199, 201, 204–205
mention of, 440, 444, 447, 467, 471
analysis, 563–564
Process variables, classification of
controllable, 53–54, 65–67
input/output, 65–67
uncontrollable, 53–54, 65–67
Production, as a fundamental process, 6, 11,
18–19, 166, 441
Production system, introduction to
benefits and burdens, 3
common purpose, expression in, 15,
502–503
common purpose, linkages within a
process-based organization, 18–21
competitive advantage/edge in, 12, 43
constituent working parts, 2–3
cooperative effort, definition of, 14–15
definition, 2–3
departmental structures, 16–18, 21
direction, associated with, 44, 502–503
linkages within and between people,
processes, and products, 13
optimization of, 15–21
people chain, 13
performance parameters, 3, 11, 48
process chain, 13
process concept, 4
process configurations, 4
process structures, 18–21
product chain, 13
as a series of processes, 4
vision, mission, core values, 15, 18, 44,
47–48, 52, 168–169, 189, 444
Production system/process enterprise models,
201, 207–209

Production systems, paradigms for
    agile, 30, 32, 36
    arrays of, 30–37
    common ground, 36–37
    craft, 30–31, 33
    evolution/transformation of, 32–33, 36
    factory, 30–31, 33–34
    lean, 30, 32, 35–36
    mass, 30–31, 34–35
Productivity
    connection to quality, 11–12
    definition, 8–9
    linkage to efficiency, 5, 11–12
    partial, 9–11
    ratios, 8–11
    total, 9, 11
P value, definition of, 93, 95, 116, 132
Pulat, P. S., 553, 560

Q

Quality, fundamentals of
    concepts, 5
    connection to productivity, 11–12
    costs in life cycle phases, 12
    creation of, 6–7
    definitions, 5–6
    experience of, 6–8, 11
    fundamental processes in creation of
        quality, 6, 11–12
    individual quality experience (IQE) model,
        6–7, 175
    linkage to effectiveness, 5, 11–12
    related to human needs, 6–7
    relationship between quality experience and
        creation, 8
    scientific view, 6
    substitute characteristics, 5–6, 47
    satisfaction 5
    true characteristics, 5–6
Quality awards, 527–530
Quality function deployment, 171–173,
    530–531

R

R^2, R^2_{max}, and R^2_{adj}. 135, 138, 141–142, 144
Radford, G. S., 5
Ramberg, J. S., 535
Random effects model/factor, definition of, 92
Random number generation, normally
    distributed, 572–573
R and S control chart mechanics and usage,
    236, 238–239, 244–248

Reengineering, 532–533
Regression (*see* Response surfaces)
Reitz, H. J., 50
Relationship diagram, 170, 449–450, 564–565
Residuals charting in SPC, 354–364
Response surfaces (*also see* Experiments,
    designs for response surfaces)
    best models, 131
    coded versus natural units of measure,
        arguments regarding, 157
    general models, 126–127
    hypothesis tests, 132–137, 141
    lack of fit, 135–136
    model assumptions, 128
    normal equations, 127–128, 137
    optimization, 155, 157
    pure error, 131–136, 146
    purpose, 126, 131, 146
    P-value, definition of, 93, 132
    model fit and adequacy, 135–136
    model justification, 133,
    model simplification, 133–135
    regression ANOVAs, 129–132, 138–139,
        141–142, 144
    residual error, 131–135
    residuals, 127–129, 131–132, 138–141, 144
    simple linear regression model, 128–129,
    sum of squares equations and terms,
        130–131, 143
Rigdon, S., 381
Risk assessment, 195, 555–556
Robust performance, 55–59
Robust design, 534–535
Rodriguez, R. N., 317
Roos, D., 35
Root cause analysis, 565–566
Ross, P. J., 124, 535, 553
Roy, R. K., 124, 535, 553
Runger, G., 381
Run length statistics and profiles for SPC
    appropriate applications, 349–354
    appropriate applications with residuals,
        356, 358- 359, 361, 363–364
Runs chart/plot, 82, 85, 566–567

S

Sample generalized variance dispersion, |S|
    chart, subgrouped data, 375–378
Sanders, M. S., 526
SAS, 104, 110, 131, 134, 142
Scatter diagram, 567–568
S^2 control chart mechanics, 258
Senge, P. M., 29, 521

Shewhart, W. A., 5, 227

Shimbun, N. K., 525

Shingo, S., 525

Siegmund, D., 296

Simons Study of Media and Markets, 169

Simulation approach in SPC performance
evaluation, 347–348

Single factor experiments (*see* Experiments,
single factor)

Sinha, N. K., 426–427

Six Sigma initiative, 309, 535–537

Smith, H., 127–129, 133–135, 157, 553

Southway, N., 518, 520

SPC/APC model combinations, 431–433

SPClab, 348

Special and common cause variation
definition of, 65
mention of, 74, 221, 227, 231, 235,
248–252, 281, 322, 328, 354

Specifications, with regard to tolerances,
47–48, 52

Spiring, F. A., 303, 317

Split-plot experimental model, 124

Statistical characterization, 73–87, 355, 360

Statistical errors in SPC
with OC curves and ARLs, 253, 255,
273–277, 347–364
type I and type II errors defined, 230,

Statistical process control (SPC), nontraditional
models
assumptions underlying SPC models, 346
residual charting, demonstration, 355–364
residual charting, regression and time series
strategies, 354, 356, 358, 361, 363,

Statistical process control (SPC), performance
assessment
appropriate applications, 347–354
appropriate applications with residuals,
356, 358- 359, 361, 363–364
inappropriate applications, 356–357,
361–362
run length statistics and profiles
appropriate applications, 349–354
appropriate applications with residuals,
356, 358–359, 361, 363–364
simulation approach in, 347–348

Storyboard structure, 487–488

Statistical process control (SPC) charting, using
attributes data
attributes control charts
C control chart,
area of opportunity for defects, 335
concepts and mechanics, 334–337
inspection unit definition, 335
usage/application, 335

Statistical process control (SPC) charting, using
attributes data *(cont' d)*
attributes control charts *(cont' d)*
C, U control chart family, 322
notation for, 325–326
P control chart
concepts and mechanics, 327–334
process improvement, relationship
to, 324, 330–332
usage/application, 329
variable sample sizes, 332–334
U control chart
concepts and mechanics, 337–340
standard area of opportunity
definition, 337
usage/application, 338
defect, definition of, 321
defective, definition of, 321
defects and defective items in, 321, 334
distributions, relationships to, 326–327
hypothesis tests, relationships to, 322
nonconforming item, definition of, 321
nonconformity, definition of, 321
nonconformity and nonconforming items
in, 321, 334
purpose, 320–321, 323
rationale, 322–325
sample size, comparison to variables
charts, 322

Statistical process control (SPC) charting using
variables, multivariate data
chi squared location chart, subgrouped data,
368
comparison to univariate SPC, 365–366,
367, 380
Hotelling T^2 location chart
individuals data, 379–380
subgrouped data, 368–375
interpretation, 380–381
joint control regions, 366–367
other charts, references for, 381
purpose, 365
sample generalized variance dispersion |S|
chart, subgrouped data, 375–378
statistical characterization of data, 365–367

Statistical process control (SPC) charting, using
variables data and individual values
common cause variation in, 281
hypothesis tests, relationships to, 281–282
natural tolerance interval, 303–304
process inertia, relationships to, 281, 300
process/production source level control,
300–302
rationale for individuals data, 281

Statistical process control (SPC) charting,
   using variables data and individual
   values *(cont'd)*
   sampling schemes for SPC
      factors involved, 300
      judgmental sampling, 299–300
      sample timing compromises, 299
      sampling strategies and intervals,
         299–300
   target-based control charts, 302–303
   variables control charts for individuals data
      CuSum control charts, 294–299
         concepts and mechanics, 294–299
         variations for standardized and FIR,
            298–299
      EWMA and EWMD control chart
         concepts and mechanics, 291–294
      moving average control chart
         mechanics, 285
      moving range control chart concepts
         and usage, 283–285
      moving range control chart mechanics,
         283–291
      notation for, 282–283
      X control chart concepts and usage,
         283–285
      X control chart mechanics, 283–291
Statistical process control (SPC) charting, using
   variables data and subgrouped values
   average run length (ARL), 253, 255,
      277–279
   benefits associated with, 235–236, 239
   central limit theorem, 242, 244, 250
   control chart limits terminology, 239, 242
   economic consequences, 244
   graphical relationship to statistical
      inference, 240
   hypothesis testing relationships, 235,
      249–251
   interpretation, 248–253
   limitations, 239
   limited duration process runs and SPC
      deviation from target control charts, 269
      standardized control charts, 269–270
      validity of models, 270
   monitoring fundamentals, 235–239
   normal model, 242
   OC (operating characteristic) curves, 253,
      255, 273–277
   process inertia relationships, 238, 240, 249
   process logs, 239
   process standard deviation estimation,
      247–248
   range definition, 236, 243

Statistical process control (SPC) charting,
   using variables data and subgrouped
   values *(cont'd)*
   runs rules, 252–253
   sampling schemes and strategy for
      subgrouped charts, 249
   Shewhart charts, 234
   subgrouping rationale, 238, 240
   trial limits, 239, 247
   type I statistical errors, 242, 250
   variables control charts for subgrouped data
      CuSum control charts, 263–268
      EWMA and EWMD control charts,
         258–263
      notation for, 240–241
      probability limits for the X-bar, R, and
         S control charts, 255–257
      R and S control charts, 244–248
      S^2 control chart, 258
      X-bar and R control chart justification,
         236, 238–239
      X-bar control chart mechanics and
         usage, 242–243, 247–248
Stratification analysis, 568–569
Stuart, B., 252
Subjective-based control models, 222, 225–228
Substitute and true quality characteristics,
   179–180, 321
Sullivan, J., 379
Sumanth, D. J., 8, 9
Sutherland, J. W., 244, 259, 264, 283, 348
Sutton, R. S., 426
Sweet, A. L., 258, 291
Systems, general classification of
   designed abstract, 27
   designed physical, 27
   hard and soft, 28
   human activity, 27
   natural, 27
   open and closed, 27

T

Taguchi, G., 5, 124, 317, 534–535, 553
Taguchi design levels, 534–535
Taguchi experimental models, 124
Targets, definition of, 47–48, 52
Taylor, J., 252
Teamwork, 483, 503–508
Theory of Constraints, 29–30, 538–539
Thinking, ways of
   adaptations of general systems thinking,
      29–30, 37–40

Thinking, ways of *(cont'd)*
    analytical thinking, 24–26, 37, 53–54
    hard and soft thinking methodologies, 28
    systems thinking, 24–30, 36–40, 53–54
Thor, C. G., 11
Tools in quality/productivity
    characteristics of, 544
    use of, 544–545
Total quality control, 539–541
Total quality management, 539–541
Tracy, N., 379, 380
True and substitute quality characteristics,
    179–180, 321
Tsui, K. 380

U

U control charts, 337–340
Ulrich, K. T., 164–165
Use/support, as a fundamental process, 6, 11,
    18–19, 166, 441

V

Value, classical definition of, 522
Value, fundamentals of
    definition, 4
    related to quality experience, 6
    value added, 13, 18–21
Value chain (PVC)
    analysis, 563–564
    definition of, 53–55, 199, 201, 204–205
    mention of, 440, 444, 447, 467, 471
Value chain (PVC) analysis, 563–564
Vance, L. C., 296
Variables control charts (*see* Statistical process
    control (SPC) charting)
Vision, mission, and core values
    applicaton in the purpose hierarchy, 47–48,
       52, 168–169, 189, 444
    definition of, 15, 18, 44
von Bertalanffy, L., 25–27

W

Walpole, R. E., 76, 79, 100, 110, 127–128,
    133–134, 250, 277, 551, 561
Warwick, K., 409
Wasserman, W., 127–129, 157, 553
Watson, G. H., 514
Whitley, D. S., et al., 426
Wierda, S., 380
Willard, M. L., 540
Wilson, B., 28
Wilson, P. F., 566
Womack, J. P., 35
Woodall, W., 379, 381, 432
Woodall, W. G., 503–504

X

X-bar control chart
    justification of, 236, 238–239
    mechanics and usage, 242–243, 247–248
X control chart mechanics and usage, 283–291

Y

Yang, Y., 428
Young, J., 379, 380

Z

Ziegler-Nichols, 407, 409